Radiomics and Radiogenomics

T0239607

Imaging in Medical Diagnosis and Therapy

Series Editors: Bruce R. Thomadsen, David W. Jordan

Beam's Eye View Imaging in Radiation Oncology
Ross I. Berbeco, Ph.D.

Principles and Practice of Image-Guided Radiation Therapy of Lung Cancer
Jing Cai, Joe Y. Chang, Fang-Fang Yin

Radiochromic Film
Role and Applications in Radiation Dosimetry
Indra J. Das

Clinical 3D Dosimetry in Modern Radiation Therapy
Ben Mijnheer

Hybrid Imaging in Cardiovascular Medicine
Yi-Hwa Liu, Albert J. Sinusas

Observer Performance Methods for Diagnostic Imaging
Foundations, Modeling, and Applications with R-Based Examples
Dev P. Chakraborty

Ultrasound Imaging and Therapy
Aaron Fenster, James C. Lacefield

Dose, Benefit, and Risk in Medical Imaging
Lawrence T. Dauer, Bae P. Chu, Pat B. Zanzonico

Big Data in Radiation Oncology
Jun Deng, Lei Xing

Radiomics and Radiogenomics
Technical Basis and Clinical Applications
Ruijiang Li, Lei Xing, Sandy Napel, Daniel L. Rubin

For more information about this series, please visit:
https://www.crcpress.com/Imaging-in-Medical-Diagnosis-and-Therapy/book-series/CRCIMAINMED

Radiomics and Radiogenomics

Technical Basis and Clinical Applications

Edited by
Ruijiang Li

Lei Xing

Sandy Napel

Daniel L. Rubin

CRC Press
Taylor & Francis Group
Boca Raton London New York

CRC Press is an imprint of the
Taylor & Francis Group, an **informa** business

CRC Press
Taylor & Francis Group
6000 Broken Sound Parkway NW, Suite 300
Boca Raton, FL 33487-2742

First issued in paperback 2020

© 2019 by Taylor & Francis Group, LLC
CRC Press is an imprint of Taylor & Francis Group, an Informa business

No claim to original U.S. Government works

ISBN-13: 978-0-8153-7585-2 (hbk)
ISBN-13: 978-0-367-77958-0 (pbk)

Library of Congress Cataloging-in-Publication Data

Names: Li, Ruijiang, editor. | Xing, Lei, editor. | Napel, Sandy, editor. |
Rubin, Daniel (Daniel L.), editor.
Title: Radiomics and radiogenomics : technical basis and clinical
applications / edited by Ruijiang Li, Lei Xing, Sandy Napel, Daniel L.
Rubin.
Other titles: Imaging in medical diagnosis and therapy.
Description: Boca Raton, FL : CRC Press, Taylor & Francis Group, [2019] |
Series: Imaging in medical diagnosis and therapy
Identifiers: LCCN 2019004822 | ISBN 9780815375852 (hardback ; alk. paper) |
ISBN 0815375859 (hardback ; alk. paper) | ISBN 9781351208277 (ebook) |
ISBN 1351208276 (ebook)
Subjects: LCSH: Cancer--Imaging. | Diagnostic imaging.
Classification: LCC RC270.3.D53 R36 2019 | DDC 616.99/40754--dc23
LC record available at https://lccn.loc.gov/2019004822

Visit the Taylor & Francis Web site at
http://www.taylorandfrancis.com

and the CRC Press Web site at
http://www.crcpress.com

Contents

Preface

This book will address two closely related emerging fields: radiomics and radiogenomics. The fields of radiomics and radiogenomics have experienced significant technical development in recent years, and have shown great promises in a growing number of clinical and translational studies, partly driven by advances in medical image analysis, increased computing power, and availability of large annotated datasets in a variety of cancer types. Currently, there is no single text available to address these developments and serve as a reference for all the issues related to *radiomics and radiogenomics*. This book will provide a comprehensive review of the field including state-of-the-art technology.

Radiomics refers to the extraction of high-throughput, quantitative features from clinical images such as CT, MRI, or positron-emission tomography (PET), while radiogenomics concerns with the study of relations between radiomic features at the tissue scale and underlying molecular features at the genomic, transcriptomic, or proteomic level. Both can be used to evaluate disease characteristics or correlate with relevant clinical outcomes such as treatment response or patient prognosis. They have the common goal of discovering useful biomarkers (diagnostic, prognostic, or predictive) to improve clinical decision-making and ultimately enable the practice of precision medicine.

In this book, we will cover in a comprehensive manner the fundamental principles, technical basis, and clinical applications of radiomics and radiogenomics, with a specific focus on oncology. Readers will benefit from a detailed understanding of the field, including the technical development and clinical relevance, and a comprehensive bibliography. This book is mainly intended for imaging scientists, medical physicists, diagnostic radiologists, and medical professionals and specialists such as radiation oncologists and medical oncologists whose patients may benefit from this research. This book may be of interest to a wide range of researchers who are affiliated with professional societies such as Radiological Society of North America (RSNA), American Association of Physicists in Medicine (AAPM), American Society for Radiation Oncology (ASTRO), and American Society of Clinical Oncology (ASCO). It will also serve as a comprehensive reference for those involved in research, teaching, management, or administration in the field of clinical oncology.

This book details the technical basis and clinical applications of two emerging and rapidly expanding fields: radiomics and radiogenomics. Radiomics has led to the discovery of promising imaging biomarkers with potential diagnostic, prognostic, or predictive value, especially in oncology, while radiogenomics can allow the identification of molecular biology behind these imaging phenotypes. The first part provides a general overview of the principles and rationale. The second part focuses on the technical basis of these two approaches and resources available to support this research. The third part is devoted to the applications of radiomics and radiogenomics in clinical oncology, organized by anatomical disease sites. The final part discusses emerging research directions and provides future outlooks including a roadmap to clinical translation.

Ruijiang Li, Lei Xing, Sandy Napel, Daniel L. Rubin
Stanford University

Acknowledgments

We wish to thank Ms. Carrie Zhang for providing tremendous administrative support throughout the editing and production of this book.

About the Editors

Ruijiang Li, PhD, is an assistant professor in the Department of Radiation Oncology at Stanford University School of Medicine. He is also an affiliated faculty member of the Integrative Biomedical Imaging Informatics at Stanford (IBIIS), a departmental section within Radiology. He has a broad background and expertise in medical imaging, with specific expertise in quantitative image analysis and machine learning as well as their applications in radiology and radiation oncology. Dr. Li has published extensively on the topics of radiomics and radiogenomics in top-tier journals such as *Radiology*, *Clinical Cancer Research*, and *Breast Cancer Research*. The goal of his research is to develop, validate, and clinically translate imaging biomarkers for predicting prognosis and therapeutic response in cancer. His research has been generously funded by the National Institute of Health (NIH)/National Cancer Institute (NCI) with K99/R00 and three R01 grants. Dr. Li received many nationally recognized awards, including the NIH Pathway to Independence Award, ASTRO Clinical/Basic Science Research Award, ASTRO Basic/Translational Science Award, American Association of Physicists in Medicine (AAPM) Science Council Research Award, etc.

Dr. Lei Xing is currently the Jacob Haimson Professor of Medical Physics and Director of Medical Physics Division of Radiation Oncology Department at Stanford University. He also holds affiliate faculty positions in the Department of Electrical Engineering, Medical Informatics, Bio-X, and Molecular Imaging Program at Stanford. Dr. Xing's research has been focused on artificial intelligence in medicine, medical imaging, treatment planning, tomographic image reconstruction, molecular imaging instrumentations, image guided interventions, nanomedicine, imaging informatics, and analysis. He has made unique and significant contributions to each of those areas. Dr. Xing is an author on more than 300 peer-reviewed publications, a co-inventor on many issued and pending patents, and a co-investigator or principal investigator on numerous NIH, Department of Defense (DOD), American Cancer Society (ACS), AAPM, RSNA and corporate grants. He is a fellow of AAPM (American Association of Physicists in Medicine) and AIMBE (American Institute for Medical and Biological Engineering). He has received numerous awards from various societies and organizations for his lab's work in artificial intelligence, medical physics, and medical imaging.

Dr. Sandy Napel's primary interests are in developing diagnostic and therapy-planning applications and strategies for the acquisition, visualization, and quantitation of multi-dimensional medical imaging data. Examples are: creation of three-dimensional images of blood vessels using CT, visualization of complex flow within blood vessels using MR, computer aided detection and characterization of lesions (e.g., colonic polyps, pulmonary nodules) from cross-sectional image data, visualization and automated assessment of 4D ultrasound data, and fusion of images acquired using different modalities (e.g., CT and MR). I have also been involved in developing and evaluating techniques for exploring cross-sectional imaging data from an internal perspective, i.e., virtual endoscopy (including colonoscopy, angioscopy, and bronchoscopy), and in the quantitation of structure parameters, e.g., volumes, lengths, medial axes, and curvatures. I am also interested in creating workable solutions to the problem of "data explosion," i.e., how to look at the thousands of images generated per examination using modern CT and MR scanners. My most recent focus includes making image features computer-accessible, to facilitate content-based retrieval of similar lesions, prediction of molecular phenotype, response to therapy, and prognosis from imaging features. I am co-director of the Radiology 3D and Quantitative Imaging Lab, providing clinical service to the Stanford and local community, and co-Director of Integrative Biomedical Imaging Informatics at Stanford (IBIIS), whose mission is to advance the clinical and basic sciences in radiology, while improving our understanding of biology and the manifestations of disease, by pioneering methods in the information sciences that integrate imaging, clinical, and molecular data.

Daniel L. Rubin, MD, MS, is a professor of radiology and medicine (Biomedical Informatics Research) at Stanford University. He is principal investigator of two centers in the National Cancer Institute's Quantitative Imaging Network (QIN), Chair of the QIN Executive Committee, Chair of the Informatics Committee of the Eastern Cooperative Oncology Group and the American College of Radiology Imaging Network (ECOG-ACRIN) cooperative group, and past Chair of the RadLex Steering Committee of the Radiological Society of North America. His NIH-funded research program focuses on quantitative imaging and integrating imaging data with clinical and molecular data to discover imaging phenotypes that can predict the underlying biology, define disease subtypes, and personalize treatment. He is a Fellow of the American College of Medical Informatics and has published over 160 scientific publications in biomedical imaging informatics and radiology.

Contributors

Zeynettin Akkus
Department of Radiology and Biomedical Engineering
Mayo Clinic
Rochester, Minnesota

Spyridon Bakas
Center for Biomedical Image Computing and Analytics (CBICA)
Department of Radiology
University of Pennsylvania
Philadelphia, Pennsylvania

Andrew Beers
Athinoula A. Martinos Center for Biomedical Imaging
Massachusetts General Hospital-Harvard Medical School
Boston, Massachusetts

Stephen R. Bowen
Departments for Radiation Oncology and Radiology
University of Washington School of Medicine
Seattle, Washington

James Brown
Athinoula A. Martinos Center for Biomedical Imaging
Massachusetts General Hospital-Harvard Medical School
Boston, Massachusetts

Guohong Cao
Department of Radiology
Shulan Hangzhou Hospital
Hangzhou, China

Ken Chang
Athinoula A. Martinos Center for Biomedical Imaging
Massachusetts General Hospital-Harvard Medical School
Boston, Massachusetts

Rhea Chitalia
Center for Biomedical Image Computing and Analytics (CBICA)
Department of Radiology
University of Pennsylvania
Philadelphia, Pennsylvania

Rivka R. Colen
Department of Diagnostic Radiology and Cancer Systems Imaging
University of Texas MD Anderson Cancer Center
Houston, Texas

Christos Davatzikos
Center for Biomedical Image Computing and Analytics (CBICA)
Department of Radiology
University of Pennsylvania
Philadelphia, Pennsylvania

Di Dong
Institute of Automation
Chinese Academy of Sciences
Beijing, China

Adrienne Dula
Department of Neurology
The University of Texas at Austin
Austin, Texas

William D. Dunn Jr.
Department of Biomedical Informatics
Emory University School of Medicine
Atlanta, Georgia

Hesham Elhalawani
Department of Radiation Oncology
University of Texas MD Anderson Cancer Center
Houston, Texas

Bradley J. Erickson
Department of Radiology
Mayo Clinic
Rochester, Minnesota

Yong Fan
Center for Biomedical Image Computing and Analytics (CBICA)
Department of Radiology
University of Pennsylvania
Philadelphia, Pennsylvania

Clifton D. Fuller
Department of Radiation Oncology, Medical Physics Program
Graduate School of Biomedical Sciences
University of Texas/MD Anderson Cancer Center
Houston, Texas

Michael Gensheimer
Department of Radiation Oncology
Stanford University
Stanford, California

Olivier Gevaert
Stanford Center for Biomedical Informatics Research
Department of Medicine
and
Department of Biomedical Data Science
Stanford University
Stanford, California

Maryellen L. Giger
Department of Radiology
University of Chicago
Chicago, Illinois

Robert Gillies
Department of Cancer Physiology
H. Lee Moffitt Cancer Center
Tampa, Florida

Assaf Hoogi
Department of Radiology
Stanford University
Stanford, California

David A. Hormuth II
Institute for Computational and Engineering Sciences
The University of Texas at Austin
Austin, Texas

Kathleen Horst
Department of Radiation Oncology
Stanford University School of Medicine
Stanford, California

Bruna Victorasso Jardim-Perassi
Department of Cancer Physiology
H. Lee Moffitt Cancer Center
Tampa, Florida

Jayashree Kalpathy-Cramer
Athinoula A. Martinos Center for Biomedical Imaging
Massachusetts General Hospital-Harvard Medical School
Boston, Massachusetts

Masoud Badiei Khuzani
Department of Radiation Oncology
Stanford University
Stanford, California

Paul E. Kinahan
Departments for Radiation Oncology and Radiology
University of Washington School of Medicine
Seattle, Washington

Timothy L. Kline
Department of Radiology
Mayo Clinic
Rochester, Minnesota

Despina Kontos
Center for Biomedical Image Computing and Analytics (CBICA)
Department of Radiology
University of Pennsylvania
Philadelphia, Pennsylvania

Panagiotis Korfiatis
Department of Radiology
Mayo Clinic
Rochester, Minnesota

Hui Li
Department of Radiology
University of Chicago
Chicago, Illinois

Ruijiang Li
Department of Radiation Oncology
Stanford University
Stanford, California

Zaiyi Liu
Department of Radiology
Guangdong Academy of Medical Sciences (Guangdong General Hospital)
Guangzhou, China

Lin Lu
Department of Radiology
Columbia University Medical Center
New York, New York

Anant Madabhushi
Department of Biomedical Engineering
Case Western Reserve University
Cleveland, Ohio

Gary Martinez
Department of Cancer Physiology
H. Lee Moffitt Cancer Center
Tampa, Florida

Matthew T. McKenna
Department of Biomedical Engineering
Vanderbilt University
Nashville, Tennessee

Michael I. Miga
Department of Biomedical Engineering, Radiology and
Radiological Sciences, and Neurological Surgery
Vanderbilt University
Nashville, Tennessee

Sandy Napel
Department of Radiology
Stanford University
Stanford, California

Issam El Naqa
Department of Radiation Oncology
University of Michigan
Ann Arbor, Michigan

Tianye Niu
Sir Run Run Shaw Hospital
Institute of Translational Medicine
Zhejiang University School of Medicine
Zhejiang University
Hangzhou, Zhejiang

Matthew J. Nyflot
Departments for Radiation Oncology and Radiology
University of Washington School of Medicine
Seattle, Washington

Kenneth Philbrick
Department of Radiology
Mayo Clinic
Rochester, Minnesota

C. Chad Quarles
Imaging Research
Barrow Neurological Research Institute
Phoenix, Arizona

Daniel L. Rubin
Integrative Biomedical Imaging Informatics
Stanford University
Stanford, California

Arvind Rao
Department of Computational Medicine and Bioinformatics
Department of Radiation Oncology
University of Michigan
Ann Arbor, Michigan

George A. Sandison
Departments for Radiation Oncology and Radiology
University of Washington School of Medicine
Seattle, Washington

Lawrence H. Schwartz
Department of Radiology
Columbia University Medical Center
New York, New York

Li Shen
Department of Biostatistics, Epidemiology, and Informatics
The Perelman School of Medicine, University of Pennsylvania
Philadelphia, Pennsylvania

Hiroki Shirato
Global Institution for Collaborative Research and Education (GI-CoRE)
Hokkaido University
Sapporo, Japan

Anna G. Sorace
Department of Biomedical Engineering and Diagnostic Medicine
The University of Texas at Austin
Austin, Texas

Ashley Stokes
Department of Neurological Surgery
Vanderbilt University
Nashville, Tennessee

Xiaoli Sun
Department of Radiation Oncology
The First Affiliated Hospital of Zhejiang University
Hangzhou, China

Khin K. Tha
Global Institution for Collaborative Research and Education (GI-CoRE)
Hokkaido University
Sapporo, Japan

Jie Tian
Institute of Automation
Chinese Academy of Sciences
Beijing, China

Harini Veeraraghavan
Department of Medical Physics
Memorial Sloan Kettering Cancer Center
New York, New York

John Virostko
Department of Diagnostic Medicine
The University of Texas at Austin
Austin, Texas

Satish E. Viswanath
Department of Biomedical Engineering
Case Western Reserve University
Cleveland, Ohio

Shuo Wang
Institute of Automation
Chinese Academy of Sciences
Beijing, China

Lise Wei
Department of Radiation Oncology
University of Michigan
Ann Arbor, Michigan

Jared A. Weis
Department of Biomedical Engineering
Wake Forest University
Winston-Salem, North Carolina

Jennifer G. Whisenant
Department of Medicine
Vanderbilt University
Nashville, Tennessee

Yan Wu
Department of Radiation Oncology
Stanford University
Stanford, California

Lei Xing
Department of Radiation Oncology
Stanford University School of Medicine
Stanford, California

Jingwen Yan
Department of BioHealth Informatics, School of Informatics and Computing
Indiana University-Purdue University
Indianapolis, Indiana

Pengfei Yang
Sir Run Run Shaw Hospital
Institute of Translational Medicine
Zhejiang University School of Medicine
Zhejiang University
Hangzhou, Zhejiang, China

Thomas E. Yankeelov
Department of Diagnostic Medicine and Oncology
Institute for Computational and Engineering Sciences, Biomedical Engineering
and
Department of Neurology
Livestrong Cancer Institute
The University of Texas at Austin
Austin, Texas

Xiaohui Yao
Department of Biostatistics, Epidemiology, and Informatics
The Perelman School of Medicine
University of Pennsylvania
Philadelphia, Pennsylvania

Binsheng Zhao
Department of Radiology
Columbia University Medical Center
New York, New York

Foreword

In radiology spanning over the last four decades, we have seen innovations in imaging transform nearly every aspect of healthcare. Yet thanks to accelerating advances in technology and the life sciences, the horizons for imaging only continue to grow. Undoubtedly, the implementation of radiomics and radiogenomics—i.e., the correlation and combination of quantitative mathematical patterns extracted from diagnostic images with clinical and genomic data—within routine clinical care would represent a major leap forward for digital medicine. However, making this leap will be a great challenge—one that will require as many capable hands on deck as possible. Happily, this book will provide members of the imaging community—from clinical radiologists, to medical physicists, and computer scientists—with the knowledge they need to engage in and strengthen the blossoming new disciplines of radiomics and radiogenomics.

Radiomics and radiogenomics clearly have vast potential to improve clinical decision support systems and aid diagnosis, prognostic assessment, and treatment selection for cancer as well as other diseases. This is especially true when radiomic and radiogenomic data are processed using machine learning techniques, such as novel types of artificial neural networks. Studies based on routinely used imaging modalities (CT, MRI, or PET) have identified sets of radiomic features predictive of factors such as tumor stage, therapeutic response, and prognosis including post-treatment recurrence and survival in patients with cancer. Radiomic and radiogenomic analyses could help guide biopsies to areas of tissue likely to contain the most clinically relevant information. Moreover, they could provide additional information that simply cannot be gleaned from biopsies. This is because, unlike biopsies, they can analyze the features of *entire* tumors or other volumes of interest at *any* location and provide a picture of tissue heterogeneity within and across multiple volumes at a single time point. This is particularly important for oncology, because tumor heterogeneity has been identified as a prognostic determinant of survival in different types of cancer, and an obstacle to cancer control.

The idea of being able to probe disease characteristics rapidly and non-invasively is, of course, tremendously exciting, and this book will give readers an overview of the many areas in which radiomics and radiogenomics studies are yielding promising results. At the same time, it will provide a solid understanding of the many technical steps involved in such studies—including image acquisition, volume-of-interest segmentation (which, today, typically requires a radiologist's participation, but may, in the future, be performed by an artificial intelligence algorithm), radiomic feature extraction, predictive modeling, and model validation, among others—and it will explain various methodologies being applied or developed for carrying out these steps. In turn, this technical understanding will help readers grasp the strengths and weaknesses of different methodologies so they can better judge the validity of published studies, participate in radiomics and radiogenomics research, and, one day, apply the tools derived from this research in clinical settings.

The greatest hurdles to making radiomics applicable in routine practice will be achieving the large-scale data-sharing and standardization necessary to ensure the validity and generalizability of radiomics-based predictive models. Indeed, even for a single task such as distinguishing between two particular lesion types, thousands of cases will be required for training and validation, and a consensus will need to be reached with regard to image acquisition protocols and radiomic feature extraction. Thankfully, a number of organizations are working to help achieve these goals. The success of their efforts will depend on the understanding, cooperation, and participation of the broader imaging community. This book will serve as an invaluable reference source for anyone wishing to help make radiomics and radiogenomics viable tools for clinical practice. Its publication could not be more timely, and we applaud the editors and contributors for their vision and hard work in creating such a wonderful, useful resource.

Marius Mayerhoefer, Hedvig Hricak
Memorial Sloan Kettering Cancer Center

INTRODUCTION

Principles and rationale of radiomics and radiogenomics

SANDY NAPEL

1.1 INTRODUCTION

Traditionally, the radiologist's job was focused on the interpretation of images, which, until recently, was facilitated entirely by their observations and supported by their training. While there are exceptions, e.g., in nuclear medicine where localized metabolic activity can be quantified as a specific uptake value (SUV), the detection and characterization of object location, shape, sharpness, and intensity, and the implications thereof, are subjective and accomplished by highly trained human observers. Today, these interpretations are critical for disease and patient management, including diagnosis, prognosis, staging, and assessment of treatment response. However, with few exceptions, this is largely a subjective operation with variable sensitivity and specificity and high inter-reader variability [1–10].

In 2007, Segal et al. showed that careful characterization of the appearances of liver lesions in contrast-enhanced CT scans could be used to infer their molecular properties [11]. Similarly, in 2008, Brown et al. showed that quantitative analysis of gray-scale image texture on magnetic resonance (MR) scans of patients with oligodendroglioma could be used to predict their genetic signatures, specifically, co-deletion of chromosomes 1p and 19q [12]. These and other early examples showed that not only could appearances on non-invasive imaging be quantified, but that these data could illuminate fundamental molecular properties of cancers [13,14]. Early examples used "semantic features," i.e., categorizations of human observations using a controlled vocabulary [15–17], and these have expanded to use "computational features" that are direct mathematical summarization of image regions. Together, semantic and computational feature classes are the basis of the new field of "radiomics" [18,19], defined as the "high-throughput extraction of quantitative features that result in the conversion of images into mineable data" [20,21] and feature prominently in what is today called "quantitative imaging" [22–24]. This chapter describes the radiomics workflow, including its strengths and its challenges, and highlights the integration of radiomics with clinical data and the molecular characterization of tissue, also known as "radiogenomics" [13,25–31], for the building of predictive models. It focuses on "conventional radiomics" (i.e., machine computation of human-engineered image features) [32]. The more recent developments in radiomics, including the use of artificial intelligence, or "deep learning," to automatically learn informative image features for linking to clinical data (e.g., expression, outcomes) from a set of suitably labeled image and clinical data [33] will be discussed in detail in the final chapter of this book. Most examples in the literature focus on cancer, i.e., imaging and analysis of medical images of tumors.

However, radiomics may be used to characterize any tissue, and so in the following we refer to the regions of imaging data to be analyzed as volumes of interest (VOIs).

1.2 PRINCIPLES AND RATIONALE OF RADIOMICS AND RADIOGENOMICS

Figure 1.1 illustrates the radiomics workflow for any imaging dataset, which may be 2D, 3D, or of higher dimension. Component parts are: (1) identification of the location of the VOI to be analyzed, (2) annotation of the tissue with semantic features, (3) VOI segmentation, i.e., identification of the entire imaged volume of tissue to be analyzed, and (4) feature computation via human-engineered image features. In some cases, "delta" features [34–38] may be computed by comparing individual feature values derived from different images acquired at different times. Collections of imaging features may be also created that combine features computed from multiple imaging methods [14,39–41]. These imaging features may be summarized in a "feature vector."

VOI identification: Each and every VOI to be processed must first be identified, either semi-automatically or manually by a radiologist or automatically using computer aided detection approaches. In cancer imaging, e.g., when multiple tumors are present in a single imaging study, human effort is generally required to identify those that are clinically relevant, as in the case of index lesions scored with RECIST ("response evaluation criteria in solid tumors"). When computing delta radiomics features, matching of tumors across observations will also be required.

Annotation with semantic features: Semantic features are descriptive observations of image content. For example, semantic features of a lung tumor might include "left lower lobe," "pleural attachment," "spiculated," "ground-glass opacity," etc. However, extraction of semantic content from unstructured radiology reports may not be appropriate due to inconsistent and/or ambiguous vocabulary across observations and observers. Further, structured reports may not support the kinds of detailed observations required for making fine distinctions among tumor characteristics that may prove useful in classification or assessing response. Advances in structured reporting include the description of semantic features using a controlled vocabulary, such as RAD-LEX [42]. However, although much effort has been expended to develop more structured reporting, it is still not common in the broad radiology community. Currently, there are at least two systems that facilitate semantic annotation of radiological images using controlled vocabularies, ePAD [43] (Figure 1.2) and the Annotated Image Markup (AIM) data service plug-in [44] to ClearCanvas [45]. Some semantic features, such as location, other morbidities, etc., are meant to be complementary to computational features; others, such as "spherical," "heterogeneous," etc., are correlated

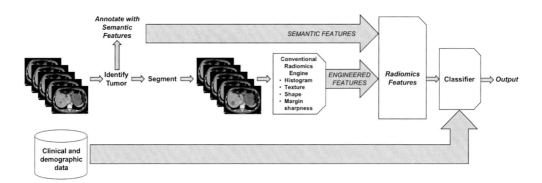

Figure 1.1 (See color insert.) Conventional radiomics workflow combining semantic and human-engineered computational features. These radiomics features can then be combined with clinical and demographic data, before the final classification stage, which generates an output such as benign/malignant, responder/non-responder, probability of 5-year survival, etc.

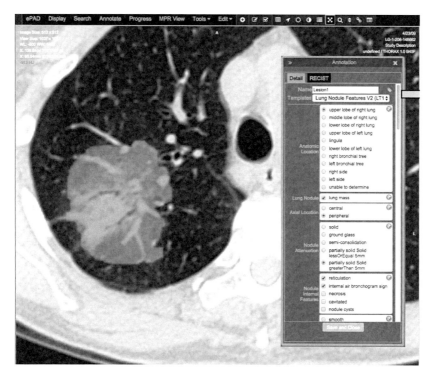

Figure 1.2 (See color insert.) Example of semantic annotation of a part solid part ground glass lung tumor using ePAD. Following tumor segmentation (green), either manually within ePAD or created via other means, an observer selects annotations using a custom template (built, e.g., using the AIM Template Builder [96]) or one of several available. The example in this figure shows a subset of the annotation topics that are required to complete this annotation.

with computational features [46,47]. One advantage of semantic annotations is that they are immediately translatable, i.e., they can be elicited in clinical environments without specialized algorithms (e.g., segmentation) or workstations. As such, they have shown interesting results in several radiogenomic studies [48–54].

VOI segmentation: Radiomics features can be extracted from arbitrary regions within the image volume: In cancer imaging, e.g., a given region may contain an entire tumor, a subset of the tumor (e.g., a habitat [55,56]; see also Chapter 7), and/or a peritumoral region [57–59] thought to be involved with or affected by the tumor. In all cases, these regions must be unambiguously identified (segmented) and input to the radiomics feature computation algorithms. This segmentation step is the single most problematic aspect of conventional radiomics workflows [60], as the features computed from tumor volumes may be extremely sensitive to the specification of the volume to be analyzed. Each combination of tumor type and image modality presents its own challenges (including volume averaging of tissues within each voxel, tumor contrast with surrounding/adjacent structures, image contrast-to-noise characteristics, and variations of image quality across vendor implementations and time). In addition, many algorithms require operator inputs, such as bounding boxes and/or seed points, and the segmentation outlines and radiomics features computed from them may be sensitive to these inputs. For example, one recent study showed wide variations in lung nodule segmentation outlines across three different algorithms and even across individual algorithms initialized with different user inputs [61]. Another study using the same dataset revealed that many radiomics features, particularly those that quantify shape and margin sharpness, are quite sensitive to segmentation [62]. Thus, the state-of-the art today is such that each segmentation must be reviewed and possibly edited by a human observer in order for the radiomics features computed from it to be trusted. One potential mitigation is to ignore segmentation altogether and to compute only features that do not require complete edge-to-edge coverage of the tumor, i.e., histogram and texture features, which may be less sensitive to the exact tumor definition, and to ignore shape and margin sharpness features, which require accurate and consistent edge delineations [63,64].

Image feature computation: Conventional or human-engineered computational image features can be divided into four classes (Figure 1.3): those that describe (1) shape [65–67], (2) margin sharpness [68,69], (3) histogram features (e.g., mean, variance, kurtosis, maximum, minimum), and (4) texture features [70–77], which describe the spatial variation of gray values within the tumor. Within each class there are hundreds to thousands of individual features, for example, some texture features quantify the spatial variation of gray values across multiple scales and orientations, and shape features can similarly quantify edge irregularity at multiple scales. It is important to recognize that many image features are inter-correlated and, as a result, not all features may add independent predictive power to radiomics models. A standard approach in many studies is to generate an autocorrelation matrix and combine correlated features into a single descriptor [78]. Several groups have made available computer code and processing pipelines for the calculation of image features from volumetric image data and segmentations (or at least volumes of interest): see, e.g., the imaging biomarker explorer [79], the Quantitative Image Feature Engine [80], and pyRadiomics [81]. Figure 1.4 shows an example of a workflow engine that allows cohorts of subjects with imaging studies and segmented VOIs to be processed, producing a vector of imaging features for each subject, and subsequently combine these with clinical data to derive associations and to build predictive models.

Challenges to radiomics: The ultimate utility of conventional radiomics workflows is to generate image features that can then be integrated with other medical data, such as, e.g., diagnosis, survival, response to therapy, mutations or genomic profiles, and demographics, for the purposes of building predictive models for one or more of the clinical variables. Space does not allow room for details, but we note that a primary

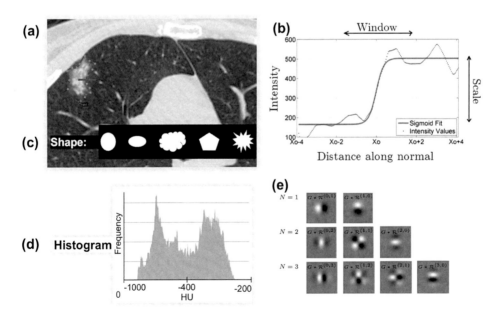

Figure 1.3 (See color insert.) Classes of conventional radiomics features. (a) CT cross-section of a chest containing a VOI of a lung tumor (outlined). (b) Edge sharpness features (window and scale) calculated by fitting intensities along normal (red line in [a]) to a sigmoid function. (c) Shape features include measures of sphericity, lobulation, spiculation, roughness, etc. (d) Gray value histogram features include statistics such as mean, maximum, standard deviation, kurtosis, etc. (e) Texture features measure statistics of spatial frequencies represented in the VOI.

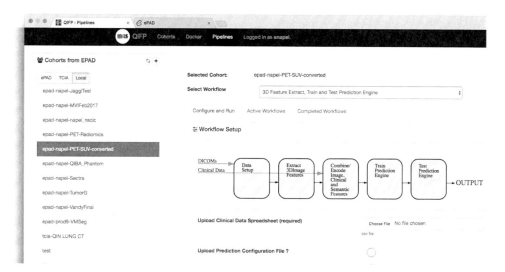

Figure 1.4 (See color insert.) Workflow of the Quantitative Image Feature Engine, which allows cohorts of subjects with imaging studies and segmented VOIs to be processed, producing a vector of imaging features for each subject, and subsequently combine these with clinical data to derive associations and to build predictive models.

concern in these analyses is overfitting, i.e., building statistically significant models based on hundreds to thousands of image features requires data from many multiples of that number of individuals. As noted, many of the features are correlated, and so feature reduction techniques based on redundancy, relevance, and/or sparse regression should be employed wherever possible [82,83]. While many investigators employ cross-validation methods to avoid training and testing on the same data, it is still important to test each predictive model on a completely independent cohort to assess generalizability.

An additional challenge is the need to standardize the methods for calculation of radiomics features so that identically intended features computed from the same data by different algorithms have the same name and values. Indeed, the same study referenced above that compared features computed from multiple segmentations also revealed that implementations from separate institutions of purportedly the same feature sometimes produced different values [62]. Much effort is underway to standardize feature naming and computation conventions, predominately led by the image biomarker standardization initiative [84] and the Quantitative Imaging Network [22–24].

In addition to the challenges raised by segmentation, a final challenge is sensitivity of radiomics features to image acquisition and reconstruction, i.e., the heterogeneity of image acquisitions [60]. Each clinical study uses their own combination of acquisition parameters, such as slice thickness, reconstruction kernel, MR pulse sequences, etc. In addition, many acquisition parameters are optimized for the particular patient under study (e.g., kilovolts [kV], milliamperes [mA] field of view). While radiologist interpretations are somewhat immune to these differences, computational radiomics features are by design sensitive to these choices. In addition, random image noise may also affect many radiomic computations [78,85]. Thus, a predictive model built for lung nodule characterization may be validated in one cohort and fail in another owing to differences in the data acquisition and reconstruction. While there have been some efforts to compensate for these differences, e.g., using phantoms and image interpolation [85–90], and Herculean efforts on protocol standardization by the quantitative imaging biomarkers initiative [91–95] are underway, much work remains before radiomics analyses can derive utility from large data bases collected by multiple institutions. Another possibility to mitigate these issues is through a branch of artificial intelligence known as deep learning, as discussed in Chapter 22.

ACKNOWLEDGMENTS

NIH/NCI U01CA187947, U24 CA180927, R01CA160251.

REFERENCES

1. Zhao B, Lee SM, Lee HJ, Tan Y, Qi J, Persigehl T, Mozley DP, Schwartz LH. Variability in assessing treatment response: Metastatic colorectal cancer as a paradigm. *Clin Cancer Res* 2014; 20:3560–3568.
2. Svahn TM, Macaskill P, Houssami N. Radiologists' interpretive efficiency and variability in true- and false-positive detection when screen-reading with tomosynthesis (3D-mammography) relative to standard mammography in population screening. *Breast* 2015; 24:687–693.
3. Pinsky PF, Gierada DS, Nath PH, Kazerooni E, Amorosa J. National lung screening trial: Variability in nodule detection rates in chest CT studies. *Radiology* 2013; 268:865–873.
4. Penn A, Ma M, Chou BB, Tseng JR, Phan P. Inter-reader variability when applying the 2013 Fleischner guidelines for potential solitary subsolid lung nodules. *Acta Radiol* 2015; 56:1180–1186.
5. Mazor RD, Savir A, Gheorghiu D, Weinstein Y, Abadi-Korek I, Shabshin N. The inter-observer variability of breast density scoring between mammography technologists and breast radiologists and its effect on the rate of adjuvant ultrasound. *Eur J Radiol* 2016; 85:957–962.
6. Hoang JK, Riofrio A, Bashir MR, Kranz PG, Eastwood JD. High variability in radiologists' reporting practices for incidental thyroid nodules detected on CT and MRI. *AJNR Am J Neuroradiol* 2014; 35:1190–1194.
7. Durao AP, Morosolli A, Pittayapat P, Bolstad N, Ferreira AP, Jacobs R. Cephalometric landmark variability among orthodontists and dentomaxillofacial radiologists: A comparative study. *Imaging Sci Dent* 2015; 45:213–220.
8. Di Grezia G, Somma F, Serra N, Reginelli A, Cappabianca S, Grassi R, Gatta G. Reducing costs of breast examination: Ultrasound performance and inter-observer variability of expert radiologists versus residents. *Cancer Invest* 2016; 34:355–360.
9. de Zwart AD, Beeres FJ, Kingma LM, Otoide M, Schipper IB, Rhemrev SJ. Interobserver variability among radiologists for diagnosis of scaphoid fractures by computed tomography. *J Hand Surg Am* 2012; 37:2252–2256.
10. Chan TY, England A, Meredith SM, McWilliams RG. Radiologist variability in assessing the position of the cavoatrial junction on chest radiographs. *Br J Radiol* 2016; 89:20150965.
11. Segal E, Sirlin CB, Ooi C, Adler AS, Gollub J, Chen X, Chan BK et al., Decoding global gene expression programs in liver cancer by noninvasive imaging. *Nat Biotechnol* 2007; 25:675–680.
12. Brown R, Zlatescu M, Sijben A, Roldan G, Easaw J, Forsyth P, Parney I et al., The use of magnetic resonance imaging to noninvasively detect genetic signatures in oligodendroglioma. *Clin Cancer Res* 2008; 14:2357–2362.
13. Rutman AM, Kuo MD. Radiogenomics: Creating a link between molecular diagnostics and diagnostic imaging. *Eur J Radiol* 2009; 70:232–241.
14. Zinn PO, Mahajan B, Sathyan P, Singh SK, Majumder S, Jolesz FA, Colen RR. Radiogenomic mapping of edema/cellular invasion MRI-phenotypes in glioblastoma multiforme. *PLoS One* 2011; 6:e25451.
15. Rubin DL, Mongkolwat P, Kleper V, Supekar K, Channin DS. Annotation and image markup: Accessing and interoperating with the semantic content in medical imaging. *IEEE Intelligent Systems* 2009; 24:57–65.
16. Channin DS, Mongkolwat P, Kleper V, Sepukar K, Rubin DL. The caBIG annotation and image markup project. *J Digit Imaging* 2009.
17. Rubin DL, Rodriguez C, Shah P, Beaulieu C. iPad: Semantic annotation and markup of radiological images. *AMIA Annu Symp Proc* 2008:626–630.
18. Lambin P, Rios-Velazquez E, Leijenaar R, Carvalho S, van Stiphout RG, Granton P, Zegers CM et al., Radiomics: Extracting more information from medical images using advanced feature analysis. *Eur J Cancer* 2012; 48:441–446.

19. Kumar V, Gu Y, Basu S, Berglund A, Eschrich SA, Schabath MB, Forster K et al., Radiomics: The process and the challenges. *Magn Reson Imaging* 2012; 30:1234–1248.

20. Gillies RJ, Kinahan PE, Hricak H. Radiomics: Images are more than pictures, they are data. *Radiology* 2016; 278:563–577.

21. Aerts HJ. The potential of radiomic-based phenotyping in precision medicine: A review. *JAMA Oncol* 2016; 2:1636–1642.

22. Clarke LP, Nordstrom RJ, Zhang H, Tandon P, Zhang Y, Redmond G, Farahani K et al., The quantitative imaging network: NCI's historical perspective and planned goals. *Transl Oncol* 2014; 7:1–4.

23. Nordstrom RJ. Special section guest editorial: Quantitative imaging and the pioneering efforts of Laurence P. Clarke. *J Med Imaging* (Bellingham) 2018; 5:011001.

24. Nordstrom RJ. The quantitative imaging network in precision medicine. *Tomography* 2016; 2:239–241.

25. Jamshidi N, Jonasch E, Zapala M, Korn RL, Brooks JD, Ljungberg B, Kuo MD. The radiogenomic risk score stratifies outcomes in a renal cell cancer phase 2 clinical trial. *Eur Radiol* 2015.

26. Kuo MD, Jamshidi N. Behind the numbers: Decoding molecular phenotypes with radiogenomics— Guiding principles and technical considerations. *Radiology* 2014; 270:320–325.

27. Aerts HJ, Velazquez ER, Leijenaar RT, Parmar C, Grossmann P, Carvalho S, Bussink J et al., Decoding tumour phenotype by noninvasive imaging using a quantitative radiomics approach. *Nat Commun* 2014; 5:4006.

28. Karlo CA, Di Paolo PL, Chaim J, Hakimi AA, Ostrovnaya I, Russo P, Hricak H, Motzer R, Hsieh JJ, Akin O. Radiogenomics of clear cell renal cell carcinoma: Associations between CT imaging features and mutations. *Radiology* 2014; 270:464–471.

29. Jamshidi N, Diehn M, Bredel M, Kuo MD. Illuminating radiogenomic characteristics of glioblastoma multiforme through integration of MR imaging, messenger RNA expression, and DNA copy number variation. *Radiology* 2014; 270:212–222.

30. Abazeed ME, Adams DJ, Hurov KE, Tamayo P, Creighton CJ, Sonkin D, Giacomelli AO, Du C, Fries DF, Wong KK, Mesirov JP, Loeffler JS, Schreiber SL, Hammerman PS, Meyerson M. Integrative radiogenomic profiling of squamous cell lung cancer. *Cancer Res* 2013; 73:6289–6299.

31. Gevaert O, Xu J, Hoang CD, Leung AN, Xu Y, Quon A, Rubin DL, Napel S, Plevritis SK. Non-small cell lung cancer: Identifying prognostic imaging biomarkers by leveraging public gene expression microarray data—Methods and preliminary results. *Radiology* 2012; 264:387–396.

32. Napel S, Giger M. Special section guest editorial: Radiomics and imaging genomics: Quantitative imaging for precision medicine. *J Med Imaging* (Bellingham) 2015; 2:041001.

33. Greenspan H, van Ginneken B, Summers RM. Deep learning in medical imaging: Overview and future promise of an exciting new technique. *IEEE Trans Med Imaging* 2016; 35:1153–1159.

34. van Timmeren JE, Leijenaar RTH, van Elmpt W, Reymen B, Lambin P. Feature selection methodology for longitudinal cone-beam CT radiomics. *Acta Oncol* 2017; 56:1537–1543.

35. Bakr S, Echegaray S, Shah R, Kamaya A, Louie J, Napel S, Kothary N, Gevaert O. Noninvasive radiomics signature based on quantitative analysis of computed tomography images as a surrogate for microvascular invasion in hepatocellular carcinoma: A pilot study. *J Med Imaging* (Bellingham) 2017; 4:041303.

36. Huang Q, Lu L, Dercle L, Lichtenstein P, Li Y, Yin Q, Zong M, Schwartz L, Zhao B. Interobserver variability in tumor contouring affects the use of radiomics to predict mutational status. *J Med Imaging* (Bellingham) 2018; 5:011005.

37. Fave X, Zhang L, Yang J, Mackin D, Balter P, Gomez DR, Followill D et al., Using pretreatment radiomics and delta-radiomics features to predict non-small cell lung cancer patient outcomes. *Int J Radiat Oncol Biol Phys* 2017; 98:249.

38. Fave X, Zhang L, Yang J, Mackin D, Balter P, Gomez D, Followill D et al., Delta-radiomics features for the prediction of patient outcomes in non-small cell lung cancer. *Sci Rep* 2017; 7:588.

39. Gevaert O, Mitchell LA, Achrol AS, Xu J, Echegaray S, Steinberg GK, Cheshier SH, Napel S, Zaharchuk G, Plevritis SK. Glioblastoma multiforme: Exploratory radiogenomic analysis by using quantitative image features. *Radiology* 2015; 276:313.

40. Itakura H, Achrol AS, Mitchell LA, Loya JJ, Liu T, Westbroek EM, Feroze AH et al., Magnetic resonance image features identify glioblastoma phenotypic subtypes with distinct molecular pathway activities. *Sci Transl Med* 2015; 7:303ra138.

41. Li Q, Bai H, Chen Y, Sun Q, Liu L, Zhou S, Wang G, Liang C, Li ZC. A fully-automatic multiparametric radiomics model: Towards reproducible and prognostic imaging signature for prediction of overall survival in glioblastoma multiforme. *Sci Rep* 2017; 7:14331.

42. Langlotz CP. RadLex: A new method for indexing online educational materials. *Radiographics* 2006; 26:1595–1597.

43. Rubin DL, Willrett D, O'Connor MJ, Hage C, Kurtz C, Moreira DA. Automated tracking of quantitative assessments of tumor burden in clinical trials. *Transl Oncol* 2014; 7:23–35.

44. Ronden MI, van Sornsen de Koste JR, Johnson C, Slotman BJ, Spoelstra FOB, Haasbeek CJA, Blom G et al., Incidence of high-risk radiologic features in patients without local recurrence after stereotactic ablative radiation therapy for early-stage non-small cell lung cancer. *Int J Radiat Oncol Biol Phys* 2018; 100:115–121.

45. Roy S, Brown MS, Shih GL. Visual interpretation with three-dimensional annotations (VITA): Three-dimensional image interpretation tool for radiological reporting. *J Digit Imaging* 2014; 27:49–57.

46. Depeursinge A, Kurtz C, Beaulieu C, Napel S, Rubin D. Predicting visual semantic descriptive terms from radiological image data: Preliminary results with liver lesions in CT. *IEEE Trans Med Imaging* 2014; 33:1669–1676.

47. Yip SSF, Liu Y, Parmar C, Li Q, Liu S, Qu F, Ye Z, Gillies RJ, Aerts H. Associations between radiologist-defined semantic and automatically computed radiomic features in non-small cell lung cancer. *Sci Rep* 2017; 7:3519.

48. Gevaert O, Echegaray S, Khuong A, Hoang CD, Shrager JB, Jensen KC, Berry GJ et al., Predictive radiogenomics modeling of EGFR mutation status in lung cancer. *Sci Rep* 2017; 7:41674.

49. Gevaert O, Xu J, Hoang CD, Leung ANC, Xu Y, Quon A, Rubin DL, Napel S, Plevritis SK. Non-small cell lung cancer: Identifying prognostic imaging biomarkers by leveraging public gene expression microarray data—Methods and preliminary results. *Radiology* 2012; 264:387–396.

50. Banerjee S, Wang DS, Kim HJ, Sirlin CB, Chan MG, Korn RL, Rutman AM et al., A computed tomography radiogenomic biomarker predicts microvascular invasion and clinical outcomes in hepatocellular carcinoma. *Hepatology* 2015; 62:792–800.

51. Jamshidi N, Jonasch E, Zapala M, Korn RL, Aganovic L, Zhao H, Tumkur Sitaram R et al., The radiogenomic risk score: Construction of a prognostic quantitative, noninvasive image-based molecular assay for renal cell carcinoma. *Radiology* 2015; 277:114–123.

52. Liu Y, Balagurunathan Y, Atwater T, Antic S, Li Q, Walker RC, Smith GT, Massion PP, Schabath MB, Gillies RJ. Radiological image traits predictive of cancer status in pulmonary nodules. *Clin Cancer Res* 2017; 23:1442–1449.

53. Liu Y, Kim J, Qu F, Liu S, Wang H, Balagurunathan Y, Ye Z, Gillies RJ. CT Features associated with epidermal growth factor receptor mutation status in patients with lung adenocarcinoma. *Radiology* 2016; 280:271–280.

54. Grossmann P, Stringfield O, El-Hachem N, Bui MM, Rios Velazquez E, Parmar C, Leijenaar RT et al. Defining the biological basis of radiomic phenotypes in lung cancer. *Elife* 2017; 6.

55. Napel S, Mu W, Jardim-Perassi B, Aerts HJWL, Gillies R. Quantitative imaging of cancer in the post-genomic era: Radio(geno)mics, deep learning and habitats. *Cancer* 2018; 124:4633–4649.

56. Gatenby RA, Grove O, Gillies RJ. Quantitative imaging in cancer evolution and ecology. *Radiology* 2013; 269:8–15.

57. Wu J, Li B, Sun X, Cao G, Rubin DL, Napel S, Ikeda DM, Kurian AW, Li R. Heterogeneous enhancement patterns of tumor-adjacent parenchyma at MR imaging are associated with dysregulated signaling pathways and poor survival in breast cancer. *Radiology* 2017; 285:401–413.

58. Prasanna P, Patel J, Partovi S, Madabhushi A, Tiwari P. Radiomic features from the peritumoral brain parenchyma on treatment-naive multi-parametric MR imaging predict long versus short-term survival in glioblastoma multiforme: Preliminary findings. *Eur Radiol* 2017; 27:4188–4197.

59. Braman NM, Etesami M, Prasanna P, Dubchuk C, Gilmore H, Tiwari P, Plecha D, Madabhushi A. Intratumoral and peritumoral radiomics for the pretreatment prediction of pathological complete response to neoadjuvant chemotherapy based on breast DCE-MRI. *Breast Cancer Res* 2017; 19:57.

60. El Naqa I, Napel S. Radiogenomics is the future of treatment response assessment in clinical oncology. *Med Phys* 2018; 45:4325–4328.

61. Kalpathy-Cramer J, Zhao B, Goldgof D, Gu Y, Wang X, Yang H, Tan Y, Gillies R, Napel S. A comparison of lung nodule segmentation algorithms: Methods and results from a multi-institutional study. *J Digit Imaging* 2016; 29:476–487.

62. Kalpathy-Cramer J, Mamomov A, Zhao B, Lu L, Cherezov D, Napel S, Echegaray S et al., Radiomics of lung nodules: A multi-institutional study of robustness and agreement of quantitative imaging features. *Tomography* 2016; 2:430–437.

63. Echegaray S, Gevaert O, Shah R, Kamaya A, Louie J, Kothary N, Napel S. Core samples for radiomics features that are insensitive to tumor segmentation: Method and pilot study using CT images of hepato-cellular carcinoma. *J Med Imaging* (Bellingham) 2015; 2:041011.

64. Echegaray S, Nair V, Kadoch M, Leung A, Rubin D, Gevaert O, Napel S. A rapid segmentation-insensitive "digital biopsy" method for radiomic feature extraction: Method and pilot study using CT images of non-small cell lung cancer. *Tomography* 2016; 2:283–294.

65. Xu J, Faruque J, Beaulieu CF, Rubin D, Napel S. A comprehensive descriptor of shape: Method and application to content-based retrieval of similar appearing lesions in medical images. *J Digit Imaging* 2012; 25:121–128.

66. Bankman IN, Spisz TS, Pavlapoilos S. Two-dimensional shape and texture quantiification. In: Bankman IN, (Ed.), *Handbook of Medical Image Processing and Analysis.* Amsterdam, the Netherlands: Elsevier/Academic Press, 2009.

67. Grove O, Berglund AE, Schabath MB, Aerts HJ, Dekker A, Wang H, Velazquez ER et al. Quantitative computed tomographic descriptors associate tumor shape complexity and intratumor heterogeneity with prognosis in lung adenocarcinoma. *PLoS One* 2015; 10:e0118261.

68. Xu J, Napel S, Greenspan H, Beaulieu CF, Agrawal N, Rubin D. Quantifying the margin sharpness of lesions on radiological images for content-based image retrieval. *Med Phys* 2012; 39:5405–5418.

69. Mu T, Nandi AK, Rangayyan RM. Classification of breast masses using selected shape, edge-sharpness, and texture features with linear and kernel-based classifiers. *J Digit Imaging* 2008; 21:153–169.

70. Zayed N, Elnemr HA. Statistical analysis of haralick texture features to discriminate lung abnormalities. *Int J Biomed Imaging* 2015; 2015:267807.

71. Yang D, Rao G, Martinez J, Veeraraghavan A, Rao A. Evaluation of tumor-derived MRI-texture features for discrimination of molecular subtypes and prediction of 12-month survival status in glioblastoma. *Med Phys* 2015; 42:6725–6735.

72. Vallieres M, Freeman CR, Skamene SR, El Naqa I. A radiomics model from joint FDG-PET and MRI texture features for the prediction of lung metastases in soft-tissue sarcomas of the extremities. *Phys Med Biol* 2015; 60:5471–5496.

73. Liu H, Tan T, van Zelst J, Mann R, Karssemeijer N, Platel B. Incorporating texture features in a computer-aided breast lesion diagnosis system for automated three-dimensional breast ultrasound. *J Med Imaging* (Bellingham) 2014; 1:024501.

74. Korfiatis P, Kline TL, Coufalova L, Lachance DH, Parney IF, Carter RE, Buckner JC, Erickson BJ. MRI texture features as biomarkers to predict MGMT methylation status in glioblastomas. *Med Phys* 2016; 43:2835–2844.

75. Depeursinge A, Yanagawa M, Leung AN, Rubin DL. Predicting adenocarcinoma recurrence using computational texture models of nodule components in lung CT. *Med Phys* 2015; 42:2054–2063.

76. Depeursinge A, Foncubierta-Rodriguez A, Van de Ville D, Muller H. Rotation-covariant texture learning using steerable riesz wavelets. *IEEE Trans Image Process* 2014; 23:898–908.

77. Wibmer A, Hricak H, Gondo T, Matsumoto K, Veeraraghavan H, Fehr D, Zheng J et al. Haralick texture analysis of prostate MRI: Utility for differentiating non-cancerous prostate from prostate cancer and differentiating prostate cancers with different Gleason scores. *Eur Radiol* 2015; 25:2840–2850.

78. Balagurunathan Y, Kumar V, Gu Y, Kim J, Wang H, Liu Y, Goldgof DB et al. Test-retest reproducibility analysis of lung CT image features. *J Digit Imaging* 2014; 27:805–823.
79. Zhang L, Fried DV, Fave XJ, Hunter LA, Yang J, Court LE. IBEX: An open infrastructure software platform to facilitate collaborative work in radiomics. *Med Phys* 2015; 42:1341–1353.
80. Echegaray S, Bakr S, Rubin DL, Napel S. Quantitative Image Feature Engine (QIFE): An open-source, modular engine for 3D quantitative feature extraction from volumetric medical images. *J Digit Imaging* 2017.
81. van Griethuysen JJM, Fedorov A, Parmar C, Hosny A, Aucoin N, Narayan V, Beets-Tan RGH, Fillion-Robin JC, Pieper S, Aerts H. Computational radiomics system to decode the radiographic phenotype. *Cancer Res* 2017; 77:e104–e107.
82. Peng HC, Long FH, Ding C. Feature selection based on mutual information: Criteria of max-dependency, max-relevance, and min-redundancy. *IEEE Trans Pattern Anal Mach Intell* 2005; 27:1226–1238.
83. Tibshirani R. Regression shrinkage and selection via the lasso. *J R Stat Soc Series B Stat Methodol* 1996; 58:267–288.
84. Zwanenburg A, Leger S, Vallières M, Löck S, Initiative ftIBS. Image biomarker standardisation initiative. arXiv preprint arXiv:1612.07003v5 2017.
85. Solomon J, Mileto A, Nelson RC, Roy Choudhury K, Samei E. Quantitative features of liver lesions, lung nodules, and renal stones at multi-detector row CT examinations: Dependency on radiation dose and reconstruction algorithm. *Radiology* 2016; 279:185–194.
86. Fave X, Mackin D, Yang J, Zhang J, Fried D, Balter P, Followill D et al., Can radiomics features be reproducibly measured from CBCT images for patients with non-small cell lung cancer? *Med Phys* 2015; 42:6784–6797.
87. Yasaka K, Akai H, Mackin D, Court L, Moros E, Ohtomo K, Kiryu S. Precision of quantitative computed tomography texture analysis using image filtering: A phantom study for scanner variability. *Medicine* (Baltimore) 2017; 96:e6993.
88. Shafiq-Ul-Hassan M, Zhang GG, Latifi K, Ullah G, Hunt DC, Balagurunathan Y, Abdalah MA et al., Intrinsic dependencies of CT radiomic features on voxel size and number of gray levels. *Med Phys* 2017; 44:1050–1062.
89. Mackin D, Fave X, Zhang L, Yang J, Jones AK, Ng CS, Court L. Harmonizing the pixel size in retrospective computed tomography radiomics studies. *PLoS One* 2017; 12:e0178524.
90. Mackin D, Fave X, Zhang L, Fried D, Yang J, Taylor B, Rodriguez-Rivera E, Dodge C, Jones AK, Court L. Measuring computed tomography scanner variability of radiomics features. *Invest Radiol* 2015; 50:757–765.
91. Sullivan DC, Obuchowski NA, Kessler LG, Raunig DL, Gatsonis C, Huang EP, Kondratovich M et al., Group R-QMW. Metrology standards for quantitative imaging biomarkers. *Radiology* 2015; 277:813–825.
92. Obuchowski NA, Buckler A, Kinahan P, Chen-Mayer H, Petrick N, Barboriak DP, Bullen J, Barnhart H, Sullivan DC. Statistical issues in testing conformance with the quantitative imaging biomarker alliance (QIBA) profile claims. *Acad Radiol* 2016; 23:496–506.
93. Kessler LG, Barnhart HX, Buckler AJ, Choudhury KR, Kondratovich MV, Toledano A, Guimaraes AR et al., The emerging science of quantitative imaging biomarkers terminology and definitions for scientific studies and regulatory submissions. *Stat Methods Med Res* 2015; 24:9–26.
94. Chen-Mayer HH, Fuld MK, Hoppel B, Judy PF, Sieren JP, Guo J, Lynch DA, Possolo A, Fain SB. Standardizing CT lung density measure across scanner manufacturers. *Med Phys* 2017; 44:974–985.
95. Chao SL, Metens T, Lemort M. TumourMetrics: A comprehensive clinical solution for the standardization of DCE-MRI analysis in research and routine use. *Quant Imaging Med Surg* 2017; 7:496–510.
96. AIM Template Builder 2.0 User's Guide. https://wiki.nci.nih.gov/display/AIM/AIM+Template+Builder+2.0+User%27s+Guide; Accessed: April 12, 2018.

PART

TECHNICAL BASIS

Imaging informatics
An overview

ASSAF HOOGI AND DANIEL L. RUBIN

Imaging has an important role in healthcare and is considered as a complementary knowledge to lab tests, patient demographic, and family information. The major topics in imaging informatics, which are also covered in this chapter, and some of them will be further detailed in later chapters, include image resolution, image enhancement, denoising, fusion, and knowledge extraction. Commonly, after the images are acquired from the imaging modalities, they are processed by these methods in a processing pipeline with the end goal of producing image features or actionable knowledge to improve healthcare. Our focus is on the image processing methods, but it is important to note that there are other informatics systems supporting routine radiology workflow, such as the picture archiving and communication systems (PACS), reporting systems (such as voice recognition and structured reporting), report information systems (RIS), and the electronic medical record system (EMR) that we do not include here, but are well covered in other works [1,2].

2.1 IMAGING MODALITIES

Imaging modalities can be divided into anatomic, functional, and molecular screening. Each of them can contribute to disease interpretation, and the combination of their extracted information can supply a significant added value [3,4]. Anatomic imaging modalities such as computerized tomography (CT) [5,6], mammography [7], and MRI [8,9] need to have high spatial-resolution to accurately identify the structure of objects of interest (e.g., organs, lesions). Other imaging techniques focus on functional imaging and tend to have a lower spatial resolution, comparing with anatomic imaging. Relying on these techniques for an accurate structural identification will not be accurate enough, though they will able to help in understanding the functional significance of the specific tissue. Example functional modalities include ultrasound Doppler that can help to analyze the blood flow and positron emission tomography [10,11] that can help in recognizing metabolic processes. Therefore, functional imaging can help in finding ischemia, inflammation, necrotic regions, and cancer tissue. Another emerging technique is molecular imaging, a sub-field of functional imaging, in which we measure the expression of particular genes [12–15]. This provides a potential platform for linking specific imaging analysis with a specific molecular gene expression pattern [16–19].

Lastly, the importance of integrating the image information from two imaging types (e.g., molecular, functional, anatomical) to detect tissue changes was recognized and became very popular over the recent years. Such a cross-modality analysis can help in understanding the role of specific genes on the tissue structure/functionality. Two approaches were introduced: (1) a single machine such as functional magnetic resonance (fMRI) that can supply both anatomical and functional analyses [20,21], and (2) fusion of the information that is obtained from two separate modalities and will be detailed below in the "Image Registration" section [22–26].

2.2 IMAGE FUSION

Image fusion is applied to construct a more detailed and representative output image by using image registration, feature extraction, and semantic information conclusion [27,28]. Image registration is the process of mapping input images with the help of reference image. Image registration is considered as an optimization problem whose goal is to maximize the similarity between the images [29]. Applications in the medical field usually include registration of anatomical modalities (e.g., CT, MRI) with functional modalities such as PET, single-photon emission computed tomography (SPECT), or f-MRI. This kind of registration supplies complementary information that can help a lot for intervention and treatment planning [30,31], computer-aided diagnosis and disease following-up [32], surgery simulation [32], atlas building and comparison [33–38], radiation therapy [39,40], assisted/guided surgery [32,41–43], anatomy segmentation [44–49], computational model building [100], and image subtraction for contrast-enhanced images [51]. Image registration can be roughly done by three different approaches. First, by measuring the intensity similarity between different pixels/voxels in the images. It can be done by applying rigid or non-rigid techniques. The main difference between them is that in case of rigid approach, we assume that the whole object is moving together and in non-rigid—different local distortions can occur in different locations within the object. Second, by detecting key points within the images and then matching those points. In this group of techniques, one can find SIFT and SURF [52–54]. These key points must be characterized by a distinction to the spatial neighbors, invariance to the original image variations, robustness against noise, and with high computational efficiency. Image descriptors are then used to represent the extracted key points. After defining the coordinates of the key points, the transformation function estimates the geometric relation between the images. The transformation functions can be selected based on the images that are needed to be registered, although, it is hard to find a single transformation function that is better for all types of images due to the strengths and weaknesses associated with each function. After registering the images, multiple image features can be extracted, and the information that was extracted from the different imaging modalities can be fused and lead to the overall clinical decision. Third, an image registration can be done by atlas-based approaches [55–57]. Figure 2.1 shows several interesting fusion examples [58].

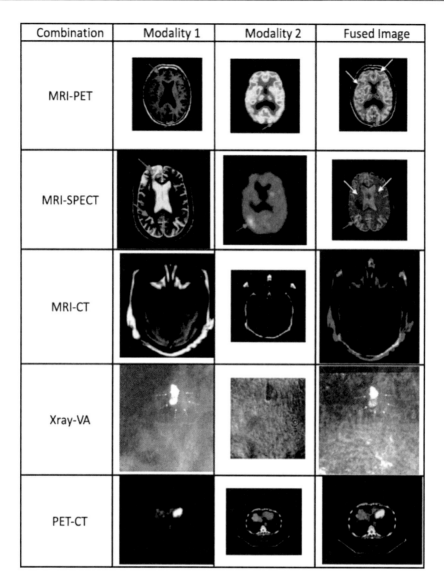

Combination	Modality 1	Modality 2	Fused Image
MRI-PET			
MRI-SPECT			
MRI-CT			
Xray-VA			
PET-CT			

Figure 2.1 (See color insert.) Examples of multimodal medical image fusion. The combination of modality 1 with modality 2 using specific image fusion techniques results in improved feature visibility for medical diagnostics and assessments as shown in the fused image column.

2.3 IMAGE RESOLUTION

Image quality depends, among all, on image resolution. Image resolution is divided into spatial, temporal, and contrast resolutions. Spatial resolution refers to the ability of differentiating between two points in the space. High spatial resolution means that we can separate well between two points that are very close to each other in the space. Temporal resolution represents the number of images that can be acquired per second. Real-time application must have a high frame rate, means high number of images per second. An optimal imaging modality would produce images with high spatial and temporal resolution, however, usually there is a trade-off between these two. There are many super-resolution methods that were developed to improve the image resolution [59–65]. Most of these techniques can be divided into four different groups—prediction models [66], edge-based models [67], image statistics [61,68,69], and patch-based models [62,63,70–75].

Contrast resolution refers to differences in spatially adjacent pixels or their local surrounding and is the basis for recognizing anatomic structures and abnormalities, which differ from adjacent regions through local differences in pixel values. Imaging contrast agents can help to increase the contrast resolution. Contrast agents have different imaging characteristics than the body tissues, and as a result, they can enhance the contrast differences between themselves and the surrounding regions. The composition of contrast agents varies according to the modality characteristics and to optimally be visible based on the physical basis of image formation. Over the recent years, advances in molecular biology have led to the ability to design contrast agents that are highly specific for individual molecules, and as a result, only the specific region of interest is highlighted in the image [76–78].

2.4 IMAGE ENHANCEMENT

The goal of image enhancement is to improve the visual appearance of the image by converting a low-quality image to a high-quality image [79] and to enable better automated image analysis such as detection, segmentation, and recognition. An optimal image enhancement technique should supply: (1) time and computational efficiency, (2) simple implementation, (3) robustness to noise to avoid noise enhancement and to different kinds of images, (4) structure preservation to keep image texture, and (5) continuity, which means that a small change in the input should cause a only a small change in the output. Enhancement procedures can be mainly divided into two classes—spatial domain methods and transform domain methods. Spatial domain methods are very popular for image enhancement, and they incorporate different histogram manipulations such as histogram equalization to automatically determine a transformation function producing an output image with a uniform histogram [80,81]. Another approach is histogram matching, wherein we generate an image that has the same intensity distribution as a predefined desired histogram. Gamma correction is another popular technique to stretch the histogram of a region of interest, separately and independently enhancing each local region. Celik and Tjahjadi proposed an adaptive image equalization algorithm using a Gaussian mixture model and a partition of the dynamic range of the image [82]. Bilateral Bezier curve (BBC) is another approach that was introduced in [83]. It divides the image into dark and bright regions, calculating the transformation function for each region separately. However, BBC often generates significant distortions in the image due to brightening and over-enhancement [84]. Tsai et al. propose a decision tree-based contrast enhancement technique, which first classifies images into six groups, and then applies a piecewise linear transformation separately for each group of the images [85]. However, the classification is done manually, by choosing six different thresholds. This is a time-consuming, tedious, and data-dependent procedure.

Transform enhancement frameworks incorporate techniques such as Fourier transforms, and as a result, the image is enhanced by modifying the frequency substance of the image [2]. Multiscale wavelet transform is also a popular method for image enhancement as is presented in [86–88].

2.5 IMAGE DENOISING

Similar to image enhancement, image denoising approaches can be categorized as spatial domain, transform domain, and dictionary learning-based approaches [53]. Spatial domain methods include local and non-local filters, which exploit the similarities between the statistics of different regions in the image. The main difference between local and non-local frameworks is the size of the surrounding region that an examined region is compared with. A large number of local filtering algorithms have been designed for noise reduction such as wavelet filter [87], Wiener filter [90], least mean squares filter, bilateral filter [91], anisotropic filtering [92], blind source separation [93], and co-occurrence filter [94]. Local methods are effective in terms of time complexity and also for considering the more relevant information within closer regions only. However, local frameworks are more sensitive than non-local ones to high amounts of image noise. Non-local frameworks filter the examined region by considering its correlation with other regions within the whole image [95–107]. Even though they are better than local filters for dealing with high noise levels, their major drawback is that they still create artifacts such as over-smoothing [108]. The second category is transform domain methods, wherein the image patches are represented by their frequency content [109–112]. These methods usually achieve better performance compared to spatial domain methods, because they have higher level properties such as sparsity and multiresolution [113,114]. The general idea of dictionary learning-based denoising methods is to learn a high-quality sparse representation of normal and noisy regions that will be used later to denoise the noisy ones [115–117].

2.6 KNOWLEDGE RETRIEVAL

Knowledge can be used in medical image analysis in two different levels/directions. On one direction, we can incorporate prior high-level knowledge in the developed models in order to improve them and obtain better image analysis. This type of information is usually a common knowledge which is relevant to the overall task (e.g., disease or the imaging modality) and can be used for registration and segmentation. On the other direction, low-level knowledge can be extracted from the specific analyzed image. This step is usually applied after segmenting the region of interest (ROI) and extracting the features inside. This retrieved knowledge can help in accurate monitoring of the patient's situation.

2.6.1 IMAGE SEGMENTATION

Image segmentation is a well-explored field and will be extensively detailed in a later chapter in the book (Chapter 6). Briefly, segmentation of images involves an automated annotation of a relevant ROI within an image. Traditional segmentation approaches mostly include thresholding [118,119], region-growing [120,121], watershed [121,122], clustering [123,124], active contours and level sets [125–127], atlas-based [128,129], and graph-based models [130,131]. Popular segmentation techniques can be divided into edge-based, regions-based, model-based or knowledge-based, and machine learning-based approaches. Edge-based segmentation relies on detecting and analyzing the boundaries of an object. However, in cases of noisy images, low contrast images, or incomplete broken boundaries, edge-based approaches will not perform well. Region-based techniques can handle better with these challenges because they consider the statistics inside a ROI, thus an object will be accurately segmented as long as the background-foreground statistics are different from each other. Both region-based and edge-based segmentation are essentially low-level techniques that only focus on local regions in the raw image data. A popular alternative method for medical image segmentation is a model-based deformable models (e.g., active contour, level set, active appearance models). These segmentation approaches have been established as one of the highly successful methods for image analysis. By developing a model that contains information about the expected shape and appearance of the structure of interest to new images, the segmentation is conducted in a top-down fashion. Due to the significant a priori information, this approach is more stable against local image artifacts and perturbations than conventional low-level algorithms that consider the image data only. Information about common variations has to be included in the model. A straight-forward approach to gather this information is to examine a number of training shapes by statistical means, leading to statistical shape models (SSMs). Well-known methods in that area are the active shape models [132,133] and active appearance models [134,135]. Figure 2.2 presents automated segmentation results, based on the method that is presented in [127].

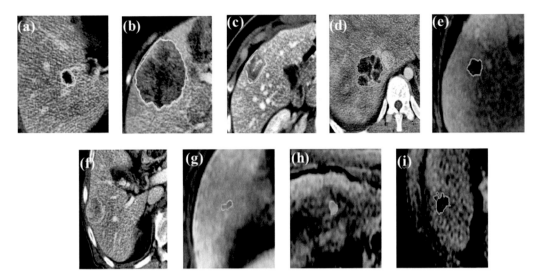

Figure 2.2 (See color insert.) Automatic segmentations (yellow) of liver lesions in CT (a–d, f) and MRI (e, g–i) images obtained using the method that is presented in [127] with the piecewise constant model (PC). (a,b) different lesion sizes (Dice of 0.88, 0.91, respectively), (b–d) heterogeneous lesions (Dice of 0.91,0.92, and 0.89, respectively), (e) homogeneous lesion (Dice of 0.97), (f,g) low contrast lesions (Dice of 0.92, and 0.96, respectively), (h,i) noisy background (Dice of 0.93,0.86 respectively). Green contours represent the manual annotations.

2.6.2 FEATURE EXTRACTION

After segmenting the desired ROI, image features can be extracted. These images can be first-order features such as pixels' intensity or higher-order features such as more complex texture features. These features can be considered as a sparse representation of the whole image data. In order to reduce the feature space dimensionality, the feature extraction procedure is usually followed by a feature selection step. Feature selection is the technique of selecting a subset of dominant features for building robust learning models by keeping the most dominant features only. Feature selection also helps people acquire better understanding about their data by telling them what are the important features and how they are related with each other and with the image itself. There are several common-used methods for features selection, including principal components analysis (PCA) [136], linear discriminant analysis (LDA) [137], least absolute shrinkage and selection operator (LASSO) [138], and generalized linear models with elastic-net penalties (GLMNET) [139] that are usually used in case that we have many more features than patients. It combines L_1 and L_2 losses by integrating LASSO and ridge regression.

2.7 DEEP LEARNING

Machine learning approaches also have been popular for medical image analysis as the strong computational resources became available. Machine learning and specifically deep learning architectures can be used in different types of tasks. However, their main limitation is the need of a lot of training labeled data. Many recent methods were developed to tackle those limitations. Generative adversarial networks (GANs), w-GANs, and stacked-GANs can be used for data augmentation and for the improvement of the image quality [140]. Convolutional neural networks such as U-Net [141,142] and V-Net [143] were designed specifically to deal with medical domain challenges such as small amount of labeled data. Autoencoders, variational autoencoders, and stacked-autoencoders can be used for image denoising and as an unsupervised features extractor [144–146]. Other methods were designed to handle with various of classifications tasks such as lesion detection [147,148], segmentation [142,149], and disease classification [150,151].

REFERENCES

1. D. S. Mendelson and D. L. Rubin, "Imaging Informatics," *Acad. Radiol.*, vol. 20, no. 10, pp. 1195–1212, 2013.

2. D. L. Rubin, H. Greenspan, and J. F. Brinkley, "Biomedical imaging informatics," in *Biomedical Informatics*, E. H. Shortliffe and J. J. Cimino (Eds.), London, UK: Springer, 2014, pp. 285–327.

3. M. R. Oliva and S. Saini, "Liver cancer imaging: Role of CT, MRI, US and PET," *Cancer Imaging*, vol. 4, no. Spec No A, pp. S42–S46, 2004.

4. Health Quality Ontario. "Ultrasound as an adjunct to mammography for breast cancer screening: A health technology assessment," *Ont. Health Technol. Assess. Ser.*, vol. 16, no. 15, pp. 1–71, 2016.

5. P. B. Bach et al., "Benefits and harms of CT screening for lung cancer: A systematic review," *JAMA*, vol. 307, no. 22, pp. 2418–2429, 2012.

6. R. Bar-Shalom et al., "Clinical performance of PET/CT in evaluation of cancer: Additional value for diagnostic imaging and patient management," *J. Nucl. Med.*, vol. 44, no. 8, pp. 1200–1209, 2003.

7. J. T. Schousboe, K. Kerlikowske, A. Loh, and S. R. Cummings, "Personalizing mammography by breast density and other risk factors for breast cancer: Analysis of health benefits and cost-effectiveness," *Ann. Intern. Med.*, vol. 155, no. 1, pp. 10–20, 2011.

8. A. R. Padhani et al., "Diffusion-weighted magnetic resonance imaging as a cancer biomarker: Consensus and recommendations," *Neoplasia N. Y. N*, vol. 11, no. 2, pp. 102–125, 2009.

9. J. H. Kim et al., "Breast cancer heterogeneity: MR imaging texture analysis and survival outcomes," *Radiology*, vol. 282, no. 3, pp. 665–675, 2016.

10. M. E. Phelps, "PET: The merging of biology and imaging into molecular imaging," *J. Nucl. Med. Off. Publ. Soc. Nucl. Med.*, vol. 41, no. 4, pp. 661–681, 2000.

11. J.-L. Alberini et al., "Single photon emission tomography/computed tomography (SPET/CT) and positron emission tomography/computed tomography (PET/CT) to image cancer," *J. Surg. Oncol.*, vol. 103, no. 6, pp. 602–606, 2011.

12. R. Popovtzer et al., "Targeted gold nanoparticles enable molecular CT imaging of cancer," *Nano Lett.*, vol. 8, no. 12, pp. 4593–4596, 2008.

13. S. S. Gambhir, "Molecular imaging of cancer with positron emission tomography," *Nat. Rev. Cancer*, vol. 2, no. 9, pp. 683–693, 2002.

14. R. Weissleder, "Molecular imaging in cancer," *Science*, vol. 312, no. 5777, pp. 1168–1171, 2006.

15. J. M. Hoffman and S. S. Gambhir, "Molecular imaging: The vision and opportunity for radiology in the future," *Radiology*, vol. 244, no. 1, pp. 39–47, 2007.

16. A. M. Rutman and M. D. Kuo, "Radiogenomics: Creating a link between molecular diagnostics and diagnostic imaging," Eur. J. Radiol., vol. 70, no. 2, pp. 232–241, 2009.

17. C. A. Karlo et al., "Radiogenomics of clear cell renal cell carcinoma: Associations between CT imaging features and mutations," *Radiology*, vol. 270, no. 2, pp. 464–471, 2013.

18. M. D. Kuo, J. Gollub, C. B. Sirlin, C. Ooi, and X. Chen, "Radiogenomic analysis to identify imaging phenotypes associated with drug response gene expression programs in hepatocellular carcinoma," *J. Vasc. Interv. Radiol.*, vol. 18, no. 7, pp. 821–830, 2007.

19. P. Kickingereder et al., "Radiogenomics of glioblastoma: Machine learning–based classification of molecular characteristics by using multiparametric and multiregional MR imaging features," *Radiology*, vol. 281, no. 3, pp. 907–918, 2016.

20. S. K. Conroy et al., "Alterations in brain structure and function in breast cancer survivors: Effect of post-chemotherapy interval and relation to oxidative DNA damage," *Breast Cancer Res. Treat.*, vol. 137, no. 2, pp. 493–502, 2013.

21. B. Cimprich et al., "Prechemotherapy alterations in brain function in women with breast cancer," *J. Clin. Exp. Neuropsychol.*, vol. 32, no. 3, pp. 324–331, 2010.

22. O. Demirci et al., "A review of challenges in the use of fMRI for disease classification/characterization and a projection pursuit application from a multi-site fMRI schizophrenia study," *Brain Imaging Behav.*, vol. 2, no. 3, pp. 207–226, 2008.

23. R. Kalash et al., "Use of functional magnetic resonance imaging in cervical cancer patients with incomplete response on positron emission tomography/computed tomography after image-based high-dose-rate brachytherapy," *Int. J. Radiat. Oncol.*, 2018.

24. M. B. de Ruiter et al., "Late effects of high-dose adjuvant chemotherapy on white and gray matter in breast cancer survivors: Converging results from multimodal magnetic resonance imaging," *Hum. Brain Mapp.*, vol. 33, no. 12, pp. 2971–2983, 2012.

25. W. A. Berg et al., "Detection of breast cancer with addition of annual screening ultrasound or a single screening MRI to mammography in women with elevated breast cancer risk," *JAMA*, vol. 307, no. 13, pp. 1394–1404, 2012.

26. P. Skaane et al., "Comparison of digital mammography alone and digital mammography plus tomosynthesis in a population-based screening program," *Radiology*, vol. 267, no. 1, pp. 47–56, 2013.

27. F. P. M. Oliveira and J. M. R. S. Tavares, "Medical image registration: A review," *Comput. Methods Biomech. Biomed. Engin.*, vol. 17, no. 2, pp. 73–93, 2014.

28. J. P. W. Pluim and J. M. Fitzpatrick, "Image registration," *IEEE Trans. Med. Imaging*, vol. 22, no. 11, pp. 1341–1343, 2003.

29. F. E.-Z. A. El-Gamal, M. Elmogy, and A. Atwan, "Current trends in medical image registration and fusion," *Egypt. Inform. J.*, vol. 17, no. 1, pp. 99–124, 2016.

30. D. T. Gering et al., "An integrated visualization system for surgical planning and guidance using image fusion and an open MR," *J. Magn. Reson. Imaging JMRI*, vol. 13, no. 6, pp. 967–975, 2001.

31. M. Staring, U. A. van der Heide, S. Klein, M. A. Viergever, and J. P. W. Pluim, "Registration of cervical MRI using multifeature mutual information," *IEEE Trans. Med. Imaging*, vol. 28, no. 9, pp. 1412–1421, 2009.

32. X. Huang, J. Ren, G. Guiraudon, D. Boughner, and T. M. Peters, "Rapid dynamic image registration of the beating heart for diagnosis and surgical navigation," *IEEE Trans. Med. Imaging*, vol. 28, no. 11, pp. 1802–1814, 2009.

33. P. A. Freeborough and N. C. Fox, "Modeling brain deformations in Alzheimer disease by fluid registration of serial 3D MR images," *J. Comput. Assist. Tomogr.*, vol. 22, no. 5, pp. 838–843, 1998.

34. K. A. Ganser, H. Dickhaus, R. Metzner, and C. R. Wirtz, "A deformable digital brain atlas system according to Talairach and Tournoux," *Med. Image Anal.*, vol. 8, no. 1, pp. 3–22, 2004.

35. S. Joshi, B. Davis, M. Jomier, and G. Gerig, "Unbiased diffeomorphic atlas construction for computational anatomy," *NeuroImage*, vol. 23, Suppl 1, pp. S151–S160, 2004.

36. A. D. Leow et al., "Longitudinal stability of MRI for mapping brain change using tensor-based morphometry," *NeuroImage*, vol. 31, no. 2, pp. 627–640, 2006.

37. C. Wu, P. E. Murtha, and B. Jaramaz, "Femur statistical atlas construction based on two-level 3D non-rigid registration," *Comput. Aided Surg. Off. J. Int. Soc. Comput. Aided Surg.*, vol. 14, no. 4–6, pp. 83–99, 2009.

38. A. Gooya, G. Biros, and C. Davatzikos, "Deformable registration of glioma images using EM algorithm and diffusion reaction modeling," *IEEE Trans. Med. Imaging*, vol. 30, no. 2, pp. 375–390, 2011.

39. W. C. Lavely et al., "Phantom validation of coregistration of PET and CT for image-guided radiotherapy," *Med. Phys.*, vol. 31, no. 5, pp. 1083–1092, 2004.

40. M. Foskey et al., "Large deformation three-dimensional image registration in image-guided radiation therapy," *Phys. Med. Biol.*, vol. 50, no. 24, pp. 5869–5892, 2005.

41. C. R. Maurer, J. M. Fitzpatrick, M. Y. Wang, R. L. Galloway, R. J. Maciunas, and G. S. Allen, "Registration of head volume images using implantable fiducial markers," *IEEE Trans. Med. Imaging*, vol. 16, no. 4, pp. 447–462, 1997.

42. A. Hurvitz and L. Joskowicz, "Registration of a CT-like atlas to fluoroscopic X-ray images using intensity correspondences," *Int. J. Comput. Assist. Radiol. Surg.*, vol. 3, no. 6, p. 493, 2008.

43. A. P. King et al., "Registering preprocedure volumetric images with intraprocedure 3-D ultrasound using an ultrasound imaging model," *IEEE Trans. Med. Imaging*, vol. 29, no. 3, pp. 924–937, 2010.

44. A. F. Frangi, M. Laclaustra, and P. Lamata, "A registration-based approach to quantify flow-mediated dilation (FMD) of the brachial artery in ultrasound image sequences," *IEEE Trans. Med. Imaging*, vol. 22, no. 11, pp. 1458–1469, 2003.

45. L. Dornheim, K. D. Tönnies, and K. Dixon, "Automatic segmentation of the left ventricle in 3D SPECT data by registration with a dynamic anatomic model," *Med. Image Comput. Comput.-Assist. Interv. MICCAI Int. Conf. Med. Image Comput. Comput.-Assist. Interv.*, vol. 8, no. Pt 1, pp. 335–342, 2005.

46. S. Martin, V. Daanen, and J. Troccaz, "Atlas-based prostate segmentation using an hybrid registration," *Int. J. Comput. Assist. Radiol. Surg.*, vol. 3, no. 6, pp. 485–492, 2008.

47. I. Isgum, M. Staring, A. Rutten, M. Prokop, M. A. Viergever, and B. van Ginneken, "Multi-atlas-based segmentation with local decision fusion—Application to cardiac and aortic segmentation in CT scans," *IEEE Trans. Med. Imaging*, vol. 28, no. 7, pp. 1000–1010, 2009.

48. Y. Gao, R. Sandhu, G. Fichtinger, and A. R. Tannenbaum, "A coupled global registration and segmentation framework with application to magnetic resonance prostate imagery," *IEEE Trans. Med. Imaging*, vol. 29, no. 10, pp. 1781–1794, 2010.

49. X. Zhuang, K. S. Rhode, R. S. Razavi, D. J. Hawkes, and S. Ourselin, "A registration-based propagation framework for automatic whole heart segmentation of cardiac MRI," *IEEE Trans. Med. Imaging*, vol. 29, no. 9, pp. 1612–1625, 2010.

50. N. M. Grosland, R. Bafna, and V. A. Magnotta, "Automated hexahedral meshing of anatomic structures using deformable registration," *Comput. Methods Biomech. Biomed. Engin.*, vol. 12, no. 1, pp. 35–43, 2009.

51. D. Maksimov et al., "Graph-matching based CTA," *IEEE Trans. Med. Imaging*, vol. 28, no. 12, pp. 1940–1954, 2009.

52. Y. Ke and R. Sukthankar, "PCA-SIFT: A more distinctive representation for local image descriptors," in *Proceedings of the 2004 IEEE Computer Society Conference on Computer Vision and Pattern Recognition, 2004. CVPR 2004*, 2004, vol. 2, pp. II-506–II-513 Vol.2.

53. H. Bay, A. Ess, T. Tuytelaars, and L. V. Gool, "Speeded-up robust features (SURF)," *Comput. Vis. Image Underst.*, vol. 110, no. 3, pp. 346–359, 2008.

54. J. Luo and G. Oubong, "A comparison of SIFT, PCA-SIFT and SURF," *Int. J. Image Process.*, vol. 3, no. 4, pp. 143–152, 2009.

55. M. Lorenzo-Valdés, G. I. Sanchez-Ortiz, R. Mohiaddin, and D. Rueckert, "Atlas-based segmentation and tracking of 3D cardiac MR images using non-rigid registration," in *Medical Image Computing and Computer-Assisted Intervention—MICCAI 2002*, Berlin, Germany: Springer, 2002, pp. 642–650.

56. R. A. Heckemann, S. Keihaninejad, P. Aljabar, D. Rueckert, J. V. Hajnal, and A. Hammers, "Improving intersubject image registration using tissue-class information benefits robustness and accuracy of multi-atlas based anatomical segmentation," *NeuroImage*, vol. 51, no. 1, pp. 221–227, 2010.

57. M. Hofmann et al., "MRI-based attenuation correction for PET/MRI: A novel approach combining pattern recognition and atlas registration," *J. Nucl. Med.*, vol. 49, no. 11, pp. 1875–1883, 2008.

58. A. P. James and B. V. Dasarathy, "Medical image fusion: A survey of the state of the art," *Inf. Fusion*, vol. 19, pp. 4–19, 2014.

59. W. T. Freeman, T. R. Jones, and E. C. Pasztor, "Example-based super-resolution," *IEEE Comput. Graph. Appl.*, vol. 22, no. 2, pp. 56–65, 2002.

60. Q. Wang, X. Tang, and H. Shum, "Patch based blind image super resolution," in *Tenth IEEE International Conference on Computer Vision (ICCV'05)*, 2005, vol. 1, pp. 709–716.

61. Q. Shan, Z. Li, J. Jia, and C.-K. Tang, "Fast image/video upsampling," *ACM Trans. Graph.*, vol. 27, no. 5, p. 1, 2008.

62. D. Glasner, S. Bagon, and M. Irani, "Super-resolution from a single image," in *2009 IEEE 12th International Conference on Computer Vision*, 2009, pp. 349–356.

63. J. Yang, J. Wright, T. S. Huang, and Y. Ma, "Image super-resolution via sparse representation," *IEEE Trans. Image Process.*, vol. 19, no. 11, pp. 2861–2873, 2010.

64. S. Yang, M. Wang, Y. Chen, and Y. Sun, "Single-image super-resolution reconstruction via learned geometric dictionaries and clustered sparse coding," *IEEE Trans. Image Process.*, vol. 21, no. 9, pp. 4016–4028, 2012.

65. C.-Y. Yang, C. Ma, and M.-H. Yang, "Single-image super-resolution: A benchmark," in *Computer Vision—ECCV 2014*, 2014, pp. 372–386.

66. M. Irani and S. Peleg, "Improving resolution by image registration," *Graph. Models Image Process.*, vol. 53, no. 3, pp. 231–239, 1991.

67. J. Sun, Z. Xu, and H.-Y. Shum, "Image super-resolution using gradient profile prior," in *2008 IEEE Conference on Computer Vision and Pattern Recognition*, 2008, pp. 1–8.

68. K. I. Kim and Y. Kwon, "Single-image super-resolution using sparse regression and natural image prior," *IEEE Trans. Pattern Anal. Mach. Intell.*, vol. 32, no. 6, pp. 1127–1133, 2010.

69. H. Zhang, J. Yang, Y. Zhang, and T. S. Huang, "Non-local kernel regression for image and video restoration," in *Computer Vision—ECCV 2010*, 2010, pp. 566–579.

70. H. Chang, D.-Y. Yeung, and Y. Xiong, "Super-resolution through neighbor embedding," in *Proceedings of the 2004 IEEE Computer Society Conference on Computer Vision and Pattern Recognition, 2004. CVPR 2004*, 2004, vol. 1, pp. I–I.

71. H. He and W. C. Siu, "Single image super-resolution using gaussian process regression," in *CVPR 2011*, 2011, pp. 449–456.

72. P. Purkait and B. Chanda, "Image upscaling using multiple dictionaries of natural image patches," in *Computer Vision—ACCV 2012*, 2012, pp. 284–295.

73. J. Yang, Z. Wang, Z. Lin, X. Shu, and T. Huang, "Bilevel sparse coding for coupled feature spaces," in *2012 IEEE Conference on Computer Vision and Pattern Recognition*, 2012, pp. 2360–2367.

74. S. Wang, L. Zhang, Y. Liang, and Q. Pan, "Semi-coupled dictionary learning with applications to image super-resolution and photo-sketch synthesis," in *2012 IEEE Conference on Computer Vision and Pattern Recognition*, 2012, pp. 2216–2223.

75. J. Yang, Z. Lin, and S. Cohen, "Fast image super-resolution based on in-place example regression," in *2013 IEEE Conference on Computer Vision and Pattern Recognition*, 2013, pp. 1059–1066.

76. L. Caschera, A. Lazzara, L. Piergallini, D. Ricci, B. Tuscano, and A. Vanzulli, "Contrast agents in diagnostic imaging: Present and future," *Pharmacol. Res.*, vol. 110, pp. 65–75, 2016.

77. C.-H. Su et al., "Nanoshell magnetic resonance imaging contrast agents," *J. Am. Chem. Soc.*, vol. 129, no. 7, pp. 2139–2146, 2007.

78. J. Weber, P. C. Beard, and S. E. Bohndiek, "Contrast agents for molecular photoacoustic imaging," *Nat. Methods*, vol. 13, no. 8, pp. 639–650, 2016.

79. S. S. Bedi and R. Khandelwal, "Various image enhancement techniques—A critical review," *IJARCET*, vol. 2, no. 3, p. 5, 2013.

80. R. A. Hummel, "Histogram modification techniques," *Comput. Graph. Image Process.*, vol. 4, no. 3, pp. 209–224, 1975.

81. S. M. Pizer et al., "Adaptive histogram equalization and its variations," *Comput. Vis. Graph. Image Process.*, vol. 39, no. 3, pp. 355–368, 1987.

82. T. Celik and T. Tjahjadi, "Automatic image equalization and contrast enhancement using Gaussian mixture modeling," *IEEE Trans. Image Process.*, vol. 21, no. 1, pp. 145–156, 2012.

83. F. C. Cheng and S. C. Huang, "Efficient histogram modification using bilateral Bezier curve for the contrast enhancement," *J. Disp. Technol.*, vol. 9, no. 1, pp. 44–50, 2013.

84. E. F. Arriaga-Garcia, R. E. Sanchez-Yanez, and M. G. Garcia-Hernandez, "Image enhancement using bi-histogram equalization with adaptive sigmoid functions," in *2014 International Conference on Electronics, Communications and Computers (CONIELECOMP)*, 2014, pp. 28–34.

85. C. M. Tsai, Z. M. Yeh, and Y. F. Wang, "Decision tree-based contrast enhancement for various color images," *Mach. Vis. Appl.*, vol. 22, no. 1, pp. 21–37, 2011.

86. X. Zong, A. F. Laine, and E. A. Geiser, "Speckle reduction and contrast enhancement of echocardiograms via multiscale nonlinear processing," *IEEE Trans. Med. Imaging*, vol. 17, no. 4, pp. 532–540, 1998.

87. A. Pizurica, W. Philips, I. Lemahieu, and M. Acheroy, "A versatile wavelet domain noise filtration technique for medical imaging," *IEEE Trans. Med. Imaging*, vol. 22, no. 3, pp. 323–331, 2003.

88. J. Zhou, A. L. Cunha, and M. N. Do, "Nonsubsampled contourlet transform: Construction and application in enhancement," in *IEEE International Conference on Image Processing 2005*, 2005, vol. 1, pp. I-469–472.

89. L. Shao, R. Yan, X. Li, and Y. Liu, "From heuristic optimization to dictionary learning: A review and comprehensive comparison of image denoising algorithms," *IEEE Trans. Cybern.*, vol. 44, no. 7, pp. 1001–1013, 2014.

90. M. Diwakar and M. Kumar, "Edge preservation based CT image denoising using wiener filtering and thresholding in wavelet domain," in *2016 Fourth International Conference on Parallel, Distributed and Grid Computing (PDGC)*, 2016, pp. 332–336.

91. C. Tomasi and R. Manduchi, "Bilateral filtering for gray and color images," in *Sixth International Conference on Computer Vision (IEEE Cat. No.98CH36271)*, 1998, pp. 839–846.

92. G. Z. Yang, P. Burger, D. N. Firmin, and S. R. Underwood, "Structure adaptive anisotropic image filtering," *Image Vis. Comput.*, vol. 14, no. 2, pp. 135–145, 1996.

93. A. M. Hasan, A. Melli, K. A. Wahid, and P. Babyn, "Denoising low-dose CT images using multi-frame blind source separation and block matching filter," *IEEE Trans. Radiat. Plasma Med. Sci.*, pp. 1–1, 2018.

94. R. J. Jevnisek and S. Avidan, "Co-occurrence filter," in *IEEE Conference on Computer Vision and Pattern Recognition (CVPR)*, 2017.

95. A. Buades, B. Coll, and J. M. Morel, "A non-local algorithm for image denoising," in *2005 IEEE Computer Society Conference on Computer Vision and Pattern Recognition (CVPR'05)*, 2005, vol. 2, pp. 60–65 vol. 2.

96. M. Mahmoudi and G. Sapiro, "Fast image and video denoising via nonlocal means of similar neighborhoods," *IEEE Signal Process. Lett.*, vol. 12, no. 12, pp. 839–842, 2005.

97. P. Coupe, P. Yger, S. Prima, P. Hellier, C. Kervrann, and C. Barillot, "An optimized blockwise nonlocal means denoising filter for 3-D magnetic resonance images," *IEEE Trans. Med. Imaging*, vol. 27, no. 4, pp. 425–441, 2008.

98. T. Thaipanich, B. T. Oh, P. H. Wu, D. Xu, and C. C. J. Kuo, "Improved image denoising with adaptive nonlocal means (ANL-means) algorithm," *IEEE Trans. Consum. Electron.*, vol. 56, no. 4, pp. 2623–2630, 2010.

99. V. Karnati, M. Uliyar, and S. Dey, "Fast non-local algorithm for image denoising," in *2009 16th IEEE International Conference on Image Processing (ICIP)*, 2009, pp. 3873–3876.

100. B. Goossens, H. Luong, R. Pižurica, and W. Philips, "An improved non-local denoising algorithm," in *Proceedings of the 2008 International Workshop on Local and Non-local Approximation in Image Processing*, 2008.

101. C. Pang, O. C. Au, J. Dai, W. Yang, and F. Zou, "A fast NL-means method in image denoising based on the similarity of spatially sampled pixels," in *2009 IEEE International Workshop on Multimedia Signal Processing*, 2009, pp. 1–4.

102. D. Tschumperlé and L. Brun, "Non-local image smoothing by applying anisotropic diffusion PDE's in the space of patches," in *2009 16th IEEE International Conference on Image Processing (ICIP)*, 2009, pp. 2957–2960.

103. S. Grewenig, S. Zimmer, and J. Weickert, "Rotationally invariant similarity measures for nonlocal image denoising," *J. Vis. Commun. Image Represent.*, vol. 22, no. 2, pp. 117–130, 2011.

104. J. Wei, "Lebesgue anisotropic image denoising," *Int. J. Imaging Syst. Technol.*, vol. 15, no. 1, pp. 64–73.

105. C. Kervrann and J. Boulanger, "Local adaptivity to variable smoothness for exemplar-based image regularization and representation," *Int. J. Comput. Vis.*, vol. 79, no. 1, pp. 45–69, 2008.

106. Y. Lou, P. Favaro, S. Soatto, and A. Bertozzi, "Nonlocal similarity image filtering," in *Image Analysis and Processing—ICIAP 2009*, 2009, pp. 62–71.

107. R. Yan, L. Shao, S. D. Cvetkovic, and J. Klijn, "Improved nonlocal means based on pre-classification and invariant block matching," *J. Disp. Technol.*, vol. 8, no. 4, pp. 212–218, 2012.

108. J. C. Brailean, S. Efstratiadis, and R. L. Lagendijk, "Noise reduction filters for dynamic image sequences: A review," *Proc. IEEE*, vol. 83, no. 9, p. 21, 1995.

109. E. P. Simoncelli and E. H. Adelson, "Noise removal via Bayesian wavelet coring," *Proc 3rd IEEE Int\l Conf on Image Proc* 1996, vol. 1, pp. 379–302.

110. J. L. Starck, E. J. Candès, and D. L. Donoho, "The curvelet transform for image denoising," *IEEE Trans. IMAGE Process.*, vol. 11, no. 6, p. 15, 2002.

111. M. N. Do and M. Vetterli, "The contourlet transform: An efficient directional multiresolution image representation," *IEEE Trans. Image Process.*, vol. 14, no. 12, pp. 2091–2106, 2005.

112. E. L. Pennec and S. Mallat, "Sparse geometric image representations with bandelets," *IEEE Trans. Image Process.*, vol. 14, no. 4, pp. 423–438, 2005.

113. A. Pizurica, A. M. Wink, E. Vansteenkiste, W. Philips, and J. Roerdink, "A review of wavelet denoising in MRI and ultrasound brain imaging," *Curr. Med. Imaging Rev.*, vol. 2, pp. 247–260, 2006.

114. R. Yan, L. Shao, and Y. Liu, "Nonlocal hierarchical dictionary learning using wavelets for image denoising," *IEEE Trans. Image Process.*, vol. 22, no. 12, pp. 4689–4698, 2013.

115. J. Mairal, F. Bach, J. Ponce, G. Sapiro, and A. Zisserman, "Non-local sparse models for image restoration," *Computer Vision, 2009 IEEE 12th International Conference on. IEEE*, 2009, pp. 2272–2279.

116. M. Elad and M. Aharon, "Image denoising via sparse and redundant representations over learned dictionaries," *IEEE Trans. Image Process.*, vol. 15, no. 12, pp. 3736–3745, 2006.

117. W. Dong, X. Li, L. Zhang, and G. Shi, "Sparsity-based image denoising via dictionary learning and structural clustering," in *CVPR 2011*, 2011, pp. 457–464.

118. N. Sharma and L. M. Aggarwal, "Automated medical image segmentation techniques," *J. Med. Phys.*, vol. 35, no. 1, p. 3, 2010.

119. A. Mustaqeem, A. Javed, and T. Fatima, "An efficient brain tumor detection algorithm using watershed & thresholding based segmentation," *Int. J. Image Graph. Signal Process. Hong Kong*, vol. 4, no. 10, pp. 34–39, 2012.

120. R. Pohle and K. D. Toennies, "Segmentation of medical images using adaptive region growing," *Proc. SPIE Med. Imag. Image Process.*, vol. 4322, pp. 1337–1346, 2001.

121. R. Rouhi, M. Jafari, S. Kasaei, and P. Keshavarzian, "Benign and malignant breast tumors classification based on region growing and CNN segmentation," *Expert Syst. Appl.*, vol. 42, no. 3, pp. 990–1002, 2015.

122. M. A. Balafar, A. R. Ramli, M. I. Saripan, and S. Mashohor, "Review of brain MRI image segmentation methods," *Artif. Intell. Rev.*, vol. 33, no. 3, pp. 261–274, 2010.

123. B. N. Li, C. K. Chui, S. Chang, and S. H. Ong, "Integrating spatial fuzzy clustering with level set methods for automated medical image segmentation," *Comput. Biol. Med.*, vol. 41, no. 1, pp. 1–10, 2011.

124. Y. Xu, J. Y. Zhu, E. Chang, and Z. Tu, "Multiple clustered instance learning for histopathology cancer image classification, segmentation and clustering," in *2012 IEEE Conference on Computer Vision and Pattern Recognition*, 2012, pp. 964–971.

125. C. Li, R. Huang, Z. Ding, J. C. Gatenby, D. N. Metaxas, and J. C. Gore, "A level set method for image segmentation in the presence of intensity inhomogeneities with application to MRI," *IEEE Trans. Image Process.*, vol. 20, no. 7, pp. 2007–2016, 2011.

126. A. Hoogi et al., "Adaptive local window for level set segmentation of CT and MRI liver lesions," *Med. Image Anal.*, vol. 37, pp. 46–55, 2017.

127. "Adaptive local window for level set segmentation of CT and MRI liver lesions." *Med. Image Anal.*, Available: https://www.medicalimageanalysisjournal.com/article/S1361-8415(17)30010-5/abstract (Accessed: June 3, 2018).

128. M. B. Cuadra, C. Pollo, A. Bardera, O. Cuisenaire, J. G. Villemure, and J. P. Thiran, "Atlas-based segmentation of pathological MR brain images using a model of lesion growth," *IEEE Trans. Med. Imaging*, vol. 23, no. 10, pp. 1301–1314, 2004.

129. P. Aljabar, R. A. Heckemann, A. Hammers, J. V. Hajnal, and D. Rueckert, "Multi-atlas based segmentation of brain images: atlas selection and its effect on accuracy," *Neuroimage*, vol. 46, pp. 726–738, 2009.

130. X. Chen, J. K. Udupa, U. Bagci, Y. Zhuge, and J. Yao, "Medical image segmentation by combining graph cut and oriented active appearance models," *IEEE Trans. Image Process.*, vol. 21, pp. 2035–2046, 2011.

131. Y. Y. Boykov and M. P. Jolly, "Interactive graph cuts for optimal boundary amp; region segmentation of objects in N-D images," in *Proceedings Eighth IEEE International Conference on Computer Vision. ICCV 2001*, 2001, vol. 1, pp. 105–112.

132. T. Shen, H. Li, and X. Huang, "Active volume models for medical image segmentation," *IEEE Trans. Med. Imaging*, vol. 30, no. 3, pp. 774–791, 2011.

133. J. Schmid, J. Kim, and N. Magnenat-Thalmann, "Robust statistical shape models for MRI bone segmentation in presence of small field of view," *Med. Image Anal.*, vol. 15, no. 1, pp. 155–168, 2011.

134. C. Petitjean and J.-N. Dacher, "A review of segmentation methods in short axis cardiac MR images," *Med. Image Anal.*, vol. 15, no. 2, pp. 169–184, 2011.

135. X. Chen, J. Udupa, U. Bagci, Y. Zhuge, and J. Yao, "Medical image segmentation by combining graph cuts and oriented active appearance models," *IEEE Trans. Image Process.*, vol. 21, no. 4, pp. 2035–2046, 2012.

136. P. G. Spetsieris, Y. Ma, V. Dhawan, and D. Eidelberg, "Differential diagnosis of parkinsonian syndromes using PCA-based functional imaging features," *NeuroImage*, vol. 45, no. 4, pp. 1241–1252, 2009.

137. Y. Sun, C. F. Babbs, and E. J. Delp, "A comparison of feature selection methods for the detection of breast cancers in mammograms: Adaptive sequential floating search vs. genetic algorithm," in *2005 IEEE Engineering in Medicine and Biology 27th Annual Conference*, 2005, pp. 6532–6535.

138. B. Mwangi, T. S. Tian, and J. C. Soares, "A review of feature reduction techniques in neuroimaging," *Neuroinformatics*, vol. 12, pp. 229–244, 2014.

139. J. Barker, A. Hoogi, A. Depeursinge, and D. L. Rubin, "Automated classification of brain tumor type in whole-slide digital pathology images using local representative tiles," *Med. Image Anal.*, vol. 30, pp. 60–71, 2016.

140. H. Zhang, T. Xu, and H. Li, "Stackgan: Text to photo-realistic image synthesis with stacked generative adversarial networks," *Proceedings of the IEEE International Conference on Computer Vision* 2017, pp. 5908–5916.

141. O. Ronneberger, P. Fischer, and T. Brox, "U-Net: Convolutional networks for biomedical image segmentation," ArXiv150504597 Cs, May 2015.

142. S. Trebeschi et al., "Deep learning for fully-automated localization and segmentation of rectal cancer on multiparametric MR," *Sci. Rep.*, vol. 7, no. 1, p. 5301, 2017.

143. F. Milletari, N. Navab, and S. A. Ahmadi, "V-Net: Fully convolutional neural networks for volumetric medical image segmentation," in *2016 Fourth International Conference on 3D Vision (3DV)*, 2016, pp. 565–571.

144. P. Vincent, H. Larochelle, I. Lajoie, Y. Bengio, and P.-A. Manzagol, "Stacked denoising autoencoders: Learning useful representations in a deep network with a local denoising criterion," *J. Mach. Learn. Res.*, vol. 11, pp. 3371–3408, 2010.

145. Y. Bengio, L. Yao, G. Alain, and P. Vincent, "Generalized denoising auto-encoders as generative models," in *Advances in Neural Information Processing Systems 26*, C. J. C. Burges, L. Bottou, M. Welling, Z. Ghahramani, and K. Q. Weinberger, (Eds.), New York: Curran Associates, 2013, pp. 899–907.

146. B. Du, W. Xiong, J. Wu, L. Zhang, L. Zhang, and D. Tao, "Stacked convolutional denoising auto-encoders for feature representation," *IEEE Trans. Cybern.*, vol. 47, no. 4, pp. 1017–1027, 2016.

147. T. Kooi, G. Litjens, B. van Ginneken, A. Gubern-Mérida, C. I. Sánchez, R. Mann, A. den Heeten, and N. Karssemeijer, "Large scale deep learning for computer aided detection of mammographic lesions," *Med. Image Anal.*, vol. 35, pp. 303–312, 2016.

148. D. Wang, A. Khosla, R. Gargeya, H. Irshad, and A. H. Beck, "Deep learning for identifying metastatic breast cancer," ArXiv160605718 Cs Q-Bio, 2016.

149. B. Kayalibay, G. Jensen, and P. van der Smagt, "CNN-based segmentation of medical imaging data," ArXiv170103056 Cs, 2017.

150. A. Esteva et al., "Dermatologist-level classification of skin cancer with deep neural networks," *Nature*, vol. 542, no. 7639, pp. 115–118, 2017.

151. J. Arevalo, F. A. González, R. Ramos-Pollán, J. L. Oliveira, and M. A. Guevara Lopez, "Representation learning for mammography mass lesion classification with convolutional neural networks," *Comput. Methods Program. Biomed.*, vol. 127, pp. 248–257, 2016.

Quantitative imaging using CT

LIN LU, LAWRENCE H. SCHWARTZ, AND BINSHENG ZHAO

3.1 INTRODUCTION

Medical imaging is one of the major technologies in medical science to assess the characteristics of human tissue non-invasively. It is widely used in clinical practice and clinical trials for diagnosis and treatment guidance [1–3]. Moreover, in recent years, medical imaging has played an increasingly important role in the context of "personalized medicine" [4,5]. However, in all imaging divisions from radiology, nuclear medicine, to pathology, the amount of quantitative analyses is still limited, and the majority of clinical decision-making is based on visual assessment and simple computation. In oncology, for example, the most common quantification for tumor response assessment is the objective response rate, i.e., the percentage change of tumor diameter from baseline as defined in the response evaluation criteria in solid tumors (RECIST) [6]. Another example is the diagnosis of pulmonary nodules using CT imaging. Nodule characteristics observed on CT such as solid tumor portion and tumor edge speculation are visually described by radiologists and used to aid the diagnosis [7]. Such qualitative nodule assessment can be affected by radiologist's experience and prone to intra and inter-observer variability [8,9].

Contrary to qualitative visual assessment, computer-aided imaging analyses use quantitative features extracted from imaging to characterize tumor phenotypes. Quantitative imaging features are computed from medical images through advanced mathematical algorithms. The conversion of medical images into arrays of mathematically computed imaging features provides the opportunity to quantitatively, objectively, and reproducibly capture tumor phenotypic information over the entire tumor and even tumor microenvironment that correlate to underlying genotypic heterogeneity and composition.

High-throughput extraction of quantitative imaging features turns the terminology *radiomics* into a reality. Radiomics, defined as the process of high-throughput mining of quantitative imaging features from medical images, is now an emerging field in oncology research [10–19]. Radiomics-based decision-support system for precision diagnosis and treatment has shown promise in predicting tumor prognosis in lung, head, neck, oropharyngeal and colorectal cancers, tumor distant metastasis in lung cancer, tumor staging in colorectal cancer, patient reaction to radiotherapy in esophageal cancer, tumor malignancy in lung screening, and patient survival in lung cancer.

A typical radiomics workflow consists of the following four procedures: (I) image acquisition, (II) tumor segmentation, (III) quantitative imaging feature extraction, and (IV) predictive model construction and validation. Figure 3.1 presents an example of utilizing the radiomic approach to investigate the association between CT imaging phenotypes and underlying somatic mutations of epidermal growth factor receptor (EGFR) in non-small cell lung cancer (NSCLC) patients treated with a targeted therapy. The radiomic workflow starts with image acquisition using a specified imaging protocol. Tumor volume is then segmented to define the

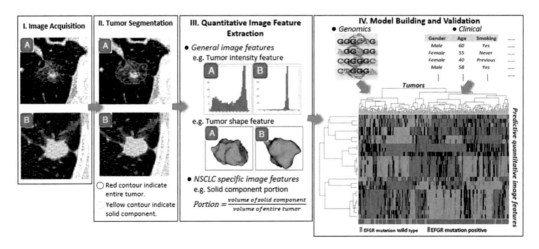

Figure 3.1 (See color insert.) A radiomic workflow to predict EGFR mutation status in lung cancer patients.

region of interest for feature extraction, which is performed using image analysis algorithms/mathematical algorithms on the original or pre-processed images. Finally, a statistical model for prediction of *EGFR* mutation status is built based on the extracted features (can also include patient demographic data) and validated.

In this chapter, we will focus on the extraction of quantitative imaging features, including feature definition and implementation. We will then address some pitfalls and challenges in CT radiomic studies. Toward the end of the chapter, we will discuss potential future research directions including harmonization of CT image acquisition and deep convolutional neural network (CNN). For constructing and validating radiomic models, readers are referred to Chapter 9, Predictive modeling, machine learning, and statistical issues.

3.2 RADIOMIC FEATURES

In this section, we will review the definitions of commonly used radiomic features, discuss choosing of feature parameters during the implementation of feature extraction, and then introduce some publicly available software for extraction of radiomic features.

3.2.1 FEATURE DEFINITION

Quantitative imaging features can be classified as statistics-based, model-based, transform-based, morphology-based, and sharpness-based, depending on their definitions and the characteristics they describe. Each class can contain several feature extraction algorithms, and each algorithm can produce a number of radiomic features (See Table 3.1).

Table 3.1 Well-defined quantitative imaging features, feature extraction algorithms, and feature classes

Feature classes	Feature extraction algorithms	Features
Statistics-based	First-order statistics	Mean, standard deviation, skewness, kurtosis
	Run-length	Short primitives emphasis, long primitives emphasis, Gray-level uniformity, primitive length uniformity, primitive percentage
	GLCM	Angular second moment, contrast, correlation, sum of squares, inverse difference moment, sum average, sum variance, sum entropy, entropy, difference variance, difference entropy, two information measurements of correlation, maximal correlation coefficient, maximum probability, cluster tendency
	Gray-Tone Difference Matrix (GTDM)	Coarseness, contrast, busyness, complexity, strength
Model-based	Fractal dimension analysis	Mean, standard deviation, lacunarity
Transform-based	Gabor transform	Average energy
	Wavelets transform	Average energies on different wavelet decomposition channels
Morphology-based	Size measurement	Uni-dimensional, bi-dimensional, volume
	Shape measurements	Eccentricity, solidity, compact-factor, round-factor
	Shape index	SI1–SI9
Sharpness-based	Sigmoid curve fitting	Slope, amplify, intensity low bounder

3.2.1.1 STATISTICS-BASED FEATURES

Statistics-based features include (but are not limited to) first-order statistics, run-length, Neighborhood Gray-Tone Difference Matrix (NGTDM) [10], and gray-level co-occurrence matrix (GLCM) [20]. The first-order statistic features are derived from density histogram. Mean, standard deviation, skewness, and kurtosis are four basic statistics-based features to describe tumor density distribution. Run-length features are used to characterize image coarseness by counting the number of maximum contiguous voxels having a constant gray-level along a line [21]. A larger number of neighboring pixels of the same gray-level represents a coarser texture, whereas a smaller number of neighboring pixels of the same gray-level indicates a fine texture. NGTDM features describe visual properties of texture based on gray-tone difference between a pixel and its neighborhood. GLCM features characterize the textures of a region of interest by creating a new matrix GLCM based on counting how often pairs of pixels with specific gray-level values in a specified spatial relationship (distance and direction) occur in the region of interest and then computing statistics from GLCM.

3.2.1.2 MODEL-BASED FEATURES

Fractal dimension [22] analysis is a technique that can provide model-based features. Fractal dimension provides a statistical index for quantifying the complexity of an image. Basically, the fractal dimension feature describes the relationship between change in a measuring scale and its resultant measurement value at that scale. The rougher the texture, the larger the fractal dimension.

3.2.1.3 TRANSFORM-BASED FEATURES

Transform-based features are extracted from transformed images. Gabor filters [23] and wavelets decomposition [24] are two important transformations in the field of digital image processing. Gabor filters are linear filters designed for edge detection. Wavelets decomposition can extract finer/coarser textures at multiple frequency scales.

3.2.1.4 MORPHOLOGY-BASED FEATURES

Morphology-based features contain information about tumor size and shape, but not tumor density. Tumor size is usually measured uni-dimensionally (Uni), bi-dimensionally (Bi), or volumetrically (Vol). Uni (maximal diameter) is defined as the longest line across a tumor on the 2D image containing the largest tumor area. Bi is defined as the product of the maximal diameter and its maximal perpendicular diameter on the same 2D image. Vol is the sum of the volumes of the tumor voxels.

Shape features can describe both global and local tumor shapes. For example, the global shape feature of eccentricity is a measure specifying how close a tumor (approximated using an ellipse) is to a circle. Solidity is the ratio of the tumor area over the area of the convex hull bounding the tumor. Compact-factor quantifies the compactness of a tumor in 3D based on its surface and volume measurements, whereas round-factor is a measure of the roundness of a tumor based on its perimeter and area.

The local shape feature class of shape index [25] captures the intuitive notion of "local surface shape" of a 3D tumor. The shape index is a numerical number ranging from −1 to 1 and can be divided into nine categories (SI1–SI9), each representing one of the following nine shapes respectively: spherical cup, trough, rut, saddle rut, saddle, saddle ridge, ridge, dome, and spherical cap.

3.2.1.5 SHARPNESS-BASED FEATURES

Sigmoid curve fitting feature [26] is used to quantify the density relationship between a tumor and its surrounding background, e.g., sharpness of the tumor margin. Sigmoid curve fitting algorithm applies sigmoid curves to fit density changes along sampling lines drawn orthogonal to the tumor surface. Each sampling line goes through one voxel on the tumor surface with specified distance inside and outside tumor margin.

3.2.2 FEATURE IMPLEMENTATION

Although feature extraction algorithms are mathematically well-defined, the computed value of a feature can vary if feature parameters are chosen differently. Variance also exists in feature implementation such as image discretization. In the next subsections, we will address these issues.

3.2.2.1 IMAGE DISCRETIZATION

Image discretization is a procedure to discretize voxel intensity within a volume of interest (VOI) to a specified value range in order to efficiently and practically compute radiomic features [27]. For instance, a CT image with an intensity range from −1024 HU to 1024 HU (i.e., total 2048 intensity bin levels) can generate a matrix size of 2048 × 2048 to compute the GLCM features, resulting in intensive computation. However, if the image density range is discretized into ten density bin levels, the corresponding GLCM matrix size can be reduced to 10 × 10 elements, which makes the computation of GLCM features efficient. Usually, there are two ways to normalize a VOI (or an image). One way is to use a pre-defined/fixed bin width to discretize image densities. The other way is to use varying bin widths to normalize a VOI. In the latter situation, the bin widths will be obtained by first computing differences between the maximum and the minimum intensity values within VOIs and then dividing the differences by 2^N (the number of bins), where N ranges from 3 to 8 per literature [10,28–30].

Attention should be paid to the selection of bin widths when normalizing VOIs, because radiomic features can be sensitive to the chosen bin widths/levels. Here is an example of calculating the GLCM features of *contrast* and *homogeneity* using two different bin levels of four and eight (Figure 3.2). We can see that *contrast* and *homogeneity* values calculated from four bins and eight bins are different. For the same image (Figure 3.2a), use of the four bins results in *contrast* = 1.09 and *homogeneity* = 0.69 (Figure 3.2b–d), while use of the eight bins results in *contrast* = 4.87 and *homogeneity* = 0.53 (Figure 3.2e–g).

3.2.2.2 SELECTION OF FEATURE PARAMETERS

Feature extraction algorithms presented in Table 3.1 involve a number of feature specified parameters. The selection of parameter values will affect the computed values of the features. For example, after image/VOI normalization, GLCM features still contain two parameters, the distance and direction of neighboring voxel pairs. As can be seen in Figure 3.3, a combination of four neighborhood distances (e.g., 1, 2, 3, 4, in pixel) and four neighborhood directions (e.g., 0, $\pi/4$, $\pi/2$, $3\pi/4$) can generate 16 groups of the GLCM feature class, each class has 17 GLCM features. When combining eight neighborhood distances and eight directions, 64 groups GLCM feature class can be reached. GLCM features generated using different bin widths/levels

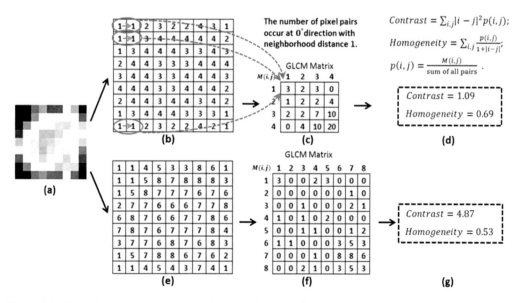

Figure 3.2 (See color insert.) Computing the GLCM features of contrast and homogeneity using different bin levels. (a) Original image. (b) Normalized image using the bin level of four. (c) GLCM matrix derived from (b). (d) Contrast and homogeneity computed from GLCM in (c). (e) Normalized image using the bin level of eight. (f) GLCM matrix derived from (e). (g) Contrast and homogeneity computed from GLCM in (f).

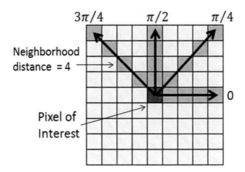

Figure 3.3 Calculation of GLCM features for a 9 × 9 2D image at four directions and a neighborhood distance equal to four.

and different neighborhood distances and directions can characterize texture patterns of different details in density and frequency. This is true for other radiomic features if the features are defined with parameters. Multiple choices of the feature parameters' values should be considered in radiomic studies.

3.2.3 FEATURE EXTRACTION SOFTWARE

Several software packages to extract radiomic features are available for conducting radiomics study, including free open source software packages such as "imaging biomarker explorer" (IBEX) [31], The University of Texas MD Anderson Cancer Center, Houston) and "Chang Gung Image Texture Analysis" (CGITA, Chang Gung University, Taiwan [32]). Commercially available packages includes RADIOMICS™ (OncoRadiomics, Maastricht, Netherlands) and TexRAD™ (Feedback plc, Cambridge, UK). In addition, the MATLAB® (The MathWorks®, Natick, Massachusetts) platform is usually used to code in-house feature extraction programs. A recent pilot study compared four quantitative texture analysis software packages (two in-house and two open source packages: MaZda and IBEX) on digital mammography and head and neck CT images and reported a large variation in the computed texture values [33].

3.3 EFFECTS OF TUMOR SEGMENTATION ON RADIOMICS

Tumor segmentation is an important step in the radiomic workflow. Tumor volumes can be obtained by manual delineation, semi-automated, and fully-automated segmentation methods. Since tumor volume specifies the region to calculate the radiomic feature, inter/intra- observer and algorithm variability in segmentation can introduce variation that affects the predictive models subsequently derived from the radiomic features [34]. In this section, we will briefly discuss tumor segmentation and segmentation's effects on radiomic features as well as on phenotype predictive models.

3.3.1 TUMOR SEGMENTATION

Manual, semi-automated, and fully automated segmentations can be used to extract radiomic features. Manual segmentation requires an operator to delineate tumor contours slice by slice using a computer mouse. Semi-automated segmentation involves human-machine interaction, for example, manual initiation by providing a seed point, a line, or a closed curve inside/outside the tumor to be segmented. Further, if a segmentation result is suboptimal, it can be modified by human. Fully automated segmentation requires no human-machine interaction at all. However, manual and fully automated segmentations are not frequently performed. Manual segmentation is a straightforward solution, but it can be time consuming and vulnerable to inter-observer variability. Though fully automated segmentation is ideal, due to the diversity and complexity of tumors, it is limited to few applications where targets are relatively simple, such as predicting malignant nodules from screening CT scans [35]. Therefore, semi-automated segmentation is the most recommended solution to radiomic studies.

3.3.2 EFFECTS OF TUMOR SEGMENTATION ON RADIOMIC FEATURES

Velazquez et al. [36] conducted the feasibility study of a semi-automated segmentation algorithm on NSCLC tumor. The semi-automated segmentation method they used was a region-growing-based algorithm integrated into the free and publicly available 3D-Slicer software platform. The authors compared the semi-automatically segmented tumor volumes obtained by three observers, who independently segmented 20 NSCLC tumors twice, to those obtained by manual delineations of five physicians. All segmented tumors were compared to the tumor macroscopic diameters in pathology, which were considered the "gold standard." The semi-automatically segmented tumors demonstrated higher agreement (overlap fractions > 0.90), lower volume variability ($p = 0.0003$), and smaller uncertainty areas ($p = 0.0002$) when compared to the manual delineations. Further, the semi-automated segmentations showed a strong correlation to pathological tumor diameters. The study demonstrated the accuracy and robustness of the semi-automated segmentation algorithm.

Using the same segmentation results mentioned above, i.e., the 20 NSCLC tumor volumes obtained semi-automatically by the three independent observers twice and five independent physicians' manual delineations, Parmar et al. [37] studied reproducibility of radiomic features. In total, 56 radiomic features quantifying phenotypes based on tumor density, shape, and texture were extracted from the segmented tumors. The intraclass correlation coefficient (ICC) [38] was used as a measurement for reproducibility. The authors reported that radiomic features extracted from semi-automated segmentations had significantly higher reproducibility (ICC = 0.8560.15, $p = 0.0009$) compared to the features extracted from the manual segmentations (ICC = 0.7760.17). Furthermore, it was found that features extracted from semi-automated segmentation were more robust, as the range was significantly smaller across observers ($p = 3.819e{-}07$), and the ranges of semi-automated extracted features overlapped with the ranges of features extracted from manual contouring (boundary lower: $p = 0.007$, higher: $p = 5.863e{-}06$).

Although semi-automated segmentation could generate more robust radiomic features than manual segmentation, inter-variability among different semi-automated segmentations should not be underestimated. Jayashree et al. [39] investigated the inter-algorithm variability by performing a multi-institutional assessment of lung tumor segmentation using CT scans of 52 lung tumors, including 40 true tumors and 12 lung phantom nodules. Three different semi-automated segmentation algorithms developed by three academic institutions were used for the comparison. Algorithm 1 [40] was based on the techniques of marker-controlled watershed, geometric active contours, and Markov random field, which required manual initiation of a region of interest that enclosed the lesion on a single image. Algorithm 2 [41] was an ensemble-based segmentation algorithm, which used the "click and grow" approach for initiation. Algorithm 3 was a region growing-based algorithm initiated with a manually placed seed "circle" inside the lesion. Proportional bias among the three algorithms were 7.9% ($p < 0.05$, algorithm 1 versus 2), 4.5% ($p = 0.100$, algorithm 3 versus 1), and -3.4% ($p = 0.266$, algorithm 3 versus 2), respectively, indicating significant differences between independent institutions.

3.3.3 EFFECTS OF SEGMENTATION ON PHENOTYPE MODEL TO PREDICT *EGFR* MUTATION STATUS

Tumor segmentation can affect the computed values of radiomic features. However, how segmentation affects the use of radiomic features to predict mutational status is unknown. Recently, Zhao et al. [42] published a pilot study to explore the effects of inter-observer variability in tumor delineation on predicting *EGFR* mutate status in NSCLC patients treated with targeted therapy. Forty-six early-stage NSCLC patients were included in the study. Using a semi-automated segmentation software, three experienced radiologists independently segmented the tumors on baseline and 3-week post-therapy thin-section CT images. Figure 3.4 shows an example of the segmented tumor contours by the three radiologists. Eighty-nine radiomic features were computed from the segmented tumors on both scan time points and feature changes (delta radiomic features) were calculated. The authors found that accuracies of the *EGFR* prediction models built upon the delta radiomic features extracted from the tumor contours delineated by the three radiologists were different,

Baseline

Follow-up

(a)

(b)

Figure 3.4 (See color insert.) An example of one tumor in the same patient, (a) one baseline, and (b) one follow-up, segmented by three radiologists independently. The segmentations of the different radiologists are indicated by contours of different color.

with the area under the curves (AUCs) of 0.87, 0.85, and 0.80, respectively. Moreover, the selected radiomic features achieving the highest AUC were also different among the three radiologists. The best radiomic features for the radiologists were delta volume, delta intensity mean, and delta compact factor, respectively.

3.3.4 TUMOR SEGMENTATION SOFTWARE

3D Slicer and ITK-SNAP are two popular segmentation software packages that are publicly available. 3D Slicer is an open source software available for multiple operating systems, including Linux, MacOSX, and Windows. It is a platform that provides image analysis (e.g., registration and interactive segmentation) and visualization (e.g., volume rendering) of medical images including CT, MRI, and positron emission tomography (PET). ITK-SNAP [43] is an easy-to-use software tool that provides semi-automatic segmentation using active contour methods, as well as manual delineation and image navigation.

3.4 EFFECTS OF IMAGING ACQUISITION ON RADIOMICS

Because radiomic features are computed based on density and/or geometry distributions of image pixels (2D)/voxels (3D), their values can be affected by imaging acquisition techniques and parameters that control the image resolutions and quality. While radiomic features have the potential to serve as imaging biomarkers for improved cancer diagnosis, prognosis, and therapeutic response assessment, the effects of the variables during imaging acquisition need to be well understood before the new imaging biomarkers can be fully utilized. In the following sections, we will review some relevant published studies on the sources of variability in radiomic features including repeat CT scans, different scanners, and scanning parameters.

3.4.1 SAME-DAY REPEAT CT SCANS

Repeat scanning is one of the sources of variability in the calculated values of radiomic features due to patient's relocation (in scanner) and organ motions caused by, for example, breathing and heart beating. In 2009, to investigate the reproducibility of tumor volume measurement in repeat CT scans, Zhao et al. [44] published the first same-day repeat CT study, also known as coffee break (test-retest) study. In the study, 32 patients underwent two non-contrast-enhanced CT scans of the chest using the same imaging protocol and the same scanner. After the first scan, the patient left the scanner, walked around, and laid back within 15 minutes for a second scan. Both scans' raw data of all patients were reconstructed into six imaging settings, i.e., combinations of three slice thicknesses (1.25 mm, 2.5 mm, and 5 mm) and two reconstruction algorithms (lung and smooth kernels). The publication reported the reproducibility of tumor volume (diameter as well) on an imaging setting of 1.25 mm slice thickness and lung reconstruction kernel, which was as low as ~15% (uni-dimensionally: ~7%). This image dataset latterly became publicly available as the Reference Image

Database to Evaluate Therapy Response (RIDER) data, which can be downloaded from the Cancer Imaging Archive (https://wiki.cancerimagingarchive.net/display/Public/RIDER+Lung+CT).

Using the downloaded test-retest dataset with the single imaging setting, Balagurunathan et al. [45] explored the reproducibility of radiomic features. The radiomic feature set included 219 features describing tumor size, shape, and texture. The authors found that 66 (30.14%) features were reproducible with concordance correlation coefficient (CCC) $\geq$ 0.90 across test-retest data.

More recently, Zhao et al. [46] studied the reproducibility of 89 commonly used radiomic features using the test-retest data of all six imaging settings. They found that the radiomic features were generally reproducible over a wide range of imaging acquisition parameters used in clinical practice and clinical trials (Figure 3.5a). However, different imaging parameters, for example, smooth and sharp reconstruction kernels, showed considerable variability especially in texture features (Figure 3.5b).

3.4.2 SCANNERS AND SCANNING PARAMETERS

The multi-brand CT scanners and different imaging acquisition parameters can inevitably introduce inter-scanner and inter-imaging parameter variability to the extraction of radiomic features. For instance, imaging acquisition parameters can be scan type, pitch, kVp, effective mAs, reconstruction kernel, slice thickness, pixel spacing, and so on. There is little knowledge about how such factors affect radiomic features.

Mackin et al. [47] scanned a specifically designed radiomic phantom (Figure 3.6) with 17 combinations of different CT scanners and scanning parameters in four different medical centers. The scans were acquired on GE, Philips, Siemens, and Toshiba CT devices using the hospitals' routine imaging parameters for thoracic studies.

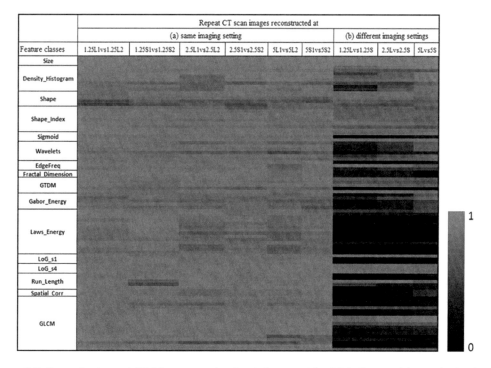

Figure 3.5 (See color insert.) CCC heat map of radiomic features. The CCCs (0 to 1) of the studied radiomic features were computed from repeat CT images reconstructed at (a) six identical imaging settings or (b) three different imaging settings. There were 89 quantitative features grouped into 15 feature classes. The brighter the red color, the higher the CCC value (i.e., the more reproducible) of a feature computed for the repeat scans. The label of "1.25LI versus 1.25L2" means both first and second scans were reconstructed at 1.25 mm slice thickness using the lung algorithm. "2.5L versus 2.5S" means both scans were reconstructed at 2.5 mm slice thickness, but using different algorithms (i.e., lung versus standard algorithms). (From Zhao, B. et al., *Sci. Rep.*, 6, 23428, 2016.)

Figure 3.6 (See color insert.) Credence cartridge radiomics (CCR) phantom. (a) CCR phantom with 10 cartridges. (b) CCR phantom set up for scanning. (c) Coronal views showing each of the 10 layers of the phantom (region of interest for analysis shown as colored squares). (From Mackin, D. et al., *Invest. Radiol.*, 50, 757–765, 2015. Acquired reuse permission from Wolters Kluwer Health, Inc.)

The phantom comprised 10 cartridges, each filled with different materials to produce a wide range of radiomic feature values (Figure 3.6a). Two classes of features (basic statistics such as mean, median, and standard deviation and neighborhood GTDM features of busyness, coarseness, contrast, etc.) calculated from NSCLC tumors in 20 patients were compared with those computed from the phantom scanned using the 17 combinations of imaging parameters. They found that the variability in the values of radiomic features calculated on CT images with different imaging parameters could be in the same order of magnitude as the variability observed in CT scans of NSCLC tumors, suggesting that the studied radiomic features were sensitive to the imaging parameters and CT scanners.

Another radiomics phantom study was published by Zhao et al. [48]. They scanned the FDA-designed anthropomorphic thorax Phantom (KyotoKagaku) under different imaging conditions to explore the effects of CT scanners and scanning parameters on radiomic features. In the published pilot study, the phantom contained 22 lesions of known sizes (10 mm and 20 mm), shapes (spherical, elliptical, lobulated, and spiculated), and densities (−630 HU, −10 HU, and +100 HU) (Figure 3.7). CT raw data were reconstructed using six imaging parameters, i.e., a combination of three slice thicknesses of 1.25 mm, 2.5 mm, and 5 mm and two reconstruction kernels of standard and lung. Fourteen representative radiomic features describing lesion size, shape, and texture were calculated and compared among the six different imaging settings. The authors found that the reconstruction kernels had little effect on the size- and shape-related features, but significantly affected the density and density texture features. Moreover, the 1.25 mm and 2.5 mm slice thicknesses were significantly better than 5 mm for volume, density mean, density standard deviation (SD), GLCM energy, and GLCM homogeneity.

3.4.3 RESPIRATORY MOTION

Fave et al. [49] examined the sensitivity of radiomic features to respiratory motion by using two phantoms scanned with cone-beam-CT (CB-CT). The first phantom was a texture phantom composed of rectangular cartridges to represent different textures (Figure 3.5). Six eight texture features from the five feature classes of histogram, GLCM, run length, NGTDM, and LoG filtered features were measured from two cartridges, shredded rubber and dense cork. The texture phantom was scanned with 19 different CB-CT imaging parameters to establish the features' inter-scanner/inter-parameter variability. The effect of respiratory motion on these features was studied using a dynamic-motion thoracic phantom, a specially designed tumor texture insert of shredded rubber material. The differences between scans acquired with different imaging parameters and different levels of motion were compared to the mean intra-patient difference from the test-retest set.

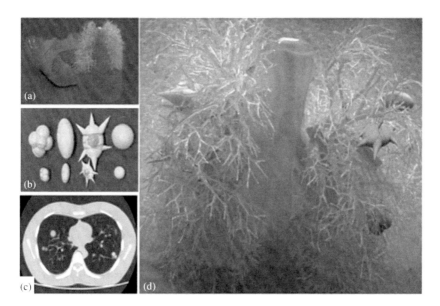

Figure 3.7 (See color insert.) The anthropomorphic thorax phantom. (a) The phantom, (b) phantom lesions of different shapes and sizes, (c) an example of a phantom CT image, and (d) an example of phantom lesions attached to vasculature. (Reprinted from *Transl. Oncol.*, 7, Zhao, B. et al., Exploring variability in CT characterization of tumors: A preliminary phantom study, 88–93, Copyright 2014, with permission from Elsevier.)

Authors found that radiomic features could be reliable as long as the imaging parameters were consistent. However, the influence of respiratory motion on some radiomic features should not be underestimated. For example, with 4 mm of motion, only 18% radiomic features (12/68) were reproducible, when motion increased to 6–8 mm, only 4% radiomic features (3/68).

3.4.4 4D-CT VERSUS 3D-CT

4D-CT is a new CT technology that can capture the location and movement of patient's lesion and organ over time. For instance, for a lung 4D-CT scanning, a number of 3D-CT images tagged with breathing signals are acquired, and each 3D-CT image corresponds to a particular breathing phase. By gathering all 3D-CT imaging, the constituted 4D-CT scanning set can cover the entire breathing cycle of the patient. Since 4D-CT is designed to alleviate the imaging distortion caused by organ motion, radiomic features extracted from 4D-CT are expected to be more reproducible than those extracted from 3D-CT.

Hunter et al. [50] explored the reproducibility of radiomic features on test-retest CT scan pairs obtained from 56 NSCLC patients imaged on three CT scanners from two institutions. One machine used average 4D-CT, one machine used end-exhale 4D-CT, and one used breath-hold helical 3D-CT. Three hundred twenty-eight radiomic features, including tumor's geometry, intensity histogram, absolute gradient image, co-occurrence matrix, and run-length matrix, were calculated on the CT images acquired from the three machines. The authors found that the number of reproducible radiomic features depended on the machine types. Compared to the end-exhale 4D-CT and breath-hold 3D-CT, the average 4D-CT could derive more reproducible radiomic features.

3.4.5 EFFECTS OF IMAGING PARAMETERS ON PHENOTYPE MODELS FOR DIAGNOSIS OF SOLITARY PULMONARY NODULE

Imaging acquisition can affect the computation of radiomic features. To determine optimal imaging acquisition parameters, there is a need to correlate imaging parameters to the performance of radiomic models. He et al. [51] published a pilot study evaluating the effects of contrast-enhancement, reconstruction slice thickness, and reconstruction kernel on the diagnostic performance of radiomics signature in solitary pulmonary nodule. Two hundred and forty patients with solitary pulmonary nodule (malignant: benign = 180: 60) underwent non-contrast CT (NECT) and contrast-enhanced CT (CECT) scans that were reconstructed using two different slice thicknesses of 1.25 mm and 5 mm and reconstruction kernels of lung (sharp kernel) and standard (smooth kernel). One hundred and fifty (250) radiomic features were extracted separately from each CT imaging series, and the diagnostic performance of each feature was assessed based on the AUC. Their results showed better discrimination capability of radiomics signature derived from NECT then from CECT in both training (AUC: 0.862 versus 0.829, $p = 0.032$) and validation cohorts (AUC: 0.750 versus 0.735, $p = 0.014$). Thin-slice CT-based radiomics signature had better diagnostic performance than thick-slice-based signature in both training (AUC: 0.862 versus 0.785, $p = 0.015$) and validation cohorts (AUC: 0.750 versus 0.725, $p = 0.025$; net reclassification index (NRI) = 0.467). Smooth kernel-based radiomics signature had better diagnostic performance than sharp kernel-based radiomics signature in both training (AUC: 0.785 versus 0.770, $p = 0.015$) and validation cohorts (AUC: 0.725 versus 0.686, $p = 0.039$). The authors concluded that the non-contrast, thin-slice, and smooth reconstruction kernel-based CT is more informative for the diagnosis of solitary pulmonary nodule.

3.5 FUTURE WORK

In the previous sections, we talked about the three key steps in radiomic studies, i.e., imaging acquisition, tumor segmentation, and feature extraction/implementation. All of these steps can introduce variability to the computed values of radiomic features and thus affect radiomics phenotype models. Imaging acquisition can be subject to heterogeneity of vendors' scanners and scanning parameters; tumor segmentation can be subject to inter-observer/software variability; and feature extraction can be subject to a wide value range in the choice of algorithm parameters. Reducing variability as much as possible will be the future aim for radiomic studies. For variability induced by imaging acquisition and tumor quantification, harmonization and standardization of the imaging parameters and tumor quantification methods can be important and efficient solutions. The recently emerging technology of deep convolutional neural networks is promising a more automated and self-learning solution.

3.5.1 HARMONIZATION OF IMAGING ACQUISITION PARAMETERS

Harmonization of imaging acquisition parameters can be fulfilled by: (1) identifying the optimal imaging parameters by correlating radiomic features with clinical outcomes and (2) defining tolerance of change to the optimal imaging parameters between different scanners and within the same scanners. However, the former is limited by the available clinical data.

To study comparability of imaging reconstruction parameters, Lu et al. [52] assessed agreements between the 89 radiomic features computed from CT images reconstructed with different slice thicknesses and kernels, using the first scan images of the same-day repeat CT dataset described in Section 4.1. As shown in Figure 3.8, the authors found that the radiomic features calculated on images reconstructed with thinner slices (1.25 mm and 2.5 mm) and smooth kernel highly agreed with each other, indicating possible interchangeable uses of these imaging settings in radiomic studies.

3.5.2 DEEP CONVOLUTIONAL NEURAL NETWORK

More recently, the deep CNN technology emerged and is developing rapidly [53]. CNN allows a mathematical model to learn semantic features automatically from images [54]. Such semantic features are not easily influenced by image quality due to heterogeneous imaging protocols. Furthermore, CNN features do not require accurate tumor segmentation, eliminating segmentation-induced variations.

There are two ways to use the CNNs: (1) use a pre-trained CNN model as a feature generator [55–58] and (2) fine-tune a pre-trained CNN model to build a new mathematical model [59–64]. For the former

Settings QIF groups	1.25S vs 2.5S	1.25L vs 2.5L	2.5S vs 5S	2.5L vs 5L	1.25S vs 5S	2.5S vs 5L	1.25S vs 5L	1.25L vs 5L	5L vs 5S	1.25S vs 2.5L	2.5L vs 2.5S	1.25L vs 1.25S	1.25L vs 2.5S	2.5L vs 5S	1.25L vs 5S	Average CCC of QIF groups
1	0.980	0.994	0.973	0.963	0.945	0.848	0.808	0.969	0.827	0.905	0.927	0.883	0.910	0.894	0.881	0.914
2	0.980	0.980	0.981	0.912	0.954	0.970	0.964	0.839	0.966	0.942	0.895	0.884	0.825	0.843	0.764	0.913
3	0.989	0.986	0.949	0.938	0.910	0.909	0.900	0.899	0.882	0.907	0.888	0.902	0.878	0.787	0.758	0.899
4	0.984	0.939	0.942	0.898	0.931	0.967	0.948	0.815	0.963	0.910	0.923	0.823	0.844	0.819	0.713	0.895
5	0.954	0.945	0.949	0.909	0.943	0.820	0.825	0.867	0.876	0.817	0.865	0.855	0.903	0.875	0.864	0.884
6	0.939	0.946	0.916	0.896	0.919	0.877	0.894	0.902	0.824	0.892	0.858	0.883	0.853	0.806	0.822	0.882
7	0.978	0.974	0.936	0.946	0.928	0.842	0.851	0.941	0.790	0.837	0.822	0.833	0.825	0.757	0.756	0.868
8	0.887	0.763	0.827	0.758	0.855	0.888	0.888	0.825	0.883	0.809	0.781	0.851	0.911	0.739	0.844	0.834
9	0.898	0.953	0.900	0.892	0.800	0.691	0.656	0.893	0.637	0.707	0.731	0.647	0.694	0.704	0.636	0.763
10	0.804	0.783	0.671	0.729	0.673	0.750	0.723	0.812	0.775	0.730	0.750	0.830	0.771	0.624	0.664	0.739
11	0.915	0.761	0.840	0.706	0.691	0.788	0.639	0.699	0.847	0.665	0.713	0.546	0.618	0.636	0.575	0.709
12	0.913	0.925	0.922	0.944	0.791	0.609	0.635	0.803	0.602	0.636	0.582	0.562	0.512	0.538	0.433	0.694
13	0.635	0.735	0.675	0.755	0.301	0.848	0.438	0.415	0.867	0.748	0.893	0.901	0.576	0.560	0.285	0.642
14	0.913	0.832	0.813	0.741	0.850	0.761	0.754	0.557	0.673	0.488	0.491	0.374	0.371	0.409	0.289	0.621
15	0.906	0.772	0.824	0.658	0.807	0.772	0.692	0.426	0.654	0.385	0.478	0.279	0.324	0.339	0.195	0.567
16	0.941	0.790	0.929	0.781	0.852	0.496	0.553	0.527	0.495	0.426	0.388	0.281	0.259	0.373	0.246	0.556
17	0.857	0.835	0.826	0.766	0.696	0.464	0.577	0.566	0.410	0.428	0.321	0.341	0.264	0.275	0.209	0.522
18	0.965	0.631	0.976	0.660	0.922	0.560	0.578	0.355	0.568	0.278	0.252	0.101	0.073	0.264	0.063	0.483
19	0.892	0.674	0.933	0.709	0.787	0.264	0.372	0.350	0.226	0.215	0.161	0.110	0.084	0.135	0.068	0.399
20	0.637	0.856	0.471	0.761	0.277	0.373	0.598	0.574	0.127	0.405	0.239	0.256	0.159	0.077	0.046	0.390
21	0.777	0.560	0.525	0.460	0.289	0.634	0.672	0.182	0.322	0.354	0.245	0.164	0.116	0.112	0.055	0.364
22	0.611	0.534	0.292	0.339	0.116	0.523	0.466	0.155	0.181	0.369	0.184	0.180	0.088	0.044	0.016	0.273
23	0.801	0.711	0.712	0.563	0.489	0.059	0.097	0.297	0.034	0.039	0.025	0.021	0.014	0.015	0.008	0.259
Average CCC of setting pairs	0.875	0.820	0.815	0.768	0.725	0.684	0.674	0.636	0.627	0.604	0.583	0.543	0.515	0.504	0.441	

Group (a) Fixing reconstruction algorithm while changing slice thickness
Group (b) Fixing slice thickness while changing reconstruction algorithm
Group (c) Smooth reconstruction algorithm (S) plus thin slice thickness versus sharp reconstruction algorithm (L) plus thick slice thickness
Group (d) Sharp reconstruction algorithm (L) plus thin slice thickness versus smooth reconstruction algorithm (S) plus thick slice thickness

1   0.5   0

Figure 3.8 (See color insert.) The CCCs of non-redundant quantitative imaging feature (QIF) groups under the 15 inter-setting comparisons. Columns are arranged in descending order according to the average CCC of the inter-setting comparisons. Rows are arranged in descending order according to average CCCs of Non-redundant QIF Groups. (From Lu, L. et al. *PLoS One*, 11, e0166550, 2016.)

situation, a pre-trained CNN model will output CNN features that can be extracted from certain layers of the network. The extracted CNN features can then be used to build new mathematical models using the machine learning methods. For instance, in [55], a pre-trained CNN model was used as a feature generator for chest pathology identification. In [58], a pre-trained CNN model was used as a feature generator for the diagnosis of breast lesion. For the latter situation, a pre-trained CNN model will be adapted and fine-tuned to build new mathematical model directly. For instance, in [62], the original top layer of a pre-trained CNN model was replaced with a new logistic layer, and then labeled data were used to only train the appended layer while keeping the rest of the network unchanged, yielding promising results for classification of unregistered multi-view mammogram. Shin In [63], fine-tuned, pre-trained CNN model was used to automatically retrieve missing or noisy cardiac acquisition plane information and predict the most common cardiac views. In [64], all layers of a pre-trained CNN model were fine tuned for automatic classification of interstitial lung diseases. The authors also suggested an attenuation rescale scheme to convert 1-channel CT images to red, green, and blue (RGB)-like 3-channel natural images, which could benefit the tuning of pre-trained model.

Two key words can be noticed in the above introduction, the "pre-trained" and the "fine-tuned." Pre-trained CNN models mean that the CNN models have already been trained on large-scale annotated datasets, e.g., ImageNet [65], an image database containing more than 14 million natural images annotated into 1000 object categories (e.g., "dog," "cat"). Popular pre-trained CNN models include Alexnet [66], GoogLeNet [67], VGG [68], and ResNet [69]. However, when applying CNN to the analysis of medical images, to obtain a comprehensively annotated dataset as ImageNet remains a challenge. Due to limited available data, fine-tuning a pre-trained CNN model is thus more efficient than training a CNN from scratch [70,71].

In summary, the future of using CNN to improve cancer diagnosis, prognosis, and response prediction/assessment is bright, and the field will be evolving rapidly.

REFERENCES

1. B. F. Kurland, E. R. Gerstner, J. M. Mountz et al. Promise and pitfalls of quantitative imaging in oncology clinical trials. *Magn Reson Imaging*, vol. 30, pp. 1301–1312, 2012.

2. A. J. Buckler, L. Bresolin, N. R. Dunnick et al. A collaborative enterprise for multi-stakeholder participation in the advancement of quantitative imaging. *Radiology*, vol. 258, pp. 906–914, 2011.

3. A. J. Buckler, L. Bresolin, N. R. Dunnick et al. Quantitative imaging test approval and biomarker qualification: Interrelated but distinct activities. *Radiology*, vol. 259, pp. 875–884, 2011.

4. L. Hood and S. H. Friend. Predictive, personalized, preventive, participatory (P4) cancer medicine. *Nat Rev Clin Oncol*, vol. 8, pp. 184–187, 2011.

5. P. Lambin, R. G. van Stiphout, M. H. Starmans et al. Predicting outcomes in radiation oncology—Multifactorial decision support systems. *Nat Rev Clin Oncol*, vol. 10, pp. 27–40, 2013.

6. E. A. Eisenhauer, P. Therasse, J. Bogaerts et al. New response evaluation criteria in solid tumours: Revised RECIST guideline (version 1.1). *Eur J Cancer*, vol. 45, pp. 228–247, 2009.

7. D. M. Hansell, A. A. Bankier, H. MacMahon et al. Fleischner society: Glossary of terms for thoracic imaging. *Radiology*, vol. 246, pp. 697–722, 2008.

8. A. McErlean, D. M. Panicek, E. C. Zabor et al. Intra- and interobserver variability in CT measurements in oncology. *Radiology*, vol. 269, pp. 451–459, 2013.

9. S. G. Armato 3rd, G. McLennan, L. Bidaut et al. The Lung Image Database Consortium (LIDC) and Image Database Resource Initiative (IDRI): A completed reference database of lung nodules on CT scans. *Med Phys*, vol. 38, pp. 915–931, 2011.

10. F. Tixier, C. C. Le Rest, M. Hatt et al. Intratumor heterogeneity characterized by textural features on baseline 18F-FDG PET images predicts response to concomitant radiochemotherapy in esophageal cancer. *J Nucl Med*, vol. 52, pp. 369–378, 2011.

11. O. Gevaert, J. Xu, C. D. Hoang et al. Non-small cell lung cancer: Identifying prognostic imaging biomarkers by leveraging public gene expression microarray data—Methods and preliminary results. *Radiology*, vol. 264, pp. 387–396, 2012.

12. V. Kumar, Y. Gu, S. Basu et al. Radiomics: The process and the challenges. *Magn Reson Imaging*, vol. 30, pp. 1234–1248, 2012.

13. P. Lambin, E. Rios-Velazquez, R. Leijenaar et al. Radiomics: Extracting more information from medical images using advanced feature analysis. *Eur J Cancer*, vol. 48, pp. 441–446, 2012.

14. H. J. Aerts, E. R. Velazquez, R. T. Leijenaar et al. Decoding tumour phenotype by noninvasive imaging using a quantitative radiomics approach. *Nat Commun*, vol. 5, p. 4006, 2014.

15. C. A. Karlo, P. L. Di Paolo, J. Chaim et al. Radiogenomics of clear cell renal cell carcinoma: Associations between CT imaging features and mutations. *Radiology*, vol. 270, pp. 464–471, 2014.

16. M. D. Kuo and N. Jamshidi. Behind the numbers: Decoding molecular phenotypes with radiogenomics—Guiding principles and technical considerations. *Radiology*, vol. 270, pp. 320–325, 2014.

17. C. Yip, D. Landau, R. Kozarski et al. Primary esophageal cancer: Heterogeneity as potential prognostic biomarker in patients treated with definitive chemotherapy and radiation therapy. *Radiology*, vol. 270, pp. 141–148, 2014.

18. T. P. Coroller, P. Grossmann, Y. Hou et al. CT-based radiomic signature predicts distant metastasis in lung adenocarcinoma. *Radiother Oncol*, vol. 114, pp. 345–350, 2015.

19. H. J. Yoon, I. Sohn, J. H. Cho et al. Decoding tumor phenotypes for ALK, ROS1, and RET fusions in lung adenocarcinoma using a radiomics approach. *Medicine (Baltimore)*, vol. 94, p. e1753, 2015.

20. R. M. Haralick, K. Shanmugam, and I. Dinstein. Textural features for image classification. *IEEE Trans Syst Man Cybern*, vol. 6, pp. 610–621, 1973.

21. M. M. Galloway. Texture analysis using gray level run lengths. *Comput Gr Image Process*, vol. 4, pp. 172–179, 1975.

22. B. Mandelbrot. How long is the coast of Britain? Statistical self-similarity and fractional dimension. *Science*, vol. 156, pp. 636–638, 1967.

23. I. Fogel and D. Sagi. Gabor filters as texture discriminator. *Biol Cybern*, vol. 61, pp. 103–113, 1989.

24. P. S. Addison, *The Illustrated Wavelet Transform HandBook*, ed: Institute of Physics, Boca Raton, FL: Taylor & Francis Group, 2002.

25. J. K. Jan and J. D. Andrea. Surface shape and curvature scales. *Image Vis Comput*, vol. 10, pp. 557–565, 1992.

26. S. A. Napel, C. F. Beaulieu, C. Rodriguez et al. Automated retrieval of CT images of liver lesions on the basis of image similarity: Method and preliminary results. *Radiology*, vol. 256, pp. 243–252, 2010.

27. R. T. Leijenaar, S. Carvalho, E. R. Velazquez et al. Stability of FDG-PET radiomics features: An integrated analysis of test-retest and inter-observer variability. *Acta Oncol*, vol. 52, pp. 1391–1397, 2013.

28. F. Orlhac, M. Soussan, J. A. Maisonobe et al. Tumor texture analysis in 18F-FDG PET: Relationships between texture parameters, histogram indices, standardized uptake values, metabolic volumes, and total lesion glycolysis. *J Nucl Med*, vol. 55, pp. 414–422, 2014.

29. G. Doumou, M. Siddique, C. Tsoumpas et al. The precision of textural analysis in (18)F-FDG-PET scans of oesophageal cancer. *Eur Radiol*, vol. 25, pp. 2805–2812, 2015.

30. W. Mu, Z. Chen, Y. Liang et al. Staging of cervical cancer based on tumor heterogeneity characterized by texture features on (18)F-FDG PET images. *Phys Med Biol*, vol. 60, pp. 5123–5139, 2015.

31. L. Zhang, D. V. Fried, X. J. Fave et al. IBEX: An open infrastructure software platform to facilitate collaborative work in radiomics. *Med Phys*, vol. 42, pp. 1341–1353, 2015.

32. Y. H. Fang, C. Y. Lin, M. J. Shih et al. Development and evaluation of an open-source software package "CGITA" for quantifying tumor heterogeneity with molecular images. *Biomed Res Int*, vol. 2014, p. 248505, 2014.

33. J. J. Foy, K. R. Mendel, H. Li et al. Variation in algorithm implementation between quantitative texture analysis software. in SPIE Medical Imaging, *RSNA*, Chicago, IL, 2017.

34. Y. Balagurunathan, Y. Gu, H. Wang et al. Reproducibility and prognosis of quantitative features extracted from CT images. *Transl Oncol*, vol. 7, pp. 72–87, 2014.

35. S. Hawkins, H. Wang, Y. Liu et al. Predicting malignant nodules from screening CT scans. *J Thorac Oncol*, vol. 11, pp. 2120–2128, 2016.

36. E. R. Velazquez, C. Parmar, M. Jermoumi et al. Volumetric CT-based segmentation of NSCLC using 3D-Slicer. *Sci Rep*, vol. 3, p. 3529, 2013.

37. C. Parmar, E. R. Velazquez, R. Leijenaar et al. Robust radiomics feature quantification using semiautomatic volumetric segmentation. *PLoS One*, vol. 9, p. e102107, 2014.

38. J. J. Bartko. The intraclass correlation coefficient as a measure of reliability. *Psychol Rep*, vol. 19, pp. 3–11, 1966.

39. J. Kalpathy-Cramer, B. Zhao, D. Goldgof et al. A comparison of lung nodule segmentation algorithms: Methods and results from a multi-institutional study. *J Digit Imaging*, vol. 29, pp. 476–487, 2016.

40. Y. Tan, L. H. Schwartz, and B. Zhao. Segmentation of lung lesions on CT scans using watershed, active contours, and Markov random field. *Med Phys*, vol. 40, p. 043502, 2013.

41. Y. Gu, V. Kumar, L. O. Hall et al. Automated delineation of lung tumors from CT images using a single click ensemble segmentation approach. *Pattern Recognit*, vol. 46, pp. 692–702, 2013.

42. Q. Huang, L. Lu, L. Dercle et al. Interobserver variability in tumor contouring affects the use of radiomics to predict mutational status. *J Med Imaging (Bellingham)*, vol. 5, p. 011005, 2018.

43. P. A. Yushkevich, J. Piven, H. C. Hazlett et al. User-guided 3D active contour segmentation of anatomical structures: Significantly improved efficiency and reliability. *Neuroimage*, vol. 31, pp. 1116–1128, 2006.

44. B. Zhao, L. P. James, C. S. Moskowitz et al. Evaluating variability in tumor measurements from same-day repeat CT scans of patients with non-small cell lung cancer. *Radiology*, vol. 252, pp. 263–272, 2009.

45. Y. Balagurunathan, V. Kumar, Y. Gu et al. Test-retest reproducibility analysis of lung CT image features. *J Digit Imaging*, vol. 27, pp. 805–823, 2014.

46. B. Zhao, Y. Tan, W. Y. Tsai et al. Reproducibility of radiomics for deciphering tumor phenotype with imaging. *Sci Rep*, vol. 6, p. 23428, 2016.

47. D. Mackin, X. Fave, L. Zhang et al. Measuring computed tomography scanner variability of radiomics features. *Invest Radiol*, vol. 50, pp. 757–765, 2015.

48. B. Zhao, Y. Tan, W. Y. Tsai et al. Exploring variability in CT characterization of tumors: A preliminary phantom study. *Transl Oncol*, vol. 7, pp. 88–93, 2014.

49. X. Fave, D. Mackin, Y. J. et al. Can radiomics features be reproducibly measured from CBCT images for patients with non-small cell lung cancer? *Med Phys*, vol. 42, pp. 6784–6797, 2015.

50. L. A. Hunter, S. Krafft, F. Stingo et al. High quality machine-robust image features: Identification in nonsmall cell lung cancer computed tomography images. *Med Phys*, vol. 40, p. 121916, 2013.

51. L. He, Y. Huang, Z. Ma, C. Liang, C. Liang, and Z. Liu. Effects of contrast-enhancement, reconstruction slice thickness and convolution kernel on the diagnostic performance of radiomics signature in solitary pulmonary nodule. *Sci Rep*, vol. 6, p. 34921, 2016.

52. L. Lu, R. C. Ehmke, L. H. Schwartz et al. Assessing agreement between radiomic features computed for multiple CT imaging settings. *PLoS One*, vol. 11, p. e0166550, 2016.

53. Y. LeCun, Y. Bengio, and G. Hinton. Deep learning. *Nature*, vol. 521, pp. 436–444, 2015.

54. Y. LeCun, L. Bottou, Y. Bengio et al. Gradient-based learning applied to document recognition. *Proc IEEE*, vol. 86, pp. 2278–2324, 1998.

55. Y. Bar, L. W. Diamant, and H. Greenspan. Deep learning with non-medical training used for chest pathology identification. *SPIE Med Imag*, 2015, vol. 9414, p. 140V.

56. B. V. Ginneken, A. A. Setio, C. Jacobs et al. Off-the-shelf convolutional neural network features for pulmonary nodule detection in computed tomography scans. *IEEE 12th Int Symp Biomed Imag,* 2015, pp. 286–289.

57. J. Arevalo, F. Gonzalez, R. Ramos-Pollan et al. Convolutional neural networks for mammography mass lesion classification. *Proc 37th Annu Int Conf IEEE EMBC*, 2015, pp. 797–800.

58. N. Antropova, B. Q. Huynh, and M. L. Giger. A deep feature fusion methodology for breast cancer diagnosis demonstrated on three imaging modality datasets. *Med Phys*, vol. 44, pp. 5162–5171, 2017.

59. T. Schlegl, J. Ofner, and G. Langs. Unsupervised pre-training across image domains improves lung tissue classification. *Presented at the Medical Computer Vision: Algorithms for Big Data. MCV 2014*, 2014.

60. H. Chen, D. Ni, J. Qin et al. Standard plane localization in fetal ultrasound via domain transferred deep neural networks. *IEEE J Biomed Health Inform*, vol. 19, pp. 1627–1636, 2015.

61. H. C. Shin, L. Lu, L. Kim et al. Interleaved text/image deep mining on a very largescale radiology database. *Presented at the Computer Vision and Pattern Recognition (CVPR), 2015 IEEE Conference on*, 2015.

62. G. Carneiro, J. Nascimento, and A. Bradley. Unregistered multiview mammogram analysis with pre-trained deep learning models. *Presented at the Medical Image Computing and Computer-Assisted Intervention-MICCAI 2015*, 2015.

63. J. Margeta, A. Criminisi, R. C. Lozoya et al. Fine-tuned convolutional neural nets for cardiac MRI acquisition plane recognition. *Computer Methods in Biomechanics and Biomedical Engineering: Imaging & Visualization*, 2015.

64. M. Gao, U. Bagci, L. Lu et al. Holistic classification of CT attenuation patterns for interstitial lung diseases via deep convolutional neural networks. *Presented at the Computer Methods in Biomechanics and Biomedical Engineering: Imaging & Visualization*, 2016.

65. O. Russakovsky, J. Deng, H. Su et al. Imagenet large scale visual recognition challenge. *Int J Comput Vision*, vol. 115, pp. 211–252, 2015.

66. A. Krizhevsky, I. Sutskever, and G. E. Hinton. Imagenet classification with deep convolutional neural networks. *Adv Neur Inf Process Syst*, 2012, pp. 1097–1105.

67. C. Szegedy, W. Liu, Y. Jia et al. Going deeper with convolutions. *Proc IEEE Conf Comput Vis Pattern Recognit*, 2015, pp. 1–9.

68. K. Simonyan and A. Zisserman. Very deep convolutional networks for large-scale image recognition. arXiv preprint arXiv:1409.1556, 2014.

69. K. He, X. Zhang, S. Ren et al. Deep residual learning for image recognition. *Proc IEEE Conf Comput Vis Pattern Recognit*, 2016, pp. 770–778.

70. N. Tajbakhsh, J. Y. Shin, S. R. Gurudu et al. Convolutional neural networks for medical image analysis: Full training or fine tuning? *IEEE Trans Med Imaging*, vol. 35, pp. 1299–1312, 2016.

71. H. C. Shin, H. R. Roth, M. Gao et al. Deep convolutional neural networks for computer-aided detection: CNN architectures, dataset characteristics and transfer learning. *IEEE Trans Med Imaging*, vol. 35, pp. 1285–1298, 2016.

4

Quantitative PET/CT for radiomics

STEPHEN R. BOWEN, PAUL E. KINAHAN, GEORGE A. SANDISON,
AND MATTHEW J. NYFLOT

The conversion of images, such as those from positron emission tomography (PET) and X-ray computed tomography (CT), into mineable high-dimensional data is known as radiomics. Radiomics is motivated by the concept that biomedical images contain information that reflects underlying physiology and molecular biology, and that these relationships can be revealed via quantitative image analyses. Thus, an understanding of quantitative PET/CT accuracy, and how variations in acquisition and image reconstruction parameters influence imaging features in the absence of true underlying biologic effects, is essential to PET/CT radiomics.

In this chapter, we first provide an overview of quantitative PET/CT, with emphasis on uncertainties and their clinical impact on quantitative precision and accuracy. We highlight ongoing efforts toward PET/CT standardization and harmonization as part of quantitative imaging biomarker quality assurance programs. With a basis in quantitative PET/CT, we introduce PET/CT radiomics, list additional sources of radiomic data processing uncertainty, propose guidelines to improve transparent reporting of PET/CT radiomic features, and emphasize the growing need for robust "omics" data analytics to help ensure reproducibility and clinical translation of scientific findings. Lastly, we summarize future directions beyond PET/CT radiomics, including hybrid PET/magnetic resonance (MR) radiomics, non-2-[^{18}F]fluoro-2-deoxy-D-glucose (FDG)-PET radiomics, normal tissue function PET radiomics, and deep learning of PET/CT images.

4.1 QUANTITATIVE PET/CT IN ONCOLOGY

Combined PET/CT imaging (1) provides superior diagnostic accuracy in many oncologic disease sites compared to either modality alone (2,3). While diagnostic evaluation historically consisted of subjective image assessment by expert clinician observers, modern diagnostics also rely on methods of objective image assessment through evaluation of quantitative PET/CT metrics. PET/CT is an inherently quantitative hybrid imaging modality (4). The known physical decay properties of positron-emitting nuclei, the production of monoenergetic 511 keV anti-parallel annihilation photons from energy and momentum conservation laws, the efficient ring configuration of coincident detection modules with precise timing and energy windows, and accurate corrections for photon attenuation as well as scattered and random coincidence events (5), all enable calibration of image intensities to absolute activity concentrations. In turn, these activity concentrations are normalized for variability in the injected activity of radiotracer and the available volume of tracer distribution in patients to produce standardized uptake values (SUV) that facilitate inter-scan and inter-patient quantitative comparisons (6).

Numerous oncologic investigations have demonstrated the clinical utility of quantitative PET/CT for risk stratification, treatment planning, treatment response assessment, and outcome prediction. They have spanned multiple disease sites, multiple PET imaging tracers, and multiple imaging time points. The most commonly reported quantitative measures in clinical PET investigations are the maximum SUV, peak SUV (7,8), the metabolic tumor volume (MTV), and the total lesion glycolysis (TLG), which is a product of MTV and SUV$_{mean}$. As it is utilized ubiquitously in oncology, we will focus primarily on applications of FDG in quantitative PET/CT (3).

4.1.1 RADIATION TREATMENT PLANNING

FDG PET/CT is indicated for radiation therapy planning in the following oncologic disease sites: lung cancer (non-small cell and small cell) (9–11), head-and-neck cancer (squamous cell carcinoma) (12), esophageal cancer (13,14), breast cancer (15), rectal cancer (16), cervical cancer (17), and hematologic malignancies (lymphoma) (18). PET/CT has been shown to better differentiate malignant tissue from atelectasis in lung cancer patients, to provide specific uptake contrast at the boundary of certain primary tumors, and to accurately identify involved nodes, which can have clinical implications for defining radiation field boundaries and prescribed radiation dosing. For example, FDG PET-positive nodes in head-and-neck cancer are typically treated as high/moderate risk targets to doses of 66–70 Gy, while PET-negative nodes are electively treated as low-risk targets to 54 Gy (12). Advances have also been made in the integration of respiratory-correlated (4D) PET/CT for lung cancer treatment planning (19), which will require patient-specific implementations for motion management during planning and image-guided treatment delivery (20). Despite progress in the integration of quantitative PET/CT into radiation therapy planning, no disease site-specific consensus guidelines exist on optimal SUV-based contouring methods, from fixed thresholds to adaptive thresholds (21) and stochastic algorithms (22). Instead, manual contouring is performed by radiation oncologists and remains susceptible to intra- and inter-observer variability, but in the future could be augmented with preliminary automatic PET segmentation that reduces contouring time (23).

4.1.2 RISK STRATIFICATION AND TREATMENT RESPONSE ASSESSMENT

In lung cancer patients, early response assessment of maximum and mean SUV (24,25), MTV (25–28), and TLG (28–32) on FDG PET/CT have predicted local failure, distant metastatic progression, and survival following concurrent or sequential chemo-radiotherapy. In esophageal cancer patients, pre-treatment and mid-treatment MTV size was associated with progression free-survival (HR > 1.2 per 10 cm^3, $p < 0.003$) and overall survival (HR > 1.2 per 10 cm^3, $p < 0.01$) (33). In head-and-neck cancer patients, a recent clinical review of quantitative FDG-PET/CT in 45 representative studies revealed that baseline volumetric parameters such as MTV and TLG were better predictors of reponse evaluation criteria in solid tumors (RECIST) status and outcome (loco-regional failure, disease progression, death from any cause) compared to SUV$_{max}$, and that MTV improved risk stratification over gross tumor volume (GTV) and conventional clinical staging criteria (34). However, the thresholds in PET metrics for patient risk stratification and outcome prediction were highly variable (e.g., pre-treatment SUV$_{max}$ > 5–12 g/mL, post-treatment SUV$_{max}$ > 3–6 g/mL) (34). This degree of variability was also observed in head-and-neck patients when comparing FDG PET SUV metrics and diffusion-weighted MRI metrics (35). In cervical cancer patients, pre-treatment SUV metrics (36) as well as mid-treatment changes in SUV metrics (37) correlated to favorable post-treatment response. Importantly, these clinical investigations have reported an array of congruent findings, contradictory findings, and "complementary" findings without consensus guidelines. Variability in the findings of clinical investigations highlights a major motivation for improving the standardized calculation and transparent reporting of quantitative PET/CT measures (38).

4.2 UNCERTAINTIES IN QUANTITATIVE PET/CT

Performance errors in quantitative PET/CT measures for clinical tasks can be framed in terms of systematic bias and random variance as confounders of accuracy and precision, respectively. Improvements in precision from lower random errors in quantitation directly translate to increases in statistical power for detecting a given effect size in clinical trials (39). Conversely, clinical studies detecting a given effect size with fixed statistical power require smaller sample size when using more quantitatively precise PET/CT parameters. Increased PET/CT sensitivity and specificity may also translate to increased effect size for treatment response assessment and clinical outcome prediction, which may further boost statistical power (39). Limits on the quantitative accuracy (bias) and precision (variance) of PET SUV metrics are governed by the following properties: stability, repeatability, and reproducibility. A quantitative PET SUV metric should be stable under consistent image formation, should be repeatable between imaging scans under consistent patient conditions, and should be reproducible between cohorts of patients under consistent patient conditions when reported in single or multi-institutions studies.

4.2.1 PHYSICAL AND TECHNICAL UNCERTAINTIES

Several sources of uncertainty contribute to both bias and variance in the estimates of PET SUV metrics, which have been summarized in dedicated review articles (40–42). These uncertainties derive from physical sources, technical sources, and biological sources. Physical sources of uncertainty include noise from scattered and random coincidence events that increase variance as a function of count rate (13), resolution loss in PET imaging systems that increase bias in SUV estimation of small objects (44,45), artifactual CT for attenuation correction (46) that is especially pronounced in 4D PET/CT (47), and reconstruction algorithms that can modify both variance and bias in PET SUV (48,49), even after normal organ or blood pool normalization techniques (50). Technical sources of uncertainty include mismatched isotope calibration between the PET scanner and dose calibrator (well counter), incorrect clock synchronization between the activity assay room (hot lab) and the PET/CT scanner computer, incorrect injected activity versus calibration time, and quality of radiotracer injection administration. Biological sources of uncertainty include variable uptake period without steady-state pharmacokinetic conditions, physiologic pulmonary/cardiac/gastrointestinal motion, and, in the case of FDG PET, differing fasting states as measured by blood glucose levels. Glucose sensitivity can have significant impact on both FDG PET/CT-based radiation treatment planning and treatment response assessment (51).

4.2.2 BIOLOGICAL UNCERTAINTIES

Additional complex biological sources of uncertainty to PET SUV include variable contribution of uptake due to tracer transport from the vasculature to the tissue of interest versus contribution of uptake due to specific retention of tracer within a tissue of interest. Accounting for this biologic source of SUV uncertainty requires modeling the transport and binding mechanisms of tracers using dynamic PET. Quantitative dynamic PET imaging provides a wealth of information on cancer phenotypes derived from tracer pharmacokinetic properties, from hypoxia to proliferation indices of tracer retention (52). However, processing dynamic PET images introduces new sources of uncertainty (53), from arterial input function definition to compartmental model choice and subsequent kinetic parameters estimation (52). Further, limited PET spatial resolution of several millimeters results in image intra-voxel heterogeneity for biologically heterogeneous tissues at the cellular scale, such as poly-phenotypic and clonally evolved solid tumors. Computational biology seeks to bridge this gap in spatial resolution by modeling underlying intra-voxel vascular distribution and cancer cell phenotypes (e.g., hypoxic fraction) to help inform on PET/CT-guided treatment strategies (54), but comes at the expense of introducing highly degenerate solutions at the cellular scale that all converge to the same image voxel representation at the millimeter scale.

4.2.3 SEGMENTATION UNCERTAINTIES

Lastly, a large source of variability that plagues all quantitative imaging modalities is region of interest definition and segmentation. Quantitative PET metrics such as SUV_{mean}, MTV, and TLG, depend on volume delineation. Many studies utilize manually delineated volumes by expert clinician observers as a gold standard, and while PET/CT has aided in radiation oncology target delineation, these volumes still display large inter-observer variability. A recent study on intra- and inter-observer manual PET segmentation variability of 806 phantom inserts and 641 patient lesions demonstrated an absolute error of >30% in phantom insert volumes and >50% in head-and-neck cancer patient lesion volumes, with repeatability of these volumes around 40% (55). Automatic segmentation using quantitative PET has been extensively researched and reviewed (22), but in general, simple fixed threshold algorithms carry higher reproducibility and bias while the more complex stochastic algorithms carry lower bias and limited reproducibility in certain cases. For example, an adaptive SUV-based segmentation algorithm (multi-level Otsu using Slicer) generated more accurate MTV contours compared to fixed SUV thresholds in 48 soft tissue sarcoma patients (intra-class correlation coefficient (ICC) = 0.93), but had variance greater than 10 times the bias in volume estimation (~500 cm³) (21). This is caused in part by a lack of ground truth validation for realistic patient tumor segmentation, in which PET segmentation algorithms are often benchmarked against manually delineated regions that carry intra- and inter-observer uncertainty. These errors in image quantification and segmentation propagate downstream in their application to radiation therapy planning, particularly in the case 4D PET/CT-guided planning for lung cancer (56).

Mitigating uncertainties in quantitative PET/CT requires a combination of existing paradigms: (1) identifying PET metrics that are insensitive to the largest sources of variability, (2) reducing the magnitude of each of the sources of variability by enforcing a fixed set of standards and allowable states in PET imaging, and (3) accounting for any residual sources of variability through transparent reporting of confidence intervals or limits on PET metrics.

4.3 STANDARDIZATION AND HARMONIZATION OF QUANTITATIVE PET/CT

Standardization of quantitative PET/CT seeks to restrict or eliminate sources of variability during patient preparation, image acquisition, image reconstruction, image post-processing, and image assessment. Harmonization of quantitative PET/CT seeks to account for the sources of variability in image formation and data analysis, with particular emphasis on establishing confidence intervals in quantitative PET metrics reported across multiple institutions (57). Such harmonization would establish limits to effect size detection and more precisely power multi-center PET/CT clinical trials. As several institutions tune their scanners to achieve harmonized reporting of quantitative PET/CT metrics, statistical power can increase at fixed patient sample size or conversely decrease the required sample size at fixed power (39).

4.3.1 QUANTITATIVE PET/CT STANDARDS

The first step toward improved quantitative PET/CT is the adoption of common standards, procedures, and metrics of performance evaluation. Standards include the widespread adoption of accreditation phantoms from several national and international organizations, such as the American College of Radiology (ACR), National Electrical Manufacturers Association (NEMA), Society of Nuclear Medicine Clinical Trials Network (SNM-CTN), and the European Association of Nuclear Medicine (EANM). NEMA/EANM/SNM-CTN phantoms are especially well-suited to detect changes in SUV and activity recovery coefficients due to changes in reconstruction settings, including number of iterations, subsets, time-of-flight kernel widths, and segmentation methods for defining SUV_{max}, SUV_{mean}, and SUV_{peak} (58).

Another class of standards for quantitative PET/CT arises during data processing in the form of synthetic phantoms or digital reference objects (DRO). These provide ground truth settings for image viewing workstations, region and volume of interest segmentation, and reporting of summary statistics. One such digital reference object, shown in Figure 4.1, was designed based on the NEMA image quality (IQ) phantom to be fully DICOM (Digital Imaging and Communications in Medicine) compliant, using a synthetic set of known PET SUV objects and corresponding CT attenuation coefficients (59). Following evaluation of the DRO by 16 institutions using 21 PET/CT display software applications, notable errors in maximum (up to 40%) and mean SUV (up to 100%) were observed. These large errors support the adoption of a digital reference standard and efforts to harmonize commercial software packages (59). The root cause of some of these errors can be linked to the structured data within DICOM formatted images and their interpretation by different DICOM readers and require an additional layer of standardization (60).

Other proposed improvements to quantitative PET standardization involved advances in scanner and injected activity calibration. One investigation used a sodium-iodide (NaI(Tl)) gamma counter with narrow energy window over 511 keV standardized by long-lived and traceable ^{68}Ge/^{68}Ga sources to achieve 0.1% stability in scanner calibration and 94% efficiency in ^{18}F activity recovery (61). Such high efficiency and narrow energy window gamma counters can provide accurate calibration over a wide range of injected volumes.

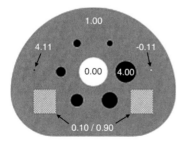 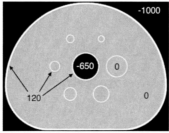 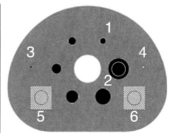

Figure 4.1 (See color insert.) A digital reference object (DRO) for quantitative PET/CT data analysis and reporting. Known PET SUV, CT HU, and ROI allow for quality assurance testing of commercial image analysis software and improved reproducibility in multi-center imaging trials. (Reproduced from Pierce, L.A. et al., *Radiology*, 277, 538–545, 2015. With permission.)

4.3.2 QUANTITATIVE PET/CT HARMONIZATION

Once physical and digital standards are adopted across institutions, they can actively seek to harmonize their PET/CT scanners for quantitative reporting through cross-calibration. One example of cross-calibration was achieved with long-lived and sealed sources within NEMA-style phantoms that can evaluate both dose calibration (well counter) and scanner calibrations. These sealed source phantoms commercially purchased as a kit and posed far fewer radiation safety challenges than conventional aqueous phantoms (62). Cross-calibration between a regional network of PET/CT scanners led to coefficients of variation in SUV mean of 1%, while bias in SUV between combinations of dose calibrators and scanners varied by 5% and fluctuated over time.

National and international efforts to promote quantitative PET/CT standardization and harmonization are ongoing. Chief among these are the Quantitative Imaging Biomarker Alliance (QIBA) under the Radiological Society of North America (RSNA) and the FDG PET/CT accreditation program under the European Association for Nuclear Medicine (EANM). QIBA has established standards for quantitative imaging biomarkers, which includes adoption of common nomenclature (e.g., standardization, harmonization, etc.) under the Metrology Working Group (63) and guidelines for appropriate statistical tests (64) (e.g., Bland-Altman, ICC, etc.) under the Technical Performance Working Group (65) of imaging metric quantitative performance in reference objects under controlled conditions. Based on their assessment of the technical performance of FDG PET imaging biomarkers, QIBA documented an FDG PET "profile" that sets a benchmark for expected measurement linearity and bias, repeatability, and reproducibility of SUV metrics within clinical trial settings (65). By meeting the criteria established by this benchmark, the repeatability of FDG-PET/CT SUV_{max} between patients should fall within 10%–12%. This framework for technical performance analysis methods and study designs spans multiple imaging modalities, with documented profiles for CT, MRI, and future profiles for single photon emission computed tomography (SPECT) and ultrasound (US).

With the benchmark FDG PET profile in place, QIBA designed a comprehensive clinical protocol to harmonize FDG PET/CT imaging, known as the Uniform Protocols for Imaging in Clinical Trials (UPICT) (66). In addition to providing minimum confidence intervals for image acquisition and reconstruction, the UPICT document reviews best practices for FDG PET/CT clinical trial site initiation, patient scheduling, patient preparation, data processing and storage, data archival and distribution, and imaging radiation risk management (66). At the heart of standardized data storage and sharing for imaging biomarker development lies common structured reporting of DICOM PET/CT data, illustrated in an example head-and-neck cancer application (60).

EANM Research Ltd. (EARL) has launched an FDG PET/CT accreditation program, which represents an international harmonization effort for quantitative imaging biomarker measurement (67). The program extends prior EANM guidelines on FDG PET/CT imaging (68) by specifying harmonization procedures for several PET/CT scanners, with variable time-of-flight (TOF) capability during image acquisition and system modeling (point spread function, PSF) for resolution recovery during image reconstruction, with the goal of reproducing treatment response assessment by the European Organization for Research and Treatment of Cancer (EORTC) or the PET Response Criteria for Solid Tumors (PERCIST). Figure 4.2 highlights the harmonized results between conventional ordered subset expectation maximization (OSEM) reconstruction and PSF + TOF when evaluating lean body mass SUV metrics (67). By post-processing the PSF + TOF to create an image equivalent to OSEM, the group was able to achieve the same response evaluation no matter what combination of pre- and post-treatment harmonized images were used for a given patient (67), illustrated in Figure 4.2. Other accreditation programs for quantitative PET imaging include the National Cancer Institute Center for Imaging Excellence, which launched in 2010 and now includes quality assurance of SUV reporting with an emphasis on evaluation of relationships between SUV errors (bias or variance) and statistical power in quantitative imaging-based trials (69). Despite improved compliance with National Cancer Institute Center for Imaging Excellence (NCI-CQIE) requirements, over 30% of PET scanners failed to pass after two requalification submissions (69).

Outside of FDG PET/CT repeatability and harmonization studies, others have characterized tracer-specific variability in quantitative metrics. One example was a multi-center study of 35 castrate-resistant metastatic prostate cancer patients undergoing test-retest ^{18}F-NaF PET/CT bone scans 3 days apart (nominal) that were

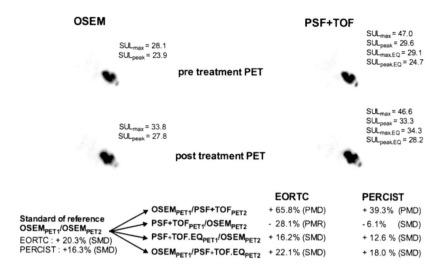

Figure 4.2 FDG PET/CT harmonization under the EANM accreditation program. Harmonized PSF + TOF PET images can be used to assess response relative to conventional OSEM PET images, producing consistent response categorization using both EORTC and PERCIST criteria. (Reproduced from Aide, N. et al., *Eur. J. Nucl. Med. Mol. Imag.*, 44, 17–31, 2017. With permission.)

analyzed for the repeatability of individual lesion SUV measures as well as the repeatability of patient aggregated SUV measures (70). Investigators found NaF SUV_{max} and SUV_{mean} coefficient of variation (CoV) to be similar between specific lesions and patient aggregates, ranging 5%–7% for SUV_{mean} and 12%–14% for SUV_{max}. The 95% limit of agreement defined an error bar ±0.3 g/mL for SUV_{max} and ±0.1 for SUV_{mean}, though authors noted larger differences in sample mean and variance of SUV measures between institutions (70).

4.3.3 PET/CT SEGMENTATION QA

Standardized and harmonized quantitative PET/CT image formation can naturally propagate to improved quality assurance of image segmentation. The importance of PET/CT image formation standardization/harmonization was exemplified by a multi-institution study that reported variability in image segmentation of up to 40% due to different scanners, 25% due to different institutional imaging protocols, 20% due to different observers, 15% due to object contrast, and 5% due to count statistics (55). Standardized segmentation can include a fixed set of volumes, such as the 1 cm diameter sphere defining peak SUV. Standard spheres placed in anatomic landmarks of the liver or mediastinum on PET/CT can be used to normalize tumor SUV for segmentation. When evaluating more complex adaptive thresholds, region growing, or stochastic clustering algorithms for segmentation, their technical performance should be reported relative to a benchmark. One such benchmark involves a PET segmentation quality assurance (QA) process from synthetically created digital reference objects (22). The QA workflow, outlined in Figure 4.3, consists of clinically realistic tumor FDG uptake distributions and corresponding CT densities that are overlaid onto a reference extended cardiac-torso (XCAT) digital phantom model. The tumor uptake is then converted to an ^{18}F activity distribution in addition to background organ activity concentration from a reference emission cardiac-torso (ECAT) digital phantom model. These sources are then passed through a virtual PET/CT scanner, such as a stochastic geant4 application for tomographic emission (GATE) Monte Carlo simulator, to produce clinically relevant virtual patient SUV distributions. Segmentation algorithms are then evaluated for their ability to accurately delineate the ground truth volumes, using performance metrics such as misclassification, spatial overlap/similarity indices, or absolute volume errors. Under different image realizations of the same ground truth tumor activity distribution, the repeatability of segmentation algorithms can be tested. Lastly, under different ground truth activity distributions from a cohort of virtual patients, the reproducibility of segmentation algorithms can be tested.

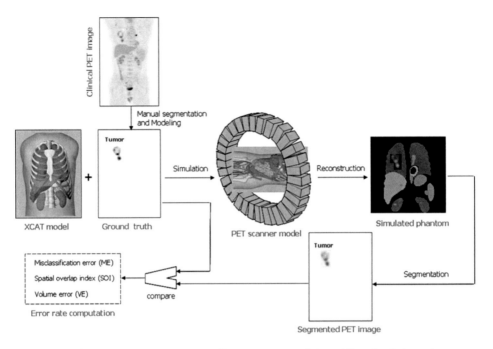

Figure 4.3 (See color insert.) Quantitative PET/CT segmentation QA workflow. Realistic patient tumors are converted to ground truth contours and activity distributions that can be used to simulate PET images and evaluate the performance of different automatic PET segmentation algorithms. (Reproduced from Zaidi, H. and El Naqa, I., *Eur. J. Nucl. Med. Mol. Imag.*, 37, 2165–2187, 2010. With permission.)

4.4 RADIOMICS OF QUANTITATIVE PET/CT

Radiomics represents large-scale mining of quantitative image features (71,72), which is well suited for application to the calibrated activity concentration and standardized uptake values inherent in modern quantitative PET/CT (73). Extracted radiomic features arise as mathematical descriptions, from statistics of the intensity histogram distribution to higher order textural features of co-occurrence, neighborhood-difference, zone-size, and run-length matrices (74). An additional strength of PET radiomics is the availability of hundreds of radiotracers which could elucidate distinct aspects of tumor phenotype. While the reduced resolution of PET relative to CT and MRI could be seen as a limitation, there is no evidence at this point that radiomics must be performed on high resolution imaging, as even MRI and CT are restricted to imaging relatively large and likely heterogeneous regions of tissue. Instead, it is likely that limited PET spatial resolution and higher image noise will impact which radiomics features are appropriate, particularly for small tumors or those with complex uptake patterns.

Several investigations have leveraged the prognostic value of FDG PET radiomics across multiple oncologic disease sites, highlighting potential clinical applications and technical challenges (75). In a series of 121 head-and-neck cancer patients and an independent validation cohort of 51 patients, the combination of FDG PET and contrast-enhanced CT texture features from zone-size matrices best predicted local tumor control (validation accuracy = 73%), with PET features reducing the rate of false-positive predictions (76). In locally advanced breast cancer, 109 FDG-PET texture features extracted at baseline in 73 patients revealed feature cluster correlations to Ki-67 expression, an immunohistochemical marker of cell proliferation, as well as pathological response status and risk of local recurrence (77). In a cohort of 66 locally advanced rectal cancer, FDG-PET texture features outperformed SUV, volume, and intensity histogram features for predicting disease-free survival and overall survival (78). As evidenced by these and other clinical investigations, high variability in the reporting and analysis of PET radiomics features persists and requires accounting of quantitative uncertainty.

As quantitative PET requires CT-based attenuation correction, multimodal feature extraction for co-localized region of interest offers an interesting advantage to single modality radiomics, by complementing low resolution functional image features with high resolution anatomic image features (73,74). Further, routine PET/CT protocols utilize contrast-free and low-dose CT scans for attenuation correction, which have reduced diagnostic quality and also higher image noise. Advances in PET detector element technology for improved spatial resolution and true signal recovery, low-dose attenuation correction CT scan processing to reduce variability in noise texture, and integration of dynamic contrast-enhanced CT or MRI acquisitions may overcome some of these limitations in the future.

4.5 UNCERTAINTIES IN PET/CT RADIOMICS

The aforementioned physical, technical, and biologic uncertainties in quantitative PET/CT data generation all propagate to impact PET/CT radiomics. As with kinetic modeling of dynamic PET data, additional sources of variability during the processing of quantitative PET/CT data for radiomic feature extraction can confound true effects.

4.5.1 IMAGE ACQUISITION AND RECONSTRUCTION UNCERTAINTY

The impact of stochastic PET image acquisition noise on radiomic features was characterized in a recent study (79). Fifty statistically independent PET images of the NEMA IQ phantom were generated from clinically representative image reconstruction settings, and a set of quantized gray-level intensity histogram, co-occurrence matrix, neighborhood-difference matrix, and zone-size matrix features were extracted. Radiomics arrays of quantitative PET metric variability under different image realizations are shown in Figure 4.4. Variability in PET radiomic features were metric- and lesion size-dependent in a manner that would be difficult to mitigate systematically. An independent study also observed FDG PET textural feature-specific variability under different image acquisition modes and reconstruction settings (80).

The repeatability of PET radiomic feature extraction under variable PET reconstruction settings has been reported by several groups in both phantom measurement and clinical patient cases. In a retrospective study

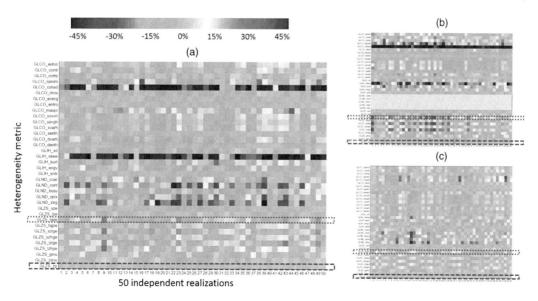

Figure 4.4 (See color insert.) Radiomics array of quantitative PET metric variability due to stochastic image noise. (a) A subset of co-occurrence matrix, (b) intensity histogram features, and (c) display variability about the mean in excess of 45%. (Reproduced from Nyflot, M.J. et al., *J. Med. Imaging (Bellingham)*, 2, 041002, 2015. With permission.)

of 20 lung cancer patients who underwent baseline FDG PET/CT scans, the effect of time-of-flight, point-spread-function modeling, iteration number, and image filtration on three SUV measures, six intensity histogram features, and 55 textural features was estimated (81). In addition to SUV_{peak} and SUV_{mean}, the PET radiomics features most robust to reconstruction settings (Coefficient of Variation (CoV) < 5%) were intensity histogram entropy, local difference entropy, inverse difference (moment) normalized, high/low gray-level run emphasis, and low gray-level zone emphasis. Other phantom and patient studies also demonstrated that fewer than half of intensity histogram features, half of textural features, and all shape-based features had CoV < 5% across different reconstruction settings (82).

Lastly, digitally reconstructed voxel size and degree of dimensional anisotropy can cause variability in PET radiomics features, particularly textural features that depend on the physical extent and shape of voxel neighborhoods. The effect of voxel size was studied in 10 phantom spheres with spatial heterogeneity patterns derived from 54 breast cancer patients and results demonstrated up 90% variation in FDG-PET texture values (83). Errors were more pronounced in homogeneous spheres using co-occurrence matrix features as opposed to zone-size matrix features. Voxel digitation may have even greater effect on imaging modalities such as MRI with high anisotropic voxel dimensions consisting of small in-plane pixel dimensions and large out-of-plane slice thickness.

4.5.2 RADIOMIC FEATURE COMPUTATION UNCERTAINTY

A key distinction of certain PET radiomics features from PET SUV summary statistics is the requirement of continuous floating-point image data transformation into quantized bins representing a discrete set of SUV levels. PET SUV data can be quantized in the intensity domain to form the same number of SUV bins for all images or to form the same interval SUV bin size for all images. The size of the PET SUV bin in a particular image will vary the number of grouped voxels and their corresponding noise properties. In oncology, a fixed number of bins that produce relative bin sizes will tend to treat tumors with different dynamic SUV ranges equally and favors relative contrast in SUV over absolute SUV scales. Under fixed number of bins, non-FDG PET avid lesions with uniformly low SUV will be susceptible to small bin sizes with high noise. On the other hand, fixed absolute SUV bin sizes will create more quantized levels in the higher dynamic range tumor, giving it greater statistical degrees of freedom compared to the lower dynamic range tumor. Other methods of image intensity discretization include intensity histogram equalization, which redistributes bin levels and sizes such that bins contain equal numbers of voxels (Hall et al. 1971). SUV quantization forms the basis for many radiomic feature calculations, including gray-level co-occurrence matrices, neighborhood-difference matrices, and zone-size matrices.

Variability in FDG PET/CT radiomics due to image intensity quantization was investigated by several repeatability studies. Maastricht University conducted a study of 35 lung cancer patients imaged before and during the second week of radiation therapy and compared the values of 44 textural features following either fixed SUV bin number (variable size) or fixed SUV bin size (variable number) across patient images (84). The effect of SUV discretization is displayed for an example patient in Figure 4.5. Fixed bin size resulted in better inter-patient and inter-scan variability, though in the absence of ground truth differences in patient rankings between discretization methods were captured as a qualitative error measure. A multi-institution investigation of test-retest FDG PET/CT datasets in 74 non-small cell lung cancer patients automatically segmented lesions on PET using a fuzzy locally adaptive Bayesian algorithm and separately semi-automatically segmented the low-dose CT using commercial software (85). Again, fixed SUV bin number (variable size) and fixed bin size (variable number) were tested prior to extraction of four shape features, ten intensity histogram features, and 26 textural features. The authors highlight that variability due to PET and CT intensity quantization was feature-dependent, and that PET features were more repeatable using fixed bin number, while low-dose CT features were more repeatable using fixed bin size. These investigations call for standardization of PET/CT image intensity discretization, though the choice of fixed bin number or size may be dependent on the patient population and specific PET/CT radiomics application.

While less well understood, it should be expected that various additional free parameters used for radiomic feature calculations affect quantitative accuracy as well. The most prominent is the choice of whether to

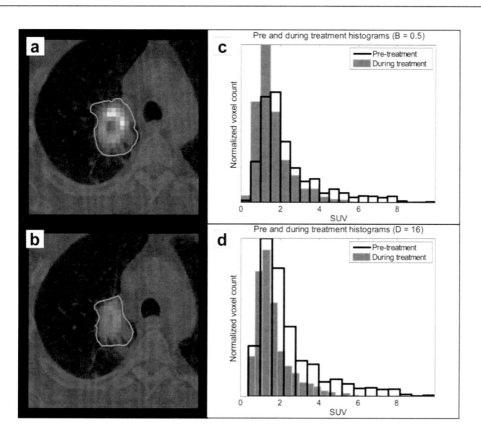

Figure 4.5 (See color insert.) Effect of SUV discretization on response assessment. Fixed SUV bin size (0.5 SUV) quantized image (a) and histogram (c) is compared against fixed SUV bin number (16 bins) quantized image (b) and histogram (d). (Reproduced from Leijenaar, R.T. et al., *Sci. Rep.* 5, 11075, 2015. With permission.)

compute radiomic features over three-dimensional image volumes, to compute on each two-dimensional image slice and average, or to compute on a single representative image slice. Whereas fully 3D calculation is information dense and therefore the least reductive technique, various existing packages use computationally efficient 2D methods, making it critical to report the radiomic feature calculation definition. Other methodological choices for feature calculation have a subtler impact. For example, the size of local neighborhoods for the neighborhood gray-tone difference matrix (NGTDM) is likely significant and may interact with the size of the filter in PET reconstruction. Preliminary studies as to the effect of filter size are mixed with no discernable trends (79,86). In another example, the gray-level co-occurrence matrix (GLCM) can be implemented by averaging 13 co-occurrence matrices computed in 13 directions, or by constructing a single matrix which incorporates the information from 13 directions (87). The relationship between feature computation and the region of interest (ROI) is also important, presenting such dilemmas as computing features at the edge of the ROI where local neighborhood features (e.g., from the NGTDM) cannot be evaluated. Potential options to overcome these technical challenges are to exclude these edge voxels, to incorporate voxels outside of the ROI, or to pad the missing data elements with zeros. Since PET images typically contain fewer voxels within the ROI than CT or MRI due to larger voxel size, this edge effect is likely amplified for PET imaging. Further work on technical image processing for radiomic feature computation is warranted.

As a result of these and other uncertainties, it is perhaps of little surprise that a recent post-treatment PET study for predicting tumor control in head-and-neck cancer patients found that 88% of 649 radiomics features were not reproducible between different radiomic data processing implementations (88). The effects of PET reconstruction settings, segmentation algorithms, and feature extraction methods on quantitative radiomics were aggregated to assess repeatability and reproducibility properties in various studies with conflicting results. A single institution study of 88 cervical cancer patient PET/CT scans tested segmentation (manual versus automatic),

SUV intensity quantization (32, 64, 128, 256 gray levels), and reconstruction algorithm (ordered subset expectation maximizatin [OSEM], Fourier rebinning maximum likelihood (ML), Fourier rebinning filtered backprojection (FBP), 3D-reprojection) variability on 79 features extracted from intensity histogram (IH), GLCM, gray level size zone matrix (GLSZM), GLRLM, NGTDM, and morphology (89). While 80% of patients had segmented volumes with similarity indices greater than 0.75, fewer than 10% of radiomics features displayed high reproducibility with Bland-Altman bounds under 30%. In a separate series of 11 non-small cell lung cancer (NSCLC) patients who underwent test-retest PET/CT scans, variable manual delineation on PET and CT images separately, and variable iterative reconstruction were applied to test the repeatability of 105 radiomic features (90). Sixty percent of these radiomic features had intra-class correlation coefficients higher than 0.9 under variable segmentation and reconstruction. In a cohort of 11 NSCLC patients, over 70% of features had high test-retest stability, while in a cohort of 23 NSCLC patients, over 90% had high inter-observer stability.

4.6 STANDARDIZATION AND HARMONIZATION OF PET/CT RADIOMICS

As with quantitative PET/CT data generation uncertainties, efforts to standardize and harmonize the extraction and reporting of PET/CT radiomic features are paramount to future clinical implementation. Specifically, mitigating uncertainties in PET/CT radiomics must involve the following: (i) selection of radiomic features that are stable under the largest sources of variability, (ii) adoption of standards for PET/CT data generation and processing, to reduce feature variability, and (iii) reporting confidence interval or limits of repeatability and reproducibility for radiomic features. These form the basis for responsible radiomics investigations that can accelerate translation of techniques into clinical practice (91).

4.6.1 PET/CT RADIOMICS STANDARDS

Criteria for radiomic feature selection include statistical independence from low-order PET metrics such as volume or mean SUV, stability under bias/variance perturbation, and sensitivity to true effect sizes measured by known changes in heterogeneity. In homogeneous and heterogeneous NEMA-like phantoms and 65 lung cancer patients, radiomic features of entropy, contrast, correlation, and coefficient of variation were volume-independent (>30 cm^3), stable across three scanners using different reconstructions (CoV $< 10\%$), and varied significantly under true changes in heterogeneity (92). Features that depend on noisy tails of PET intensity distributions, such as SUV$_{max}$, IH skewness, and IH kurtosis, or depend on isotropic voxel dimensions and voxel neighborhood size such as cluster shade and zone percentage, should be utilized with care following stricter protocol standardization. Consideration should also be given to independent feature selection methods for PET and low-dose CT, in order to properly account for differing sources of uncertainty. Under test-retest repeatability and inter-observer reproducibility of PET radiomics features in lung cancer patients, reported in Figure 4.6, simple SUV measures were most repeatable and reproducible, while intensity histogram features were least repeatable and reproducible (93). Among PET textural metrics, co-occurrence matrix features were most variable between repeat scans and observers, while zone-size matrix features were least variable.

PET/CT radiomics guidelines on image acquisition and reconstruction follow from quantitative PET/CT standardization initiatives (patient preparation, scanner calibration, post-injection timing) and harmonization initiatives (scanner reconstruction algorithm, data correction methods, image filtration). Following image formation, post-processing should be standardized prior to radiomics feature extraction. Digital voxel sampling of PET and low-dose CT images should be consistent in size across imaging protocols and minimize the degree of voxel anisotropy. Smaller voxels provide better spatial sampling of radiomics features, but increase variance due to image noise, while larger voxels provide better statistical sampling, but reduced spatial sampling. Highly anisotropic voxels from large CT or PET slice thickness relative to in-plane resolution lead to imbalanced spatial sampling over voxel neighborhoods, with larger physical distances spanned between slices. Neighborhood-difference and zone-size textural features are particularly susceptible to digital voxel size and voxel dimension anisotropy. Recommended isotropic voxel sizes for PET are 3–5 mm, while those for low-dose CT are 1–2 mm.

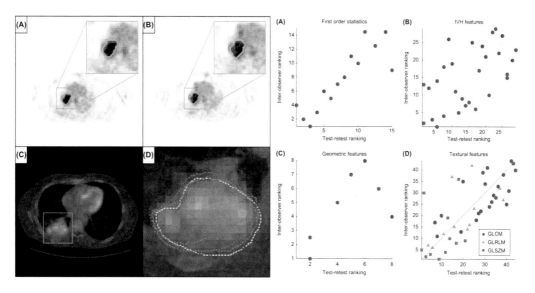

Figure 4.6 (See color insert.) Test-retest stability and inter-observer variability of PET radiomics. PET SUV summary statistics and zone-size matrix (GLSZM) features (a and c) achieved high ranks in both test-retest and inter-observer variability, while intensity histogram (IH) and co-occurrence matrix (GLCM) features (b and d) had unfavorable trade-offs in variability. (Reproduced from Leijenaar, R.T. et al., *Acta Oncol.*, 52, 1391–1397, 2013. With permission.)

Linear interpolation is preferred for high resolution CT data, while cubic spline interpolation is preferred for low resolution PET data. Inspection of resampled PET/CT data is critical and should confirm that the variance in the data is minimally perturbed.

Following PET/CT image digitation and resampling if necessary, ROI should be carefully defined for post-processing and radiomics feature extraction. Every ROI should exceed a minimum size that is dependent on the voxel neighborhood and zone size. Typical neighborhoods are computed over $3 \times 3 \times 3$ or $4 \times 4 \times 4$ voxel sliding windows, though the selection of voxel neighborhoods over which to define radiomics features should follow from correlation with ROI volume, which is particularly critical for smaller volumes. The ROI boundary definition should be consistent and, in the conversion between high-dimensional polygon meshes/contours to low-dimensional Cartesian grid masks, a reasonable methodology includes weighting edge voxel contributions by the fractional volume contained inside the polygon mesh. Low-dimensional image modalities with edge blurring and partial volume effects such as PET tend to require more inclusive ROI boundaries than high-dimensional image modalities such as CT, which would warrant construction of independent PET and low-dose CT ROI for radiomics applications. ROI volumes should ideally remain contiguous, but in cases of multifocal regions, feature extraction should be performed separately on each contiguous volume. Lastly, the ROI boundary shape and morphology should be considered when extracting radiomics features. In areas of high ROI concavity and complexity, textural features that depend on voxel neighborhoods and zone sizes become less stable. Irregular boundary shapes, particularly on segmented PET images, should be carefully inspected and if necessary modified when extracting certain radiomics features of image texture or shape. Other low-order PET SUV features that are not as sensitive to ROI morphology may not require such inspection.

Following ROI definition, textural and wavelet features, among others, may require data normalization such as z-score transformation (i.e., mean subtraction and standard deviation division) and discretization. PET SUV and CT HU discretization into bins should be standardized to best reflect the underlying voxel-level variance in the ROI. For example, in populations with high inter-patient variability in tumor SUV or HU dynamic range, such as lung cancer, fixed quantized bin size may be used to maximize the number of bins in highly heterogeneous tumors. In other populations with similar patient SUV or HU dynamic range, such as head-and-neck cancer, fixed bin number may be used to best represent normalized heterogeneity. With lower spatial resolution, higher noise, and lower dynamic SUV range, PET data are frequently quantized into fewer bins (e.g., SUV levels ≤ 64) than diagnostic CT data (e.g., HU levels ≥ 128). Conversely, noise-limited PET data are often quantized above a minimum size (e.g., ΔSUV bin ≥ 0.5), whereas CT data are quantized below a maximum size (e.g., ΔHU bin ≤ 10 HU), though larger bin size may be required for low-dose CT scans. The exact discretization scheme will depend on the patient population, the imaging technique, and the effect size being investigated with PET/CT radiomics.

Lastly, standards should be established for mathematically calculating PET/CT radiomics features. An example review of the equations that define these features is presented as part of the Image Biomarker Standardization Initiative (IBSI), which could be adopted as a template for future feature calculation definitions. With a larger and growing number of PET/CT radiomics features, it becomes important to group them into families with common definitions and mathematical foundations. Improved formalisms will become increasingly important as new image features specific to PET imaging are defined. It is important to distinguish suitable PET radiomics features from CT radiomics features for a particular application, following the previously listed criteria. In general, PET radiomics features should be extracted over the 3D ROI rather than over individual 2D slice ROI, especially when computing neighborhood-difference or zone-size matrices.

4.6.2 PET/CT RADIOMICS QA PROGRAMS AND HARMONIZATION

Improving the quality of PET/CT radiomics investigations will rely on the establishment of harmonization guidelines and QA programs, which are born out of existing initiatives that promote quantitative PET/CT QA. For instance, an extension of the cross-calibration kits used between different PET/CT scanners could be modified for QA of PET/CT radiomics analysis software. Such a radiomics QA kit would require consistent image digitation, data transformation, ROI definition, image quantization, and image feature extraction procedures. The QA program would first test radiomics software on a PET/CT digital reference object or geometric phantom of spatial heterogeneity, with known intensity, shape, and texture properties. The second test would be conducted on a lesion with patient-derived texture in a digital anthropomorphic phantom. Lastly, QA would be performed on radiomics analysis of patients to meet the following measures of robustness:

- Feature extraction *stability* under image formation and image processing
- Feature extraction *repeatability* following test-retest images in the same patient under biologically invariant conditions
- Feature extraction *reproducibility* following longitudinal images in the same patient under biologically variant conditions
- Feature extraction *harmonization* between cohorts of patients at different institutions

An example of a PET radiomics harmonization methodology was proposed for clinical investigations of non-small cell lung cancer patients (94). Another harmonization approach involved feature selection to simultaneously maximize prediction accuracy for a clinical task while minimizing variability between different PET scanners in a cohort of 118 cervical patients, which allowed a radiomic signature made up of eight texture features to be trained on a cohort of patients using a single scanner and validated on a different cohort using a second scanner (95). The validated and harmonized multivariate PET radiomic signature demonstrated better prediction of local recurrence compared to SUV$_{max}$, which may be due to increased variability of SUV$_{max}$ across different scanners. Following the blueprint of quantitative imaging biomarker and accreditation initiatives, future PET/CT radiomics QA programs and harmonization efforts could also be sponsored by RSNA, EANM, and National Clinical Trials Network (NCTN) Imaging Radiation Oncology Core (IROC), among others.

PET/CT radiomics QA is a foundational aspect for responsible radiomics research, as suggested in the editorial by Vallières and colleagues (91). They recommend that standardized details on experimental design, image acquisition, image reconstruction, image feature selection, model assessment, clinical implications, and open-source/data access be provided in every radiomics study. We now consider the role of machine learning for PET/CT radiomics prediction modeling, its effect on reporting results in radiomics investigations, and suggest analytical best practices.

4.7 MACHINE LEARNING ANALYTICS OF PET/CT RADIOMICS

Extraction of large-scale PET/CT radiomics features, from simple SUV metrics to high-order textural and wavelet features, requires comprehensive decision support systems for clinical applications such as patient classification, risk stratification, and outcome prediction. The advent of machine learning algorithms to handle high-dimensional data analytics are well-suited to select PET/CT radiomics features for integration into clinical decision support. Adoption of the appropriate machine learning algorithm depends on the clinical task and more specifically the target variable being predicted by radiomics (96). Binary or categorical target variables require machine learning for classification tasks, such as oncological staging, treatment selection, or treatment response assessment. Machine learning algorithms for classification range from simple logistic regression, to non-linear support vector machines and random forests of logistic regression trees. The performance of these algorithms can be stated in terms of classification accuracy, such as area under the receiver-operating characteristic curve (AUC) and truth table metrics (sensitivity, specificity, positive predictive value (PPV), negative predictive value (NPV)). Time-to-event target variables require machine learning to predict survival or toxicity. Machine learning algorithms for survival prediction range from penalized/boosted Cox proportional hazard regression to random survival forests of Cox regression trees (96). Performance of time-to-event risk prediction models should be evaluated with concordance indices or Brier scores that account for data censoring.

Several best practices should be followed when developing and evaluating prediction models based on PET/CT radiomics. First, PET/CT radiomic feature distributions across patient cohorts should be examined for skewness or outliers and normalized to a common scale. Continuous radiomics features based on PET SUV can be standardized by subtracting the population mean and dividing by the population standard deviation. Second, high-dimensional PET/CT radiomic feature sets for predicting imbalanced target variable classes or few events should be reduced prior to machine learning model design. Dimensionality reduction techniques include principal component analysis (PCA), variance inflation factorization (VIF), and hierarchical clustering, among many other supervised and unsupervised techniques, which seek to weight variables by their degree of statistical variation and thereby minimize cross-correlation that does not add independent information. All remaining univariate comparisons between PET/CT features and outcome variables should incorporate some form of false discovery rate correction (97). Third, machine learning model performance should be reported either on independent datasets when available or on internally validated datasets using cross-validation, optimism adjustment, or bootstrapping. These techniques guard against model overfitting to a training dataset in order to maintain prediction accuracy on independent datasets, ideally across multiple institutions. Best-practices for PET/CT radiomics modeling broadly parallel Institute of Medicine (IOM) committee recommendations for robust analytics of translational omics investigations (98) and are encompassed by the checklists for transparent reporting of a multivariable prediction model for individual prognosis or diagnosis (TRIPOD) (99). IOM and TRIPOD reports promote strategies to guard against overfitting models of large feature sets relative to single biomarker tests (98), to facilitate translation of complex databases and computational models for verification of conclusions (98), and to establish transparent multivariate analytical workflows that achieve consensus standards of statistical rigor (99).

PET/CT radiomics feature extraction and machine learning prediction models have been successfully integrated into clinical investigations. A secondary analysis of ACRIN 6668/RTOG 0235 extracted 43 textural features from pre-treatment FDG-PET/CT images in 201 NSCLC patients was conducted to predict overall survival (100). The study utilized the least absolute shrinkage operator (LASSO) to select features that were independent of MTV for predicting overall survival (OS) and recursive partitioning to identify thresholds in continuous variables. This procedure allowed construction of a bivariate MTV and texture feature (SumMean) model for defining three independent risk groups with median 23, 20, and 6 months OS (log rank $p < 0.001$). In head-and-neck cancer, 1615 radiomic features from pre-treatment FDG-PET/CT images of 300 patients from four cohorts were extracted to predict risk of locoregional recurrence and distant metastasis (101). The workflow for the study is illustrated in Figure 4.7, in which radiomic prediction models were designed using random forests for feature selection, bootstrapping for cohort and event proportion balancing, and Cox regression to independently train and validate datasets. They achieved high accuracy for predicting distant

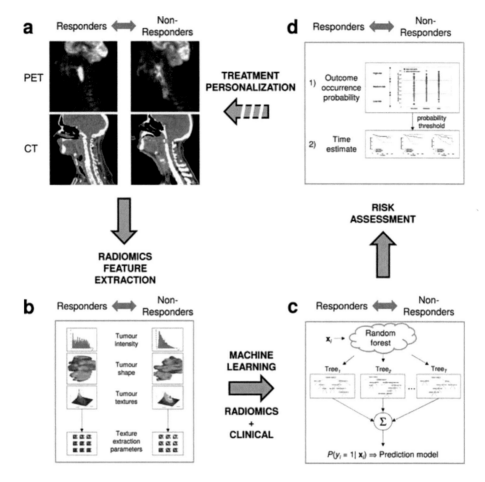

Figure 4.7 (See color insert.) Workflow for combining PET/CT (a) radiomics feature extraction (b) and machine learning analytics (c) for predicting risk and clinical outcome (d) in head-and-neck cancer patients. (Reproduced from Vallieres, M. et al., *Sci. Rep.*, 7, 10117, 2017. With permission.)

metastases (c-index = 0.88) and moderate accuracy for predicting local failure (c-index = 0.67). In esophageal cancer, a series of 174 stage III–IV patients was mined for FDG-PET radiomic features to predict incidence of local failure and distant metastases (102). Prediction models consisted of machine learning feature selection algorithms coupled to conventional logistic regression. Under 5-folds internal cross-validation and independent testing, the local failure prediction model retained the highest level of significance (AUC = 0.68, $p = 0.03$).

4.8 FUTURE DIRECTIONS

In this age of precision medicine and big healthcare data analytics, several exciting future directions lie beyond PET/CT radiomics. Here, we focus on the following frontiers: PET/MR radiomics, non-FDG PET radiomics, normal tissue PET radiomics, and deep learning PET/CT radiomics.

4.8.1 BEYOND PET/CT: PET/MR RADIOMICS

PET/MR radiomics is as a natural extension to quantitative PET/MR. A novel joint FDG PET/MR radiomics model was designed with optimized feature extraction methods to predict presence of lung metastases from soft-tissue sarcomas in a cohort of 51 patients (103). Multivariate logistic regression from features

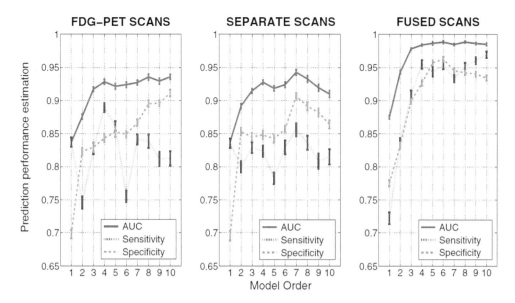

Figure 4.8 (See color insert.) Joint FDG-PET/MR radiomics model for predicting lung metastatic incidence from primary sarcoma. Prediction performance improves when operating jointly on fused PET/MR images (right) relative to individual PET images (left) or separate PET and MR images (middle). (Reproduced from Vallieres, M. et al., *Phys. Med. Biol.*, 60, 5471–5496, 2015. With permission.)

extracted on wavelet-transformed fused FDG-PET/MR in primary sarcoma ROI outperformed models from FDG-PET or MRI alone (Figure 4.8), with 1000 bootstrap-sample-adjusted AUC = 0.90 (*p* = 0.004, sensitivity = 0.80, specificity = 0.81). Interestingly, all FDG PET features in the joint PET/MR model were drawn from the zone-size or run-length texture family, which reinforces modality-specific utility of certain radiomics features.

While the quantitative challenges with PET/MR, such as accurate attenuation correction, are beyond the scope of this chapter, it is clear that the standardization and harmonization paradigms from quantitative PET and quantitative MR should be integrated to define PET/MR biomarkers. Likewise, guidelines for PET/CT radiomics QA can be adopted by PET/MR radiomics, with the understanding that some guidelines will be modality specific in order to leverage differing strengths. For example, optimal PET/MR radiomics in certain clinical applications may benefit from PET discretization into fixed SUV bin sizes due to absolute calibration of activity concentrations, coupled with MR discretization into fixed intensity bin number due to arbitrary unit scales. PET/MR radiomics sensitivity was recently studied (103), with findings that isotropic voxel size and quantization method were the largest source of outcome prediction variability. The same group later found that optimizing PET/MR acquisition parameters, beyond identifying stable PET/MR radiomics features, can further improve outcome prediction (104).

In addition to risk stratification and outcome prediction, multimodality PET/MR/CT radiomics can be leveraged toward improved definition of therapeutic targets. In radiation oncology, target definition has evolved from a planning target volume on CT receiving uniform dose to biological target volumes on PET receiving non-uniform adaptively escalated doses. Comprehensive radiomic feature extraction on CT, MRI, and PET may reveal differential response to treatment between different target volumes, while texture and wavelet feature maps may identify variable radiation resistance between intra-tumoral regions. This so-called radiomic target volume (RTV) is conceptually defined in Figure 4.9, including both regions for dose escalation and dose de-escalation. In practice, targets for dose painting based on PET may be defined on both SUV and radiomic feature images.

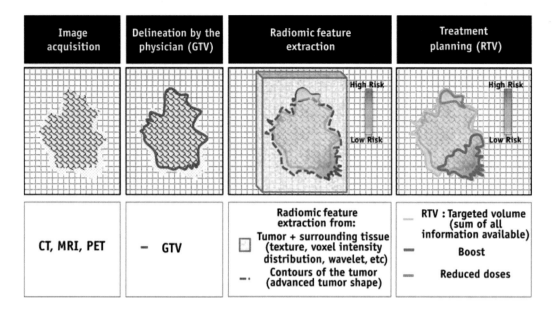

Figure 4.9 (See color insert.) Concept of the radiomic target volume (RTV) based on feature extraction and parametric maps derived from CT, MRI, and PET. Spatially distinct risk regions with the RTV can guide dose escalation or dose de-escalation strategies. (Reproduced from Sun, R. et al., *Int. J. Radiat. Oncol. Biol. Phys.*, 95, 1544–1545, 2016. With permission.)

4.8.2 BEYOND FDG PET: HYPOXIA, PROLIFERATION, AND AMINO ACID METABOLISM PET RADIOMICS

As quantitative PET standardization and harmonization initiatives expand beyond FDG, other oncologic PET tracer images will be increasingly mined for radiomic features. Tracers with high SUV ranges and heterogeneous tumor distribution patterns are more likely to yield useful radiomics analysis. This has implications in the extraction of textural features from hypoxia PET images, where, for example, high tumor-to-background fluoroazomycinarabinoside (FAZA) or fluoronitroimidazol-methyltriazolpropanol (HX4) PET features may provide increased utility compared to low tumor-to-background fluoromisonidazole (FMISO) features. Other PET tracers such as NaF PET and prostate specific membrane antigen (PSMA) PET for metastatic prostate cancer can be utilized for radiomics of both intra-lesion and inter-lesion feature extraction. Radiomic signatures from a spectrum of metastatic lesions in each patient, or from a primary tumor index, may improve the selection and combination of therapies tailored to individuals.

The stability and sensitivity of fluorothymidine (FLT) PET/diffusion-weighted MRI radiomics for response assessment in renal cell carcinoma patients was evaluated (106), revealing that a radiomics signature of FLT PET SUV and apparent diffusion coefficient (ADC) textural features was stable under test/retest imaging and most sensitive to response to tyrosine kinase inhibitor therapy. Amino acid imaging with fluoroethyltyrosine (FET) PET can be processed for kinetic parameters from dynamic acquisitions or mined for textural features from static acquisitions. A recent investigation in 47 brain patients with 54 contrast-enhancing lesions reported that FET PET radiomics on static images achieved higher diagnostic accuracy for discriminating post-radiation injury from disease recurrence compared to FET PET kinetic parameters on dynamic image series (85% versus 70% accuracy) (107). Importantly, FET PET radiomics analysis does not depend on resource-intensive dynamic image acquisition.

4.8.3 BEYOND ONCOLOGY PET: NORMAL TISSUE PET RADIOMICS

While many PET radiomics investigations have focused on cancer phenotyping, applications extend beyond oncology. Pulmonary ventilation and perfusion imaging with [68Ga]Galligas and [68Ga]macro-aggregated albumin (MAA), respectively, could be utilized for PET radiomic feature extraction and correlation with pulmonary toxicity risk following surgery or radiation therapy in lung cancer patients. These high-risk patients could then be selected for normal tissue sparing therapeutic strategies. Cardiac PET radiomics may improve diagnosis of defects or risk of cardiac toxicity and perhaps reduce the need for complex cardiac-gated dynamic acquisitions through feature extraction on simple static images. FDG PET or FLT PET radiomics of bone marrow may better identify patients who are at risk for hematopoietic toxicity and require marrow-sparing therapeutic strategies. Radiomics of beta amyloid PET, tau PET, and other brain PET imaging may identify early predictive signatures of Alzheimer's and other neurodegenerative diseases. While not an exhaustive list, these conceptual examples demonstrate the enormous clinical potential of normal tissue function radiomics combined with data analytics.

4.8.4 ARTIFICIAL INTELLIGENCE AND DEEP LEARNING OF PET/CT

Advances in artificial intelligence and the subsequent application of machine learning algorithms are having a profound impact on applications in healthcare, from service line operation optimization to clinical decision making. The aforementioned machine learning algorithms all provide decision support based on pre-defined PET/CT radiomics features. However, a class of machine learning, so-called deep learning, has emerged in medical imaging applications from innovations in artificial intelligence and computer vision. Deep learning through artificial neural networks mimics the human neural architecture by building many layers of partially or fully connected synthetic neurons through convolution or dimensionality reduction operations. For example, each subsequent layer in an imaging convolution neural network (CNN) has reduced spatial dimension (i.e., fewer voxels), but increased feature dimension (i.e., more gray levels in each voxel), which is a trade-off between breadth and depth. Deep learning algorithms can operate on quantitative PET/CT images without pre-defining region of interest, discretizing SUV into pre-determined bins, or extracting a fixed set of intensity, textural, wavelet, and shape features. They accomplish this feat by first training on very large imaging databases, such as ImageNet, to accurately segment and label different image regions. Pre-trained deep learning algorithms are then applied to PET/CT images by modifying only the last few layers, so-called transfer learning, which tunes the prediction model to a particular task without overfitting the data.

There are clear advantages and limitations to deep learning of PET/CT images. Transfer learning of pre-trained and commonly used artificial neural networks reduce sources of variability from user definitions that are required for traditional radiomics analysis. Definitions in PET voxel size, voxel anisotropy, SUV quantization, and feature calculation are all absorbed into convolution and spatial dimensionality reduction steps. Deep learning allows for simple data augmentation techniques, such as rotation, scaling, and cropping of PET/CT images, to improve robustness of the model prediction. While deep-learning algorithms continue to impress by setting new standards in image classification performance, they are plagued by weak model interpretability and transparency. Many of the hidden layers in a deep neural network applied to medical images do not provide intuitive rationale behind predicting outcomes or making decisions, which has negative clinical and even legal ramifications. The evolution of deep learning in combination with other artificial intelligence approaches may overcome this deficiency by adding transparency to hidden neural network layers or combining artificial neural networks with simpler rule-based decision trees. By highlighting areas in the original PET/CT image that were weighted most for decision making, such as subclinical disease surrounding a primary tumor and risk of metastatic progression, investigators can add transparency and invaluable clinical feedback beyond predictive performance accuracy. One study utilized deep learning for feature extraction and coupled it to a transparent support vector machine algorithm for classification (108), thus allowing greater control over model interpretation (Figure 4.10).

Another interesting application of deep learning in PET is for attenuation correction from MR images rather than gold-standard CT images. A recent study demonstrated that a deep convolution auto-encoder network could automatically delineate air, bone, and soft tissue regions on routine spin-lattice relaxation time-weighed (T1) MR images to generate synthetic attenuation maps with 1% error, which was superior to a pseudo CT atlas-based approach with 5% error (109). Through transfer learning, the algorithm could train on a modestly sized cohort of 30 patients with T1-weighted images and then predict the attenuation maps of 10 new patients. While deep learning for MR-based attenuation correction of quantitative PET via automatic tissue type segmentation is promising, it is still currently limited to intracranial sites that are less susceptible to attenuation correction errors and artifacts.

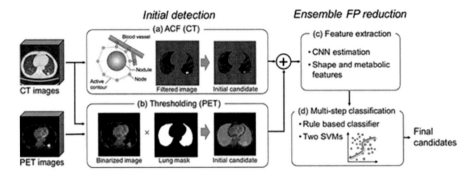

Figure 4.10 (See color insert.) Deep learning algorithm for PET/CT feature extraction combined with conventional support vector machine for pulmonary lesion detection and classification. The convolution neural network (CNN) is made more transparent by replacing the prediction model with a simple SVM classifier, allowing greater user control and interpretability. (Reproduced from Teramoto, A. et al., *Med. Phys.*, 43, 2821–2827, 2016. With permission.)

4.9 CONCLUSIONS

Quantitative PET/CT is an established imaging modality with tremendous clinical utility for diagnosis, staging, treatment planning, treatment response assessment, and outcome prediction. National and international organizations have committed to building quality assurance tools and accreditation programs that standardize and harmonize the reporting of quantitative PET/CT metrics as imaging biomarkers. When extending quantitative PET/CT analysis to extraction of high-dimensional sets of intensity, textural, and shape features, additional general and modality-specific data processing standards and best practices should be followed. Machine learning techniques can facilitate PET/CT radiomic feature selection by optimizing trade-offs in feature stability under variation and sensitivity to true effects. Adhering to consensus guidelines and recommendations on robust data analytics promotes transparency, reproducibility, and validity of radiomics investigations. Emergent research directions in radiomics of PET/MR and several PET tracers, as well as increased integration of artificial intelligence for medical imaging decision support, offer promising areas of innovation.

REFERENCES

1. Beyer T, Townsend DW, Brun T et al. A combined PET/CT scanner for clinical oncology. *J Nucl Med* 2000;41:1369–1379.
2. Weber WA, Grosu AL, Czernin J. Technology insight: Advances in molecular imaging and an appraisal of PET/CT scanning. *Nat Clin Pract Oncol* 2008;5:160–170.
3. Fletcher JW, Djulbegovic B, Soares HP et al. Recommendations on the use of 18F-FDG PET in oncology. *J Nucl Med* 2008;49:480–508.
4. Alessio AM, Kinahan PE, Cheng PM et al. PET/CT scanner instrumentation, challenges, and solutions. *Radiol Clin North Am* 2004;42:1017–1032, vii.
5. Kinahan PE, Townsend DW, Beyer T et al. Attenuation correction for a combined 3D PET/CT scanner. *Med Phys* 1998;25:2046–2053.
6. Thie JA. Understanding the standardized uptake value, its methods, and implications for usage. *J Nucl Med* 2004;45:1431–1434.
7. Joo Hyun O, Lodge MA, Wahl RL. Practical PERCIST: A simplified guide to PET response criteria in solid tumors 1.0. *Radiology* 2016;280:576–584.
8. Wahl RL, Jacene H, Kasamon Y et al. From RECIST to PERCIST: Evolving considerations for PET response criteria in solid tumors. *J Nucl Med* 2009;50 Suppl 1:122S–150S.

9. Dahele MR, Ung YC. 18F-FDG PET in planning radiation treatment of non-small cell lung cancer: Where exactly is the tumor? *J Nucl Med* 2007;48:1403.

10. Shirai K, Nakagawa A, Abe T et al. Use of FDG-PET in radiation treatment planning for thoracic cancers. *Int J Mol Imaging* 2012;2012:609545.

11. Zheng Y, Sun X, Wang J et al. FDG-PET/CT imaging for tumor staging and definition of tumor volumes in radiation treatment planning in non-small cell lung cancer. *Oncol Lett* 2014;7:1015–1020.

12. Arens AI, Troost EG, Schinagl D et al. FDG-PET/CT in radiation treatment planning of head and neck squamous cell carcinoma. *Q J Nucl Med Mol Imaging* 2011;55:521–528.

13. Muijs CT, Beukema JC, Woutersen D et al. Clinical validation of FDG-PET/CT in the radiation treatment planning for patients with oesophageal cancer. *Radiother Oncol* 2014;113:188–192.

14. Shimizu S, Hosokawa M, Itoh K et al. Can hybrid FDG-PET/CT detect subclinical lymph node metastasis of esophageal cancer appropriately and contribute to radiation treatment planning? A comparison of image-based and pathological findings. *Int J Clin Oncol* 2009;14:421–425.

15. Davidson T, Ben-David M, Galper S et al. Use of 18F-FDG PET-CT imaging to determine internal mammary lymph node location for radiation therapy treatment planning in breast cancer patients. *Pract Radiat Oncol* 2016;7:373–381.

16. Braendengen M, Hansson K, Radu C et al. Delineation of gross tumor volume (GTV) for radiation treatment planning of locally advanced rectal cancer using information from MRI or FDG-PET/CT: A prospective study. *Int J Radiat Oncol Biol Phys* 2011;81:e439–e445.

17. Haynes-Outlaw ED, Grigsby PW. The role of FDG-PET/CT in cervical cancer: Diagnosis, staging, radiation treatment planning and follow-Up. *PET Clin* 2010;5:435–446.

18. Terezakis SA, Schoder H, Kowalski A et al. A prospective study of (1)(8)FDG-PET with CT coregistration for radiation treatment planning of lymphomas and other hematologic malignancies. *Int J Radiat Oncol Biol Phys* 2014;89:376–383.

19. Aristophanous M, Berbeco RI, Killoran JH et al. Clinical utility of 4D FDG-PET/CT scans in radiation treatment planning. *Int J Radiat Oncol Biol Phys* 2012;82:e99–e105.

20. Bowen SR, Nyflot MJ, Gensheimer M et al. Challenges and opportunities in patient-specific, motion-managed and PET/CT-guided radiation therapy of lung cancer: Review and perspective. *Clin Transl Med* 2012;1:18.

21. Lee I, Im HJ, Solaiyappan M et al. Comparison of novel multi-level Otsu (MO-PET) and conventional PET segmentation methods for measuring FDG metabolic tumor volume in patients with soft tissue sarcoma. *EJNMMI Phys* 2017;4:22.

22. Zaidi H, El Naqa I. PET-guided delineation of radiation therapy treatment volumes: A survey of image segmentation techniques. *Eur J Nucl Med Mol Imag* 2010;37:2165–2187.

23. HM TT, Devakumar D, Sasidharan B et al. Hybrid positron emission tomography segmentation of heterogeneous lung tumors using 3D Slicer: Improved GrowCut algorithm with threshold initialization. *J Med Imaging (Bellingham)* 2017;4:011009.

24. van Elmpt W, Ollers M, Dingemans AM et al. Response assessment using 18F-FDG PET early in the course of radiotherapy correlates with survival in advanced-stage non-small cell lung cancer. *J Nucl Med* 2012;53:1514–1520.

25. Ohri N, Piperdi B, Garg MK et al. Pre-treatment FDG-PET predicts the site of in-field progression following concurrent chemoradiotherapy for stage III non-small cell lung cancer. *Lung Cancer* 2015;87:23–27.

26. Ohri N, Duan F, Machtay M et al. Pretreatment FDG-PET metrics in stage III non-small cell lung cancer: ACRIN 6668/RTOG 0235. *J Natl Cancer Inst* 2015;107.

27. Gensheimer MF, Hong JC, Chang-Halpenny C et al. Mid-radiotherapy PET/CT for prognostication and detection of early progression in patients with stage III non-small cell lung cancer. *Radiother Oncol* 2017;125:338–343.

28. Chung HW, Lee KY, Kim HJ et al. FDG PET/CT metabolic tumor volume and total lesion glycolysis predict prognosis in patients with advanced lung adenocarcinoma. *J Cancer Res Clin Oncol* 2014;140:89–98.

29. Chen HH, Chiu NT, Su WC et al. Prognostic value of whole-body total lesion glycolysis at pretreatment FDG PET/CT in non-small cell lung cancer. *Radiology* 2012;264:559–566.

30. Park SY, Cho A, Yu WS et al. Prognostic value of total lesion glycolysis by 18F-FDG PET/CT in surgically resected stage IA non-small cell lung cancer. *J Nucl Med* 2015;56:45–49.

31. Vu CC, Matthews R, Kim B et al. Prognostic value of metabolic tumor volume and total lesion glycolysis from (1)(8)F-FDG PET/CT in patients undergoing stereotactic body radiation therapy for stage I non-small-cell lung cancer. *Nucl Med Commun* 2013;34:959–963.

32. Yildirim F, Yurdakul AS, Ozkaya S et al. Total lesion glycolysis by 18F-FDG PET/CT is independent prognostic factor in patients with advanced non-small cell lung cancer. *Clin Respir J* 2017;11:602–611.

33. Pollom EL, Song J, Durkee BY et al. Prognostic value of midtreatment FDG-PET in oropharyngeal cancer. *Head Neck* 2016;38:1472–1478.

34. Castelli J, De Bari B, Depeursinge A et al. Overview of the predictive value of quantitative 18 FDG PET in head and neck cancer treated with chemoradiotherapy. *Crit Rev Oncol Hematol* 2016;108:40–51.

35. Varoquaux A, Rager O, Lovblad KO et al. Functional imaging of head and neck squamous cell carcinoma with diffusion-weighted MRI and FDG PET/CT: Quantitative analysis of ADC and SUV. *Eur J Nucl Med Mol Imaging* 2013;40:842–852.

36. Kidd EA, El Naqa I, Siegel BA et al. FDG-PET-based prognostic nomograms for locally advanced cervical cancer. *Gynecol Oncol* 2012;127:136–140.

37. Bowen SR, Yuh WTC, Hippe DS et al. Tumor radiomic heterogeneity: Multiparametric functional imaging to characterize variability and predict response following cervical cancer radiation therapy. *J Magn Reson Imaging* 2017;47:1388–1396.

38. Kinahan PE, Mankoff DA, Linden HM. The value of establishing the quantitative accuracy of PET/CT imaging. *J Nucl Med* 2015;56:1133–1134.

39. Doot RK, Thompson T, Greer BE et al. Early experiences in establishing a regional quantitative imaging network for PET/CT clinical trials. *Magn Reson Imaging* 2012;30:1291–1300.

40. Boellaard R. Standards for PET image acquisition and quantitative data analysis. *J Nucl Med* 2009;50(Suppl 1):11S–20S.

41. Kinahan PE, Fletcher JW. Positron emission tomography-computed tomography standardized uptake values in clinical practice and assessing response to therapy. *Semin Ultrasound CT MR* 2010;31:496–505.

42. Adams MC, Turkington TG, Wilson JM et al. A systematic review of the factors affecting accuracy of SUV measurements. *AJR Am J Roentgenol* 2010;195:310–320.

43. Macdonald LR, Schmitz RE, Alessio AM et al. Measured count-rate performance of the Discovery STE PET/CT scanner in 2D, 3D and partial collimation acquisition modes. *Phys Med Biol* 2008;53:3723–3738.

44. Samuraki M, Matsunari I, Chen WP et al. Partial volume effect-corrected FDG PET and grey matter volume loss in patients with mild Alzheimer's disease. *Eur J Nucl Med Mol Imaging* 2007;34:1658–1669.

45. Soret M, Bacharach SL, Buvat I. Partial-volume effect in PET tumor imaging. *J Nucl Med* 2007;48:932–945.

46. Ay MR, Zaidi H. Assessment of errors caused by X-ray scatter and use of contrast medium when using CT-based attenuation correction in PET. *Eur J Nucl Med Mol Imaging* 2006;33:1301–1313.

47. Nyflot MJ, Lee TC, Alessio AM et al. Impact of CT attenuation correction method on quantitative respiratory-correlated (4D) PET/CT imaging. *Med Phys* 2015;42:110–120.

48. Brendle C, Kupferschlager J, Nikolaou K et al. Is the standard uptake value (SUV) appropriate for quantification in clinical PET imaging?—Variability induced by different SUV measurements and varying reconstruction methods. *Eur J Radiol* 2015;84:158–162.

49. Takahashi Y, Oriuchi N, Otake H et al. Variability of lesion detectability and standardized uptake value according to the acquisition procedure and reconstruction among five PET scanners. *Ann Nucl Med* 2008;22:543–548.

50. Kuhnert G, Boellaard R, Sterzer S et al. Impact of PET/CT image reconstruction methods and liver uptake normalization strategies on quantitative image analysis. *Eur J Nucl Med Mol Imaging* 2016;43:249–258.

51. Wong CY, Thie J, Gaskill M et al. Addressing glucose sensitivity measured by F-18 FDG PET in lung cancers for radiation treatment planning and monitoring. *Int J Radiat Oncol Biol Phys* 2006;65:132–137.

52. Muzi M, O'Sullivan F, Mankoff DA et al. Quantitative assessment of dynamic PET imaging data in cancer imaging. *Magn Reson Imaging* 2012;30:1203–1215.

53. Svensson PE, Olsson J, Engbrant F et al. Characterization and reduction of noise in dynamic PET data using masked volumewise principal component analysis. *J Nucl Med Technol* 2011;39:27–34.

54. Petit SF, Dekker AL, Seigneuric R et al. Intra-voxel heterogeneity influences the dose prescription for dose-painting with radiotherapy: A modelling study. *Phys Med Biol* 2009;54:2179–2196.

55. Beichel RR, Smith BJ, Bauer C et al. Multi-site quality and variability analysis of 3D FDG PET segmentations based on phantom and clinical image data. *Med Phys* 2017;44:479–496.

56. Bowen SR, Nyflot MJ, Herrmann C et al. Imaging and dosimetric errors in 4D PET/CT-guided radiotherapy from patient-specific respiratory patterns: A dynamic motion phantom end-to-end study. *Phys Med Biol* 2015;60:3731–3746.

57. Boellaard R. The engagement of FDG PET/CT image quality and harmonized quantification: from competitive to complementary. *Eur J Nucl Med Mol Imaging* 2016;43:1–4.

58. Makris NE, Huisman MC, Kinahan PE et al. Evaluation of strategies towards harmonization of FDG PET/CT studies in multicentre trials: comparison of scanner validation phantoms and data analysis procedures. *Eur J Nucl Med Mol Imaging* 2013;40:1507–1515.

59. Pierce LA, 2nd, Elston BF, Clunie DA et al. A digital reference object to analyze calculation accuracy of PET standardized uptake value. *Radiology* 2015;277:538–545.

60. Fedorov A, Clunie D, Ulrich E et al. DICOM for quantitative imaging biomarker development: A standards based approach to sharing clinical data and structured PET/CT analysis results in head and neck cancer research. *Peer J* 2016;4:e2057.

61. Lodge MA, Holt DP, Kinahan PE et al. Performance assessment of a NaI(Tl) gamma counter for PET applications with methods for improved quantitative accuracy and greater standardization. *EJNMMI Phys* 2015;2:11.

62. Byrd DW, Doot RK, Allberg KC et al. Evaluation of cross-calibrated 68Ge/68Ga phantoms for assessing PET/CT measurement bias in oncology imaging for single- and multicenter trials. *Tomography* 2016;2:353–360.

63. Sullivan DC, Obuchowski NA, Kessler LG et al. Metrology standards for quantitative imaging biomarkers. *Radiology* 2015;277:813–825.

64. Obuchowski NA, Buckler A, Kinahan P et al. Statistical issues in testing conformance with the quantitative imaging biomarker alliance (QIBA) profile claims. *Acad Radiol* 2016;23:496–506.

65. Raunig DL, McShane LM, Pennello G et al. Quantitative imaging biomarkers: A review of statistical methods for technical performance assessment. *Stat Methods Med Res* 2015;24:27–67.

66. Graham MM, Wahl RL, Hoffman JM et al. Summary of the UPICT protocol for 18F-FDG PET/CT imaging in oncology clinical trials. *J Nucl Med* 2015;56:955–961.

67. Aide N, Lasnon C, Veit-Haibach P et al. EANM/EARL harmonization strategies in PET quantification: From daily practice to multicentre oncological studies. *Eur J Nucl Med Mol Imaging* 2017;44:17–31.

68. Boellaard R, Delgado-Bolton R, Oyen WJ et al. FDG PET/CT: EANM procedure guidelines for tumour imaging: Version 2.0. *Eur J Nucl Med Mol Imaging* 2015;42:328–354.

69. Scheuermann JS, Reddin JS, Opanowski A et al. Qualification of national cancer institute-designated cancer centers for quantitative PET/CT imaging in clinical trials. *J Nucl Med* 2017;58:1065–1071.

70. Lin C, Bradshaw T, Perk T et al. Repeatability of quantitative 18F-NaF PET: A multicenter study. *J Nucl Med* 2016;57:1872–1879.

71. Gillies RJ, Kinahan PE, Hricak H. Radiomics: Images are more than pictures, they are data. *Radiology* 2016;278:563–577.

72. Lambin P, Leijenaar RTH, Deist TM et al. Radiomics: The bridge between medical imaging and personalized medicine. *Nat Rev Clin Oncol* 2017;14:749.

73. Hatt M, Tixier F, Visvikis D et al. Radiomics in PET/CT: More than meets the eye? *J Nucl Med* 2017;58:365–366.

74. Hatt M, Tixier F, Pierce L et al. Characterization of PET/CT images using texture analysis: The past, the present... any future? *Eur J Nucl Med Mol Imaging* 2017;44:151–165.

75. Yip SS, Aerts HJ. Applications and limitations of radiomics. *Phys Med Biol* 2016;61:R150–R166.

76. Bogowicz M, Riesterer O, Stark LS et al. Comparison of PET and CT radiomics for prediction of local tumor control in head and neck squamous cell carcinoma. *Acta Oncol* 2017:1–6.

77. Ha S, Park S, Bang JI et al. Metabolic radiomics for pretreatment 18F-FDG PET/CT to characterize locally advanced breast cancer: Histopathologic characteristics, response to neoadjuvant chemotherapy, and prognosis. *Sci Rep* 2017;7:1556.

78. Lovinfosse P, Polus M, Van Daele D et al. FDG PET/CT radiomics for predicting the outcome of locally advanced rectal cancer. *Eur J Nucl Med Mol Imaging* 2017;45:365–375.

79. Nyflot MJ, Yang F, Byrd D et al. Quantitative radiomics: Impact of stochastic effects on textural feature analysis implies the need for standards. *J Med Imaging (Bellingham)* 2015;2:041002.

80. Galavis PE, Hollensen C, Jallow N et al. Variability of textural features in FDG PET images due to different acquisition modes and reconstruction parameters. *Acta Oncol* 2010;49:1012–1016.

81. Yan J, Chu-Shern JL, Loi HY et al. Impact of image reconstruction settings on texture features in 18F-FDG PET. *J Nucl Med* 2015;56:1667–1673.

82. Shiri I, Rahmim A, Ghaffarian P et al. The impact of image reconstruction settings on 18F-FDG PET radiomic features: Multi-scanner phantom and patient studies. *Eur Radiol* 2017;27:4498–4509.

83. Orlhac F, Nioche C, Soussan M et al. Understanding changes in tumor texture indices in PET: A comparison between visual assessment and index values in simulated and patient data. *J Nucl Med* 2017;58:387–392.

84. Leijenaar RT, Nalbantov G, Carvalho S et al. The effect of SUV discretization in quantitative FDG-PET radiomics: The need for standardized methodology in tumor texture analysis. *Sci Rep* 2015;5:11075.

85. Desseroit MC, Tixier F, Weber WA et al. Reliability of PET/CT shape and heterogeneity features in functional and morphologic components of non-small cell lung cancer tumors: A repeatability analysis in a prospective multicenter cohort. *J Nucl Med* 2017;58:406–411.

86. Doumou G, Siddique M, Tsoumpas C et al. The precision of textural analysis in (18)F-FDG-PET scans of oesophageal cancer. *Eur Radiol* 2015;25:2805–2812.

87. Hatt M, Majdoub M, Vallieres M et al. 18F-FDG PET uptake characterization through texture analysis: Investigating the complementary nature of heterogeneity and functional tumor volume in a multi-cancer site patient cohort. *J Nucl Med* 2015;56:38–44.

88. Bogowicz M, Leijenaar RTH, Tanadini-Lang S et al. Post-radiochemotherapy PET radiomics in head and neck cancer—The influence of radiomics implementation on the reproducibility of local control tumor models. *Radiother Oncol* 2017;125:385–391.

89. Altazi BA, Zhang GG, Fernandez DC et al. Reproducibility of F18-FDG PET radiomic features for different cervical tumor segmentation methods, gray-level discretization, and reconstruction algorithms. *J Appl Clin Med Phys* 2017;18:32–48.

90. van Velden FH, Kramer GM, Frings V et al. Repeatability of radiomic features in non-small-cell lung cancer [(18)F]FDG-PET/CT studies: Impact of reconstruction and delineation. *Mol Imaging Biol* 2016;18:788–795.

91. Vallieres M, Zwanenburg A, Badic B et al. Responsible radiomics research for faster clinical translation. *J Nucl Med* 2018;59:189–193.

92. Forgacs A, Pall Jonsson H, Dahlbom M et al. A study on the basic criteria for selecting heterogeneity parameters of F18-FDG PET images. *PLoS One* 2016;11:e0164113.

93. Leijenaar RT, Carvalho S, Velazquez ER et al. Stability of FDG-PET radiomics features: An integrated analysis of test-retest and inter-observer variability. *Acta Oncol* 2013;52:1391–1397.

94. Sollini M, Cozzi L, Antunovic L et al. PET radiomics in NSCLC: State of the art and a proposal for harmonization of methodology. *Sci Rep* 2017;7:358.

95. Reuze S, Orlhac F, Chargari C et al. Prediction of cervical cancer recurrence using textural features extracted from 18F-FDG PET images acquired with different scanners. *Oncotarget* 2017;8:43169–43179.

96. Leger S, Zwanenburg A, Pilz K et al. A comparative study of machine learning methods for time-to-event survival data for radiomics risk modelling. *Sci Rep* 2017;7:13206.

97. Chalkidou A, O'Doherty MJ, Marsden PK. False discovery rates in PET and CT studies with texture features: A systematic review. *PLoS One* 2015;10:e0124165.

98. Micheel CM, Nass SJ, Omenn GS, editors. *Evolution of Translational Omics: Lessons Learned and the Path Forward*. Washington, DC: National Academies Press; 2012.

99. Moons KG, Altman DG, Reitsma JB et al. New guideline for the reporting of studies developing, validating, or updating a multivariable clinical prediction model: The TRIPOD statement. *Adv Anat Pathol* 2015;22:303–305.

100. Ohri N, Duan F, Snyder BS et al. Pretreatment 18F-FDG PET textural features in locally advanced non-small cell lung cancer: Secondary analysis of ACRIN 6668/RTOG 0235. *J Nucl Med* 2016;57:842–848.

101. Vallieres M, Kay-Rivest E, Perrin LJ et al. Radiomics strategies for risk assessment of tumour failure in head-and-neck cancer. *Sci Rep* 2017;7:10117.

102. Folkert MR, Setton J, Apte AP et al. Predictive modeling of outcomes following definitive chemoradiotherapy for oropharyngeal cancer based on FDG-PET image characteristics. *Phys Med Biol* 2017;62:5327–5343.

103. Vallieres M, Freeman CR, Skamene SR et al. A radiomics model from joint FDG-PET and MRI texture features for the prediction of lung metastases in soft-tissue sarcomas of the extremities. *Phys Med Biol* 2015;60:5471–5496.

104. Vallieres M, Laberge S, Diamant A et al. Enhancement of multimodality texture-based prediction models via optimization of PET and MR image acquisition protocols: A proof of concept. *Phys Med Biol* 2017;62:8536–8565.

105. Sun R, Orlhac F, Robert C et al. In regard to Mattonen et al. *Int J Radiat Oncol Biol Phys* 2016;95:1544–1545.

106. Antunes J, Viswanath S, Rusu M et al. Radiomics analysis on FLT-PET/MRI for characterization of early treatment response in renal cell carcinoma: A proof-of-concept study. *Transl Oncol* 2016;9:155–162.

107. Lohmann P, Stoffels G, Ceccon G et al. Radiation injury vs. recurrent brain metastasis: Combining textural feature radiomics analysis and standard parameters may increase 18F-FET PET accuracy without dynamic scans. *Eur Radiol* 2017;27:2916–2927.

108. Teramoto A, Fujita H, Yamamuro O et al. Automated detection of pulmonary nodules in PET/CT images: Ensemble false-positive reduction using a convolutional neural network technique. *Med Phys* 2016;43:2821–2827.

109. Liu F, Jang H, Kijowski R et al. Deep learning MR imaging-based attenuation correction for PET/MR imaging. *Radiology* 2017:170700.

Quantitative imaging using MRI

DAVID A. HORMUTH II, JOHN VIROSTKO, ASHLEY STOKES, ADRIENNE DULA, ANNA G. SORACE, JENNIFER G. WHISENANT, JARED A. WEIS, C. CHAD QUARLES, MICHAEL I. MIGA, AND THOMAS E. YANKEELOV

5.1 INTRODUCTION

In this section, we introduce several quantitative methods of magnetic resonance imaging that can provide the input data for radiomic and/or radiogenomic analysis. For clarity, we divide the chapter into two sections. The first focused on characterizing tissue properties with endogenous contrast mechanisms (i.e., those for which an injection of any material is not required), while the second is focused on characterizing tissue properties with exogenous contrast agents. For each subsection, we first provide a qualitative overview of the technique, then a more quantitative description, then a discussion of applications, repeatability/reproducibility of the technique, and clinical examples. Given the space limitations, we will not be able to provide comprehensive introductions to all methods presented. For a more complete description of most of the topics in this chapter, the interested reader is referred to [1].

5.1.1 CHARACTERIZING TISSUE PROPERTIES WITH ENDOGENOUS CONTRAST MECHANISMS

5.1.1.1 QUANTITATIVE MEASUREMENT OF T_1, T_2, T_2^*, AND PROTON DENSITY

The contrast between tissues on conventional magnetic resonance imaging (MRI) stems primarily from the water content of tissue and its inherent relaxation processes following radiofrequency excitation.

Chief among these are the recovery of longitudinal magnetization following excitation, which is quantified by the time constant T_1, and the decay of transverse magnetization, which is quantified by the time constant T_2. Longitudinal relaxation is also known as spin-lattice relaxation, as it reflects the loss of spin excitation as thermal energy into the surrounding environment. Transverse relaxation is also known as spin-spin relaxation, as it is caused primarily by interaction with neighboring spins leading to dephasing. In practical settings, transverse relaxation occurs more quickly than predicted from T_2 due to the presence of inhomogeneity in the main magnetic field. This dephasing due to inhomogeneity is captured by the relaxation time T_2^*, which incorporates the effect of inhomogeneity as well as T_2 relaxation. In contrast with the aforementioned relaxation times, proton density reflects the number of protons (i.e., water content) in the tissue in the absence of relaxation.

MR images are frequently referred to as T_1, T_2, T_2^*, or proton density weighted, in which acquisition parameters are set so that the corresponding contrast mechanism dominates the image intensity. However, these "weighted" images are typically mixed measures in which the intensity reflects aspects of water content and all relaxation mechanisms. Techniques have been developed for quantifying relaxation times and mapping them on a voxel-by-voxel basis. For example, a variable flip angle spoiled gradient echo approach is commonly used in the clinical setting to measure T_1 [2]. The signal intensity for a spoiled gradient echo sequence is shown in Eq. (5.1):

$$S = S_0 \frac{\left(\sin\alpha \cdot \left(1 - e^{-TR/T_1} \right) \right) \cdot e^{-TE/T_2^*}}{\left(1 - \cos\alpha \cdot e^{-TR/T_1} \right)}, \tag{5.1}$$

where S is the measured signal intensity, S_0 is a constant describing the proton density, α is the flip angle, TR is the repetition time, and TE is the echo time. TR, TE, and α are selected by the investigator, while T_1 and T_2^* are the measured properties. For measuring T_1, several spoiled gradient echo images are acquired with a varied α, while keeping TE short to reduce T_2^* effects. Eq. (5.1) can then be fit to the acquired gradient echo data to estimate T_1 on a voxel-wise basis. While the variable flip angle approach is a fast means to estimate T_1, its accuracy can suffer from inhomogeneity in the radiofrequency field ($\textbf{B}_1$), particularly at high magnetic field strengths [1]. Measurement of the $\textbf{B}_1$ field can be performed to correct for this inhomogeneity [3]. Inversion or saturation recovery approaches for measuring T_1 are more accurate in the presence of $\textbf{B}_1$ inhomogeneity, but require substantially longer acquisition times [4], frequently precluding them from clinical application. The signal equation for an inversion recovery spin-echo is shown below:

$$S = S_0 \left(1 - 2 \cdot e^{-TI/T_1} + e^{-TR/T_1} \right), \tag{5.2}$$

where TI is the inversion time. To measure T_1 with this approach, spin echo images are acquired with different TI (typically logarithmically spaced), where TR is typically selected to be equal to five times the maximum T_1 of the sample. T_2 can be measured by acquiring a spin echo image at multiple TEs. The signal equation for a spin echo sequence is shown in Eq. (5.3):

$$S = S_0 \left(1 - \exp(-TR/T_1) \right) \exp(-TE/T_2) \tag{5.3}$$

A multiple spin-echo sequence is typically used to acquire a series of echoes at different TEs to adequately sample the decay curve for different T_2's. T_2^* can be calculated using a gradient-echo sequence (e.g., Eq. [5.1]) at varied TEs and fitting the logarithm of the signal intensity versus the echo time. Similarly, proton density can be quantified using sequences with minimal T_1 and T_2 contributions, followed by analytical weighting to remove residual T_1 and T_2 effects, and comparing the intensity with a standard of known proton density [5].

While seminal work in the development of MRI proposed altered relaxation times in tumors [6], neither native relaxation times nor proton density have been established as clinical biomarkers in oncology though some efforts have been made. For example, a study of anti-angiogenic therapy in colorectal cancer found reduced tumor T_1 after treatment, although the presumed alterations in tumor structure resulting in the underlying changes in T_1 were not fully explored [7]. A study of T_2^* in breast adenocarcinoma found higher T_2^* in tumors than surrounding breast tissue and a decrease in T_2^* following neoadjuvant therapy that correlated with treatment response [8]. The T_2^* value of tumors likely reflects a mixed measure incorporating edema, vascular effects, and hemorrhage which likely hinders straightforward interpretation of these results.

As T_1, T_2, and T_2^* maps are often required for quantitative analysis of cancer using other MRI techniques (e.g., dynamic contrast-enhanced MRI (DCE-MRI) [9], dynamic susceptibility MRI [10], MR thermometry [11], and oxygen enhanced MRI [12]—see Section 5.2 below), repeatability and reproducible measurement techniques are desired. While the commonly used techniques are typically quite repeatable and reproducible, the accurate measurement of relaxometric properties can be affected by temperature, B_1 and B_0 inhomogeneity, or changes in physiology (e.g., increased blood flow, oxygenation). One study on the reproducibility of T_1 measurements in myocardial tissue using an inversion recovery sequence [13] observed a reproducibility coefficient of variation of 1.5%. Another study on the reproducibility of T_2 measurements acquired in human knee cartilage [14] observed high repeatability of T_2 measurements within a given study. Briefly, the investigators observed a long-term repeatability coefficient of variation less than 5.4%, while a short-term reproducibility of less than 3.9% was observed. A reproducibility study of T_2^* mapping in human cortex demonstrated a high level reproducibility (average coefficient of variation = 1.79%) [15] using a multi-echo fast low angle shot (FLASH) MRI approach.

5.1.1.2 DIFFUSION-WEIGHTED MRI

Diffusion-weighted imaging (DWI) is an MRI technique that is sensitive to the microscopic, thermally induced self-diffusion of water molecules within a system [16,17]. In biological tissues, water movement is neither completely free nor random due to the constant physical and chemical interactions that water molecules have with other macromolecules, intracellular organelles, cell membranes, and vascular structures. The differing patterns of water diffusion in various tissues give rise to the specific image contrast in a diffusion-weighted image. For example, necrotic tissues that have lost cellular integrity allow for greater mobility of water molecules and will exhibit lower signals on diffusion-weighted images. Conversely, tissues with increased cellularity (e.g., tumors) have numerous boundaries that impede the diffusion of water and will exhibit high signal intensities. A quantitative analysis of DWI data returns estimates of the apparent diffusion coefficient (ADC), and in well-controlled situations, the ADC has been shown to correlate inversely with tissue cellularity [18,19]. Thus, DWI provides indirect functional information about the cytoarchitecture of specific biological tissues, which could aid in distinguishing between normal and diseased tissues.

A diffusion-weighted image is typically obtained by applying two diffusion-sensitizing gradients with amplitude G and duration δ separated by the diffusion time Δ on either side of a $180°$ refocusing radiofrequency (RF) pulse (Figure 5.1A). The total diffusion weighting is indicated by the "b-value," which is calculated as:

$$b = \gamma^2 \delta^2 G^2 (\Delta - \delta/3) \tag{5.4}$$

where γ is the gyromagnetic ratio of the proton. The signal intensity from a standard spin echo pulse sequence with diffusion weighting depends on ρ, T_1, T_2, and b:

$$S = S_0 \rho \cdot (1 - e^{TR/T_1}) \cdot e^{-TE/T_2} \cdot e^{-ADC \cdot b} \tag{5.5}$$

where TR and TE are the repetition and echo times, respectively. By acquiring multiple diffusion-weighted images with different amplitudes and durations of the diffusion gradients (i.e., by varying the b-value), the resulting imaging series can be fit to Eq. (5.5) for each image voxel to calculate an ADC parametric map (Figure 5.1B-E).

To allow appropriate study design, it is imperative to demonstrate the reproducibility of ADC measurements, thereby providing information on individual and intergroup variability [20]. Koh et al. and Kim et al. used a Bland-Altman analysis [21] to calculate the coefficient of reproducibility (r) in patients with advanced solid tumors [22] and malignant hepatic tumors [23], respectively. (Note that r reflects the magnitude of change in ADC that can be confidently detected in a single individual and is presented below as the 95%

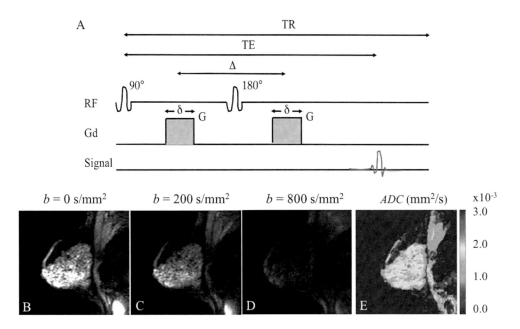

Figure 5.1 (See color insert.) An example of DWI acquisition and analysis. (A) Representative schematic of a typical diffusion-weighted spin-echo sequence where diffusion sensitizing gradients with amplitude G and duration δ separated by the diffusion time Δ, are applied symmetrically around the 180° refocusing pulse. Panel B shows an image of the normal breast without diffusion weighting (b-value = 0), whereas panels C and D show diffusion-weighted images with b-values equal to 200 s/mm^2 and 800 s/mm^2, respectively. Image voxels from panels B, C, and D are fit exponentially to the diffusion-weighted spin-echo signal equation (Eq. [5.2]) to generate an ADC parametric map (E).

confidence interval of the mean ADC.) Koh et al. acquired DWI data under free-breathing and observed good reproducibility of ADC measurements using multiple b-values ($r = 13.3\%$; $b = 0, 50, 100, 250, 500$, and 750 s/mm^2) and ADC high b-values ($r = 14.1\%$; $b = 100-750$ s/mm^2), but poor reproducibility with ADC low b-values ($r = 62.5\%$; $b = 0-100$ s/mm2) [22]. Kim et al. triggered DWI acquisition on respiration and observed similar reproducibility of ADC measurements using multiple b-values ($r = 15.0\%$; $b = 0, 50, 300, 500$, and 1000 s/mm^2), but lower reproducibility with a two b-value technique ($r = 28.0\%$; $b = 0$ and 1000 s/mm^2) [23]. Both studies reported reproducible ADC measurements, yet the magnitude of error was different among acquisition techniques, ADC calculation methods, and selection of b-values, thus emphasizing the need for DWI standardization. Investigators within the Quantitative Imaging Network (QIN) have evaluated multisite reproducibility of ADC measurements in ice water phantoms, where the ADC value is known and can be controlled, and in the breast of women with cancer [24,25].

Academic investigators have partnered with The American College of Radiology Imaging Network (ACRIN) to perform multi-center trials of quantitative DWI. ACRIN 6698 evaluated DWI as a treatment response metric for breast cancer patients undergoing neoadjuvant chemotherapy [26]. Recent results presented at the 2017 American Society of Clinical Oncology meeting showed that DWI data reflect cytotoxic effects of chemotherapy and mid-treatment ADC was predictive of pathological complete response [27]. ACRIN 6702 evaluated the utility of DWI for detection and diagnosis of breast cancer in combination with current standard-of-care MRI techniques. Results from this study are not yet published, however, preliminary data suggest that DWI, when added to a conventional breast MRI exam, could reduce the number of unnecessary biopsies without missing any malignant lesions [28].

5.1.1.3 MAGNETIZATION TRANSFER MRI

Biological tissue includes freely moving water molecules as well as a milieu of proteins and other large macromolecules. Whereas freely moving water is readily imaged using traditional MRI techniques, the spins of large macromolecules dephase too quickly to be detected directly. However, magnetization transfer

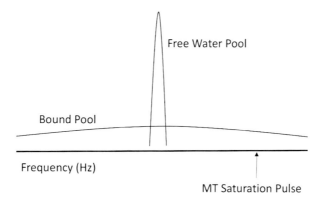

Figure 5.2 Cartoon showing the application of the MT pulse several hundred to several thousand hertz away from the free water frequency to saturate the bound pool.

MRI (MT-MRI) indirectly probes the macromolecular pool in tissue by exploiting its interaction with the surrounding free water pool. Protons exhibit dipolar cross-relaxation and chemical exchange between the "bound pool," which consists of protons in macromolecules and closely associated water, and the free water pool. MT-MRI interrogates the bound pool indirectly by examining the transfer of energy from the free water pool to the bound pool.

Magnetization transfer MRI was first described by Wolff and Balaban [29], who found that application of a saturation pulse off resonance from the free water frequency reduced the net magnetization of the free water. Further investigation showed that saturation of the bound water pool, which has a high degree of coupling and thus absorbs a broad range of frequencies, leads to transfer of magnetization from the free water pool to the bound water pool (Figure 5.2). The resulting decrease in the net magnetization of the free water can be quantified by the MTR which is calculated as in Eq. (5.6):

$$MTR = (S_0 - S_{MT})/S_0, \tag{5.6}$$

in which S_{MT} represents the signal associated with the MT saturation, and S_0 is the signal with no MT pulse. Importantly, MTR is not an intrinsic property of tissue, rather, it is influenced by the pulse sequence used for acquisition [30]. Attempts to acquire and analyze magnetization transfer MRI data in a manner which is pulse sequence independent are known collectively as quantitative magnetization transfer (qMT). A number of qMT approaches are being developed [31–35] which differ primarily in terms of saturation technique and the presence or absence of steady state magnetization. Common to each qMT technique is that all models calculate a set of quantitative parameters rather than a single MTR value. For instance, a two-pool qMT model calculates six total parameters: the longitudinal and transverse relaxation rates for both the free and bound water pools as well as the exchange rates between these two pools [32]. qMT methods can also be used to quantify the pool size ratio, which yields a ratio of the number of bound macromolecular protons to those in free water.

Measurements of MTR in the peripheral nerve have indicated good repeatability, with a repeatability coefficient of 1.3% at a single site [36]. Despite the use of a standardized MRI protocol, a multi-center study found a large degree of variation in MTR measured in the brain [37]. Test-retest of qMT parameters in breast tissue indicates that measurements are repeatable (repeatability value = 2.39%) [38]. Multi-site repeatability of qMT parameters have not yet been established as of 2018.

The most successful clinical application of MT-MRI has occurred in neurology, owing to the sensitivity of MT to myelin and its use in interrogated demyelinating disease [39]. Correspondingly, MT-MRI has been applied clinically in neuro-oncology. Studies of tumor grading have indicated that MTR is higher in high grade astrocytomas [40] and gliomas [41] versus low grade tumors. In these studies, the MTR correlated with the collagen content of meningiomas [40], suggesting that magnetization transfer may probe the tumor

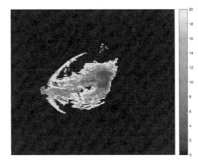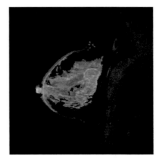

Figure 5.3 (See color insert.) Quantitative magnetization transfer image of an estrogen and progesterone receptor positive human breast tumor (3.7 cm diameter) reveals a low pool size ratio in the tumor compared with the surrounding fibroglandular tissue. Shown are the contrast-enhanced anatomical image (left), pool size ratio map (middle), and pool size ratio map overlaid with the anatomical image (right).

extracellular matrix. Given the recent interest in the role of the extracellular matrix in tumor progression and its identification as a druggable target [42], *MT-MRI* may prove useful in future oncological studies assessing therapeutic response. *MT-MRI* has also been performed in breast cancer with initial preliminary indications that *MTR* may be useful in differentiated benign and malignant lesions [43,44] (Figure 5.3).

5.1.1.4 CHEMICAL EXCHANGE SATURATION TRANSFER MRI

Chemical exchange saturation transfer (CEST) enables indirect detection of tissue metabolites with exchangeable protons rendering MRI a non-invasive molecular imaging tool sensitive to the concentrations of endogenous metabolites as well as their environment [45]. Endogenous metabolites with exchangeable protons, including mobile proteins with amide protons (e.g., glycosaminoglycans, glycogen, myo-inositol, glutamate, and creatine), have been identified as sources of endogenous CEST contrast. These metabolites can be exploited as non-invasive and non-ionizing biomarkers of disease diagnosis [46] and treatment monitoring [47,48]. Of particular interest for cancer imaging are the amide protons which can provide the amide proton transfer (APT) metric, reflective of the concentration of amide protons and their exchange rate with the free proton pool [49]. This APT metric has been used to assess physical and physiological characteristics of the tissue microenvironment such as temperature [50], pH [51–53], and metabolite concentration [54–56].

Image contrast in CEST MRI is generated by selectively saturating exchangeable protons on molecules of the solute resonating at frequencies different ($\Delta\omega$ = frequency shift relative to Larmor frequency of water) from bulk water, $\Delta\omega = 0$ [57], followed by a delayed detection of the water signal, which is affected by the transfer of this saturation. The saturated solute species interact with the magnetization of the bulk water through direct chemical exchange reducing the observed water signal [58] as seen in Figure 5.4. A detectable CEST effect requires that a discrete chemical shift difference ($\Delta\omega$) between water and the exchangeable proton on the solute is preserved, and the exchange rate, k_{sw}, has to fulfill the slow to intermediate exchange

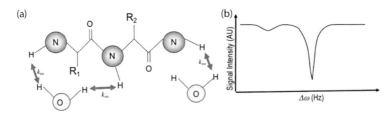

Figure 5.4 (See color insert.) Panel (a) displays an example macromolecular nitrogen (N) backbone with exchangeable hydrogens (H) demonstrating direct chemical exchange with the water protons as denoted by the red arrows. Panel (b) presents an example z-spectrum with the signal intensity shown as a function of saturation offset frequency, $\Delta\omega$, demonstrating a slight signal decrease with the CEST effect and nearly complete saturation around the water frequency.

condition on the time scale defined as $k_{sw} \leq \Delta\omega$ in order for the exchange of saturated protons to sufficiently affect the bulk water signal [57]. This $\Delta\omega$ increases linearly with static field strength, therefore CEST imaging greatly benefits from higher magnetic fields. The chemical exchange rate, k_{sw}, depends mainly on the environment such as the tissue pH [54,56,59,60].

The CEST effect depends on many factors such as field strength (B_0), concentration of exchanging spins, exchange rate, as well as field homogeneities (both B_0 and B_1), T_1 of water protons, and RF saturation pulse duration and amplitude. Thus, these factors must be addressed to achieve the optimal CEST effect from a given metabolite. The typical saturation pulse is never perfectly frequency-selective, resulting in direct saturation effects. Additionally, saturation pulses *in vivo* will result in magnetization transfer (MT) effects due to water bound to larger macromolecules in solid or semi-solid phases. These MT effects also decrease the bulk water signal. These CEST, direct saturation, and MT effects can be evaluated using a z-spectrum which depicts the normalized signal intensity of bulk water as a function of saturation pulse offset frequency [61] as seen in panel B of Figure 5.5. The z-spectrum is characterized by a symmetric direct saturation around the water frequency, and aberrations from this symmetry at the resonances of the exchangeable protons, particularly due to the CEST effect. Since the direct saturation and MT effects are roughly symmetric about the water frequency, an asymmetry analysis can be used to isolate the CEST effect by directly comparing the opposing sides of the centered z-spectra with respect to the water frequency [62] using:

$$CEST_{asym}(\Delta\omega) = [S(-\Delta\omega) - S(\Delta\omega)]/S_0, \tag{5.7}$$

where S_0 is the signal intensity of water in the absence of saturation.

Amide protons exist in abundance along the backbone of mobile proteins and peptides and produce a CEST effect at 3.5 ppm downfield from the bulk water resonance [54]. APT can be used to examine the z-spectra asymmetry caused by the APT ($\Delta\omega$ = 3.5 ppm), termed APT_{asym}, where $APT_{asym} = S(\Delta\omega = -3.5$ ppm$)/S_0 - S(\Delta\omega = 3.5$ ppm$)/S_0$. The APT_{asym} was initially applied in humans to assess amide proton content and pH in brain tumors at 3T [46]. This study, as well as others [63], including migration to 7T [64,65], demonstrate that APT_{asym} is increased in glioma relative to surrounding tissue. This increase in APT contrast is hypothesized to be a result of tumor cells accumulating defective proteins at a higher rate than normal while also experiencing alterations in pH due to hypoxia [66]. This increase in mobile macromolecular protons in brain tumors relative to normal white matter has been detected by MR spectroscopy and found to be concentration dependent [67]. Additional comparisons to MR spectroscopy have indicated potential to estimate tumor proliferation index [68]. The extracellular acidosis has been detected using pH-sensitive CEST MRI in human glioblastoma [67]. Pre-clinical [69] and clinical [46,63] studies indicate the ability to distinguish tumor from edema, tumor grading, as well as prediction of treatment response [63,70,71]. These capabilities

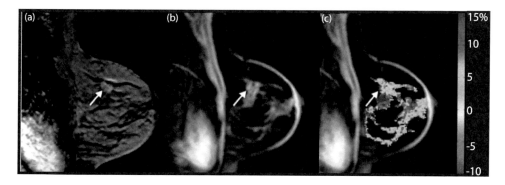

Figure 5.5 (See color insert.) CEST imaging of breast cancer patient at 3 Tesla. (a) Contrast-enhanced image with arrow indicating lesion location, (b) S_0 image where $\Delta\omega$ = 3.5 ppm, (c) APT_{asym} map demonstrating a higher APT within the lesion relative to surrounding fibroglandular tissue indicating increased concentration of proteins/peptides or tissue pH. The heterogeneity seen in Panel C is due to the low SNR that is inherent to the CEST technique.

have potential to contribute complimentary information to existing clinical imaging tools. CEST MRI has also been applied to various other cancers including bladder [72] and breast [48,73–76].

CEST MRI of the APT as well as glycosaminoglycans CEST (gagCEST) was shown to be feasible and reproducible for examination of breast cancer patients at 7T [74] with measured standard deviations of CEST effects of 1%. Application at clinical field strength of 3T [48] resulted in a repeatability of 1.91 with 95% confidence interval limits of $\pm$ 0.70% (α = 0.05). Further repeatability has been demonstrated in muscles and glands in the head and neck for tumor detection [77] and the cervical spinal cord [78] with repeatability values of 0.0241 and intraclass correlation coefficient of 0.82, respectively. However, this technology is not without its limitations. For example, CEST imaging *in vivo* is a complex technique because of interferences with direct water saturation (spillover effect, [79]), the involvement of other exchanging pools [80], in particular MT [80], and nuclear Overhauser effects [81]. Moreover, there is a strong dependence of the many effects on the employed acquisition parameters of radiofrequency irradiation for selective saturation which renders quantification difficult to reproduce between laboratories [82].

APT_{asym} is a unique contrast offering complementary information to that provided by standard clinical MRI measures, but there are several barriers currently preventing general clinical use. The complexity of the CEST spectrum and its dependence on acquisition parameters have precluded standardization of the use as a biomarker for clinical trials. Additionally, the ability to sweep a saturation pulse through frequencies is not possible using FDA-approved scanning software. Despite these hurdles, an early phase clinical trial is underway evaluating the feasibility in women with early stage breast cancer (39). This trial will generate preliminary data regarding the feasibility of this imaging technique.

5.1.2 CHARACTERIZING TISSUE PROPERTIES WITH EXOGENOUS CONTRAST AGENTS

5.1.2.1 DYNAMIC CONTRAST-ENHANCED MRI

DCE-MRI utilizes the acquisition of heavily T_1-weighted images acquired before, during, and after the injection of a paramagnetic contrast agent. The injection of the contrast agent results in an increase in signal intensity in regions where the contrast agent accumulates. DCE-MRI is a term used to describe a wide range of dynamic MRI techniques and analytic approaches including qualitative, semi-quantitative, and quantitative methods applied to consecutive contrast-enhanced MRI data acquired at high or low temporal resolution [83–85]. Qualitative assessment of DCE-MRI data provides information regarding location, malignancy, and extent of disease burden, while quantitative analysis of DCE-MRI can yield parametric maps of physiological characteristics that are directly related to the vascularity and cellularity of the tissue of interest [86]. In the clinical setting, great emphasis is placed on obtaining DCE-MRI data at high spatial resolution, in order to appropriately visualize the extent of disease [96–99]. Using common, clinically available methods, high spatial resolution necessitates lower temporal resolution, therefore the resulting signal intensity time series can only be analyzed qualitatively or semi-quantitatively to characterize general curve shape features (e.g., washout, plateau, or persistent [87,88]) or signal enhancement ratio (SER) [89,90]. Conversely, high temporal resolution DCE-MRI data are required for pharmacokinetic quantitative analysis [91]. With constant improvements in clinical imaging hardware and new technologies, the acquisition of both high spatial and high temporal MRI is becoming feasible. Therefore, there are increasing opportunities to utilize DCE-MRI to characterize tumor biology, distinguish benign and malignant lesions, and evaluate treatment response [92–94].

DCE-MRI begins with measurements of the baseline, or native, tissue T_1 for each voxel in the field of view. Following the T_1 mapping protocol, a dynamic acquisition of serial 3D T_1-weighted spoiled gradient echo sequences are then acquired before, during, and after the intravascular injection of the low molecular weight, gadolinium-based contrast agent. The initial pass of the contrast reflects perfusion, while subsequent imaging (typically 3 minutes–10 minutes) allows for characterizing the passage of the contrast agent into the extravascular space [95]. By considering each voxel in each image over the course of time, a time-concentration curve can be measured and then analyzed in a semi-quantitative fashion or with quantitative analysis to quantify a range of physiological parameters within the tissue [85]. Quantitative parameters

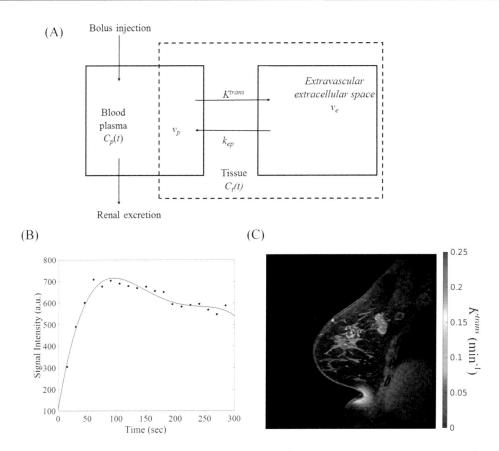

Figure 5.6 (See color insert.) DCE-MRI can be quantified using the two-compartment Tofts-Kety model (panel A) evaluating the measured voxel-based time-intensity curve resulting from sequential T_1-weighted images acquired before, during, and after contrast injection (panel B). Panel C displays an example of a parametric map of the DCE-MRI pharmacokinetic parameter K^{trans} is shown for a malignant breast lesion overlaid on a cross-sectional high-resolution MRI.

are most commonly derived in the Tofts-Kety pharmacokinetic model [84,98], which is a two compartment model of the contrast agent distribution within the body [99]. One compartment represents the vascular space, and the other represents the tissue compartment (Figure 5.6). To perform quantitative DCE-MRI, the following measurements are required: the baseline T_1 map and the dynamic T_1-weighted data introduced above, estimation of the time rate of change of the concentration of contrast agent in the blood plasma (i.e., the arterial input function, AIF), and a pharmacokinetic model to analyze the resulting data. An AIF can be measured directly from a large vessel within an individual subject [100,101], a population average [102,103], or alternatively through a reference region model [101,102]. DCE-MRI analysis is then completed by fitting the data to the pharmacokinetic model (region of interest or voxel basis) to estimate the physiological parameters, as seen in Eq. (5.8):

$$C_t(T) = K^{trans} \int_0^T C_p(u) \bullet \exp\left(-\left(\frac{K^{trans}}{v_e}\right)(T-u)\right)du + v_p C_p(T), \tag{5.8}$$

where $C_t(T)$ is the time course of the concentration of contrast agent in the tissue compartment, $C_p(u)$ is the time course of the concentration of the contrast agent in the plasma (AIF), K^{trans} is the volume transfer rate between blood plasma and the tissue extravascular extracellular space, v_e is the extravascular extracellular volume fraction, v_p is the plasma volume fraction, and k_{ep} (= K^{trans}/v_e) is the efflux constant [104].

An example of a DCE-MRI parametric map of K^{trans} overlaid on an anatomical image is seen in Figure 5.6. K^{trans} (typically expressed in units of mL(blood)/mL(tissue)/time) is the product of the blood flow and capillary leakage, and therefore correlates with the initial slope ("wash-in" rate) of the time-intensity curve. v_e (a volume fraction between 0 and 1) correlates with the peak height and time-to-peak and relates to the available space within the tissue interstitial space for an accumulating contrast agent. k_{ep} (units of 1/time) is related to the "wash-out" of the extravascular extracellular space back into the vascular space and controls the shape of the curve. Finally, v_p (a volume fraction between 0 and 1), which is included in the extended Tofts-Kety model, results in a higher and earlier peak height [91].

Quantifying the repeatability and reproducibility of DCE-MRI (including T_1 mapping and AIF characterization) is important to establish confidence in the technique [105]. Unfortunately, as it is difficult to perform test-retest analysis of DCE-MRI in patients due to complications associated with multiple contrast injections, the majority of repeatability and reproducibility analysis has been completed in the pre-clinical setting [106–109]. In the clinical setting, it has been shown that inter- and intra-observer variability of the placement of the region of interest plays a significant role in the quantification of DCE-MRI pharmacokinetic parameters [110]. Additionally, the method of AIF selection (i.e., population average AIF versus an individual AIF) established the strengths and weaknesses of various approaches [111]. Reproducibility of quantitative DCE-MRI of normal human tissues was evaluated in human pelvis tissues and revealed that both K^{trans} and v_e were reproducible in many non-cancerous tissues, including muscle and ischium [112]. Furthermore, reproducibility of DCE-MRI was quantified from paired MRI scans for K^{trans}, k_{ep}, v_e, and semi-quantitative parameters in normal and tumor tissue in 16 patients with a solid tumor of various cancer origin. The parameter v_e, enhancement, and area under the curve (AUC) were shown to be highly reproducible DCE-MRI parameters, whereas K^{trans} and k_{ep} have greater variability [113].

There are numerous single-center clinical trials that have employed DCE-MRI for evaluation of tissue response [94,114–120]. One example is Chung et al. who observed a decrease in K^{trans} in tumors that exhibited eventual response to radiosurgery in patient with brain metastases [116]. Another common application of DCE-MRI is in conjunction with a phase I/II trial in which the imaging biomarker might be used to obtain approval of the therapy, such as an anti-angiogenic agent, for phase III trials [114,121,122]. In a study of metastatic renal carcinoma response to sorafenib, patients with a high baseline of K^{trans} revealed overall better progression free survival [114]. Additionally, despite the challenges in streamlining acquisition and analysis across multiple centers, several multi-site clinical trials have employed DCE-MRI to evaluate the cellular or vascular changes in response to various anti-cancer treatments [123,124]. One noteworthy, large multi-center quantitative imaging trial was the ACRIN 6655 I-SPY TRIAL, "Investigation of Serial Studies to Predict Your Therapeutic Response with Imaging And moLecular Analysis," which evaluated quantitative MRI in 216 patients for early prediction of response to neoadjuvant therapy in locally advanced breast cancer [123]. This collaborative study evaluated lesion size, shape, extent, distribution, kinetics, and morphologic pattern through up to four DCE-MRIs (and other quantitative MRI methods) during standard-of-care neoadjuvant treatment. Early imaging biomarkers of response were correlated with pathological complete response at the time of surgery and a multivariable model including both MRI and clinical measurements achieved the highest predictive value of response [123].

5.1.2.2 DYNAMIC SUSCEPTIBILITY MRI

Hemodynamic abnormalities are common characteristics that accompany many neurologic disorders and brain cancer. Dynamic susceptibility contrast (DSC)-MRI is one of the most widely used techniques to non-invasively interrogate such abnormalities and enables patient-specific diagnosis, prognostication, and evaluation of therapeutic response. A typical DSC-MRI scan consists of the sequential acquisition of T_2^*- or T_2-weighted MR images before and after the injection of an intravenous contrast agent. The passage of the contrast agent through tissue induces a concentration-dependent change in the local relaxation times (T_2^* and T_2) and the associated dynamically acquired signals, enabling the estimation of contrast agent concentration time profiles. Pharmacokinetic modeling is applied to such data in order to extract maps of cerebral blood volume (CBV), cerebral blood flow (CBF), and mean transit time (MTT) [124–126]. In practice, DSC-MRI exhibits high specificity for the distinctive vascular characteristics exhibited in glioblastoma [127–130], sensitivity to treatment induced changes [131–134], identification of salvageable penumbra tissue following acute stroke [135], prediction of disease severity in multiple sclerosis [136], and robust repeatability [137].

Before kinetic modeling can be used for hemodynamic mapping, the acquired DSC-MRI signals must be converted into concentration of contrast agent time courses. The observed decrease in tissue T_2^* and T_2 during passage of the contrast agent bolus reflects the loss of phase coherence as water protons diffuse through magnetic field perturbations between intravoxel tissue structures (e.g., between vessels and the extravascular space) with differing susceptibilities that arise due to compartmentalization of the contrast agent [144]. It is typically assumed that there exists a linear relationship between the tissue concentration of contrast agent and the change in the transverse relaxation rate:

$$\Delta R_2^*(t) = r_2^* \, C_t(t),$$ (5.9)

where $C_t(t)$ is the tissue concentration of contrast agent, and r_2^* is the effective relaxivity of the contrast agent. By considering DSC-MRI signals before and during the passage of the contrast agent, voxel-wise ΔR_2^* time profiles can be analytically derived from the gradient echo signal equation [138]. Note R_2^* is equal to $1/T_2^*$. Note that without an independent measure, or assumed value of the effective contrast agent relaxivity, the derived time curves (and, hence, the computed kinetic parameters) are considered relative and are often presented as normalized to those values found in contralateral brain tissue.

The computation of relative CBV, CBF, and MTT using DSC-MRI relies on the principles of tracer dilution theory [139,140]. As indicated above, CBV is the most widely used DSC-MRI parameter for brain tumor evaluation and can be estimated using the integral of the ΔR_2^* time profiles in each voxel. Estimating CBF requires introduction of the residue function, $R(t)$, which measures the fraction of contrast agent remaining in the vasculature after its injection, and knowledge of the arterial input function, $C_a(t)$. By definition, the residue function at $t = 0$ equals 1 and is a decreasing function of time. The local tissue contrast agent (CA) concentration can then be expressed as the convolution of the residue function and the arterial input function:

$$C_t(t) = CBF \, C_a(t) \otimes R(t).$$ (5.10)

Deconvolution is used to derive $CBF \cdot R(t)$ and, since $R(0) = 1$, CBF is determined as the initial (or maximum) value of this product. Finally, MTT can be computed according to the central volume theorem [141], MTT = CBV/CBF. A representative CBV map derived from DSC-MRI in a patient with a grade IV glioblastoma multiforme (GBM) is illustrated in Figure 5.7. As compared to normal appearing gray and white matter, the hemodynamic parameters found in GBMs are substantially elevated.

A double baseline study, where DSC-MRI data were collected twice within 8 days, was completed in 33 treatment-naïve glioblastoma patients to characterize the repeatability of the technique and to establish the minimum sample size required to detect an assumed CBV change of 10% or 20% [137]. Importantly, this study also compared the repeatability of CBV mapping methods that incorporated different post-processing strategies, such as CA leakage correction techniques [142,143], the use of the DSC-MRI signal, rather than the derived as ΔR_2^* described above, to estimate CBV, and the use of standardization, rather than normalization, of the derived CBV maps. ("Standardization" transforms CBV maps to a standardized intensity scale, meaning that a given tissue type should have the same intensity across subjects and does not require drawing region of interest in a reference tissue [144]). In general, the lowest within-subject coefficient of variation (<20%) was found for post-processing techniques that utilize contrast leakage correction, integration of ΔR_2^*, and CBV standardization. With this degree of variation, the minimum number of patients needed to detect a 10% and 20% change in CBV would be 109 and 28, respectively.

While numerous studies have investigated the potential of DSC-MRI as a prognostic biomarker for survival at individual sites, ACRIN 6677 companion study to Radiation Therapy Oncology Group (RTOG) 0625, a phase II trial of bevacizumab with irinotecan or temozolomide in recurrent GBM, was the first performed in a multi-center trial setting [134]. Patients underwent DSC-MRI scans before, 2, 8, and/or 16 weeks after the initiation of therapy, and the potential of CBV maps to predict overall survival at 1 year post-therapy was assessed.

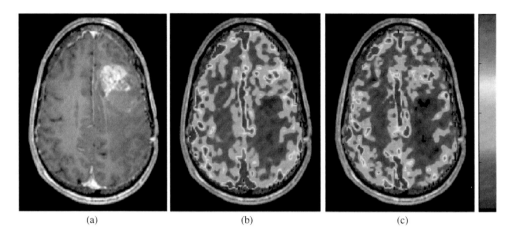

(a)   (b)   (c)

Figure 5.7 (See color insert.) An in vivo example of a DSC-MRI experiment in a patient with a glioblastoma multiforme (GBM). (a) Post-contrast T_1-weighted image exhibits the characteristic enhancement accompanying a disrupted blood brain barrier. DSC-MRI derived cerebral blood volume (CBV) and cerebral blood flow (CBF) maps (b and c, respectively) reveal high vascular density and perfusion within the enhancing tumor, with values substantially higher than those observed in gray or white matter. Also note that normal appearing gray matter CBV and CBF values are roughly 1.5–2.5 times higher than those found in white matter.

On average, the difference in *CBV* prior to and 2 weeks after therapy in patients who were alive at 1 year was −52.9%, while those who did not survive had an increase in CBV of 5.2% ($p = 0.04$). The receiver operating characteristic area under the curve at week 2 and week 16 were 0.85 and 0.9. Kaplan-Meier analysis of *CBV* data before and 2 weeks after therapy revealed that patients with a negative change in *CBV* had a significantly longer survival as compared to those with an increase in *CBV* ($p = 0.0015$), indicating that early decreases in *CBV* can be used as a prognostic marker of improved survival in recurrent GBM patients treated with bevacizumab.

5.1.2.3 IMAGING TISSUE OXYGENATION WITH MRI

Hypoxia (the state of insufficient oxygen) results from an imbalance between tissue oxygenation and metabolic consumption of oxygen. Although hypoxia has implications in multiple pathologies, in the case of cancer, hypoxia is known to lead to more aggressive tumor phenotypes and poor therapeutic response to conventional treatments [145–148]. More specifically, an oxygen partial pressure (pO_2) in tumors greater than 7–10 mmHg is typically considered the critical threshold to support tissue metabolic function [149]. The ability to quantitatively measure tissue oxygenation is currently an unmet clinical need, as such a method would facilitate personalized medicine. For example, knowledge of tumor oxygenation status may aid in selecting patients that would benefit from hypoxia-targeted treatments. Historically, positron emission tomography-based methods have dominated clinical strategies to image hypoxia [146], but no consensus currently exists on the optimal hypoxia imaging method. MRI-based hypoxia-imaging methods may provide several advantages over PET approaches, including being more widely available, able to provide repeated measures, wide dynamic sensitivity to varying levels of hypoxia, and improved spatial resolution to quantify tumor heterogeneity [145,146,150]. Two of the most promising MRI-based methods exploit the MR signal sensitivity to blood and tissue oxygenation levels. The first is based on the blood oxygen level dependent (BOLD) method and exploits the sensitivity of R_2^* ($R_2^* = 1/T_2^*$)" to paramagnetic deoxyhemoglobin in blood vessels and nearby tissues [151]. Because BOLD contrast is influenced by multiple factors (including blood volume, magnetic field inhomogeneities, and diffusion), the quantitative BOLD (qBOLD) method was developed to separate these effects and provide quantitative metrics of tissue oxygenation [152]. The second method, oxygen-enhanced MRI (OE-MRI), exploits the sensitivity of R_1 ($R_1 = 1/T_1$) to paramagnetic molecular oxygen dissolved in blood plasma or interstitial tissue. In this method, a measurable R_1 change is induced by a hyperoxic gas challenge [153].

Given the complex relationship between MR signal and oxygenation, no direct MRI-based approach exists for quantifying tissue oxygenation. Multiple studies have shown that both R_2^* and R_1 lack a direct correlation with oxygenation levels, as each are impacted by multiple potential confounding factors [154–156]. In the case of R_2^*, these factors include macroscopic field inhomogeneities, R_2 ($R_2 = 1/T_2$), water diffusion, and vascular characteristics. Most of the early implementations of the qBOLD method involve fitting multi-echo (with both gradient- and spin-echo sensitivities) data to complex BOLD signal models [151,152,157–160] that isolate the relevant tissue oxygenation parameters, such as oxygen extraction fraction (OEF) and oxygen saturation level (SO_2). The utility of this method for measuring SO_2 was demonstrated in human [157–160] and rodent [157] brains and has been validated using direct measurements of blood oxygenation levels [157]. One caveat to this multi-parameter fitting approach is that high signal-to-noise ratio (SNR) is needed to minimize the resulting parameter uncertainty. An alternative approach, called multi-parametric qBOLD and proposed by Christen et al. [161], is to simplify the fitting procedure by acquiring independent measures of certain model parameters. More specifically, separate measurements of blood volume fraction (BV_f), R_2, and B_0 field inhomogeneities are acquired, reducing the number of parameters to be fit as well as the high SNR requirements. In this method, the gradient echo signal can be expressed as a function of the proportionality constant (Cte), static magnetic field inhomogeneities, $F(t)$, as well as R_2, and R_2':

$$s(t) = Cte\, F(t) \exp(-R_2 t)\exp(-R_2' t), \tag{5.11}$$

where

$$R_2' = \left(BV_f \frac{4}{3}\pi\Delta\chi_0 Hct(1-LSO_2)\gamma B_0 \right). \tag{5.12}$$

In Eq. (5.12), R_2' is a function of BV_f, the susceptibility difference between oxy- and deoxyhemoglobin ($\Delta\chi_0$), hematocrit (Hct), local blood oxygen saturation (LSO_2), gyromagnetic ratio (γ), and B_0. This latter multi-parametric qBOLD method has been demonstrated in both humans [162] and rodents [155,161], with good correlation between MR-based oxygen saturation and blood gas analysis. Figure 5.8 shows an example qBOLD LSO_2 map in a C6 rat tumor model, showing reduced tumor oxygenation relative to contralateral tissue. One potential limitation for the multi-parametric qBOLD method is that the separate blood volume measures are

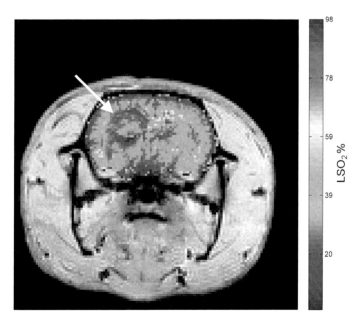

Figure 5.8 (See color insert.) An example LSO_2 map obtained with the multiparametric qBOLD protocol in a C6 rat glioma model. The LSO_2 values in tumor (indicated by the arrow) are much lower than those found in the surrounding normal-appearing tissue.

typically obtained using an exogenously administered contrast agent (e.g., gadolinium (Gd)-based agents in human studies and iron-oxide-based agents in pre-clinical studies). Overall, the qBOLD approach shows promise for quantitatively mapping tissue oxygenation using MRI.

The OE-MRI method exploits the paramagnetic nature of molecular oxygen to induce an increase in R_1, however, this effect is small due to the low *in vivo* concentration of oxygen dissolved in plasma. Due to the high affinity of hemoglobin to oxygen, OE-MRI involves inhalation of hyperoxic gas to effectively saturate hemoglobin, yielding a measurable R_1 change induced by the remaining dissolved molecular oxygen. These changes in R_1 (i.e., ΔR_1) should be directly related to the change in oxygen concentration, thereby providing a method to assess tissue oxygenation. While most applications of OE-MRI thus far have focused on pulmonary function [163–165], several studies have demonstrated promising results in tumors [153,166–168]. In the general OE-MRI method, an R_1 map is acquired prior to gas challenge, followed by a dynamic acquisition of R_1-weighted images during which the hyperoxic gas challenge is initiated. The resulting ΔR_1 can be positive, negative, or unchanged (refractory). A negative ΔR_1 was shown to correlate with (*ex vivo*) pimonidazole staining [166]. More recently, O'Connor utilized a voxel classification system based on information from both the hyperoxic challenge and DCE-MRI to distinguish perfused hypoxic regions from non-perfused and perfused normoxic regions [153]. In this method, the fraction of perfused hypoxic voxels, defined as those refractory to hyperoxic challenge, was found to be a robust biomarker of hypoxia as determined by (*ex vivo*) pimonidazole staining. This promising method is both cost-effective and clinically translatable [150], though further studies are needed to assess whether R_1 changes in OE-MRI can be quantitative biomarkers of tissue oxygenation across tumor types and treatment conditions.

For the fitting-based qBOLD method, low test-retest variation (one-hour span) was observed in three subjects [160]. In the multi-parametric qBOLD method, less than 5% variability was observed in LSO_2 estimates, suggesting good reproducibility of the method [161]. Using OE-MRI for pulmonary function showed good test-retest reproducibility (one-month span) [164]. In tumor models, the OE-MRI measurements were shown to be accurate (as assessed with an R_1 phantom), precise (within-scan coefficient of variation for R_1 of 0.41% in vivo), and stable (no significant changes in R_1 while breathing air) [153]. In the future, the repeatability of these methods needs to be established in patients with cancer.

While a limited number of clinical trials have incorporated PET-based methods to assess hypoxia, to the best of our knowledge, there are no examples of clinical trials incorporating MRI-based measures of hypoxia. This will likely change in the future, as the ability to quantitatively measure tissue oxygenation becomes more robust and widespread. In particular, hypoxia imaging methods may improve clinical trial outcomes by selecting patients who would benefit from treatments designed to overcome the limitations hypoxia imposes on particular treatments. This is especially important for hypoxia-activated drugs that are able to tailor treatment to hypoxia and for assessing treatment strategies that aim to reduce the extent of hypoxic disease.

5.1.2.4 MAGNETIC RESONANCE ELASTOGRAPHY

Magnetic resonance elastography (MRE) is a quantitative imaging technique that uses functional mechanical elasticity information as a source of image contrast. While its use does not specifically require administration of a "contrast agent" in the traditional sense, MRE relies on the application of an exogenous mechanical stimulation to generate spatial estimates of mechanical stiffness for image contrast. MRE is a member of the broader field of mechanical elasticity imaging methods known generally as the field of "elastography," in which other imaging modalities can be similarly used to generate images containing mechanical elasticity information based on externally applied (or internally induced) mechanical stimulation. These include ultrasound, where the general method of elastography was first developed [169], in addition to computed tomography and optical methods [169–172]. Interestingly, the richness of tissue material characterization in studying disease has begun to even outgrow its namesake. Whereas the conceptualization of the field owes its origins to determining the "elasticity" of tissue, in recent years, we have seen that characterization expand to addressing the nature of tissue compressibility, anisotropy, viscoelasticity, and poro-elasticity as a means to further differentiate tissue, one might label as a mechanical phenotype [172–178].

With this understanding, MRE has become a valuable diagnostic and therapy response tool for the quantitative assessment of tissue stiffness in cancer, inflammation, and fibrosis [179–185]. For example, early work has demonstrated that mechanical elasticity is directly relatable and a functional surrogate for fibrosis and stromal extracellular matrix structural organization within tissue. Disease progression is often accompanied by significant additional fibrosis, extracellular matrix deposition, and desmoplasia [186–190], and therefore MRE is well positioned to monitor this phenomenon by quantifying tissue elasticity to assess disease progression and therapy response. Best known for pioneering work in the assessment of liver fibrosis [179], MRE has also been successfully applied to quantify tissue mechanical elasticity in a multitude of other sites, including brain, breast, lung, and prostate cancers, among others in both clinical and pre-clinical domains [185,191–195].

Principal to the acquisition of MRE data is the external application of mechanical stimulation to the body, which is used to generate internal deformation within tissue. This mechanical stimulus can be either dynamic or quasi-static, depending on the specific elasticity imaging technique. In the case of dynamic MRE, mechanical shear waves are induced within tissue by externally applying a mechanical vibrational source, usually via an acoustic or piezoelectric driver. The induced tissue displacement is then acquired via phase-contrast MR imaging where oscillating motion-sensitized gradient sequences are synchronized to the vibrational frequency of the mechanical stimulus. Spatial maps of mechanical elasticity are generated from deformation images by applying numerical post-processing using assumptions of mechanical constitutive models, usually linear, in either model-based or direct inversion techniques. For a more thorough discussion of dynamic MRE, the interested reader is directed to [185]. In the case of quasi-static MRE, conventional anatomical MR images are acquired in both a baseline state and a deformed state. Deformation is externally applied, usually through the use of a pneumatic-driven displacement source/device, to cause internal tissue deformation. Spatial maps of tissue mechanical properties are then generated depending on the specific technique, either MR spin-tagging (or strain-encoding)-based techniques [196–199] or modality independent elastography (MIE) [171,200–202]. MIE utilizes non-rigid image registration methodologies within a mechanical model-based inverse problem framework that generates a best-fit estimation of the spatial distribution of mechanical elasticity that causes the baseline image to match the applied-deformation image, given a description of the forcing boundary conditions. An example resulting elasticity map is shown in Figure 5.9 for quasi-static MRE (MIE) in the breast.

As the field of MRE has matured, several recent studies have established the repeatability and reproducibility of the technique across a range of conditions with generally satisfactory results, comparable to many common quantitative imaging methodologies. In a study of hepatic MRE in nine healthy volunteers using different mechanical deformation drivers and breath-hold phase cycles, Wang et al. demonstrated intraclass

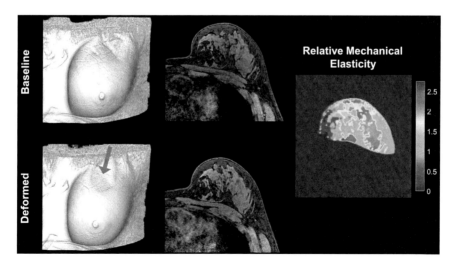

Figure 5.9 (See color insert.) An example of quasi-static MRE in the breast. Conventional anatomical MR image volumes are acquired in both a baseline and a deformed state and biomechanical model-based reconstruction techniques are used to generate a map of mechanical elasticity within the tissue. Note that the red arrow denotes the location of the external deformation source.

correlation coefficients (ICC) that ranged from 0.71 to 0.96 for repeatability studies examining measurements under identical conditions [203]. However, this study also found moderate bias when comparing MRE measurements across different breath-hold phases in reproducibility studies, with concordance correlation coefficients (CCC) that ranged from 0.61 to 0.94, indicating the need for additional studies to examine the variability of measurements throughout the breath-hold phase cycle. For MRE measurements of mechanical elasticity in breast tissue at 3 Tesla field strength, Hawley et al. found satisfactory reproducibility results with CCC that ranged between 0.87 and 0.91 in a study of 22 healthy volunteers in a test-retest study (36). For the case of quasi-static MRE in a murine pre-clinical breast cancer system, Weis et al. established a range of ICC of 0.75–0.99 at the bulk level and 0.70 at the voxel level for quasi-static MRE (MIE) of murine tumor [202].

As the repeatability and reproducibility of MRE becomes more established, MRE measurements of mechanical elasticity have begun to appear in clinical trial studies in which the diagnostic accuracy and efficacy of MRE is tested in a rigorous manner. There are several examples of clinical trials examining MRE that demonstrate good diagnostic capability, especially for the assessment of hepatic fibrosis [182,205,206]. In an important clinical study by Siegmann et al. in breast cancer lesion characterization, the diagnostic value of adding MRE measurements to standard contrast-enhanced MRI was examined [207]. This study presented an initial examination in 57 suspected breast lesions and was the first to demonstrate the additive value of MRE measurements beyond that of standard contrast-enhanced imaging alone. When assessing lesions for malignancy, when only using contrast-enhanced MRI, sensitivity and specificity for breast cancer detection was 97.3% and 55%, respectively, with AUC of 0.93. When MRE was combined with contrast-enhanced MRI, diagnostic accuracy was improved, with AUC of 0.96 and specificity elevated from 75% to 90%. While preliminary, this clinical study shows the complementary value of MRE in addition to more standard quantitative imaging methodologies.

5.2 SUMMARY

While the initial efforts in radiomics and radiogenomics typically employed methods to analyze non-specific imaging data types that were acquired with an emphasis on characterizing anatomy and morphology, there are a range of imaging parameters that have the ability to report on more specific aspects of cancer biology. In this chapter, we focused on eight such techniques that are available from magnetic resonance imaging that have shown the ability to accurately report on such varied characteristics as cellularity, protein content, perfusion, oxygen status, and tissue mechanics. Given the success of these techniques in isolation, and their increasing accessibility, it is likely that they will have the ability to enhance the diagnostic and prognostic abilities of radiomics and radiogenomics.

REFERENCES

1. Yankeelov TE, Pickens, DR, Price RR *Quantitative MRI in Cancer.* Hendee WR, editor. Boca Raton, FL: Taylor & Francis Group, LLC; 2012.
2. Fram EK et al. Rapid calculation of T1 using variable flip angle gradient refocused imaging. *Magn Reson Imaging.* 1987;5(3):201–208.
3. Whisenant JG et al. Bloch-Siegert B1-mapping improves accuracy and precision of longitudinal relaxation measurements in the breast at 3 T. *Tomography.* 2016;2(4):250–259.
4. Crawley AP et al. A comparison of one-shot and recovery methods in T1 imaging. *Magn Reson Med.* 1988;7(1):23–34.
5. Wehrli FW et al. Quantification of contrast in clinical MR brain imaging at high magnetic field. *Invest Radiol.* 1985;20(4):360–369.
6. Damadian R. Tumor detection by nuclear magnetic resonance. *Science* (80-). 1971;171(3976):1151–1153.
7. O'Connor JPB et al. Quantifying antivascular effects of monoclonal antibodies to vascular endothelial growth factor: Insights from imaging. *Clin Cancer Res.* 2009;15(21):6674–6682.

8. Li SP et al. Primary human breast adenocarcinoma: Imaging and histologic correlates of intrinsic susceptibility-weighted MR imaging before and during chemotherapy. *Radiology.* 2010;257(3):643–652.

9. Yankeelov TE et al. Dynamic contrast enhanced magnetic resonance imaging in oncology: Theory, data acquisition, analysis, and examples. *Curr Med Imaging Rev.* 2009;3(2):91–107.

10. Boxerman JL et al. Longitudinal DSC-MRI for distinguishing tumor recurrence from pseudoprogression in patients with a high-grade glioma. *Am J Clin Oncol.* 2017;40.3:228–234.

11. Rieke V et al. MR thermometry. *J Magn Reson Imaging.* 2008;27(2):376–390.

12. O'Connor JPB et al. Oxygen-enhanced MRI accurately identifies, quantifies, and maps tumor hypoxia in preclinical cancer models. *Cancer Res.* 2016;76(4):787–795.

13. Pica S et al. Reproducibility of native myocardial T1 mapping in the assessment of Fabry disease and its role in early detection of cardiac involvement by cardiovascular magnetic resonance. *J Cardiovasc Magn Reson.* 2014;16(1):99.

14. Hannila I et al. The repeatability of T2 relaxation time measurement of human knee articular cartilage. *Magn Reson Mater Physics, Biol Med.* 2015;28(6):547–553. doi:10.1007/s10334-015-0494-3

15. Govindarajan ST et al. Reproducibility of T2* mapping in the human cerebral cortex in vivo at 7 tesla MRI. *J Magn Reson Imaging.* 2015;42(2):290–296. doi:10.1002/jmri.24789.

16. Carr HY, Purcell EM. Effects of diffusion on free precession in nuclear magnetic resonance experiments. *Phys Rev* 1954;94(3):630–630.

17. Hahn EL. Spin echoes. *Phys Rev* 1950;80(4):580–594.

18. Anderson AW, Xie J, Pizzonia J, Bronen RA, Spencer DD, Gore JC. Effects of cell volume fraction changes on apparent diffusion in human cells. *Magn Reson Imaging* 2000;18(6):689–695.

19. Barnes SL, Sorace AG, Loveless ME, Whisenant JG, Yankeelov TE. Correlation of tumor characteristics derived from DCE-MRI and DW-MRI with histology in murine models of breast cancer. *NMR Biomed* 2015;28(10):1345–1356.

20. Padhani AR, Liu G, Koh DM et al. Diffusion-weighted magnetic resonance imaging as a cancer biomarker: Consensus and recommendations. *Neoplasia* 2009;11(2):102–125.

21. Bland JM, Altman DG. Measuring agreement in method comparison studies. *Stat Methods Med Res* 1999;8(2):135–160.

22. Koh DM, Blackledge M, Collins DJ et al. Reproducibility and changes in the apparent diffusion coefficients of solid tumours treated with combretastatin A4 phosphate and bevacizumab in a two-centre phase I clinical trial. *Eur Radiol* 2009;19(11):2728–2738.

23. Kim SY, Lee SS, Park B et al. Reproducibility of measurement of apparent diffusion coefficients of malignant hepatic tumors: Effect of DWI techniques and calculation methods. *J Magn Reson Imaging* 2012;36(5):1131–1138.

24. Malyarenko D, Galban CJ, Londy FJ, Meyer CR, Johnson TD, Rehemtulla A, Ross BD, Chenevert TL. Multi-system repeatability and reproducibility of apparent diffusion coefficient measurement using an ice-water phantom. *J Magn Reson Imaging* 2013;37(5):1238–1246.

25. Newitt DC, Zhang Z, Gibbs J, Partridge SC, Chenevert TL, Bolan PJ, Rosen M, Marques H, Hylton NM. Reproducibility of ADC measures by breast DWI: Results of the ACRIN 6698 Trial. *Proc Intl Soc Mag Reson Med* (ISMRM) 2017;25(Abst# 0949).

26. Galban CJ, Ma B, Malyarenko D et al. Multi-site clinical evaluation of DW-MRI as a treatment response metric for breast cancer patients undergoing neoadjuvant chemotherapy. *PLoS One* 2015;10(3):e0122151.

27. Partridge SC, Zhang Z, Newitt DC et al. ACRIN 6698 trial: Quantitative diffusion-weighted MRI to predict pathologic response in neoadjuvant chemotherapy treatment of breast cancer. *J Clin Onco* 2017;35(15 suppl):11520.

28. Partridge SC, DeMartini WB, Kurland BF, Eby PR, White SW, Lehman CD. Quantitative diffusion-weighted imaging as an adjunct to conventional breast MRI for improved positive predictive value. *AJR Am J Roentgenol* 2009;193(6):1716–1722.

29. Wolff SD, Balaban RS. Magnetization transfer contrast (MTC) and tissue water proton relaxation in vivo. *Magn Reson Med* 1989;10(1):135–144.

30. Eng J, Ceckler TL, Balaban RS. Quantitative 1H magnetization transfer imaging in vivo. *Magn Reson Med* 1991;17(2):304–314.

31. Gochberg DF, Kennan RP, Robson MD, Gore JC. Quantitative imaging of magnetization transfer using multiple selective pulses. *Magn Reson Med* 1999;41(5):1065–1072.

32. Henkelman RM, Huang X, Xiang QS, Stanisz GJ, Swanson SD, Bronskill MJ. Quantitative interpretation of magnetization transfer. *Magn Reson Med* 1993;29(6):759–766.

33. Ropele S, Seifert T, Enzinger C, Fazekas F. Method for quantitative imaging of the macromolecular 1H fraction in tissues. *Magn Reson Med* 2003;49(5):864–871.

34. Sled JG, Pike GB. Quantitative interpretation of magnetization transfer in spoiled gradient echo MRI sequences. *J Magn Reson* 2000;145(1):24–36.

35. Yarnykh VL. Pulsed Z-spectroscopic imaging of cross-relaxation parameters in tissues for human MRI: Theory and clinical applications. *Magn Reson Med* 2002;47(5):929–939.

36. Dortch RD, Dethrage LM, Gore JC, Smith SA, Li J. Proximal nerve magnetization transfer MRI relates to disability in Charcot-Marie-Tooth diseases. *Neurology* 2014;83(17):1545–1553.

37. Berry I, Barker GJ, Barkhof F, Campi A, Dousset V, Franconi JM, Gass A, Schreiber W, Miller DH, Tofts PS. A multicenter measurement of magnetization transfer ratio in normal white matter. *J Magn Reson Imaging* 1999;9(3):441–446.

38. Arlinghaus LR, Dortch RD, Whisenant JG, Kang H, Abramson RG, Yankeelov TE. Quantitative magnetization transfer imaging of the breast at 3.0°T: Reproducibility in healthy volunteers. *Tomography* 2016;2(4):260–266.

39. Petrella JR, Grossman RI, McGowan JC, Campbell G, Cohen JA. Multiple sclerosis lesions: Relationship between MR enhancement pattern and magnetization transfer effect. *AJNR Am J Neuroradiol* 1996;17(6):1041–1049.

40. Lundbom N. Determination of magnetization transfer contrast in tissue: An MR imaging study of brain tumors. *AJR Am J Roentgenol* 1992;159(6):1279–1285.

41. Kurki T, Lundbom N, Kalimo H, Valtonen S. MR classification of brain gliomas: Value of magnetization transfer and conventional imaging. *Magn Reson Imaging* 1995;13(4):501–511.

42. Venning FA, Wullkopf L, Erler JT. Targeting ECM disrupts cancer progression. *Front Oncol* 2015;5:224.

43. Bonini RH, Zeotti D, Saraiva LA, Trad CS, Filho JM, Carrara HH, de Andrade JM, Santos AC, Muglia VF. Magnetization transfer ratio as a predictor of malignancy in breast lesions: Preliminary results. *Magn Reson Med* 2008;59(5):1030–1034.

44. Heller SL, Moy L, Lavianlivi S, Moccaldi M, Kim S. Differentiation of malignant and benign breast lesions using magnetization transfer imaging and dynamic contrast-enhanced MRI. *J Magn Reson Imaging* 2013;37(1):138–145.

45. van Zijl PC, Yadav NN. Chemical exchange saturation transfer (CEST): What is in a name and what isn't? *Magn Reson Med* 2011;65(4):927–948.

46. Jones CK, Schlosser MJ, van Zijl PC, Pomper MG, Golay X, Zhou J. Amide proton transfer imaging of human brain tumors at 3T. *Magn Reson Med* 2006;56(3):585–592.

47. Zhou J, Tryggestad E, Wen Z et al. Differentiation between glioma and radiation necrosis using molecular magnetic resonance imaging of endogenous proteins and peptides. *Nat Med* 2011;17(1):130–134.

48. Dula AN, Arlinghaus LR, Dortch RD, Dewey BE, Whisenant JG, Ayers GD, Yankeelov TE, Smith SA. Amide proton transfer imaging of the breast at 3 T: Establishing reproducibility and possible feasibility assessing chemotherapy response. *Magn Reson Med* 2013;70(1):216–224.

49. van Zijl PC, Zhou J, Mori N, Payen JF, Wilson D, Mori S. Mechanism of magnetization transfer during on-resonance water saturation. A new approach to detect mobile proteins, peptides, and lipids. *Magn Reson Med* 2003;49(3):440–449.

50. McVicar N, Li AX, Suchy M, Hudson RH, Menon RS, Bartha R. Simultaneous in vivo pH and temperature mapping using a PARACEST-MRI contrast agent. *Magn Reson Med* 2013;70(4):1016–1025.

51. Aime S, Delli Castelli D, Terreno E. Novel pH-reporter MRI contrast agents. *Angew Chem Int Ed Engl* 2002;41(22):4334–4336.

52. Delli Castelli D, Ferrauto G, Cutrin JC, Terreno E, Aime S. In vivo maps of extracellular pH in murine melanoma by CEST-MRI. *Magn Reson Med* 2014;71(1):326–332.

53. Sun PZ, Sorensen AG. Imaging pH using the chemical exchange saturation transfer (CEST) MRI: Correction of concomitant RF irradiation effects to quantify CEST MRI for chemical exchange rate and pH. *Magn Reson Med* 2008;60(2):390–397.

54. Zhou J, Payen JF, Wilson DA, Traystman RJ, van Zijl PC. Using the amide proton signals of intracellular proteins and peptides to detect pH effects in MRI. *Nat Med* 2003;9(8):1085–1090.

55. Sun PZ, Zhou J, Sun W, Huang J, van Zijl PC. Detection of the ischemic penumbra using pH-weighted MRI. *J Cereb Blood Flow Metab* 2007;27(6):1129–1136.

56. Ward KM, Balaban RS. Determination of pH using water protons and chemical exchange dependent saturation transfer (CEST). *Magn Reson Med* 2000;44(5):799–802.

57. Ward KM, Aletras AH, Balaban RS. A new class of contrast agents for MRI based on proton chemical exchange dependent saturation transfer (CEST). *J Magn Reson* 2000;143(1):79–87.

58. Wolff SD, Balaban RS. Magnetization transfer contrast (MTC) and tissue water proton relaxation in vivo. *Magn Reson Med* 1989;10(1):135–144.

59. Liepinsh E, Otting G. Proton exchange rates from amino acid side chains—Implications for image contrast. *Magn Reson Med* 1996;35(1):30–42.

60. Sun PZ, Murata Y, Lu J, Wang X, Lo EH, Sorensen AG. Relaxation-compensated fast multislice amide proton transfer (APT) imaging of acute ischemic stroke. *Magn Reson Med* 2008;59(5):1175–1182.

61. Bryant RG. The dynamics of water-protein interactions. *Annu Rev Biophys Biomol Struct* 1996;25:29–53.

62. Guivel-Scharen V, Sinnwell T, Wolff SD, Balaban RS. Detection of proton chemical exchange between metabolites and water in biological tissues. *J Magn Reson* 1998;133(1):36–45.

63. Zhou J, Blakeley JO, Hua J, Kim M, Laterra J, Pomper MG, van Zijl PC. Practical data acquisition method for human brain tumor amide proton transfer (APT) imaging. *Magn Reson Med* 2008;60(4):842–849.

64. Jones CK, Polders D, Hua J, Zhu H, Hoogduin HJ, Zhou J, Luijten P, van Zijl PC. In vivo three-dimensional whole-brain pulsed steady-state chemical exchange saturation transfer at 7 T. *Magn Reson Med* 2012;67(6):1579–1589.

65. Mougin OE, Coxon RC, Pitiot A, Gowland PA. Magnetization transfer phenomenon in the human brain at 7 T. *Neuroimage* 2010;49(1):272–281.

66. Salhotra A, Lal B, Laterra J, Sun PZ, van Zijl PC, Zhou J. Amide proton transfer imaging of 9L gliosarcoma and human glioblastoma xenografts. *NMR Biomed* 2008;21(5):489–497.

67. Howe FA, Barton SJ, Cudlip SA et al. Metabolic profiles of human brain tumors using quantitative in vivo 1H magnetic resonance spectroscopy. *Magn Reson Med* 2003;49(2):223–232.

68. Park JE, Kim HS, Park KJ, Kim SJ, Kim JH, Smith SA. Pre- and posttreatment glioma: Comparison of amide proton transfer imaging with MR spectroscopy for biomarkers of tumor proliferation. *Radiology* 2016;278(2):514–523.

69. Zhou J, Lal B, Wilson DA, Laterra J, van Zijl PC. Amide proton transfer (APT) contrast for imaging of brain tumors. *Magn Reson Med* 2003;50(6):1120–1126.

70. Wen Z, Hu S, Huang F, Wang X, Guo L, Quan X, Wang S, Zhou J. MR imaging of high-grade brain tumors using endogenous protein and peptide-based contrast. *Neuroimage* 2010;51(2):616–622.

71. Sagiyama K, Mashimo T, Togao O et al. In vivo chemical exchange saturation transfer imaging allows early detection of a therapeutic response in glioblastoma. *Proc Natl Acad Sci USA* 2014;111(12):4542–4547.

72. Whyard T, Waltzer WC, Waltzer D, Romanov V. Metabolic alterations in bladder cancer: Applications for cancer imaging. *Exp Cell Res* 2016;341(1):77–83.

73. Donahue MJ, Donahue PC, Rane S, Thompson CR, Strother MK, Scott AO, Smith SA. Assessment of lymphatic impairment and interstitial protein accumulation in patients with breast cancer treatment-related lymphedema using CEST MRI. *Magn Reson Med* 2016;75(1):345–355.

74. Klomp DW, Dula AN, Arlinghaus LR et al., Amide proton transfer imaging of the human breast at 7T: Development and reproducibility. *NMR Biomed* 2013;26(10).1271 1277.

75. Cai K, Xu HN, Singh A, Moon L, Haris M, Reddy R, Li LZ. Breast cancer redox heterogeneity detectable with chemical exchange saturation transfer (CEST) MRI. *Mol Imaging Biol* 2014;16(5):670–679.

76. Schmitt B, Trattnig S, Schlemmer HP. CEST-imaging: A new contrast in MR-mammography by means of chemical exchange saturation transfer. *Eur J Radiol* 2012;81 Suppl 1:S144–S146.

77. Yuan J, Chen S, King AD, Zhou J, Bhatia KS, Zhang Q, Yeung DK, Wei J, Mok GS, Wang YX. Amide proton transfer-weighted imaging of the head and neck at 3 T: A feasibility study on healthy human subjects and patients with head and neck cancer. *NMR Biomed* 2014;27(10):1239–1247.

78. By S, Barry RL, Smith AK, Lyttle BD, Box BA, Bagnato FR, Pawate S, Smith SA. Amide proton transfer CEST of the cervical spinal cord in multiple sclerosis patients at 3T. *Magn Reson Med* 2018;79:806–814.

79. Mulkern RV, Williams ML. The general solution to the Bloch equation with constant RF and relaxation terms: Application to saturation and slice selection. *Med Phys* 1993;20(1):5–13.

80. van Zijl PC, Jones CK, Ren J, Malloy CR, Sherry AD. MRI detection of glycogen in vivo by using chemical exchange saturation transfer imaging (glycoCEST). *Proc Natl Acad Sci USA* 2007;104(11):4359–4364.

81. Li H, Zu Z, Zaiss M, Khan IS, Singer RJ, Gochberg DF, Bachert P, Gore JC, Xu J. Imaging of amide proton transfer and nuclear Overhauser enhancement in ischemic stroke with corrections for competing effects. *NMR Biomed* 2015;28(2):200–209.

82. Sun PZ, van Zijl PC, Zhou J. Optimization of the irradiation power in chemical exchange dependent saturation transfer experiments. *J Magn Reson* 2005;175(2):193–200.

83. Jackson A. Analysis of dynamic contrast enhanced MRI. *Br J Radiol* 2004;77 Spec No 2:S154-166.

84. Padhani AR. Dynamic contrast-enhanced MRI in clinical oncology: Current status and future directions. *J Magn Reson Imaging* 2002;16(4):407–422.

85. Yankeelov TE, Gore JC. Dynamic contrast enhanced magnetic resonance imaging in oncology: Theory, data acquisition, analysis, and examples. *Curr Med Imaging Rev* 2007;3(2):91–107.

86. Jackson A, Buckley DL, Parker GJ. *Dynamic Contrast-Enhanced Magnetic Resonance Imaging in Oncology.* New York: Springer; 2004.

87. Verma S, Turkbey B, Muradyan N, Rajesh A, Cornud F, Haider MA, Choyke PL, Harisinghani M. Overview of dynamic contrast-enhanced MRI in prostate cancer diagnosis and management. *AJR Am J Roentgenol* 2012;198(6):1277–1288.

88. Schnall MD, Blume J, Bluemke DA et al. Diagnostic architectural and dynamic features at breast MR imaging: Multicenter study. *Radiology* 2006;238(1):42–53.

89. Abe H, Mori N, Tsuchiya K, Schacht DV, Pineda FD, Jiang Y, Karczmar GS. Kinetic analysis of benign and malignant breast lesions with ultrafast dynamic contrast-enhanced MRI: Comparison with standard kinetic assessment. *AJR Am J Roentgenol* 2016:1–8.

90. Arasu VA, Chen RC, Newitt DN, Chang CB, Tso H, Hylton NM, Joe BN. Can signal enhancement ratio (SER) reduce the number of recommended biopsies without affecting cancer yield in occult MRI-detected lesions? *Acad Radiol* 2011;18(6):716–721.

91. Tofts PS. T1-weighted DCE imaging concepts: Modelling, acquisition and analysis. *MAGNETOM Flash* 2010;3:30–35.

92. Huang W, Tudorica LA, Li X, Thakur SB, Chen Y, Morris EA, Tagge IJ, Korenblit ME, Rooney WD, Koutcher JA, Springer CS, Jr. Discrimination of benign and malignant breast lesions by using shutter-speed dynamic contrast-enhanced MR imaging. *Radiology* 2011;261(2):394–403.

93. Schabel MC, Morrell GR, Oh KY, Walczak CA, Barlow RB, Neumayer LA. Pharmacokinetic mapping for lesion classification in dynamic breast MRI. *J Magn Reson Imaging* 2010;31(6):1371–1378.

94. Li X, Kang H, Arlinghaus LR, Abramson RG, Chakravarthy AB, Abramson VG, Farley J, Sanders M, Yankeelov TE. Analyzing spatial heterogeneity in DCE- and DW-MRI parametric maps to optimize prediction of pathologic response to neoadjuvant chemotherapy in breast cancer. *Transl Oncol* 2014;7(1):14–22.

95. Heisen M, Fan X, Buurman J, van Riel NA, Karczmar GS, ter Haar Romeny BM. The influence of temporal resolution in determining pharmacokinetic parameters from DCE-MRI data. *Magn Reson Med* 2010;63(3):811–816.

96. Abramson RG, Li X, Hoyt TL, Su PF, Arlinghaus LR, Wilson KJ, Abramson VG, Chakravarthy AB, Yankeelov TE. Early assessment of breast cancer response to neoadjuvant chemotherapy by semi-quantitative analysis of high-temporal resolution DCE-MRI: Preliminary results. *Magn Reson Imaging* 2013;31(9):1457–1464.

97. Karahaliou A, Vassiou K, Arikidis NS, Skiadopoulos S, Kanavou T, Costaridou L. Assessing heterogeneity of lesion enhancement kinetics in dynamic contrast-enhanced MRI for breast cancer diagnosis. *Br J Radiol* 2010;83(988):296–309.

98. Tofts PS. Modeling tracer kinetics in dynamic Gd-DTPA MR imaging. *J Magn Reson Imaging* 1997;7(1):91–101.

99. Tofts PS, Berkowitz B, Schnall MD. Quantitative analysis of dynamic Gd-DTPA enhancement in breast tumors using a permeability model. *Magn Reson Med* 1995;33(4):564–568.

100. Fedorov A, Fluckiger J, Ayers GD, Li X, Gupta SN, Tempany C, Mulkern R, Yankeelov TE, Fennessy FM. A comparison of two methods for estimating DCE-MRI parameters via individual and cohort based AIFs in prostate cancer: A step towards practical implementation. *Magn Reson Imaging* 2014;32(4):321–329.

101. Li X, Welch EB, Arlinghaus LR et al. A novel AIF tracking method and comparison of DCE-MRI parameters using individual and population-based AIFs in human breast cancer. *Phys Med Biol* 2011;56(17):5753–5769.

102. Parker GJ, Roberts C, Macdonald A, Buonaccorsi GA, Cheung S, Buckley DL, Jackson A, Watson Y, Davies K, Jayson GC. Experimentally-derived functional form for a population-averaged high-temporal-resolution arterial input function for dynamic contrast-enhanced MRI. *Magn Reson Med* 2006;56(5):993–1000.

103. Shukla-Dave A, Lee N, Stambuk H, Wang Y, Huang W, Thaler HT, Patel SG, Shah JP, Koutcher JA. Average arterial input function for quantitative dynamic contrast enhanced magnetic resonance imaging of neck nodal metastases. *BMC Med Phys* 2009;9:4.

104. Tofts PS, Brix G, Buckley DL et al. Estimating kinetic parameters from dynamic contrast-enhanced T1-weighted MRI of a diffusable tracer: Standardized quantities and symbols. *J Magn Reson Imaging* 1999;10:223–232.

105. Yankeelov TE, Mankoff DA, Schwartz LH et al. Quantitative imaging in cancer clinical trials. *Clin Cancer Res* 2016;22(2):284–290.

106. Barnes SL, Whisenant JG, Loveless ME, Yankeelov TE. Practical dynamic contrast enhanced MRI in small animal models of cancer: Data acquisition, data analysis, and interpretation. *Pharmaceutics* 2012;4(3):442–478.

107. Barnes SL, Whisenant JG, Loveless ME, Ayers GD, Yankeelov TE. Assessing the reproducibility of dynamic contrast enhanced magnetic resonance imaging in a murine model of breast cancer. *Magn Reson Med* 2013;69(6):1721–1734.

108. Yankeelov TE, DeBusk LM, Billheimer DD, Luci JJ, Lin PC, Price RR, Gore JC. Repeatability of a reference region model for analysis of murine DCE-MRI data at 7T. *J Magn Reson Imaging* 2006;24(5):1140–1147.

109. Yankeelov TE, Cron GO, Addison CL, Wallace JC, Wilkins RC, Pappas BA, Santyr GE, Gore JC. Comparison of a reference region model with direct measurement of an AIF in the analysis of DCE-MRI data. *Magn Reson Med* 2007;57(2):353–361.

110. Beresford MJ, Padhani AR, Taylor NJ, Ah-See ML, Stirling JJ, Makris A, d'Arcy JA, Collins DJ. Inter- and intraobserver variability in the evaluation of dynamic breast cancer MRI. *J Magn Reson Imaging* 2006;24(6):1316–1325.

111. Rata M, Collins DJ, Darcy J, Messiou C, Tunariu N, Desouza N, Young H, Leach MO, Orton MR. Assessment of repeatability and treatment response in early phase clinical trials using DCE-MRI: Comparison of parametric analysis using MR- and CT-derived arterial input functions. *Eur Radiol* 2016;26(7):1991–1998.

112. Padhani AR, Hayes C, Landau S, Leach MO. Reproducibility of quantitative dynamic MRI of normal human tissues. *NMR Biomed* 2002;15(2):143–153.

113. Galbraith SM, Lodge MA, Taylor NJ, Rustin GJ, Bentzen S, Stirling JJ, Padhani AR. Reproducibility of dynamic contrast-enhanced MRI in human muscle and tumours: Comparison of quantitative and semi-quantitative analysis. *NMR Biomed* 2002;15(2):132–142.

114. Hahn OM, Yang C, Medved M, Karczmar G, Kistner E, Karrison T, Manchen E, Mitchell M, Ratain MJ, Stadler WM. Dynamic contrast-enhanced magnetic resonance imaging pharmacodynamic biomarker study of sorafenib in metastatic renal carcinoma. *J Clin Oncol* 2008;26(28):4572–4578.

115. Mross K, Drevs J, Muller M et al. Phase I clinical and pharmacokinetic study of PTK/ZK, a multiple VEGF receptor inhibitor, in patients with liver metastases from solid tumours. *Eur J Cancer* 2005;41(9):1291–1299.

116. Chung C, Driscoll B, Gorjizadeh A, Foltz W, Lee S, Menard C, Coolens C. Early detection of tumor response using 4D DCE-CT and DCE-MRI in patients treated with radiosurgery for brain metastases. *Pract Radiat Oncol* 2013;3(2 Suppl 1):S17–S18.

117. Esserman L, Hylton N, George T, Weidner N. Contrast-enhanced magnetic resonance imaging to assess tumor histopathology and angiogenesis in breast carcinoma. *Breast J* 1999;5(1):13–21.

118. Tudorica A, Oh KY, Chui SY et al. Early prediction and evaluation of breast cancer response to neoadjuvant chemotherapy using quantitative DCE-MRI. *Transl Oncol* 2016;9(1):8–17.

119. Kuchcinski G, Le Rhun E, Cortot AB et al. Dynamic contrast-enhanced MR imaging pharmacokinetic parameters as predictors of treatment response of brain metastases in patients with lung cancer. *Eur Radiol* 2017;27(9):3733–3743.

120. Li L, Wang K, Sun X, Wang K, Sun Y, Zhang G, Shen B. Parameters of dynamic contrast-enhanced MRI as imaging markers for angiogenesis and proliferation in human breast cancer. *Med Sci Monit* 2015;21:376–382.

121. Ashton E, Riek J. Advanced MR techniques in multicenter clinical trials. *J Magn Reson Imaging* 2013;37(4):761–769.

122. O'Connor JP, Jackson A, Parker GJ, Jayson GC. DCE-MRI biomarkers in the clinical evaluation of antiangiogenic and vascular disrupting agents. *Br J Cancer* 2007;96(2):189–195.

123. Hylton NM, Blume JD, Bernreuter WK et al. Locally advanced breast cancer: MR imaging for prediction of response to neoadjuvant chemotherapy—Results from ACRIN 6657/I-SPY TRIAL. *Radiology* 2012;263(3):663–672.

124. Aronen H, Cohen M, Belliveau J, Fordham J, Rosen B. Ultrafast imaging of brain tumors. *J Magn Reson Imaging* 1993;5:14–24.

125. Ostergaard L, Weisskoff RM, Chesler DA, Gyldensted C, Rosen BR. High resolution measurement of cerebral blood flow using intravascular tracer bolus passages. Part I: Mathematical approach and statistical analysis. *Magn Reson Med* 1996;36:715–725.

126. Willats L, Calamante F. The 39 steps: Evading error and deciphering the secrets for accurate dynamic susceptibility contrast MRI. *NMR Biomed* 2012. doi:10.1002/nbm.2833.

127. Aronen H, Gazit I, Louis D, Buchbinder B, Pardo F, Weisskoff R, Harsh G, Cosgrove G, Halpern E, Hochberg F. Cerebral blood volume maps of gliomas: Comparison with tumor grade and histologic findings. *Radiology* 1994;191:41–51.

128. Donahue KM, Krouwer HG, Rand SD, Pathak AP, Marszalkowski CS, Censky SC, Prost RW. Utility of simultaneously acquired gradient-echo and spin-echo cerebral blood volume and morphology maps in brain tumor patients. *Magn Reson Med* 2000;43:845–853.

129. Schmainda K, Rand S, Joseph A, Lund R, Ward B, Pathak A, Ulmer J, Badruddoja M, Krouwer H. Characterization of a first-pass gradient-echo spin-echo method to predict brain tumor grade and angiogenesis. *Am J Neuroradiol* 2004;25:1524–1532.

130. Hakyemez B, Erdogan C, Ercan I, Ergin N, Uysal S, Atahan S. High-grade and low-grade gliomas: Differentiation by using perfusion MR imaging. *Clin Radiol* 2005;60:493–502.

131. Fuss M, Wenz F, Essig M, Muenter M, Debus J, Herman TS, Wannenmacher M. Tumor angiogenesis of low-grade astrocytomas measured by dynamic susceptibility contrast-enhanced MRI (DSC-MRI) is predictive of local tumor control after radiation therapy. *Int J Radiat Oncol Biol Phys* 2001;51:478–482.

132. Weber M, Thilmann C, Lichy M et al. Assessment of irradiated brain metastases by means of arterial spin-labeling and dynamic susceptibility-weighted contrast-enhanced perfusion MRI: Initial results. *Invest Radiol* 2004;39:277–287.

133. Boxerman JL, Ellingson BM, Jeyapalan S, Elinzano H, Harris RJ, Rogg JM, Pope WB, Safran H. Longitudinal DSC-MRI for distinguishing tumor recurrence from pseudoprogression in patients with a high-grade glioma. *Am J Clin Oncol* 2014. doi:10.1097/COC.0000000000000156.

134. Schmainda KM, Zhang Z, Prah M, Snyder BS, Gilbert MR, Sorensen AG, Barboriak DP, Boxerman JL. Dynamic susceptibility contrast MRI measures of relative cerebral blood volume as a prognostic marker for overall survival in recurrent glioblastoma: Results from the ACRIN 6677/RTOG 0625 multicenter trial. *Neuro-Oncology* 2015. doi:10.1093/neuonc/nou364.

135. Livne M, Kossen T, Madai VI, Zaro-Weber O, Moeller-Hartmann W, Mouridsen K, Heiss W-D, Sobesky J. Multiparametric model for penumbral flow prediction in acute stroke. *Stroke* 2017;48:1849–1854. doi:10.1161/STROKEAHA.117.016631.

136. Sowa P, Nygaard GO, Bjornerud A, Celius EG, Harbo HF, Beyer MK. Magnetic resonance imaging perfusion is associated with disease severity and activity in multiple sclerosis. *Neuroradiology* 2017. doi:10.1007/s00234-017-1849-4.

137. Prah MA, Stufflebeam SM, Paulson ES, Kalpathy-Cramer J, Gerstner ER, Batchelor TT, Barboriak DP, Rosen BR, Schmainda KM. Repeatability of standardized and normalized relative CBV in patients with newly diagnosed glioblastoma. *Am J Neuroradiol* 2015. doi:10.3174/ajnr. A4374.

138. Ostergaard L, Sorensen AG, Kwong KK, Weisskoff RM, Gyldensted C, Rosen BR. High resolution measurement of cerebral blood flow using intravascular tracer bolus passages. Part II: Experimental comparison and preliminary results. *Magn Reson Med* 1996;36:726–736.

139. Zierler K. Theoritcal basis of indicator-dilution methods for measuring blood flow and volume. *Circ Res* 1965;10:393–407.

140. Zierler K. Equations for measuring blood flow by external monitoring of radioisotopes. *Circ Res* 1965;16:309–321.

141. Meier P, Zierler KL. On the theory of the indicator-dilution method for measurement of blood flow and volume. *J Appl Physiol* 1954;6:731–744.

142. Boxerman JL, Prah DE, Paulson ES, Machan JT, Bedekar D, Schmainda KM. The role of preload and leakage correction in gadolinium-based cerebral blood volume estimation determined by comparison with MION as a criterion standard. *Am J Neuroradiol* 2012. doi:10.3174/ajnr. A2934.

143. Paulson ES, Schmainda KM. Comparison of dynamic susceptibility-weighted contrast-enhanced MR methods: Recommendations for measuring relative cerebral blood volume in brain tumors. *Radiology* 2008;249:601–613. doi:10.1148/radiol.2492071659.

144. Bedekar D, Jensen T, Schmainda KM. Standardization of relative cerebral blood volume (rCBV) image maps for ease of both inter- and intrapatient comparisons. *Magn Reson Med* 2010;64:907–913. doi:10.1002/mrm.22445.

145. Krohn KA, Link JM, Mason RP. Molecular imaging of hypoxia. *J Nucl Med* 2008;49(Suppl 2):129S–148S.

146. Tatum JL, Kelloff GJ, Gillies RJ et al. Hypoxia: Importance in tumor biology, noninvasive measurement by imaging, and value of its measurement in the management of cancer therapy. *Int J Radiat Biol* 2006;82(10):699–757.

147. Shannon AM, Bouchier-Hayes DJ, Condron CM, Toomey D. Tumour hypoxia, chemotherapeutic resistance and hypoxia-related therapies. *Cancer Treat Rev* 2003;29(4):297–307.

148. Hammond EM, Asselin MC, Forster D, O'Connor JP, Senra JM, Williams KJ. The meaning, measurement and modification of hypoxia in the laboratory and the clinic. *J Clin Oncol* 2014;26(5):277–288.

149. Höckel M, Vaupel P. Tumor hypoxia: Definitions and current clinical, biologic, and molecular aspects. *J Natl Cancer Inst* 2001;93(4):266–276.

150. Dewhirst MW, Birer SR. Oxygen-enhanced MRI is a major advance in tumor hypoxia imaging. *Cancer Res* 2016;76(4):769–772.

151. Yablonskiy DA, Haacke EM. Theory of NMR signal behavior in magnetically inhomogeneous tissues: The static dephasing regime. *Magn Reson Med* 1994;32(6):749–763.

152. Yablonskiy DA, Sukstanskii AL, He X. Blood oxygenation level-dependent (BOLD)-based techniques for the quantification of brain hemodynamic and metabolic properties—Theoretical models and experimental approaches. *NMR Biomed* 2013;26(8):963–986.

153. O'Connor JP, Boult JK, Jamin Y et al. Oxygen-enhanced MRI accurately identifies, quantifies, and maps tumor hypoxia in preclinical cancer models. *Cancer Res.* 2016;76(4):787–795.

154. O'Connor JP, Jackson A, Buonaccorsi GA et al. Organ-specific effects of oxygen and carbogen gas inhalation on tissue longitudinal relaxation times. *Magn Reson Med* 2007;58(3):490–496.

155. Christen T, Lemasson B, Pannetier N, Farion R, Remy C, Zaharchuk G, Barbier EL. Is T2* enough to assess oxygenation? Quantitative blood oxygen level–dependent analysis in brain tumor. *Radiology* 2012;262(2):495–502.

156. Burrell JS, Walker-Samuel S, Baker LC, Boult JK, Jamin Y, Halliday J, Waterton JC, Robinson SP. Exploring deltaR(2) * and deltaR(1) as imaging biomarkers of tumor oxygenation. *J Magn Reson Imaging* 2013;38(2):429–434.

157. He X, Zhu M, Yablonskiy DA. Validation of oxygen extraction fraction measurement by qBOLD technique. *Magn Reson Med* 2008;60(4):882–888.

158. An H, Lin W. Quantitative measurements of cerebral blood oxygen saturation using magnetic resonance imaging. *J Cereb Blood Flow Metab* 2000;20(8):1225–1236.

159. An H, Lin W. Cerebral oxygen extraction fraction and cerebral venous blood volume measurements using MRI: Effects of magnetic field variation. *Magn Reson Med* 2002;47(5):958–966.

160. He X, Yablonskiy DA. Quantitative BOLD: Mapping of human cerebral deoxygenated blood volume and oxygen extraction fraction: Default state. *Magn Reson Med* 2007;57(1):115–126.

161. Christen T, Lemasson B, Pannetier N, Farion R, Segebarth C, Remy C, Barbier EL. Evaluation of a quantitative blood oxygenation level-dependent (qBOLD) approach to map local blood oxygen saturation. *NMR Biomed* 2011;24(4):393–403.

162. Christen T, Schmiedeskamp H, Straka M, Bammer R, Zaharchuk G. Measuring brain oxygenation in humans using a multiparametric quantitative blood oxygenation level dependent MRI approach. *Magn Reson Med* 2012;68(3):905–911.

163. Kindvall SSI, Diaz S, Svensson J, Wollmer P, Olsson LE. The change of longitudinal relaxation rate in oxygen enhanced pulmonary MRI depends on age and BMI but not diffusing capacity of carbon monoxide in healthy never-smokers. *PloS One* 2017;12(5):e0177670.

164. Zhang WJ, Niven RM, Young SS, Liu YZ, Parker GJ, Naish JH. Dynamic oxygen-enhanced magnetic resonance imaging of the lung in asthma—Initial experience. *Eur J Radiol* 2015;84(2):318–326.

165. Edelman RR, Hatabu H, Tadamura E, Li W, Prasad PV. Noninvasive assessment of regional ventilation in the human lung using oxygen-enhanced magnetic resonance imaging. *Nat Med* 1996;2(11):1236–1239.

166. Linnik IV, Scott ML, Holliday KF et al. Noninvasive tumor hypoxia measurement using magnetic resonance imaging in murine U87 glioma xenografts and in patients with glioblastoma. *Magn Reson Med* 2013; doi:10.1002/mrm.24826.

167. O'Connor JP, Naish JH, Parker GJ et al. Preliminary study of oxygen-enhanced longitudinal relaxation in MRI: A potential novel biomarker of oxygenation changes in solid tumors. *Int J Radiat Oncol Biol Phys* 2009;75(4):1209–1215.

168. Matsumoto K-i, Bernardo M, Subramanian S, Choyke P, Mitchell JB, Krishna MC, Lizak MJ. MR assessment of changes of tumor in response to hyperbaric oxygen treatment. *Magn Reson Med* 2006;56(2):240–246.

169. Ophir J, Cespedes I, Ponnekanti H, Yazdi Y, Li X. Elastography: A quantitative method for imaging the elasticity of biological tissues. *Ultrason Imaging* 1991;13(2):111–134.

170. Miga MI. A new approach to elastography using mutual information and finite elements. *Phys Med Biol* 2003;48(4):467–480.

171. Pheiffer TS, Ou JJ, Ong RE, Miga MI. Automatic generation of boundary conditions using demons nonrigid image registration for use in 3-D modality-independent elastography. *IEEE Trans Biomed Eng* 2011;58(9):2607–2616.

172. Han ZL, Li JS, Singh M, Wu C, Liu CH, Raghunathan R, Aglyamov SR, Vantipalli S, Twa MD, Larin KV. Optical coherence elastography assessment of corneal viscoelasticity with a modified Rayleigh-Lamb wave model. *J Mech Behav Biomed Mater* 2017;66:87–94.

173. Budelli E, Brum J, Bernal M, Deffieux T, Tanter M, Lema P, Negreira C, Gennisson JL. A diffraction correction for storage and loss moduli imaging using radiation force based elastography. *Phys Med Biol* 2017;62(1):91–106.

174. Hossain MM, Moore CJ, Gallippi CM. Acoustic radiation force impulse-induced peak displacements reflect degree of anisotropy in transversely isotropic elastic materials. *IEEE Trans Ultrason Ferroelectr Freq* 2017;64(6):989–1001.

175. Righetti R, Ophir J, Srinivasan S, Krouskop TA. The feasibility of using elastography for imaging the Poisson's ratio in porous media. *Ultrasound Med Biol* 2004;30(2):215–228.

176. Righetti R, Srinivasan S, Kumar AT, Ophir J, Krouskop TA. Assessing image quality in effective Poisson's ratio elastography and poroelastography: I. *Phys Med Biol* 2007;52(5):1303–1320.

177. Tan LK, McGarry MDJ, Van Houten EEW, Ji M, Solamen L, Weaver JB, Paulsen KD. Gradient-based optimization for poroelastic and viscoelastic MR elastography. *IEEE Trans Med Imaging* 2017;36(1):236–250.

178. Zhao JX, Zhai F, Cheng J, He Q, Luo JW, Yang XP, Shao JH, Xing HC. Evaluating the significance of viscoelasticity in diagnosing early-stage liver fibrosis with transient elastography. *Plos One* 2017;12(1).

179. Yin M, Talwalkar JA, Glaser KJ, Manduca A, Grimm RC, Rossman PJ, Fidler JL, Ehman RL. Assessment of hepatic fibrosis with magnetic resonance elastography. *Clin Gastroenterol Hepatol* 2007;5(10):1207–1213 e1202.

180. Mariappan YK, Glaser KJ, Ehman RL. Magnetic resonance elastography: A review. *Clin Anat* 2010;23(5):497–511.

181. Leitao HS, Doblas S, Garteiser P, d'Assignies G, Paradis V, Mouri F, Geraldes CF, Ronot M, Van Beers BE. Hepatic fibrosis, inflammation, and steatosis: Influence on the MR viscoelastic and diffusion parameters in patients with chronic liver disease. *Radiology* 2017;283(1):98–107.

182. Thompson SM, Wang J, Chandan VS, Glaser KJ, Roberts LR, Ehman RL, Venkatesh SK. MR elastography of hepatocellular carcinoma: Correlation of tumor stiffness with histopathology features-Preliminary findings. *Magn Reson Imaging* 2017;37:41–45.

183. Yin M, Glaser KJ, Manduca A et al. Distinguishing between hepatic inflammation and fibrosis with MR elastography. *Radiology* 2017;284(3):694–705.

184. Zhu B, Wei L, Rotile N, Day H, Rietz T, Farrar CT, Lauwers GY, Tanabe KK, Rosen B, Fuchs BC, Caravan P. Combined magnetic resonance elastography and collagen molecular magnetic resonance imaging accurately stage liver fibrosis in a rat model. *Hepatology* 2017;65(3):1015–1025.

185. Glaser KJ, Manduca A, Ehman RL. Review of MR elastography applications and recent developments. *J Magn Reson Imaging* 2012;36(4):757–774.

186. Butcher DT, Alliston T, Weaver VM. A tense situation: Forcing tumour progression. *Nat Rev Cancer* 2009;9(2):108–122.

187. Lu P, Weaver VM, Werb Z. The extracellular matrix: A dynamic niche in cancer progression. *J Cell Biol* 2012;196(4):395–406.

188. Paszek MJ, Weaver VM. The tension mounts: Mechanics meets morphogenesis and malignancy. *J Mammary Gland Biol Neoplasia* 2004;9(4):325–342.

189. Paszek MJ, Zahir N, Johnson KR et al. Tensional homeostasis and the malignant phenotype. *Cancer Cell* 2005;8(3):241–254.

190. Pickup MW, Mouw JK, Weaver VM. The extracellular matrix modulates the hallmarks of cancer. *EMBO Rep* 2014;15(12):1243–1253.

191. Li J, Jamin Y, Boult JK, Cummings C, Waterton JC, Ulloa J, Sinkus R, Bamber JC, Robinson SP. Tumour biomechanical response to the vascular disrupting agent ZD6126 in vivo assessed by magnetic resonance elastography. *Br J Cancer* 2014;110(7):1727–1732.

192. Mariappan YK, Glaser KJ, Levin DL, Vassallo R, Hubmayr RD, Mottram C, Ehman RL, McGee KP. Estimation of the absolute shear stiffness of human lung parenchyma using (1) H spin echo, echo planar MR elastography. *J Magn Reson Imaging* 2014;40(5):1230–1237.

193. Murphy MC, Huston J, 3rd, Jack CR, Jr., Glaser KJ, Senjem ML, Chen J, Manduca A, Felmlee JP, Ehman RL. Measuring the characteristic topography of brain stiffness with magnetic resonance elastography. *PLoS One* 2013;8(12):e81668.

194. Weis JA, Kim DK, Yankeelov TE, Miga MI. Validation and reproducibility assessment of modality independent elastography in a pre-clinical model of breast cancer. 2014. *Medical Imaging 2014: Biomedical Applications in Molecular, Structural, and Functional Imaging, 90381I.* Vol. 9038. Bellingham, WA: International Society for Optics and Photonics.

195. Good DW, Stewart GD, Hammer S, Scanlan P, Shu W, Phipps S, Reuben R, McNeill AS. Elasticity as a biomarker for prostate cancer: A systematic review. *BJU Int* 2014;113(4):523–534.

196. Osman NF, Sampath S, Atalar E, Prince JL. Imaging longitudinal cardiac strain on short-axis images using strain-encoded MRI. *Magn Reson Med* 2001;46(2):324–334.

197. Pan L, Stuber M, Kraitchman DL, Fritzges DL, Gilson WD, Osman NF. Real-time imaging of regional myocardial function using fast-SENC. *Magn Reson Med* 2006;55(2):386–395.

198. Harouni AA, Gharib AM, Osman NF, Morse C, Heller T, Abd-Elmoniem KZ. Assessment of liver fibrosis using fast strain-encoded MRI driven by inherent cardiac motion. *Magn Reson Med* 2015;74:106–114.

199. Watanabe H, Kanematsu M, Kitagawa T et al. MR elastography of the liver at 3 T with cine-tagging and bending energy analysis: Preliminary results. *Eur Radiol* 2010;20(10):2381–2389.

200. Ou JJ, Ong RE, Yankeelov TE, Miga MI. Evaluation of 3D modality-independent elastography for breast imaging: A simulation study. *Phys Med Biol* 2008;53(1):147–163.

201. Washington CW, Miga MI. Modality independent elastography (MIE): A new approach to elasticity imaging. *IEEE Trans Med Imaging* 2004;23(9):1117–1128.

202. Weis JA, Flint KM, Sanchez V, Yankeelov TE, Miga MI. Assessing the accuracy and reproducibility of modality independent elastography in a murine model of breast cancer. *J Med Imaging* (Bellingham) 2015;2(3):036001.

203. Wang K, Manning P, Szeverenyi N et al. Repeatability and reproducibility of 2D and 3D hepatic MR elastography with rigid and flexible drivers at end-expiration and end-inspiration in healthy volunteers. *Abdom Radiol* 2017;42(12):2843–2854.

204. Hawley JR, Kalra P, Mo X, Raterman B, Yee LD, Kolipaka A. Quantification of breast stiffness using MR elastography at 3 Tesla with a soft sternal driver: A reproducibility study. *J Magn Reson Imaging* 2017;45(5):1379–1384.

205. Yin M, Glaser KJ, Talwalkar JA, Chen J, Manduca A, Ehman RL. Hepatic MR elastography: Clinical performance in a series of 1377 consecutive examinations. *Radiology* 2016;278(1):114–124.

206. Loomba R, Wolfson T, Ang B et al. Magnetic resonance elastography predicts advanced fibrosis in patients with nonalcoholic fatty liver disease: A prospective study. *Hepatology* 2014;60(6):1920–1928.

207. Siegmann KC, Xydeas T, Sinkus R, Kraemer B, Vogel U, Claussen CD. Diagnostic value of MR elastography in addition to contrast-enhanced MR imaging of the breast-initial clinical results. *Eur Radiol* 2010;20(2):318–325.

Tumor segmentation

SPYRIDON BAKAS, RHEA CHITALIA, DESPINA KONTOS, YONG FAN, AND CHRISTOS DAVATZIKOS

6.1 OVERVIEW

Over the last years it has become more evident that cancer comes with highly heterogeneous molecular and radiographic profiles, both in the spatial and temporal domain [1–15]. This heterogeneity hinders successful treatment, while making it impossible to fully assess the tumor's spectrum of molecular characteristics from a single tissue specimen, as depending on the analyzed specimen different characteristics may be identified. Furthermore, multiple/repeated surgeries and biopsies that could evaluate existing or emerged molecular changes during treatment are typically neither considered nor performed due to their invasive nature and potential complications. Moreover, the tumor's biopsy/resection might not always be possible, such as in cases of inoperable and deep-seated tumors, and consequently tissue for histopathological and molecular assessment and analysis is unavailable, or limited. Finally, molecular tissue characterization is limited to reference centers, as it requires costly and not widely-available equipment for tissue-based molecular testing.

Furthermore, the Response Evaluation Criteria In Solid Tumors (RECIST and the Response Assessment in Neuro-Oncology (RANO) working group (also known as the Macdonald) criteria consider 1D or 2D measurements, such as the major axis of a tumor or the products of perpendicular major and minor axes [16–18] on an subjectively/arbitrarily selected slice. Although the RANO criteria are widely used due to their applicability, their subjective nature hinders their repeatability and reproducibility, especially for irregular-shaped tumors [19].

Non-invasive, longitudinal measures capturing the complete spatial tumor extent and peritumoral tissue would allow for more comprehensive, hence accurate, characterization of the heterogeneity of the tumor, which would subsequently influence personalized prognosis and treatment. Imaging is a non-invasive and widely available method for assessing tumor status *in vivo* macroscopically, and hence, has a very promising role toward these goals. In recent years, imaging features describing the texture (also known as radiomics) have been included in diagnostic imaging reporting and data systems (IRADS) for (i) breast (i.e., BI-RADS) [20], (ii) prostate (i.e., PI-RADS) [21], and (iii) lung (i.e., LI-RADS) [22]. Furthermore, there has been mounting evidence that computational analysis of quantitative imaging phenomic features (a group of which are often called radiomic features) extracted from multi-parametric radiographic imaging modalities can characterize tumors comprehensively and provide critical information about various biological processes in the tumor microenvironment, as well as yield associations with the underlying molecular characteristics of the cancer (often referred to as radiogenomics) [23–38]. Since cancer patients are routinely scanned radiographically (typically before surgery and at multiple time points during treatment and follow-up), the availability of radiomic and radiogenomic imaging biomarkers would help evaluate the spatial and temporal heterogeneity of tumors.

The foundation for extracting quantitative imaging phenomic (QIP) features (beyond radiomics) from specific anatomical regions would be based on producing accurate segmentation labels of the respective various tumor sub-regions and surrounding anatomy. Note that tumor heterogeneity will be discussed in detail in Chapter 7, feature extraction in Chapter 8, and radiogenomics in Chapters 10, 11, and 21.

6.2 METHODOLOGICAL APPROACHES

Tissue segmentation is an essential task for clinical assessment and evaluation of medical images, irrespective of whether it refers to the involuntary partitioning of an image that an expert radiologist conceptually performs to assess a clinical image and create a report for a patient or to the actual delineation of the various structures appearing in a clinical radiographic scan. Example applications of both personalized and populations-based studies that require the segmentation of a clinically acquired scan into various components include, but not limited to, surgical planning, delivery of radiation therapy, cortical structure analysis and thickness estimation (in brain scans), quantification of extent of resection and subsequent longitudinal evaluation of tumor growth, extraction of QIP features, tissue characterization, monitoring tumor changes in response to neo-adjuvant chemotherapy, and identification of association with molecular characteristics (radiogenomic research). The tedious and time-consuming manual processing steps involved in such a process, as well as the human rater variability that potentially impede not only the further analyses, but also their repeatability and reproducibility, assisted in making apparent the importance of automated approaches for the segmentation task, leading to various computer-aided segmentation methodologies. A stratification schema is hereby introduced assessing and describing methodological principles based on patient-wise or population-based knowledge (i.e., generative, discriminative, hybrid), as well as based on using information derived from registration techniques.

6.2.1 GENERATIVE METHODS

These methods use patient-wise prior domain knowledge, provided in the form of initialization within the patient's image(s) either manually from an operator or from another automated method, as well as a given number of distinct classes that are expected to appear within the image(s). The term generative is used to characterize them as they "generate" results, based on the given initialization and the information available on the rest of the image, typically in the form of a membership criterion for each of the given classes. Such prior domain knowledge might describe: (i) statistical/parametric models of intensity distributions across initialized region of interest (ROI)/classes or even (ii) information from biophysical tumor growth models to drive the segmentation labels [39–44].

6.2.2 DISCRIMINATIVE METHODS

These methodologies apply knowledge learned/modeled from population-based assessments to new images and "discriminate"/classify parts of an image into the various modeled classes [45–51]. Examples of conventional discriminative methods include support vector machines, boosting, and random forests.

The disadvantages of these methods lay in handling parts of the image that do not match any of the profiles included in the population-based assessment and in being dependent on ample/sufficient and well-annotated data.

6.2.2.1 DEEP LEARNING (CONVOLUTIONAL NEURAL NETWORK)

Deep learning methods may also be included in the category of discriminative methods. Although the learning of their models is based on neural networks introduced in the 1940s [52], they were recently given the term "deep learning" due to their "deep" stacking of multiple layers of processed input data. Referring to the architecture of convolutional neural networks (CNNs), this processing typically includes the convolution of the input data with a number of local kernels followed by a non-linear transformation on each layer, the output of which is sent (optionally down-sampled) on the following layer. This processing increases the effectiveness of the analysis of the assessed input data, while produces inferences invariant to translation (i.e., increases the field of view of each neuron). A CNN, during training, tries to minimize a cost function (e.g., Kullback-Leibler divergence [53]), which usually describes the variance between the labels predicted from its last layer and the ones provided as the ground-truth (GT) of the training dataset.

Since most of the patch-based tumor segmentation methods typically solve the tumor segmentation problem as a voxel-wise classification problem, lacking constraints that encourage spatial and appearance consistency in the segmentation result, they may obtain degraded segmentation results due to imaging noise. To improve the segmentation's robustness to imaging noise, different strategies have been proposed, such as building a cascaded architecture of CNNs [54], training CNNs using multi-scale image patches [45,54], and building fully Convolutional Neural Networks (FCNNs), i.e., fully connected layers as convolutions with 12 or 13 kernels [55]. Promising segmentation performance has also been achieved by integrating CNNs with Markov random fields or conditional random fields (CRFs), either used as a post-process step of CNNs [45,56] or formulated as neural networks [57,58]. In particular, CRFs can be implemented as recurrent neural networks (RNNs) [58], referred to as CRF-RNNs hereafter, so that both CNNs and CRF-RNNs can be trained end-to-end with back-propagation algorithms. The FCNNs have been integrated with CRF-RNNs in a unified deep learning-based tumor segmentation model to achieve end-to-end learning.

6.2.3 HYBRID METHODS

Hybrid methods combining generative and discriminative methods have shown promise, in terms of their segmentation label accuracy, as they first "generate" results taking advantage of initializations incorporating the experts' knowledge, and then correct systematic errors on specific classes based on population based studies [59–63].

6.2.4 REGISTRATION-DRIVEN METHODS

Radiomic analysis of heterogeneous tumor disease often requires the consolidation of patient information from multiple modalities, or different time points, for accurate analysis. To this end, image registration can be used to align medical scans of different modalities or different time points of the same patient, bringing them into a comparable domain prior to segmentation [64]. Registration-driven methods can allow for the combined analysis of multi-modality images, as well as for longitudinal analysis of tumor development, or response to therapy.

Alignment of multiple modalities is also used as a pre-processing step for the previously described methods (i.e., generative, discriminative, hybrid), allowing them to determine more accurately tumor boundaries by leveraging complementary information provided in multiple modalities capturing the entire disease burden [42]. Incorporation of such extensive/combined information in their prior domain knowledge leads to a more comprehensive and superior segmentation performance.

1. *Multi-modality segmentation*

   Building upon the accuracy of predictive and prognostic models utilizing QIP features, joint models combining features from multiple imaging modalities can provide valuable information for tumor characterization. This can be seen in positron emission tomography (PET) where tumor visualization is often limited by the spatial distribution of radiotracer uptake. Segmentation of tumors from PET images often requires registration and fusion with computed tomography (CT) images to capture the entire tumor boundary [65].

   Multi-spectral MRI can provide information about the dynamics and presence of tissue water, contrast agents, or selected metabolites. When registered together, these images provide the boundary of the whole tumor extent, while accounting for intra-tumor heterogeneity, resulting in accurate segmentation [66,67]. This is particularly useful when segmenting tumors from non-deformable structures such as the brain. For example, gliomas describe a group of central nervous system tumors with various degrees of aggressiveness, each comprised of heterogeneous imaging phenotypes when visualized using multi-modality MRI. Registering these multi-modal MRI sequences allows for an accurate segmentation of the various sub-regions of each glioma, which can then be further quantified.

2. *Use in longitudinal segmentation*

   Monitoring the progression of a tumor and its response to therapy requires longitudinal imaging of the same patient, from which tumor morphology and radiomic features can be quantified after segmenting the tumor on each of these imaging scans. For meaningful comparisons between these features, images must be harmonized and co-registered prior to tumor segmentation. Previous studies have utilized registered longitudinal scans to extract radiomic and morphologic features from segmented tumors to assess response to therapy [68]. Ou et al. [69] implemented an attribute-based deformable registration algorithm, DRAMMS [70] to compare longitudinal breast MRI over the course of neoadjuvant chemotherapy. Identifying tumor regions on these registered images allowed for the quantification of heterogeneous tumor changes, showing the potential to quantitatively monitor response to therapy.

3. *Atlas-based segmentation*

   When differentiating tumors from background structures, segmentation by labeling voxels based on intensity value alone is often inadequate, as multiple structures comprising of the same tissue may be indistinguishable. As a result, incorporating the spatial information from neighboring voxels can add value to tumor segmentations. Atlas-based segmentations utilize previously confirmed image information (e.g., expert annotated) as reference information for tumor segmentation in new/unseen images. An atlas consists of images with prior knowledge of the desired anatomical structure, often pooled from population data [71]. Atlases can be generated by manually segmenting anatomical locations in multiple scans or by registering multiple segmented images together, often of different individuals, to integrate information from each image and each patient. An input, non-segmented image can then be mapped to the atlas, with unknown labels for each voxel determined by assessing the label of the corresponding location in the atlas image. This approach is well-established and has been utilized for tumor segmentations [72–74].

6.3 EVALUATION

The performance of computer-aided segmentation labels, both for cancerous and healthy tissue, is evaluated against labels delineated manually by expert radiologists. Following the widely accepted high inter- and intra-rater variability in such delineations [75–78], studies have shown that fusion of such labels reduces such variability [5]. Therefore, qualitative evaluation of segmentation labels produced by different automated methods is not expected to suffice toward judging the performance of each method. To quantitatively evaluate the performance of these methods, statistical principles (adapted from set-theory) are employed to compare the predicted labels (sets) with the GT labels (sets).

6.3.1 PERFORMANCE EVALUATION METRICS

The effectiveness of automated segmentation methods is quantitatively evaluated at the pixel level, by comparing the overlap between the manually annotated GT of the ROIs (i.e., ROI_{GT}) with the segmentation area predicted from the automated method (i.e., ROI_{Pred}). The overlap measurement defined in Figure 6.1, are the ones widely accepted and used by the scientific community to evaluate the performance of segmentation labels, i.e., Jaccard similarity coefficient, Precision (or Positive Predictive Value), Recall (or Sensitivity), Specificity, Dice coefficient, and the Hausdorff distance. Set theoretic notation is employed hereafter to mathematically define these measurements, as well as confusion matrix values (i.e., True Positives, False Positives, True Negatives, and False Negatives) and related graphical representations.

6.3.1.1 JACCARD SIMILARITY COEFFICIENT (OR JACCARD INDEX)

The Jaccard index (J) includes information only from pixels $p \in (ROI_{GT} \cap ROI_{Pred})$, but penalizes pixels misclassified as either ROI or non-ROI, i.e., $p \in (ROI_{GT} \Delta ROI_{Pred})$, where "$\Delta$" represents the symmetric different of ROI_{GT} and ROI_{Pred}, i.e., $p \in (ROI_{GT} \cup ROI_{Pred})\backslash(ROI_{GT} \cap ROI_{Pred})$. In terms of its interpretation, high J means that most pixels returned by the method were also included in the GT.

$$J = |ROI_{GT} \cap ROI_{Pred}|/|ROI_{GT} \cup ROI_{Pred}|$$

$$J = TP/(TP + FP + FN)$$

6.3.1.2 PRECISION (OR POSITIVE PREDICTIVE VALUE)

The Precision (P) for an ROI includes information from pixels $p \in (ROI_{GT} \cap ROI_{Pred})$, but its definition differs from that for J, as pixels $p \in (ROI_{Pred}^C \cap ROI_{GT})$, whenever ROI_{Pred} is a subset of ROI_{GT}, are not penalized. This would allow (without penalizing) the ROI_{Pred} to be much smaller than, but within ROI_{GT}. In other words

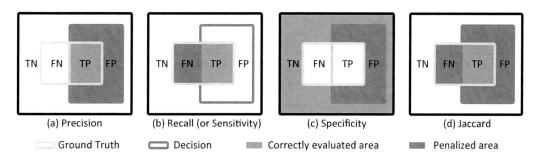

Figure 6.1 **(See color insert.)** Visual representation of the performance evaluation metrics of (a) Precision, (b) Recall, (c) Specificity, and (d) Jaccard.

P would not penalize undersegmented ROIs. In terms of the application, high P means that the segmentation method returned substantially more pixels of the ROI (relevant/correct pixels) than pixels outside the ROI (irrelevant pixels).

$$P = |ROI_{GT} \cap ROI_{Pred}|/|ROI_{Pred}|$$

$$P = TP/(TP + FP)$$

6.3.1.3 RECALL (OR SENSITIVITY, OR TRUE POSITIVE RATE)

The measurement of "Recall" (R) evaluates the proportion of the GT label that was correctly predicted. Specifically, R, includes information from pixels $p \in (ROI_{GT} \cap ROI_{Pred})$, but its definition complements that of P, as pixels $p \in (ROI_{Pred}{}^C \cap ROI_{GT})$, whenever ROI_{Pred} is a subset of ROI_{GT}, are penalized, while pixels $p \in (ROI_{GT}{}^C \cap ROI_{Pred})$, are not. High recall means that the automated method correctly returned most of the relevant pixels (pixels of the ROI).

$$R = Sensitivity = TPR = |ROI_{GT} \cap ROI_{Pred}|/|ROI_{GT}|$$

$$R = Sensitivity = TPR = TP/(TP + FN)$$

6.3.1.4 SPECIFICITY (OR TRUE NEGATIVE RATE)

The Specificity, S, for an ROI includes information from pixels $p \in (ROI_{GT} \cup ROI_{Pred})^C$, and its definition complements that for R as it accounts for the proportion of pixels that do not belong to the ROI, but are correctly classified. High Specificity means that the evaluated method correctly returned most of the irrelevant pixels (pixels outside of the ROI).

$$Specificity = TNR = |ROI_{GT} \cup ROI_{Pred}|^C/|ROI_{Pred}{}^C \cap ROI_{GT}|$$

$$Specificity = TNR = TN/(TN + FP)$$

6.3.1.5 DICE COEFFICIENT

The "Dice Coefficient" (D) describes the similarity between ROI_{GT} and ROI_{Pred}.

$$D = 2|ROI_{GT} \cap ROI_{Pred}|/|ROI_{GT}| + |ROI_{Pred}|$$

6.3.1.6 HAUSDORFF DISTANCE

Hausdorff Distance (H) is mathematically defined as the maximum distance of a set to the nearest point in the other set [79]. In other words, it measures how close the boundaries of ROI_{Pred} are to those of ROI_{GT}. H is used to assess the alignment between the contours of the segmentations and is defined as:

$$H(ROI_{GT}, ROI_{Pred}) = \max\{\min\{d(ROI_{GT}, ROI_{Pred})\}\}$$

6.3.2 REPRODUCIBILITY

Inability of reproducing research findings published in scientific manuscripts is a well-known issue for studies using data acquired in a single institution. The main reason hindering such reproducibility is the variability across acquisition equipment. Over recent years, taking into consideration the need for open data availability, and in consistency with the findable, accessible, interoperable, reusable (FAIR) principle [80], there have been collective efforts toward making available benchmark datasets acquired from multiple institutions, using varying equipment and acquisition protocols, thereby allowing for reproducible findings.

An example of such efforts is the International Brain Tumor Segmentation (BraTS) Challenge (www.braintumorsegmentation.org) [5,59,81], which has been offering a growing database of multi-institutional data accompanied with expert ground-truth segmentation labels, for benchmarking automated methods on the task of segmenting gliomas from multiple modalities. Substantial efforts have also been made by the National Cancer Institute (NCI) of the National Institutes of Health (NIH), and specifically by The Cancer Imaging Archive (TCIA), in releasing primary data for various cancer scans [82]. Note that further research resources, as well as the scientific benefit of sharing data and computational challenges will be described in more detail in Chapter 12.

6.4 APPLICATION EXAMPLES

In the following two subsections, there is the description of current trends and successful approaches for the segmentation of breast and lung cancer.

6.4.1 BRAIN CANCER

Although brain is a well-defined anatomical structure with *a priori* structural knowledge, brain tumors (and particularly gliomas) are challenging to segment, primarily due to their intrinsic histologic heterogeneity, which is portrayed in their radiographic phenotype (e.g., appearance and shape). Multi-modal assessment of such tumors is considered essential, for their accurate segmentation, as their various histologically distinct sub-regions are represented by varying intensity profiles disseminated across multiple modalities, reflecting varying biological properties. Such multiple modalities typically comprise native (T1) and post-contrast T1-weighted (T1Gd), T2-weighted (T2), and T2 fluid attenuated inversion recovery (FLAIR) volumes. Additional factors that make the task of brain tumor segmentation challenging, include the deformation of tissues surrounding the tumor (also known as "mass-effect") and the location of the radiofrequency (RF)coils (and the imperfection of the RF pulses) that may introduce non-uniformity in MR images (also known as "bias field").

Various automated methods have been proposed for this task that span across the whole spectrum of the methodological stratification described above [5,40–42,45,47–51,59–63,67,73,74,83–90]. However, differences across private datasets, such as: (i) the exact modalities included in the acquisition protocol, (ii) the type of the tumor (i.e., glioblastoma or lower grade glioma, primary or secondary tumor, solid or infiltratively growing), and (iii) the state of disease during acquisition (i.e., pre-operative scans acquired prior to treatment, versus postoperative scans including treatment effects and resection cavities), make the comparison of different segmentation strategies challenging. Therefore, the BraTS challenge was formed since 2012, with the intention of evaluating the performance of automated computational methods in the BraTS benchmark dataset. The BraTS dataset forms benchmark data of routine clinically acquired multi modal MRI scans, acquired in various institutions.

Example methods for brain tumor segmentation include applications based on: outlier detection [73], support vector machines [67], stacked denoising autoencoders [87], expectation-maximization [40,42], hybrid combination of generative and discriminative approaches [59–61,63], restricted Boltzmann machines [41], CNN [48–50,90], and 3D CNN with fully connected conditional random field [45]. Methods that have been proposed for segmenting brain tumors and being ranked within the top-performing methods during the last few years include (i) semi-automatic generative methods [40,42], (ii) semi-automatic and fully automatic discriminative methods [41,45,48–50,90], as well as (iii) hybrid generative-discriminative methods [59–61]. Specifically, both the semi-automatic generative methods [40,42] follow an expectation-maximization framework to segment the brain scans into tumor, as well as healthy tissue labels, and register a healthy population probabilistic atlas to glioma patients' brain scans. However, the method described in [40] incorporates a glioma growth model (based on a reaction diffusion-advection model) [91–93] to account for mass-effects. Semi-automatic and fully automatic discriminative methods that have been proposed [45,48–50,86,90] describe deep learning techniques of various architectures and have shown a lot of promise for the task of segmentation as they tend to be the top-ranked algorithms during the last two instances of BraTS. Finally, a hybrid generative-discriminative method [59–61] has also managed to be the top-ranked method during the BraTS 2015. Its generative part is based on GLioma Image SegmenTation and Registration (GLISTR) [40] (described above), and its discriminative part is based on a gradient boosting multi-class classification scheme [94,95], which was trained on the data of BraTS'15 challenge [5], to refine tumor labels based on information from multiple patients. Lastly, a Bayesian strategy [96] is employed to further refine and finalize the tumor segmentation labels based on patient-specific intensity statistics from the multiple available MRI modalities.

Although there has been a great effort on developing individual algorithms for the accurate segmentation of brain tumors, comprehensive studies have shown that consensus-based segmentation produces results of better performance, as well as of less variation across subjects, when compared to manually annotated segmentation labels from expert radiologists [5].

6.4.2 BREAST CANCER

The greater availability of mammography images taken during breast cancer screenings has allowed for the development of automated segmentation-based software tools to identify risk factors for breast cancer development [97,98]. Similarly, there are many developed methodologies for the segmentation of well-defined anatomical structures such as the brain, due to *a priori* structural knowledge allowing for many of the initialization steps to be automated under the assumption of a base control point in shape. However, this poses a challenge for the segmentation of arbitrarily shaped structures such as breast lesions. Despite this, multiple groups have developed both semi-automated and fully automated techniques for the segmentation of breast lesions from multi-modality images, as manual segmentation is often time consuming and subject to inter-observer bias.

The most commonly implemented segmentation technique utilizes clustering-based approaches due to its simplicity. Chen et al. developed a fuzzy C-means (FCM)-based approach for automated segmentation of 3D breast lesions from dynamic contrast enhanced (DCE) MRI [99]. Using a clinical dataset consisting of 121 MRI scans of primary mass lesions, they correctly segmented 97% of the lesions using their proposed FCM method, as compared to ground-truth manual segmentations provided by an expert radiologist. The FCM approach was extended by Lee et al. to incorporate the pharmacokinetic activity of each pixel in a DCE-MRI scan. To do so, a scalar signal called the variance enhancement slope was determined using the variance of pixel intensity differences across post-contrast MRI scans. FCM was then used to segment regions with similar variance enhancement slopes.

Beyond clustering based techniques, Ashraf et al. presented a method using a multi-channel extension of Markov random fields to utilize features derived from the imaging data specific to DCE-MRI [100]. This algorithm achieved a receiver operating characteristic area under curve (AUC) of 0.97 for tumor segmentation when using ground truth manual segmentation.

Methodologies have also been developed for use in PET images, with specific applications in PET-based breast cancer treatment planning. Examples of these techniques include thresholding, vibrational approaches, learning-based methods, and stochastic models [101]. The challenges of applying these various segmentation methodologies to PET images are the low spatial resolution and high noise. Additionally, breast lesion visualization in PET is dependent on the injected radiotracer and can lead to inaccurate segmentation of only the radiotracer bound tumor region.

Overall, there is a great effort in developing semi-automated or fully automated techniques for the segmentation of breast tumors [102–105]. Accurate segmentation of lesions from surrounding breast tissue can allow for radiomic feature extraction to characterize the tumor and assess relationships to gene expression, breast cancer molecular markers, and ultimately breast cancer recurrence [12,37,100].

6.4.3 LUNG CANCER

Early prediction of treatment response and survival is important in terms of optimal treatment planning and prognosis [106,107]. Imaging techniques, particularly PET and CT, have been playing important roles in diagnosis, staging, and predicting treatment response in lung cancer, particularly non-small cell lung cancer (NSCLC) [108–112]. Diagnostic and prognostic value of tumor metabolic burden and morphological information, such as standard uptake value, tumor size, and texture features derived from PET/CT imaging data, have been demonstrated in radiomic studies [111–115].

As the first step in radiomic studies of the lung cancer [113,114], lung tumor segmentation plays an important role [116,117]. Since manual segmentation of the lung tumors is laborious and often shows inter-rater variability, an enormous effort has been devoted to development of semi-automatic or fully automatic tumor segmentation methods in order to achieve reliable and accurate tumor segmentation [5,45,54,73,83,84,101,116,118–124]. The automatic segmentation of lung tumors typically follows a two-step procedure: lung field segmentation and lung nodule detection [125,126]. Particularly, graph-based methods, including graph-cut and random walk techniques, have been widely adopted in lung tumor segmentation studies for their flexibility to be used as either a semi-automatic method by manually providing tumor seed points or a fully automatic method by automatically identifying tumor seed points [118,127].

More recently, deep learning techniques have been adopted in tumor segmentation studies following their success in general image analysis field, such as images classification [128], objects detection [129], and semantic segmentation [55,57,58]. Most of the deep learning-based tumor segmentation methods are built upon CNNs [47,54,86,130] and convolutional restricted Boltzman machines [85]. Although a variety of network structures have been proposed for tumor segmentation [47,54,130], most of these tumor segmentation methods train CNNs in a pattern classification setting using image patches, i.e., local regions in images. These methods classify each image patch into healthy or tumor tissues, and the classification result of each image patch is used to label its center voxel for achieving the tumor segmentation. Recent studies have demonstrated that such deep learning-based method could achieve promising lung nodule segmentation performance [131,132].

6.5 FUTURE DIRECTIONS

Considering the amount of methods and analyses conducted in radiographic images, including radiomic and radiogenomic research, it is evident that this field could take advantage of analyses integrating radiographic and histologic imaging during its immediate future, revealing a new field of "integrated diagnostics" that could shed light toward understanding more mechanistically the source and potential evolution of individual tumors, thereby facilitating more aggressive yet targeted and personalized treatment regimens, increasing the potential for cure, and decreasing the side-effects of the current "one-size-fits-all" approaches.

REFERENCES

1. R. Verhaak, K. Hoadley, E. Purdom, V. Wang, Y. Qi, M. Wilkerson et al., "Integrated genomic analysis identifies clinically relevant subtypes of glioblastoma characterized by abnormalities in PDGFRA, IDII1, EGFR, and NF1," *Cancer Cell*, vol. 17, pp. 98–110, 2010.
2. C. W. Brennan, R. G. W. Verhaak, A. McKenna, B. Campos, H. Noushmehr, S. R. Salama et al., "The somatic genomic landscape of glioblastoma," *Cell*, vol. 155, pp. 462–477, 2013.
3. A. Sottoriva, I. Spiterib, S. G. M. Piccirillo, A. Touloumis, V. P. Collins, J. C. Marioni et al., "Intratumor heterogeneity in human glioblastoma reflects cancer evolutionary dynamics," *Proceedings of the National Academy of Sciences USA*, vol. 110, pp. 4009–4014, 2013.
4. J.-M. Lemee, A. Clavreul, and P. Menei, "Intratumoral heterogeneity in glioblastoma: Don't forget the peritumoral brain zone," *Neuro-Oncology*, vol. 17, pp. 1322–1332, 2015.
5. B. H. Menze, A. Jakab, S. Bauer, J. Kalpathy-Cramer, K. Farahani, J. Kirby et al., "The multimodal brain tumor image segmentation benchmark (BRATS)," *IEEE Transactions on Medical Imaging*, vol. 34, pp. 1993–2024, 2015.
6. H. Akbari, L. Macyszyn, X. Da, R. Wolf, M. Bilello, R. Verma et al., "Pattern analysis of dynamic susceptibility contrast-enhanced MR imaging demonstrates peritumoral tissue heterogeneity," *Radiology*, vol. 273, pp. 502–510, 2014.

7. M. Bilello, H. Akbari, X. Da, J. M. Pisapia, S. Mohan, R. L. Wolf et al., "Population-based MRI atlases of spatial distribution are specific to patient and tumor characteristics in glioblastoma," *NeuroImage: Clinical*, vol. 12, pp. 34–40, 2016.

8. S. P. Niclou, "Gauging heterogeneity in primary versus recurrent glioblastoma," *Neuro-Oncology*, vol. 17, pp. 907–909, 2015.

9. M. J. v. d. Bent, Y. Gao, M. Kerkhof, J. M. Kros, T. Gorlia, K. v. Zwieten et al., "Changes in the EGFR amplification and EGFRvIII expression between paired primary and recurrent glioblastomas "*Neuro-Oncology*, vol. 17, pp. 935–941, 2015.

10. P. C. Gedeon, B. D. Choi, J. H. Sampson, and D. D. Bigner, "Rindopepimut: Anti-EGFRvIII peptide vaccine, oncolytic," *Drugs Future*, vol. 38, pp. 147–155, 2013.

11. D. O'Rourke, A. Desai, J. Morrissette, M. Martinez-Lage, M. Nasrallah, S. Brem et al., "Pilot study of T cells redirected to EGFRvIII with a chimaric antigen receptor in patients with EGFRvIII+ glioblastoma," *Neuro-Oncology*, vol. 17, pp. v110–v111, 2015.

12. A. Ashraf, B. Gaonkar, C. Mies, A. DeMichele, M. Rosen, C. Davatzikos et al., "Breast DCE-MRI kinetic heterogeneity tumor markers: Preliminary associations with neoadjuvant chemotherapy response," *Translational Oncology*, vol. 8, pp. 154–162, 2015.

13. J. P. B. O'Connor, C. J. Rose, J. C. Waterton, R. A. D. Carano, G. J. M. Parker, and A. Jackson, "Imaging intratumor heterogeneity: Role in therapy response, resistance, and clinical outcome," *Clinical Cancer Research*, vol. 21, pp. 249–257, 2015.

14. R. Fisher, L. Pusztai, and C. Swanton, "Cancer heterogeneity: Implications for targeted therapeutics," *British Journal of Cancer*, vol. 108, pp. 479–485, 2013.

15. D. Zardavas, A. Irrthum, C. Swanton, and M. Piccart, "Clinical management of breast cancer heterogeneity," *Nature Reviews Clinical Oncology*, vol. 12, p. 381, 2015.

16. P. Therasse, S. G. Arbuck, E. A. Eisenhauer, J. Wanders, R. S. Kaplan, L. Rubinstein et al., "New guidelines to evaluate the response to treatment in solid tumors. European Organization for Research and Treatment of Cancer, National Cancer Institute of the United States, National Cancer Institute of Canada," *Journal of the National Cancer Institute*, vol. 92, pp. 205–216, 2000.

17. D. R. Macdonald, T. L. Cascino, S. C. Schold, Jr., and J. G. Cairncross, "Response criteria for phase II studies of supratentorial malignant glioma," *Journal of Clinical Oncology*, vol. 8, pp. 1277–1280, 1990.

18. P. Y. Wen, D. R. Macdonald, D. A. Reardon, T. F. Cloughesy, A. G. Sorensen, E. Galanis et al., "Updated response assessment criteria for high-grade gliomas: Response assessment in neuro-oncology working group," *Journal of Clinical Oncology*, vol. 28, pp. 1963–1972, 2010.

19. M. J. Vos, B. M. Uitdehaag, F. Barkhof, J. J. Heimans, H. C. Baayen, W. Boogerd et al., "Interobserver variability in the radiological assessment of response to chemotherapy in glioma," *Neurology*, vol. 60, pp. 826–830, 2003.

20. E. S. Burnside, E. A. Sickles, L. W. Bassett, D. L. Rubin, C. H. Lee, D. M. Ikeda et al., "The ACR BI-RADS(®) experience: Learning from history," *Journal of the American College of Radiology: JACR*, vol. 6, pp. 851–860, 2009.

21. J. G. R. Bomers and J. O. Barentsz, "Standardization of multiparametric prostate MR imaging using PI-RADS," *BioMed Research International*, vol. 2014, p. 9, 2014.

22. E. A. Kazerooni, M. R. Armstrong, J. K. Amorosa, D. Hernandez, L. A. Liebscher, H. Nath et al., "ACR CT accreditation program and the lung cancer screening program designation," *Journal of the American College of Radiology*, vol. 12, pp. 38–42, 2016.

23. C. C. Jaffe, "Imaging and genomics: Is there a synergy?" *Radiology*, vol. 264, pp. 329–331, 2012.

24. C. Proud, "Radiogenomics: The promise of personalized treatment in radiation oncology?" *Clinical Journal of Oncology Nursing*, vol. 18, p. 185, 2014.

25. B. S. Rosenstein, C. M. West, S. M. Bentzen, J. Alsner, C. N. Andreassen, D. Azria et al., "Radiogenomics: Radiobiology enters the era of big data and team science," *International Journal of Radiation Oncology, Biology, Physics*, vol. 89, pp. 709–713, 2014.

26. A. M. Rutman and M. D. Kuo, "Radiogenomics: Creating a link between molecular diagnostics and diagnostic imaging," *European Journal of Radiology*, vol. 70, pp. 232–241, 2009.

27. H. L. Aerts, "The potential of radiomic-based phenotyping in precision medicine: A review," *JAMA Oncology*, vol. 2, pp. 1636–1642, 2016.

28. B. M. Ellingson, "Radiogenomics and imaging phenotypes in glioblastoma: Novel observations and correlation with molecular characteristics," *Current Neurology and Neuroscience Reports*, vol. 15, p. 506, 2015.

29. O. Gevaert, L. A. Mitchell, A. S. Achrol, J. Xu, S. Echegaray, G. K. Steinberg et al., "Glioblastoma multiforme: Exploratory radiogenomic analysis by using quantitative image features," *Radiology*, vol. 273, pp. 168–174, 2014.

30. B. J. Gill, D. J. Pisapia, H. R. Malone, H. Goldstein, L. Lei, A. Sonabend et al., "MRI-localized biopsies reveal subtype-specific differences in molecular and cellular composition at the margins of glioblastoma," *Proceedings of the National Academy of Sciences USA*, vol. 111, pp. 12550–12555, 2014.

31. D. A. Gutman, L. A. D. Cooper, S. N. Hwang, C. A. Holder, J. Gao, T. D. Aurora et al., "MR imaging predictors of molecular profile and survival: Multi-institutional study of the TCGA glioblastoma data set," *Radiology*, vol. 267, pp. 560–569, 2013.

32. R. Jain, L. M. Poisson, D. Gutman, L. Scarpace, S. N. Hwang, C. A. Holder et al., "Outcome prediction in patients with glioblastoma by using imaging, clinical, and genomic biomarkers: Focus on the nonenhancing component of the tumor," *Radiology*, vol. 272, pp. 484–493, 2014.

33. M. A. Mazurowski, "Radiogenomics: What it is and why it is important," *Journal of the American College of Radiology*, vol. 12, pp. 862–866, 2015.

34. M. A. Mazurowski, A. Desjardins, and J. M. Malof, "Imaging descriptors improve the predictive power of survival models for glioblastoma patients," *Neuro-Oncology*, vol. 15, pp. 1389–1394, 2013.

35. M. Mahrooghy, A. B. Ashraf, D. Daye, E. S. McDonald, M. Rosen, C. Mies et al., "Pharmacokinetic tumor heterogeneity as a prognostic biomarker for classifying breast cancer recurrence risk," *IEEE Transactions on Biomedical Engineering*, vol. 62, pp. 1585–1594, 2015.

36. M. Mahrooghy, A. B. Ashraf, D. Daye, C. Mies, M. Feldman, M. Rosen et al., "Heterogeneity wavelet kinetics from DCE-MRI for classifying gene expression based breast cancer recurrence risk," *International Conference on Medical Image Computing and Computer-Assisted Intervention*, vol. 16, pp. 295–302, 2013.

37. H. Li, Y. Zhu, E. S. Burnside, E. Huang, K. Drukker, K. A. Hoadley et al., "Quantitative MRI radiomics in the prediction of molecular classifications of breast cancer subtypes in the TCGA/TCIA data set," *NPJ Breast Cancer*, vol. 2, 2016.

38. H. Li, Y. Zhu, E. S. Burnside, K. Drukker, K. A. Hoadley, C. Fan et al., "MR imaging radiomics signatures for predicting the risk of breast cancer recurrence as given by research versions of MammaPrint, oncotype DX, and PAM50 gene assays," *Radiology*, vol. 281, pp. 382–391, 2016.

39. A. Gooya, G. Biros, and C. Davatzikos, "Deformable registration of glioma images using EM algorithm and diffusion reaction modeling," *IEEE Transactions on Medical Imaging*, vol. 30, pp. 375–390, 2011.

40. A. Gooya, K. M. Pohl, M. Bilello, L. Cirillo, G. Biros, E. R. Melhem et al., "GLISTR: Glioma image segmentation and registration," *IEEE Transactions on Medical Imaging*, vol. 31, pp. 1941–1954, 2012.

41. M. Agn, O. Puonti, P. M. af Rosenschöld, I. Law, and K. V. Leemput, "Brain tumor segmentation using a generative model with an RBM prior on tumor shape," *Brainlesion: Glioma, Multiple Sclerosis, Stroke and Traumatic Brain Injuries*, vol. 9556, pp. 168–180, 2016.

42. B. H. Menze, K. van Leemput, D. Lashkari, M.-A. Weber, N. Ayache, and P. Golland, "A generative model for brain tumor segmentation in multi-modal images," in *Medical Image Computing and Computer-Assisted Intervention—MICCAI 2010: 13th International Conference, Beijing, China, September 20–24, 2010, Proceedings, Part II*, T. Jiang, N. Navab, J. P. W. Pluim, and M. A. Viergever, Eds., Berlin, Germany: Springer, 2010, pp. 151–159.

43. K. R. Swanson, E. C. Alvord, and J. D. Murray, "A quantitative model for differential motility of gliomas in grey and white matter," *Cell Proliferation*, vol. 33, pp. 317–329, 2000.

44. K. R. Swanson, C. Bridge, J. D. Murray, and E. C. Alvord, "Virtual and real brain tumors. Using mathematical modeling to quantify glioma growth and invasion," *Journal of the Neurological Sciences*, vol. 216, pp. 1–10, 2003.

45. K. Kamnitsas, C. Ledig, V. F. J. Newcombe, J. P. Simpson, A. D. Kane, D. K. Menon et al., "Efficient multi-scale 3D CNN with fully connected CRF for accurate brain lesion segmentation," *Medical Image Analysis*, vol. 36, pp. 61–78, 2017.

46. Y. Freund, R. Schapire, and N. Abe, "A short introduction to boosting," *Journal-Japanese Society for Artificial Intelligence*, vol. 14, p. 1612, 1999.

47. P. Dvorak and B. H. Menze, "Structured prediction with convolutional neural networks for multimodal brain tumor segmentation," *Presented at the in Proceedings of the Multimodal Brain Tumor Image Segmentation Challenge held in conjunction with MICCAI 2015 (MICCAI-BRATS 2015)*, Technische Universität München (T.U.M.), Munich, Germany, 2015.

48. M. Havaei, F. Dutil, C. Pal, H. Larochelle, and P.-M. Jodoin, "A convolutional neural network approach to brain tumor segmentation," *Brainlesion: Glioma, Multiple Sclerosis, Stroke and Traumatic Brain Injuries*, vol. 9556, pp. 195–208, 2016.

49. S. Pereira, A. Pinto, V. Alves, and C. A. Silva, "Deep convolutional neural networks for the segmentation of gliomas in multi-sequence MRI," *Brainlesion: Glioma, Multiple Sclerosis, Stroke and Traumatic Brain Injuries*, vol. 9556, pp. 131–143, 2016.

50. S. Pereira, A. Pinto, V. Alves, and C. A. Silva, "Brain tumor segmentation using convolutional neural networks in MRI images," *IEEE Transactions on Medical Imaging*, vol. 35, pp. 1240–1251, 2016.

51. V. Rao, M. S. Sarabi, and A. Jaiswal, "Brain tumor segmentation with deep learning," *Presented at the in Proceedings of the Multimodal Brain Tumor Image Segmentation Challenge held in conjunction with MICCAI 2015 (MICCAI-BRATS 2015)*, Technische Universität München (T.U.M.), Munich, Germany, 2015.

52. W. S. McCulloch and W. Pitts, "A logical calculus of the ideas immanent in nervous activity," *Bulletin of Mathematical Biophysics*, vol. 5, p. 115, 1943.

53. S. Kullback and R. A. Leibler, "On information and sufficiency," *The Annals of Mathematical Statistics*, vol. 22, pp. 79–86, 1951.

54. M. Havaei, A. Davy, D. Warde-Farley, A. Biard, A. Courville, Y. Bengio et al., "Brain tumor segmentation with deep neural networks," *Medical Image Analysis*, vol. 35, pp. 18–31, 2017.

55. J. Long, E. Shelhamer, and T. Darrell, "Fully convolutional networks for semantic segmentation," *Proceedings of the IEEE Conference on Computer Vision and Pattern Recognition*, pp. 3431–3440, 2015.

56. L.-C. Chen, G. Papandreou, I. Kokkinos, K. Murphy, and A. L. Yuille, "Deeplab: Semantic image segmentation with deep convolutional nets, atrous convolution, and fully connected crfs," arXiv preprint arXiv:1606.00915, 2016.

57. Z. Liu, X. Li, P. Luo, C.-C. Loy, and X. Tang, "Semantic image segmentation via deep parsing network," *Proceedings of the IEEE International Conference on Computer Vision*, pp. 1377–1385, 2015.

58. S. Zheng, S. Jayasumana, B. Romera-Paredes, V. Vineet, Z. Su, D. Du et al., "Conditional random fields as recurrent neural networks," *Proceedings of the IEEE International Conference on Computer Vision*, pp. 1529–1537, 2015.

59. S. Bakas, H. Akbari, A. Sotiras, M. Bilello, M. Rozycki, J. S. Kirby et al., "Advancing the cancer genome atlas glioma MRI collections with expert segmentation labels and radiomic features," *Nature Scientific Data*, vol. 4, pp. 170117, 2017.

60. S. Bakas, K. Zeng, A. Sotiras, S. Rathore, H. Akbari, B. Gaonkar et al., "Segmentation of gliomas in multimodal magnetic resonance imaging volumes based on a hybrid generative-discriminative framework," *Presented at the in Proceedings of the Multimodal Brain Tumor Image Segmentation Challenge held in conjunction with MICCAI 2015 (MICCAI-BRATS 2015)*, Technische Universität München (T.U.M.), Munich, Germany, 2015.

61. S. Bakas, K. Zeng, A. Sotiras, S. Rathore, H. Akbari, B. Gaonkar et al., "GLISTRboost: Combining multimodal MRI segmentation, registration, and biophysical tumor growth modeling with gradient boosting machines for glioma segmentation," *Brainlesion: Glioma, Multiple Sclerosis, Stroke and Traumatic Brain Injuries*, vol. 9556, pp. 144–155, 2016.

62. K. Zeng, S. Bakas, A. Sotiras, H. Akbari, M. Rozycki, S. Rathore et al., "Segmentation of gliomas in pre-operative and post-operative multimodal magnetic resonance imaging volumes based on a hybrid generative-discriminative framework," *Brainlesion: Glioma, Multiple Sclerosis, Stroke and Traumatic Brain Injuries,* vol. 10154, pp. 184–194, 2017.

63. B. H. Menze, K. V. Leemput, D. Lashkari, T. Riklin-Raviv, E. Geremia, E. Alberts et al., "A generative probabilistic model and discriminative extensions for brain lesion segmentation–With application to tumor and stroke," *IEEE Transactions on Medical Imaging,* vol. 35, pp. 933–946, 2016.

64. L. G. H. Derek, G. B. Philipp, H. Mark, and J. H. David, "Medical image registration," *Physics in Medicine & Biology,* vol. 46, p. R1, 2001.

65. R. Bar-Shalom, N. Yefremov, L. Guralnik, D. Gaitini, A. Frenkel, A. Kuten, et al., "Clinical performance of PET/CT in evaluation of cancer: Additional value for diagnostic imaging and patient management," *Journal of Nuclear Medicine,* vol. 44, pp. 1200–1209, 2003.

66. S. Ozer, D. L. Langer, X. Liu, M. A. Haider, T. H. van der Kwast, A. J. Evans et al., "Supervised and unsupervised methods for prostate cancer segmentation with multispectral MRI," *Medical Physics,* vol. 37, pp. 1873–1883, 2010.

67. S. Ruan, S. Lebonvallet, A. Merabet, and J. M. Constans, "Tumor segmentation from a multispectral MRI images by using support vector machine classification," *2007 4th IEEE International Symposium on Biomedical Imaging: From Nano to Macro,* pp. 1236–1239, 2007.

68. M. Mamede, P. Abreu-e-Lima, M. R. Oliva, V. Nosé, H. Mamon, and V. H. Gerbaudo, "FDG-PET/CT tumor segmentation-derived indices of metabolic activity to assess response to neoadjuvant therapy and progression-free survival in esophageal cancer: Correlation with histopathology results," *American Journal of Clinical Oncology,* vol. 30, pp. 377–388, 2007.

69. Y. Ou, S. P. Weinstein, E. F. Conant, S. Englander, X. Da, B. Gaonkar et al., "Deformable registration for quantifying longitudinal tumor changes during neoadjuvant chemotherapy," *Magnetic Resonance in Medicine,* vol. 73, pp. 2343–2356, 2015.

70. Y. Ou, A. Sotiras, N. Paragios, and C. Davatzikos, "DRAMMS: Deformable registration via attribute matching and mutual-saliency weighting," *Medical Image Analysis,* vol. 15, pp. 622–639, 2011.

71. D. L. Pham, C. Xu, and J. L. Prince, "Current methods in medical image segmentation," *Annual Review of Biomedical Engineering,* vol. 2, pp. 315–337, 2000.

72. L. J. Stapleford, J. D. Lawson, C. Perkins, S. Edelman, L. Davis, M. W. McDonald et al., "Evaluation of automatic atlas-based lymph node segmentation for head-and-neck cancer," *International Journal of Radiation Oncology, Biology, Physics,* vol. 77, pp. 959–966, 2010.

73. M. Prastawa, E. Bullitt, S. Ho, and G. Gerig, "A brain tumor segmentation framework based on outlier detection," *Medical Image Analysis,* vol. 8, pp. 275–283, 2004.

74. M. Prastawa, E. Bullitt, N. Moon, K. Van Leemput, and G. Gerig, "Automatic brain tumor segmentation by subject specific modification of atlas priors," *Academic Radiology,* vol. 10, pp. 1341–1348, 2003.

75. E. R. Velazquez, H. J. Aerts, Y. Gu, D. B. Goldgof, D. De Ruysscher, A. Dekker et al., "A semiautomatic CT-based ensemble segmentation of lung tumors: Comparison with oncologists' delineations and with the surgical specimen," *Radiotherapy and Oncology,* vol. 105, pp. 167–173, 2012.

76. I. E. van Dam, J. R. v. S. de Koste, G. G. Hanna, R. Muirhead, B. J. Slotman, and S. Senan, "Improving target delineation on 4-dimensional CT scans in stage I NSCLC using a deformable registration tool," *Radiotherapy and Oncology,* vol. 96, pp. 67–72, 2010.

77. M. A. Deeley, A. Chen, R. Datteri, J. H. Noble, A. J. Cmelak, E. F. Donnelly et al., "Comparison of manual and automatic segmentation methods for brain structures in the presence of space-occupying lesions: A multi-expert study," *Physics in Medicine and Biology,* vol. 56, pp. 4557–4577, 2011.

78. G. P. Mazzara, R. P. Velthuizen, J. L. Pearlman, H. M. Greenberg, and H. Wagner, "Brain tumor target volume determination for radiation treatment planning through automated MRI segmentation," *International Journal of Radiation Oncology, Biology, Physics,* vol. 59, pp. 300–312, 2004.

79. G. Rote, "Computing the minimum Hausdorff distance between two point sets on a line under translation," *Information Processing Letters,* vol. 38, pp. 123–127, 1991.

80. M. D. Wilkinson, M. Dumontier, I. J. Aalbersberg, G. Appleton, M. Axton, A. Baak et al., "The FAIR guiding principles for scientific data management and stewardship," *Scientific Data*, vol. 3, p. 160018, 2016.

81. S. Bakas, M. Reyes, A. Jakab, S. Bauer, M. Rempfler, A. Crimi et al., "Identifying the best machine learning algorithms for brain tumor segmentation, progression assessment, and overall survival prediction in the BRATS challenge," *arXiv preprint arXiv:1811.02629*, 2018.

82. K. Clark, B. Vendt, K. Smith, J. Freymann, J. Kirby, P. Koppel et al., "The Cancer Imaging Archive (TCIA): Maintaining and operating a public information repository," *Journal of Digital Imaging*, vol. 26, pp. 1045–1057, 2013.

83. D. Cobzas, N. Birkbeck, M. Schmidt, M. Jagersand, and A. Murtha, "3D variational brain tumor segmentation using a high dimensional feature set," *Computer Vision, IEEE 11th International Conference on ICCV*, pp. 1–8, 14–21, 2007.

84. S. Reza and K. Iftekharuddin, "Improved brain tumor tissue segmentation using texture features," *MICCAI-BRATS Challenge on Multimodal Brain Tumor Segmentation*, 2014.

85. M. Agn, O. Puonti, I. Law, P. M. af Rosenschold, and K. V. Leemput, "Brain tumor segmentation by a generative model with a prior on tumor shape," *Presented at the in Proceedings of the Multimodal Brain Tumor Image Segmentation Challenge held in conjunction with MICCAI 2015 (MICCAI-BRATS 2015)*, Technische Universität München (T.U.M.), Munich, Germany, 2015.

86. M. Havaei, F. Dutil, C. Pal, H. Larochelle, and P.-M. Jodoin, "A convolutional neural network approach to brain tumor segmentation," *Presented at the in Proceedings of the Multimodal Brain Tumor Image Segmentation Challenge held in conjunction with MICCAI 2015 (MICCAI-BRATS 2015)*, Technische Universität München (T.U.M.), Munich, Germany, 2015.

87. K. Vaidhya, R. Santhosh, S. Thirunavukkarasu, V. Alex, and G. Krishnamurthi, "Multi-modal brain tumor segmentation using stacked denoising autoencoders," *Presented at the in Proceedings of the Multimodal Brain Tumor Image Segmentation Challenge held in conjunction with MICCAI 2015 (MICCAI-BRATS 2015)*, Technische Universität München (T.U.M.), Munich, Germany, 2015.

88. R. Meier, V. Karamitsou, S. Habegger, R. Wiest, and M. Reyes, "Parameter learning for CRF-based tissue segmentation of brain tumors," *Presented at the in Proceedings of the Multimodal Brain Tumor Image Segmentation Challenge held in conjunction with MICCAI 2015 (MICCAI-BRATS 2015)*, Technische Universität München (T.U.M.), Munich, Germany, 2015.

89. R. Meier, S. Bauer, J. Slotboom, R. Wiest, and M. Reyes, "Appearance- and context-sensitive features for brain tumor segmentation," *Presented at the in Proceedings of the Multimodal Brain Tumor Image Segmentation Challenge held in conjunction with MICCAI 2014 (MICCAI-BRATS 2014)*, 2014.

90. P. Chang, "Fully convolutional neural networks with hyperlocal features for brain tumor segmentation," *Presented at the in Proceedings of the Multimodal Brain Tumor Image Segmentation Challenge held in conjunction with MICCAI 2015 (MICCAI-BRATS 2016)*, Athens, Greece, 2016.

91. C. Hogea, G. Biros, F. Abraham, and C. Davatzikos, "A robust framework for soft tissue simulations with application to modeling brain tumor mass effect in 3D MR images," *Physics in Medicine and Biology*, vol. 52, p. 6893, 2007.

92. C. Hogea, C. Davatzikos, and G. Biros, "Brain-tumor interaction biophysical models for medical image registration," *SIAM Journal on Scientific Computing*, vol. 30, pp. 3050–3072, 2008.

93. C. Hogea, C. Davatzikos, and G. Biros, "An image-driven parameter estimation problem for a reaction–diffusion glioma growth model with mass effects," *Journal of Mathematical Biology*, vol. 56, pp. 793–825, 2008.

94. J. H. Friedman, "Greedy function approximation: A gradient boosting machine," *Annals of Statistics*, pp. 1189–1232, 2001.

95. J. H. Friedman, "Stochastic gradient boosting," *Computational Statistics & Data Analysis*, vol. 38, pp. 367–378, 2002.

96. S. Bakas, K. Chatzimichail, G. Hunter, B. Labbe, P. S. Sidhu, and D. Makris, "Fast semi-automatic segmentation of focal liver lesions in contrast-enhanced ultrasound, based on a probabilistic model," *TCIV Computer Methods in Biomechanics and Biomedical Engineering: Imaging & Visualization*, vol. 5, pp. 329–338, 2017.

97. B. M. Keller, J. Chen, D. Daye, E. F. Conant, and D. Kontos, "Preliminary evaluation of the publicly available laboratory for breast radiodensity assessment (LIBRA) software tool: Comparison of fully

automated area and volumetric density measures in a case–control study with digital mammography," *Breast Cancer Research*, vol. 17, p. 117, 2015.

98. B. M. Keller, D. L. Nathan, Y. Wang, Y. Zheng, J. C. Gee, E. F. Conant et al., "Estimation of breast percent density in raw and processed full field digital mammography images via adaptive fuzzy c-means clustering and support vector machine segmentation," *Medical Physics*, vol. 39, pp. 4903–4917, 2012.

99. W. Chen, M. L. Giger, and U. Bick, "A Fuzzy C-Means (FCM)-based approach for computerized segmentation of breast lesions in dynamic contrast-enhanced MR images," *Academic Radiology*, vol. 13, pp. 63–72, 2006.

100. A. B. Ashraf, S. C. Gavenonis, D. Daye, C. Mies, M. A. Rosen, and D. Kontos, "A multichannel Markov random field framework for tumor segmentation with an application to classification of gene expression-based breast cancer recurrence risk," *IEEE Transactions on Medical Imaging*, vol. 32, pp. 637–648, 2013.

101. H. Zaidi and I. El Naqa, "PET-guided delineation of radiation therapy treatment volumes: A survey of image segmentation techniques," *European Journal of Nuclear Medicine and Molecular Imaging*, vol. 37, pp. 2165–2187, 2010.

102. X. Xi, H. Shi, L. Han, T. Wang, H. Y. Ding, G. Zhang et al., "Breast tumor segmentation with prior knowledge learning," *Neurocomputing*, vol. 237, pp. 145–157, 2017.

103. N. Yu, J. Wu, S. P. Weinstein, B. Gaonkar, B. M. Keller, A. B. Ashraf et al., "A superpixel-based framework for automatic tumor segmentation on breast DCE-MRI," in *SPIE Medical Imaging*, Orlando, FL, 2015, pp. 94140O-94140O-7, 2015.

104. Y. C. Lin, Y. L. Huang, and D. R. Chen, "Breast tumor segmentation based on level-set method in 3d sonography," in *2013 Seventh International Conference on Innovative Mobile and Internet Services in Ubiquitous Computing*, pp. 637–640, 2013.

105. H. M. Moftah, A. T. Azar, E. T. Al-Shammari, N. I. Ghali, A. E. Hassanien, and M. Shoman, "Adaptive k-means clustering algorithm for MR breast image segmentation," *Neural Computing and Applications*, vol. 24, pp. 1917–1928, 2014.

106. P. M. Rothwell, "Prognostic models," *Practical Neurology*, vol. 8, pp. 242–253, 2008.

107. H. Hemingway, P. Croft, P. Perel, J. A. Hayden, K. Abrams, A. Timmis et al., "Prognosis research strategy (PROGRESS) 1: A framework for researching clinical outcomes," *BMJ*, vol. 346, p. e5595, 2013.

108. S. Chicklore, V. Goh, M. Siddique, A. Roy, P. K. Marsden, and G. J. Cook, "Quantifying tumour heterogeneity in 18F-FDG PET/CT imaging by texture analysis," *European Journal of Nuclear Medicine and Molecular Imaging*, vol. 40, pp. 133–140, 2013.

109. D. Lardinois, W. Weder, T. F. Hany, E. M. Kamel, S. Korom, B. Seifert et al., "Staging of non–small-cell lung cancer with integrated positron-emission tomography and computed tomography," *New England Journal of Medicine*, vol. 348, pp. 2500–2507, 2003.

110. H. Y. Lee, H. J. Lee, Y. T. Kim, C. H. Kang, B. G. Jang, D. H. Chung et al., "Value of combined interpretation of computed tomography response and positron emission tomography response for prediction of prognosis after neoadjuvant chemotherapy in non-small cell lung cancer," *Journal of Thoracic Oncology*, vol. 5, pp. 497–503, 2010.

111. Z. D. Horne, D. A. Clump, J. A. Vargo, S. Shah, S. Beriwal, S. A. Burton et al., "Pretreatment SUV max predicts progression-free survival in early-stage non-small cell lung cancer treated with stereotactic body radiation therapy," *Radiation Oncology*, vol. 9, p. 41, 2014.

112. A. Depeursinge, A. Foncubierta-Rodriguez, D. Van De Ville, and H. Müller, "Three-dimensional solid texture analysis in biomedical imaging: Review and opportunities," *Medical Image Analysis*, vol. 18, pp. 176–196, 2014.

113. G. Lee, H. Y. Lee, H. Park, M. L. Schiebler, E. J. van Beek, Y. Ohno et al., "Radiomics and its emerging role in lung cancer research, imaging biomarkers and clinical management: State of the art," *European Journal of Radiology*, vol. 86, pp. 297–307, 2017.

114. M. Scrivener, E. E. de Jong, J. E. van Timmeren, T. Pieters, B. Ghaye, and X. Geets, "Radiomics applied to lung cancer: A review," *Translational Cancer Research*, vol. 5, pp. 398–409, 2016.

115. H. Zhang, K. Wroblewski, S. Liao, R. Kampalath, D. C. Penney, Y. Zhang et al., "Prognostic value of metabolic tumor burden from 18 F-FDG PET in surgical patients with non–small-cell lung cancer," *Academic Radiology*, vol. 20, pp. 32–40, 2013.

116. B. Foster, U. Bagci, A. Mansoor, Z. Xu, and D. J. Mollura, "A review on segmentation of positron emission tomography images," *Computers in Biology and Medicine*, vol. 50, pp. 76–96, 2014.

117. A. Mansoor, U. Bagci, B. Foster, Z. Xu, G. Z. Papadakis, L. R. Folio et al., "Segmentation and image analysis of abnormal lungs at CT: Current approaches, challenges, and future trends," *RadioGraphics*, vol. 35, pp. 1056–1076, 2015.

118. U. Bagci, J. K. Udupa, N. Mendhiratta, B. Foster, Z. Xu, J. Yao et al., "Joint segmentation of anatomical and functional images: Applications in quantification of lesions from PET, PET-CT, MRI-PET, and MRI-PET-CT images," *Medical Image Analysis*, vol. 17, pp. 929–945, 2013.

119. D. Ciardo, M. A. Gerardi, S. Vigorito, A. Morra, V. Dell'acqua, F. J. Diaz et al., "Atlas-based segmentation in breast cancer radiotherapy: Evaluation of specific and generic-purpose atlases," *The Breast*, vol. 32, pp. 44–52, 2017.

120. A.-S. Dewalle-Vignion, N. Betrouni, C. Baillet, and M. Vermandel, "Is STAPLE algorithm confident to assess segmentation methods in PET imaging?" *Physics in Medicine and Biology*, vol. 60, p. 9473, 2015.

121. A. R. Eldesoky, E. S. Yates, T. B. Nyeng, M. S. Thomsen, H. M. Nielsen, P. Poortmans et al., "Internal and external validation of an ESTRO delineation guideline–dependent automated segmentation tool for loco-regional radiation therapy of early breast cancer," *Radiotherapy and Oncology*, vol. 121, pp. 424–430, 2016.

122. C.-J. Tao, J.-L. Yi, N.-Y. Chen, W. Ren, J. Cheng, S. Tung et al., "Multi-subject atlas-based auto-segmentation reduces interobserver variation and improves dosimetric parameter consistency for organs at risk in nasopharyngeal carcinoma: A multi-institution clinical study," *Radiotherapy and Oncology*, vol. 115, pp. 407–411, 2015.

123. S. Ruan, S. Lebonvallet, A. Merabet, and J.-M. Constans, "Tumor segmentation from a multispectral MRI images by using support vector machine classification," in Biomedical Imaging: From Nano to Macro, 2007. ISBI 2007. 4th IEEE International Symposium on, pp. 1236–1239, 2007.

124. J. Sykes, "Reflections on the current status of commercial automated segmentation systems in clinical practice," *Journal of Medical Radiation Sciences*, vol. 61, pp. 131–134, 2014.

125. S. L. A. Lee, A. Z. Kouzani, and E. J. Hu, "Automated detection of lung nodules in computed tomography images: A review," *Machine Vision and Applications*, vol. 23, pp. 151–163, 2012.

126. F. Shaukat, G. Raja, A. Gooya, and A. F. Frangi, "Fully automatic and accurate detection of lung nodules in CT images using a hybrid feature set," *Medical Physics*, vol. 44, pp. 3615–3629, 2017.

127. Q. Song, J. Bai, D. Han, S. Bhatia, W. Sun, W. Rockey et al., "Optimal co-segmentation of tumor in PET-CT images with context information," *IEEE Transactions on Medical Imaging*, vol. 32, pp. 1685–1697, 2013.

128. A. Krizhevsky, I. Sutskever, and G. E. Hinton, "ImageNet classification with deep convolutional neural networks," in *Advances in Neural Information Processing Systems*, pp. 1097–1105, 2012.

129. R. Girshick, J. Donahue, T. Darrell, and J. Malik, "Rich feature hierarchies for accurate object detection and semantic segmentation," in *Proceedings of the IEEE Conference on Computer Vision and Pattern Recognition*, pp. 580–587, 2014.

130. S. Pereira, A. Pinto, V. Alves, and C. A. Silva, "Deep convolutional neural networks for the segmentation of gliomas in multi-sequence MRI," *Presented at the in Proceedings of the Multimodal Brain Tumor Image Segmentation Challenge held in conjunction with MICCAI 2015 (MICCAI-BRATS 2015)*, Technische Universität München (T.U.M.), Munich, Germany, 2015.

131. H. Jiang, H. Ma, W. Qian, M. Gao, and Y. Li, "An automatic detection system of lung nodule based on multi-group patch-based deep learning network," *IEEE Journal of Biomedical and Health Informatics*, vol. 22, pp. 1227–1237, 2017.

132. S. Wang, M. Zhou, Z. Liu, Z. Liu, D. Gu, Y. Zang et al., "Central focused convolutional neural networks: Developing a data-driven model for lung nodule segmentation," *Medical Image Analysis*, vol. 40, pp. 172–183, 2017.

Habitat imaging of tumor evolution by magnetic resonance imaging (MRI)

BRUNA VICTORASSO JARDIM-PERASSI, GARY MARTINEZ, AND ROBERT GILLIES

Magnetic Resonance Imaging (MRI) offers images with exquisite soft tissue contrast provided by multiple contrast mechanisms. These yield insight into biophysical properties, which in turn report on the physiological status of the tissue in question. Multiple MR imaging pulse-sequences simultaneously sample information about the tissue, and when used in combination, may be capable of discerning subtle physiological differences. Habitat imaging (HI) is a technique that aspires to capture these subtle differences in tumors that can be related back to histopathological differences in viability and immunohistochemical (IHC) differences, such as staining for hypoxia or acidosis. As such, it can give a real-time and longitudinal view of tumor physiology and its changes with growth or response to therapy. In order to achieve this, anatomical and functional imaging sequence-datasets are simultaneously applied and incorporated into an analytical framework to separate subdomains of a tumor. These different regions experience disparate evolutionary selection pressures, which have been the subject of numerous hypotheses (Gatenby and Gillies, 2004, 2008; Gerlinger, 2012; Gerlinger and Swanton, 2010; Merlo et al., 2006; Nowell, 1976). The fact that a quantitative view of tumor evolution may be possible is encouraging and may lay the groundwork for more advanced metrics of tumor response. An especially useful example in this regard is in a recent therapeutic regimen that relies upon Darwinian dynamics to inform the drug dosing schedules (Enriquez-Navas et al., 2016; Zhang et al., 2017). Habitats reflect subtly different regions in a tumor and distinction between these should be made with caution. Therefore, multiparametric MRI (mpMRI) leveraged with the ground truth provided by histology/IHC may be the minimum essential components needed to discriminate habitats.

In routine clinical oncological and pre-clinical imaging studies, a variety of pulse sequences are frequently used to generate basic anatomic contrast, whereas more advanced functional techniques, such as diffusion and perfusion, may add significant value to imaging (Hyare et al., 2017; Marino et al., 2017; O'Connor et al., 2015). The apparent diffusion coefficient (ADC) map is calculated using multiple diffusion weightings (b-values). It reflects water motion, which can be restricted by cellular membranes and hence can reflect tumor cellularity. In the general collective, it has been postulated that ADC values can increase as the cell density decreases (Marino et al., 2017; Winfield et al., 2016). However, mean ADC does not necessarily reflect cellularity in all tumors (Surov et al., 2017). For example, LaViolette et al. (2014) found both hypercellularity and necrosis in diffusion-restricted regions in glioma, which was confirmed as coagulative necrosis in co-registered histology and IHC for vascular endothelial growth factor (VEGF), likely due to extreme hypoxia. Other histological features may also influence the ADC, such as nuclear volume, extracellular matrix/parenchyma ratio, and the density of microvessels (Surov et al., 2017).

Other useful functional parameters may be obtained by dynamic contrast-enhanced (DCE) MRI, which consists of obtaining spin-lattice relaxation time (T1)-weighted images in a time series before and after contrast agent administration, and allows assessment of blood flow and vascular permeability (Price et al., 2013). Different parameters can be obtained from DCE-MRI (Cabarrus and Westphalen, 2017; Price et al., 2013; Winfield et al., 2016), which may be informative of angiogenesis and hypoxia in solid tumors (Brindle, 2008; Chang et al., 2017). Clinical (Newbold et al., 2009) and pre-clinical studies (Egeland et al., 2012) have correlated DCE-MRI parameters with intratumoral hypoxia, as detected in histological sections using a hypoxia marker (pimonidazole) (Jackson et al., 2007).

Thus, instead of the tumor size measurement using anatomic T2-weighted images, different MRI sequences can be used to assess tissue metabolism and physiology, potentially revealing additional information relevant to therapy response and cancer progression (Gillies and Beyer, 2016; O'Connor et al., 2015). In habitat imaging, multiple MRI parameter maps are combined and simultaneously classified, on a pixel-by-pixel basis, to identify different tumor cell subpopulations, rather than conventional analysis that averages MRI parameter for an entire region. By corroborating functional imaging to histopathology, a clear method toward identification of habitats may allow for the eventual application of this technique to evaluate evolutionary dynamics within a tumor.

The idea of using multiple MRI parameters has been around for quite some time. In the 1980s, mpMRI data were first segmented into tissue populations by adapting techniques used for satellite image analyses (Vannier et al., 1985, 1987, 1991). More recently, tumors have been segmented into subregions using specific MRI sequences or different mpMRI combinations (Carano et al., 2004; Henning et al., 2007a, 2007b).

Studies based on DCE-MRI have also identified noteworthy tumor subregions (Chang et al., 2017; Chaudhury et al., 2015; Han et al., 2017; Stoyanova et al., 2012). As tumor oxygenation is dependent on microcirculation, regions with inadequate perfusion could be indicative of hypoxic areas. Hypoxia also increases vascular endothelial growth factor which is also known as vascular permeability factor, which will increase contrast extravasation (Bhujwalla et al., 2001). Thus, tumors may be segmented into well-perfused, hypoxic, and necrotic subregions (Chang et al., 2017). Since hypoxia is associated with more aggressive tumor phenotypes, the detection of hypoxic regions *in vivo* is of great relevance to determinate prognosis and therapeutic outcomes, including response to hypoxia-activated pro-drugs, HAPs. Features extracted from these subregions with distinct blood flow patterns showed additional prognostic value combined with other clinicopathologic factors (Chang et al., 2016; Wu et al., 2017). With regard to anti-angiogenic therapy response, spatially heterogeneous changes that were evaluated by DCE-MRI were also shown to be informative compared to the averaged values of the whole tumor (Longo et al., 2015). Recently, Featherstone and co-workers used a data-driven method to cluster the tumors of two xenograft models into six subregions based on perfusion (DCE-MRI) and tissue oxygenation (oxygen-enhanced [OE]-MRI). Authors inferred these clusters as necrotic, hypoxic, and viable subregions, however, they assumed that biological interpretation regarding these subregions are compromised by the lack of corroboration through histological information (Featherstone et al., 2018).

Low perfusion regions could indicate a habitat of cells that are adapted to survive in hypoxic and acidic conditions. These regions may be inferred by different mpMRI combinations, for example, as shown in Figure 7.1, where a habitat map was created by clustering T2 map, T2* map, diffusion-weighted imaging (DWI), and DCE-MRI data. Application of pixel-by-pixel clustering, in a multidimensional space, leverages the information content that is present in the individual pulse-sequence images. These clustering algorithms reveal the complexity of spatial heterogeneity, and the co-registration of MRI-derived habitat maps with histology is crucial to confirm the cellular phenotypes. In the context of intratumoral heterogeneity, clustering algorithms that allow the mixtures of clusters (habitats), such as Gaussian mixture model, are generally more informative than hard-clustering algorithms (Figure 7.2). Figures 7.1 and 7.2 show four distinct habitats detected by mpMRI, which were spatially corroborated by habitats identified histologically in a pre-clinical model of breast cancer. These maps identified both low and high ADC values in regions of necrosis.

Other groups using different pre-clinical models have identified multiple tumor regions using k-means clustering of ADC, T2, and proton density (M0), which was correlated with histology (Barck et al., 2009; Carano et al., 2004; Henning et al., 2007a). In these studies, two viable and two necrotic regions were identified. Both necrotic regions showed high ADC values, but differed in the T2 values. Areas of viable tumor cells exhibited low ADC values, as well as areas containing subcutaneous adipose tissue, which were differentiated by higher M0 values. These regions were differentiated in well-oxygenated or hypoxic areas, based on the hypoxia-inducible factor (HIF-1) staining in histology, and responded differently to radiotherapy (Henning et al., 2007b). The capacity of mpMRI to detect tumor regions with differential responses was also demonstrated in another study, in which a quantitative parameter, K^{trans}, was derived from DCE-MRI. It showed reduction in tumor regions previously classified as viable by clustering DWI-MRI, T2, and proton density, while it did not change in the necrotic regions following antiangiogenic therapy (Berry et al., 2008). More recently, Divine et al. (2016) showed the spatial-temporal evolution of viable and necrotic tumor subregions combining 18F-fluorodeoxyglucose (FDG)-Positron Emission Tomography (PET) (18F-FDG-PET) and DWI-MRI data in a xenograft model of lung cancer, which were co-registered with regions manually classified on histological hematoxylin and eosin (H&E) staining.

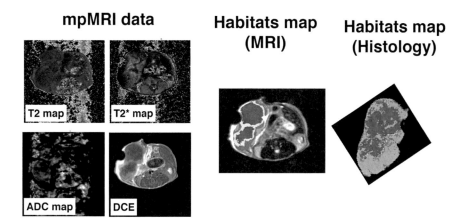

Figure 7.1 (See color insert.) Multiparametric MRI data were clustered to create a habitats map, which is corroborated by histology in a pre-clinical model of breast cancer (4T1). Histological images (H&E and immunohistochemical sections stained with a hypoxia marker pimonidazole) were segmented to generate a habitat map at histological level. Blue represents necrotic areas and green represents viable tumor cells; yellow represents a perinecrotic region stained for hypoxia, while pink represents regions of hypoxic viable tumor cells.

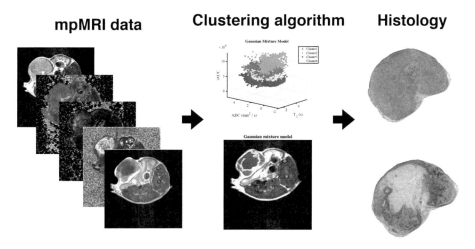

Figure 7.2 (See color insert.) Multiparametric MRI was used as raw data of different clustering algorithms to create habitats maps, which each cluster is represented by different colors. Evaluation of the corresponding histological slices is crucial to confirm the cellular phenotype of the habitats and determinate the best image segmentation algorithm. Example of a murine breast cancer (4T1) allograft in a mouse model. In the Gaussian mixture model, cluster green represents regions of viable tumor cells; clusters yellow and magenta represent regions with moderate enhancement in DCE-MRI and are confirmed as hypoxic in hypoxia (positive for pimonidazole); cluster blue represents region with non-viable tumor cells.

A series of studies have evaluated tumor subregions in glioblastoma (GBM) patients by MRI and correlated specific features of MRI-defined subregions with survival and molecular characteristics (Lee et al., 2015a, 2015b; Zhou et al., 2014, 2017). By using T1 post-gadolinium and either T2-weighted or fluid attenuated inversion recovery (FLAIR) sequences, GBM tumors can be segmented into different subregions. In particular, subregions with low blood flow (low enhancement in T1-post) and low cell density (high FLAIR signals) were mainly found in a group of patients with short-term survival (Zhou et al., 2014). Similarly, tumor subregions showing high T1-contrast-enhanced and low ADC values were characterized as high-risk intratumoral subregions and associated with survival in glioblastoma patients (Cui et al., 2017). In addition, Caulo et al. (2014) proposed a glioma-grading index, based on specific parameters assessed in different tumor regions identified by mpMRI. The index showed better sensitivity and specificity to differentiate low- and high-grade glioma, than each of the parameters individually, evidencing, once again, the importance to evaluate the tumor taking into account the intratumoral spatial heterogeneity.

Following these studies, Dextraze et al. (2017) used four MRI-sequences (pre-contrast T1-weighted, post-contrast T1, T2-weighted, and T2 FLAIR images) to segment into 16 clusters, based on high and low signal for each of the four sequences using Kmeans clustering. Three subregions were identified as showing association with overall survival, after adjusting for the clinical covariates, and called as "relevant habitats." In the last instance, a pathological interpretation was inferred for these habitats, based on the MRI data combinations and molecular pathways analyzed from genomic data. Heterogeneity features of different combinations of mpMRI were associated with histopathology and gene expression data in patients with hepatocellular carcinoma and with biological changes in response to treatment (Hectors et al., 2017). Interestingly, Yip et al. (2017) showed a reduction in the spatial heterogeneity after therapy with trastuzumab in a human epidermal growth factor receptor 2 (HER-2) positive esophageal tumor xenograft, while neither MRI nor histological parameters showed significant differences between the control and treated groups when evaluated across the whole tumor rather than subregions.

The advantage of HI is that it may have the capability to distinguish the aforementioned subtle differences because multiple independent MRI parameters are simultaneously used, whereas a single parameter, such as ADC or T2, may be insufficient. In addition, HI can cast these differences within a framework of Darwinian evolution. Therefore, HI has two essential components: (1) it requires that some form of classification or segmentation of regions, based on mpMRI data, be performed and (2) that these regions be relatable to tumor physiological habitats. Without (2), this would simply be mpMRI. Without (1), (2) may not be possible.

In summary, HI is the segmentation of regions within a tumor that experience different physiological microenvironments, which are corroborated by histopathology and immunohistochemistry. The work done so far shows that it possesses great promise as a way to discern subtle tumor physiological states and changes in state determined longitudinally. Furthermore, these are anticipated to provide dramatic improvements in the ability to observe the emergence of resistant populations within tumors.

REFERENCES

Barck, K. H., Willis, B., Ross, J., French, D. M., Filvaroff, E. H., and Carano, R. A. (2009). Viable tumor tissue detection in murine metastatic breast cancer by whole-body MRI and multispectral analysis. *Magn Reson Med 62*, 1423–1430.

Berry, L. R., Barck, K. H., Go, M. A., Ross, J., Wu, X., Williams, S. P., Gogineni, A., Cole, M. J., Van Bruggen, N., Fuh, G. et al. (2008). Quantification of viable tumor microvascular characteristics by multispectral analysis. *Magn Reson Med 60*, 64–72.

Bhujwalla, Z. M., Artemov, D., Natarajan, K., Ackerstaff, E., and Solaiyappan, M. (2001). Vascular differences detected by MRI for metastatic versus nonmetastatic breast and prostate cancer xenografts. *Neoplasia 3*, 143–153.

Brindle, K. (2008). New approaches for imaging tumour responses to treatment. *Nat Rev Cancer 8*, 94–107.

Cabarrus, M. C., and Westphalen, A. C. (2017). Multiparametric magnetic resonance imaging of the prostate-a basic tutorial. *Transl Androl Urol 6*, 376–386.

Carano, R. A., Ross, A. L., Ross, J., Williams, S. P., Koeppen, H., Schwall, R. H., and Van Bruggen, N. (2004). Quantification of tumor tissue populations by multispectral analysis. *Magn Reson Med 51*, 542–551.

Caulo, M., Panara, V., Tortora, D., Mattei, P. A., Briganti, C., Pravata, E., Salice, S., Cotroneo, A. R., and Tartaro, A. (2014). Data-driven grading of brain gliomas: A multiparametric MR imaging study. *Radiology 272*, 494–503.

Chang, R. F., Chen, H. H., Chang, Y. C., Huang, C. S., Chen, J. H., and Lo, C. M. (2016). Quantification of breast tumor heterogeneity for ER status, HER2 status, and TN molecular subtype evaluation on DCE-MRI. *Magn Reson Imaging 34*, 809–819.

Chang, Y. C., Ackerstaff, E., Tschudi, Y., Jimenez, B., Foltz, W., Fisher, C., Lilge, L., Cho, H., Carlin, S., Gillies, R. J., et al. (2017). Delineation of tumor habitats based on dynamic contrast enhanced MRI. *Sci Rep 7*, 9746.

Chaudhury, B., Zhou, M., Goldgof, D. B., Hall, L. O., Gatenby, R. A., Gillies, R. J., Patel, B. K., Weinfurtner, R. J., and Drukteinis, J. S. (2015). Heterogeneity in intratumoral regions with rapid gadolinium washout correlates with estrogen receptor status and nodal metastasis. *J Magn Reson Imaging 42*, 1421–1430.

Cui, Y., Ren, S. J., Tha, K. K., Wu, J., Shirato, H., and Li, R. J. (2017). Volume of high-risk intratumoral subregions at multi-parametric MR imaging predicts overall survival and complements molecular analysis of glioblastoma. *Eur Radiol 27*, 3583–3592.

Dextraze, K., Saha, A., Kim, D., Narang, S., Lehrer, M., Rao, A., Narang, S., Rao, D., Ahmed, S., Madhugiri, V. et al. (2017). Spatial habitats from multiparametric MR imaging are associated with signaling pathway activities and survival in glioblastoma. *Oncotarget 8*, 112992–113001.

Divine, M. R., Katiyar, P., Kohlhofer, U., Quintanilla-Martinez, L., Pichler, B. J., and Disselhorst, J. A. (2016). A population-based gaussian mixture model incorporating 18F-FDG PET and diffusion-weighted MRI quantifies tumor tissue classes. *J Nucl Med 57*, 473–479.

Egeland, T. A., Gulliksrud, K., Gaustad, J. V., Mathiesen, B., and Rofstad, E. K. (2012). Dynamic contrast-enhanced-MRI of tumor hypoxia. *Magn Reson Med 67*, 519–530.

Enriquez-Navas, P. M., Kam, Y., Das, T., Hassan, S., Silva, A., Foroutan, P., Ruiz, E., Martinez, G., Minton, S., Gillies, R. J., and Gatenby, R. A. (2016). Exploiting evolutionary principles to prolong tumor control in preclinical models of breast cancer. *Sci Transl Med 8*, 327ra324.

Featherstone, A. K., O'Connor, J. P. B., Little, R. A., Watson, Y., Cheung, S., Babur, M., Williams, K. J., Matthews, J. C., and Parker, G. J. M. (2018). Data-driven mapping of hypoxia-related tumor heterogeneity using DCE-MRI and OE-MRI. *Magn Reson Med 79*, 2236–2245.

Gatenby, R. A., and Gillies, R. J. (2004). Why do cancers have high aerobic glycolysis? *Nat Rev Cancer 4*, 891–899.

Gatenby, R. A., and Gillies, R. J. (2008). A microenvironmental model of carcinogenesis. *Nat Rev Cancer 8*, 56–61.

Gerlinger, M. (2012). Intratumor heterogeneity and branched evolution revealed by multiregion sequencing (vol 366, pg 883, 2012). *New Engl J Med 367*, 976–976.

Gerlinger, M., and Swanton, C. (2010). How Darwinian models inform therapeutic failure initiated by clonal heterogeneity in cancer medicine. *Brit J Cancer 103*, 1139–1143.

Gillies, R. J., and Beyer, T. (2016). PET and MRI: Is the whole greater than the sum of its parts? *Cancer Res 76*, 6163–6166.

Han, S., Stoyanova, R., Lee, H., Carlin, S. D., Koutcher, J. A., Cho, H., and Ackerstaff, E. (2017). Automation of pattern recognition analysis of dynamic contrast-enhanced MRI data to characterize intratumoral vascular heterogeneity. *Magn Reson Med 79*, 1736–1744.

Hectors, S. J., Wagner, M., Bane, O., Besa, C., Lewis, S., Remark, R., Chen, N., Fiel, M. I., Zhu, H., Gnjatic, S. et al. (2017). Quantification of hepatocellular carcinoma heterogeneity with multiparametric magnetic resonance imaging. *Sci Rep 7*, 2452.

Henning, E. C., Azuma, C., Sotak, C. H., and Helmer, K. G. (2007a). Multispectral quantification of tissue types in a RIF-1 tumor model with histological validation. Part I. *Magn Reson Med 57*, 501–512.

Henning, E. C., Azuma, C., Sotak, C. H., and Helmer, K. G. (2007b). Multispectral tissue characterization in a RIF-1 tumor model: Monitoring the ADC and T2 responses to single-dose radiotherapy. Part II. *Magn Reson Med 57*, 513–519.

Hyare, H., Thust, S., and Rees, J. (2017). Advanced MRI techniques in the monitoring of treatment of gliomas. *Curr Treat Options Neurol 19*, 11.

Jackson, A., O'Connor, J. P., Parker, G. J., and Jayson, G. C. (2007). Imaging tumor vascular heterogeneity and angiogenesis using dynamic contrast-enhanced magnetic resonance imaging. *Clin Cancer Res 13*, 3449–3459.

LaViolette, P. S., Mickevicius, N. J., Cochran, E. J., Rand, S. D., Connelly, J., Bovi, J. A., Malkin, M. G., Mueller, W. M., and Schmainda, K. M. (2014). Precise ex vivo histological validation of heightened cellularity and diffusion-restricted necrosis in regions of dark apparent diffusion coefficient in 7 cases of high-grade glioma. *Neuro-Oncology 16*, 1599–1606.

Lee, J., Narang, S., Martinez, J., Rao, G., and Rao, A. (2015a). Spatial habitat features derived from multiparametric magnetic resonance imaging data are associated with molecular subtype and 12-Month survival status in glioblastoma multiforme. *PLoS One 10*, e0136557.

Lee, J., Narang, S., Martinez, J. J., Rao, G., and Rao, A. (2015b). Associating spatial diversity features of radio-logically defined tumor habitats with epidermal growth factor receptor driver status and 12-month sur-vival in glioblastoma: Methods and preliminary investigation. *J Med Imaging* (Bellingham) *2*, 041006.

Longo, D. L., Dastru, W., Consolino, L., Espak, M., Arigoni, M., Cavallo, F., and Aime, S. (2015). Cluster analysis of quantitative parametric maps from DCE-MRI: Application in evaluating heterogeneity of tumor response to antiangiogenic treatment. *Magn Reson Imaging 33*, 725–736.

Marino, M. A., Helbich, T., Baltzer, P., and Pinker-Domenig, K. (2017). Multiparametric MRI of the breast: A review. *J Magn Reson Imaging 47*, 301–315.

Merlo, L. M. F., Pepper, J. W., Reid, B. J., and Maley, C. C. (2006). Cancer as an evolutionary and ecological process. *Nat Rev Cancer 6*, 924–935.

Newbold, K., Castellano, I., Charles-Edwards, E., Mears, D., Sohaib, A., Leach, M., Rhys-Evans, P., Clarke, P., Fisher, C., Harrington, K., and Nutting, C. (2009). An exploratory study into the role of dynamic contrast-enhanced magnetic resonance imaging or perfusion computed tomography for detection of intratumoral hypoxia in head-and-neck cancer. *Int J Radiat Oncol Biol Phys 74*, 29–37.

Nowell, P. C. (1976). The clonal evolution of tumor cell populations. *Science 194*, 23–28.

O'Connor, J. P., Rose, C. J., Waterton, J. C., Carano, R. A., Parker, G. J., and Jackson, A. (2015). Imaging intratumor heterogeneity: Role in therapy response, resistance, and clinical outcome. *Clin Cancer Res 21*, 249–257.

Price, J. M., Robinson, S. P., and Koh, D. M. (2013). Imaging hypoxia in tumours with advanced MRI. *Q J Nucl Med Mol Imaging 57*, 257–270.

Stoyanova, R., Huang, K., Sandler, K., Cho, H., Carlin, S., Zanzonico, P. B., Koutcher, J. A., and Ackerstaff, E. (2012). Mapping tumor gypoxia in vivo using pattern recognition of dynamic contrast-enhanced MRI data. *Transl Oncol 5*, 437–447.

Surov, A., Meyer, H. J., and Wienke, A. (2017). Correlation between apparent diffusion coefficient (ADC) and cellularity is different in several tumors: A meta-analysis. *Oncotarget 8*, 59492–59499.

Vannier, M. W., Butterfield, R. L., Jordan, D., Murphy, W. A., Levitt, R. G., and Gado, M. (1985). Multispectral analysis of magnetic resonance images. *Radiology 154*, 221–224.

Vannier, M. W., Butterfield, R. L., Rickman, D. L., Jordan, D. M., Murphy, W. A., and Biondetti, P. R. (1987). Multispectral magnetic resonance image analysis. *Crit Rev Biomed Eng 15*, 117–144.

Vannier, M. W., Pilgram, T. K., Speidel, C. M., Neumann, L. R., Rickman, D. L., and Schertz, L. D. (1991). Validation of magnetic resonance imaging (MRI) multispectral tissue classification. *Comput Med Imaging Graph 15*, 217–223.

Winfield, J. M., Payne, G. S., Weller, A., and deSouza, N. M. (2016). DCE-MRI, DW-MRI, and MRS in Cancer: Challenges and advantages of implementing qualitative and quantitative multi-parametric imaging in the clinic. *Top Magn Reson Imaging 25*, 245–254.

Wu, J., Cui, Y., Sun, X., Cao, G., Li, B., Ikeda, D. M., Kurian, A. W., and Li, R. (2017). Unsupervised clustering of quantitative image phenotypes reveals breast cancer subtypes with distinct prognoses and molecular pathways. *Clin Cancer Res 23*, 3334–3342.

Yip, C., Weeks, A., Shaw, K., Siddique, M., Chang, F., Landau, D. B., Cook, G. J., and Goh, V. (2017). Magnetic resonance imaging (MRI) of intratumoral voxel heterogeneity as a potential response biomarker: Assessment in a HER2+ esophageal adenocarcinoma xenograft following trastuzumab and/or cisplatin therapy. *Transl Oncol 10*, 459–467.

Zhang, J., Cunningham, J. J., Brown, J. S., and Gatenby, R. A. (2017). Integrating evolutionary dynamics into treatment of metastatic castrate-resistant prostate cancer. *Nat Commun 8*, 1816.

Zhou, M., Chaudhury, B., Hall, L. O., Goldof, D. B., Gillies, R. J., and Gatenby, R. A. (2017). Identifying spatial imaging biomarkers of glioblastoma multiforme for survival group prediction. *J Magn Reson Imaging 46*, 115–123.

Zhou, M., Hall, L., Goldof, D., Russo, R., Balagurunathan, Y., Gillies, R., and Gatenby, R. (2014). Radiologically defined ecological dynamics and clinical outcomes in glioblastoma multiforme: Preliminary results. *Transl Oncol 7*, 5–13.

Feature extraction and qualification

LISE WEI AND ISSAM EL NAQA

8.1 INTRODUCTION

"Feature", also known as a variable, an attribute, is a commonly used concept in different fields, such as machine learning, image processing, and pattern recognition. In the context of computer vision and image processing, *it means a descriptor of an image that is relevant for dealing with some tasks related to certain applications,* e.g., object detection, face recognition. The idea of imaging features for computer vision appeared as early as the 1960s, using edges and simple shapes as visual features (Roberts 1963). However, the systematic application in medical imaging didn't show up until the 1980s, with computer aided diagnosis (CAD) being developed for quantitative image analysis (Doi 2007). Recently, the development of medical imaging techniques, new hardware, and standardized protocols further pushed the field stepping into the quantitative imaging analysis era. Varieties of modalities such as computed tomography (CT), positron emission tomography (PET), and magnetic resonance imaging (MRI) are widely used in the diagnosis, treatment planning, and endpoint prediction, providing us with huge amount of data. Therefore, extracting useful information from these images becomes necessary. Considering the complex procedures for clinicians to read the medical images and make a decision, they are also extracting information, however, in a relatively qualitative way, for example, the boundary of the tumor, the heterogeneity, etc. Fortunately, with the development of pattern recognition techniques and statistical learning tools, high-throughput extraction of quantitative features that can convert digital medical images into mineable high-dimensional data addresses the problem and shows great potential for precision medicine. The conversion of medical images into a large amount of advanced quantitative features and the subsequent analysis that relates these features with biological endpoints and clinical outcomes is referred to the field of "radiomics".

It is interesting to consider the origin of the word—radiomics. "Radio" comes from radiology, which refers to radiology images, e.g., CT, MRI, and PET. "-omics" stands for the technologies that aim at providing collective and quantitative features for the entire system and explore the underlying mechanisms. It is widely used in biology, such as in the study of genes (genomics), proteins (proteomics), and metabolites (metabolomics) (Gillies et al. 2016). The origin of radiomics is medical images, without all these modalities providing images we cannot do anything. The goal of radiomics is to take advantage of these resources and develop diagnostic, predictive, or prognostic radiomic models to understand the biological processes underneath, support personalized clinical decisions, and optimize individualized treatment planning. The core of radiomics is the extraction of quantitative features, with which we can apply all the advanced machine learning algorithms and build the models to bridge between images and biological and clinical endpoints. One point we have to mention is the hypothesis of radiomic analysis, that these imaging features are able to capture distinct phenotypic differences, like the genetics and proteomics patterns or other clinical outcomes so that we can infer the endpoints using different models. This hypothesis has been proven recently by many research. Segal et al. showed that the dynamic imaging traits (features) in CT systematically correlate with the global gene expression programs of primary human cancer (Segal et al. 2007). Aerts et al. found that a large number of radiomic features extracted from CT images have prognostic power in independent datasets of lung and head and neck cancer patients (Aerts et al. 2014). It is interesting that it identifies a general prognostic phenotype in both lung and head and neck cancer. Vallières et al. extracted features from fluorodeoxyglucose (FDG)-PET and CT images and did the risk assessment of locoregional recurrences (LR) and distant metastases (DM) in head and neck cancer (Vallières et al. 2017). These studies assured the potential of radiomic features for analyzing the properties of specific tumors. Given the examples stated above, the main idea is how can we obtain the "hidden" information from those pixels (voxels) in the images. The answer is imaging features should be generated that reflect the characteristics and complexity of the region of interest (ROIs) from different modalities and not to be redundant or correlated with each other. Thus, there are two aspects that we should consider in this process: (1) what kind of features exist for us to extract and (2) avoidance of correlated and redundancy of the extracted features.

The aspects of ROIs we are interested in are spatially and temporally heterogeneous. It was shown that features capturing these intra-tumor heterogeneities were strongly powerful in prognostic modeling of aggressive tumors. These changes in 4D space play an important role in the analysis to monitor the disease status (Aerts et al. 2014). Based on this fact, intuitively, it is natural to divide radiomic features into two types: spatial (static) and temporal (dynamic) features (El Naqa 2014). Static features are based on intensity, shape, size or volume, and texture, while dynamic features are based on kinetic analysis using time-varying acquisition protocols, such as dynamic PET or MRI. Both of these features offer information on tumor phenotype and its microenvironment (habitat). We can also categorize the static features into semantic and agnostic features (Gillies et al. 2016). Semantic features are those commonly used in the radiology lexicon describing ROIs, such as shape, vascularity, etc. Agnostic features are quantitative statistical image descriptors that are generally histogram-based, including first-, second-, or higher-order statistical metrics. In this chapter, we will first talk about different features based on two categories: static and dynamic features, with intensity and first-order histograms descriptors, intensity-volume histograms, morphological (shape), and texture features in the static case and with kinetic models and rate coefficients in the dynamic features case. In the third section, we will discuss how can we select the most powerful subset of features to get good predicting performance as well as stability by getting rid of redundant and correlated features.

8.2 TYPES OF FEATURES

Radiological imaging techniques are powerful tools that can provide useful information for the detection, diagnosis, and treatment plan response for personalized patient care. Different modalities capture different characteristics (e.g., contrast) of the target tissue. Previously, clinicians exploit these characteristics to reach a diagnosis. However, we can extract more information from these images and facilitate the clinicians' effort to make decisions more accurately and efficiently. One thing we can do is to quantify the characteristic that a clinician is looking at, such as image contrast, heterogeneity, etc. Moreover, extraction of information that is "hidden" and not visible to the clinician could be of greater interest for revealing unknown biological mechanisms and may have practical application to precision medicine. Another advantage is that radiomic approaches have the potential to extract data over the whole ROIs compared with the current biopsy approach, which additionally is invasive and could be inaccurate due to random sampling and is limited in terms of its reach, due to risks of pain and injury to the patients. While obtaining such information is a challenging task in the era of "big data". To deal with this challenge, varieties of radiomic features have been developed to decode the tissue pathology and translate the wholesome images into high-dimensional mineable data.

8.2.1 PRE-PROCESSING

One important prerequisite procedure is the pre-processing of imaging data before the extraction of features, which aims at enhancing image quality, obtaining region of interest, thus enabling the repeatable and comparable radiomic analysis. The starting point is the image stack (e.g., readouts from Digital Imaging and Communications in Medicine (DICOM) files). For some imaging modalities, such as PET, the images should be converted to a more meaningful representation (standardized uptake value, SUV) to take into account varying biodistributions related to the injected amount of the tracer, its activity, and patient's weight. Voxel size resampling is a vital pre-processing step for datasets that have variable voxel sizes (Thibault et al. 2014). Specifically, isotropic voxel size is required for some texture feature extraction. Down-sampling to larger dimension will lead to information loss. While, up-sampling may add artificial information. It is still unclear which one is better and perhaps the best way to evaluate this is to base the judgement on the model's performance. There are two main categories of interpolation algorithms: polynomial and spline interpolation. Nearest neighbor is a zero-order polynomial method that assigns gray-level values of the nearest neighbor to the interpolated point. Bilinear or trilinear interpolation and bicubic or tricubic interpolation are often used for 2D in-plane interpolation or 3D cases. Cubic spline and convolution interpolation are third-order polynomial methods that interpolate smoother surface than linear methods, while being slower in implementation. Linear interpolation is a rather commonly used algorithm, since it neither leads to the rough blocking artifacts images that are generated by nearest neighbors, nor will it cause out-of-range gray levels that might be produced by higher order interpolation (Zwanenburg et al. 2016).

Next step is to obtain the ROIs, such as the tumor or organs at risk, so that features can be extracted from these volumes. These are obtained by segmentation methods of the whole region or sampling within such region, i.e., digital biopsy (Echegaray et al. 2016). Typically, ROIs will be represented by a mask that equals 1 inside the volume and 0 outside. Usually, ROIs are contoured by experienced clinicians or automatically (fully or semi) by computer algorithms.

Before the calculation of radiomic features, especially texture features, the gray levels should be discretized or quantized to make the computation tractable. For example, if an image has 512 discrete intensity values, it will yield $2^{512} \times 2^{512} \times 2^{512}$ gray-level co-occurrence matrix (GLCM), which would be computationally too expensive to implement (Yip and Aerts 2016). There are two ways to do the discretization: fixed bin number N and fixed bin width B. For fixed bin number, we first decide a fixed number of N bins, and the gray levels will be discretized into these bins using the formula below:

$$I_N(x) = \begin{cases} 1 & I(x) = I_{min} \\ N \times \dfrac{I(x) - I_{min}}{I_{max} - I_{min}} & otherwise \end{cases} \tag{8.1}$$

The intensity resolution equals $(I_{max} - I_{min})/N$.

For fixed bin width, the discretized gray levels are calculated as follow:

$$I_B(x) = \begin{cases} 1 & I(x) = I_{min} \\ \dfrac{I(x) - I_{min}}{B} & otherwise \end{cases} \tag{8.2}$$

The fixed bin number method is better when the modality used is not well calibrated. It maintains the contrast and makes the images of different patients comparable. But it loses the relationship between image intensity. While fixed bin size method keeps the direct relationship with the original scale. Some investigations about the effect of both methods have shown that fixed bin size method gave better repeatability and thus may be suitable for intra- and inter-patient studies, however, this remains the subject of ongoing research (van Velden et al. 2016, Leijenaar et al. 2015).

After the pre-processing of original images, in this section, we discuss some fundamental radiomic features from two categories: static and dynamic features. We divide the static features into two main categories: morphological features and statistical features, describing the geometric property and the distribution of intensities of the ROIs in relation to its spatial distribution. Among the statistical features, absolute intensity statistical features, intensity histogram features, intensity-volume histogram features, and texture features are contained. Absolute intensity statistical features, intensity histogram features, and intensity-volume histogram features are labelled as first-order features, and texture features are labelled as second-order features. The set of features we summarized here is built on the common feature sets described by many reports, but primarily in the comprehensive works of Aerts et al. (2014), Zwanenburg et al. (2016), and Kickingereder et al. (2016). A lot of them are also based on earlier efforts, which were referenced somewhere else.

8.2.2 STATIC FEATURES

8.2.2.1 MORPHOLOGICAL (SHAPE) FEATURES

Morphological (or volume and shape) features, which is a kind of external representation of a region, characterize the shape, size, and surface information of the ROIs according to some metrics which will be described below (Jain 1989). The volume of ROI is denoted to be V, surface area A (Aerts et al. 2014, Vaidya et al. 2012).

1. **Compactness 1:**

$$compactness\,1 = \frac{V}{\sqrt{\pi} * A^{\frac{2}{3}}} \tag{8.3}$$

2. **Compactness 2:**

$$compactness\,2 = 36\pi \frac{A^2}{V^3} \tag{8.4}$$

The compactness is a quantity representing how compact the region is and is independent of scale and orientation, is dimensionless numbers, and is not overly dependent on one or two extreme points in the shape (Gillman 2002). The most compact shape is a perfect sphere.

3. **Maximum 3D diameter:** The maximum Euclidean distance between any two voxels on the surface of the ROI.

4. **Spherical disproportion:**

$$\text{spherical disproportion} = \frac{A}{4\pi R^2} = \frac{A}{\left(6\sqrt{\pi} * V\right)^{\frac{2}{3}}},$$ (8.5)

where R is the radius of a sphere with the same volume as the ROI. It is the ratio of the surface area of the ROI to the surface area of a sphere with the same volume as the ROI.

5. **Sphericity:**

$$\text{sphericity} = \frac{\left(6\pi^2 V\right)^{\frac{2}{3}}}{A}$$ (8.6)

Sphericity measures the roundness of the shape of the ROI relative to a sphere. $0 < \text{sphericity} \leq 1$, where 1 indicates a perfect sphere.

6. **Surface area:** First of all, a mesh is generated by dividing the surface into connected triangles, then area can be calculated as below:

$$A = \sum_{i=1}^{N} \frac{1}{2} |a_i b_i \times a_i c_i|,$$ (8.7)

where N is the total number of triangles in the surface and a, b, c are edge vectors of the triangles.

7. **Volume:** The volume can be determined by counting the number of voxels in the ROI and timing the voxel size (voxel based) or by summing the volume of the mesh elements.

8. **Eccentricity (EC):**

$$EC = 1 - \sqrt{\frac{b^2}{a^2}}$$ (8.8)

9. **Surface to volume ratio:**

$$\text{surface to volume ratio} = \frac{A}{V}$$ (8.9)

This feature is dimensionless, and the lower the value is, the more compact the region is.

8.2.2.2 FIRST-ORDER FEATURES

First-order features are based on first-order histograms that show the distribution of the voxel intensities in the ROIs. Depending on how we process the intensity values of an ROI, we can further divide the first-order features into absolute intensity statistical features, intensity histogram features, and intensity-volume histogram features. These features help to reduce the large number of voxel values in ROIs to single values, such as mean, minimum, etc.

8.2.2.2.1 Absolute intensity features

We use the raw unprocessed absolute intensity values for calculating these features. Let I denote the intensity values in region of interest with N voxels, I_{whole} the intensity values in the whole image, and $\bar{I}$ the average intensity in the ROIs.

Then, the first-order histogram features can be expressed as:

1. **Mean:**

$$\text{mean} = \frac{1}{N}\sum_{i}^{N} I(i) \tag{8.10}$$

2. **Variance:**

$$\text{variance} = \frac{1}{N-1}\sum_{i=1}^{N}\left(I(i)-\bar{I}\right)^{2} \tag{8.11}$$

3. **Standard deviation (SD):**

$$\text{standard deviation} = \left(\frac{1}{N-1}\sum_{i=1}^{N}\left(I(i)-\bar{I}\right)^{2}\right)^{1/2} \tag{8.12}$$

4. **Skewness:**

$$\text{skewness} = \frac{\frac{1}{N}\sum_{i=1}^{N}\left(I(i)-\bar{I}\right)^{3}}{\left(\sqrt{\frac{1}{N}\sum_{i=1}^{N}\left(I(i)-\bar{I}\right)^{2}}\right)^{3}} \tag{8.13}$$

5. **Kurtosis:**

$$kurtosis = \frac{\frac{1}{N}\sum_{i=1}^{N}\left(I(i)-\bar{I}\right)^{4}}{\left(\sqrt{\frac{1}{N}\sum_{i=1}^{N}\left(I(i)-\bar{I}\right)^{2}}\right)^{2}} \tag{8.14}$$

Kurtosis is a measure of the "tailedness" of the intensity distribution. Higher kurtosis is the result of infrequent extreme deviations (or outliers), as opposed to the frequent modestly deviations.

6. **Median:** The median intensity value of I.
7. **Maximum:** The maximum intensity value of I.
8. **90th percentile:** P_{90} is the 90th percentile of I. It is more robust to outliers than the maximum.
9. **Minimum:** The minimum intensity value of I.
10. **10th percentile:** P_{10} is the 10th percentile of I. It is more robust to outliers than the minimum.
11. **Interquartile range:** The interquartile range is defined as:

$$\text{interquartile range} = P_{75} - P_{25} \tag{8.15}$$

P_{75} and P_{25} are the 75th and 25th percentile of I.

12. **Range:**

$$\text{range} = \max(I) - \min(I) \tag{8.16}$$

13. **Covered image intensity range:**

$$\text{intensity range} = \frac{\max I - \min I}{\max I_{whole} - \min I_{whole}} \tag{8.17}$$

14. **No. of voxels:** The number of voxels in I.

15. **Mean absolute deviation:**

$$\text{mean absolute deviation} = \frac{1}{N}\sum_{i}^{N}\left|I(i)-\bar{I}\right| \tag{8.18}$$

16. **Robust mean absolute deviation:** Considering the existence of outliers, a more robust way to get the mean absolute deviation is to restrict the gray levels closer to the center of the distribution, e.g., between P_{10} and P_{90}. I_{10-90} stands for the voxel values that lie between the 10th and 90th percentile.

$$\text{robust mean absolute deviation} = \frac{1}{N_{10-90}}\sum_{i=1}^{N_{10-90}}\left|I_{10-90}-\overline{I_{10-90}}\right| \tag{8.19}$$

where N_{10-90} denotes the voxels with intensities between $P_{10}(I)$ and $P_{90}(I)$, $\overline{I_{10-90}}$ is the sample mean of I_{10-90}.

17. **Median absolute deviation:** Similar with the mean absolute deviation, but it measures the deviation from the median instead of mean.

$$\text{median absolute deviation} = \frac{1}{N}\sum_{i}^{N}\left|I(i)-M\right| \tag{8.20}$$

where M denotes the median of I.

18. **Coefficient of variation (CV):**

$$\text{coefficient of variation} = \frac{SD}{\text{mean}} \tag{8.21}$$

It shows the extent of variability in relation to the mean of the population.

19. **Root mean square (RMS):**

$$RMS = \sqrt{\frac{\sum_{i}^{N}I(i)^{2}}{N}} \tag{8.22}$$

20. **Sum of intensities:**

$$\text{sum of intensities} = \sum_{i=1}^{N}I(i) \tag{8.23}$$

21. **Energy:**

$$\text{energy} = \sum_{i}^{N}I(i)^{2}\,N \tag{8.24}$$

Mean, variance, skewness, and kurtosis are commonly used statistical moments that describe the dispersion of a distribution. The skewness characterizes the degree of asymmetry around its mean. The kurtosis measures the relative peakedness or flatness of a distribution. Entropy and uniformity describe the heterogeneity of the images (Aerts et al. 2014, Parekh and Jacobs 2016, Zwanenburg et al. 2016, Avanzo et al. 2017). Note that some of the metrics may have similar names across the different feature categories such as Energy for instance redefined below, therefore, it may be useful to add a prefix/suffix, to clarify that when reporting. Efforts to standardize and harmonize radiomics terminology are currently being pursued by several organizations in medical physics.

8.2.2.2.2 Intensity histogram features

If we extract features based on the intensity histogram, which is generated by discretizing the absolute intensity values into gray level bins, these features will fall into the category of intensity histogram features. Comparing with the absolute features, these features can reduce noise and limit the number of values for the intensity. Let I_d be the discretized intensity levels of N voxels in the ROIs, and P the first-order histogram with L intensity levels and the center gray values with C. A normalized first-order histogram (P_N) is the percentage of voxels that fall into a certain intensity bin if we divide the intensities of an image into L equally spaced bins. The calculation of features from intensity histogram is similar with those described above, except that the discretized gray intensities (I_d) are used instead of I, and the normalized histogram P_N can be used to provide another equivalent formulation. Thus, we will not provide the formula for these repeated features, such as mean, variance. Readers can refer to the first part or reference (Zwanenburg et al. 2016) to check. Except for these features mentioned in absolute features above, several other features are listed below.

1. **Intensity histogram mode:** The mode of I_d is the bin for which the count is maximal. There are cases that the mode is not unique, e.g., if multiple bins have the same maximal count, then the bin closest to the mean will be the mode.
2. **Intensity histogram entropy:**

$$\text{entropy} = -\sum_{i=1}^{L} P_N(i)\log_2 P_N(i) \tag{8.25}$$

Entropy is the measure of uncertainty or inherent randomness. It is a concept from thermodynamics, representing the disorder within a macroscopic system. Later, Claude Shannon developed the entropy in the field of information theory in his paper *A mathematical Theory of Communication* in 1948. If the histogram (P_N) is uniformly distributed, entropy will achieve the maximum value of $\log_2 P(i)$.

3. **Intensity histogram uniformity:**

$$\text{uniformity} = \sum_{i}^{L} P_N(i)^2 \tag{8.26}$$

4. **Maximum histogram gradient:** Calculation of histogram gradient is:

$$P' = \left[P(2)-P(1),\ldots,\frac{P(i+1)-P(i-1)}{2},\ldots,P(L)-P(L-1) \right] \tag{8.27}$$

Actually, there are different ways to calculate the gradient, the method presented above is quite simple to implement and generates a gradient with the same number of elements as the original histogram. Thus, maximum histogram gradient $=\max(P')$ (van Dijk et al. 2017).

5. **Maximum histogram gradient gray level:** It is the discretized gray level corresponding to the maximum histogram gradient.
6. **Minimum histogram gradient:** *minimum histogram gradient* $=\min(P')$
7. **Minimum histogram gradient gray level:** It is the discretized gray level corresponding to the minimum histogram gradient.

8.2.2.2.3 Intensity volume histogram (IVH) features

Intensity volume histogram (IVH) features are extracted from first-order histogram (El Naqa et al. 2009). IVH is analogous to the dose volume histogram (DVH) used in the field of radiotherapy treatment planning, which summarizes the complex 3D imaging information into one single curve. Similarly, there are differential and cumulative IVH, displaying the relationship between the gray level i and the volume fraction v that has the gray level of i and that has at least gray level i or higher, respectively. Here, we only discuss the cumulative IVH for extraction of features. Each point in the IVH curve represents the absolute or relative volume of the ROI that exceeds a variable intensity threshold as a percentage of the maximum intensity. Comparing with the histogram we used in the first-order features above, IVH describes the fractional volume for the ROIs versus the intensity gray levels, while the intensity histogram above is the number of voxels versus intensity. A discretized gray level ROI is necessary for the calculation of IVH features. For images which are originally discretized (e.g., CT), we can directly use the original intensity values. While for images without discretized features, we should first quantize or discretize the intensities. Details about gray level discretization can be found above.

Followed by the acquisition of the discretized gray level ROI image I_d with level number L, IVH can be calculated by first obtaining the fraction volume corresponding to certain gray level i. The point on the curve (v_i, i) could be computed by counting the voxels having gray level smaller than i and dividing by the total voxel number N:

$$v_i = 1 - \frac{1}{N}\sum_{j=1}^{N}\delta(I_d(j) < i),\tag{8.28}$$

δ is the Kronecker delta.

The gray level fraction γ for discrete gray level i is:

$$\gamma_i = \frac{i - \min(I_d)}{\max(I_d) - \min(I_d)}.$$

IVH enables the extraction of several metrics for further modeling. Similar with the above two first-order histograms, statistical features like mean, variance can be extracted from IVH. Moreover, some specific features for IVH will be shown below:

1. **Volume at intensity fraction V_x:** It is the largest volume that has an intensity fraction γ of at least $x\%$.
2. **Intensity at volume fraction I_x:** It is the smallest intensity gray level i that exists in at most $x\%$ of the volume.
3. **Volume at intensity fraction difference:** the difference between the volume fractions at two different intensity fractions, e.g., $V_{10-90} = V_{10} - V_{90}$.
4. **Intensity at volume fraction difference:** the difference between the intensity levels at two different volume fractions, e.g., $I_{10-90} = I_{10} - I_{90}$.
5. **Area under IVH curve:** van Velden et al. defined the area under the IVH curve (van Velden et al. 2011). They showed that the area under the curve is a quantitative index of uptake heterogeneity, where lower values correspond to increased heterogeneity.

In summary, all these IVH features can be used for analyzing heterogeneity in images (El Naqa et al. 2009). There are reported examples on the use of the IVH approach for predicting local control in lung cancer from PET/CT images (Vaidya et al. 2012). The intensity gray levels here are either Hounsfield units (HU) for a CT scan or SUVs for a FDG-PET scan.

Though these first-order histogram features simplify and summarize the 3D imaging into 1D curves and extract useful predictors to build predictive models, they ignored the spatial and topological information (Kumar et al. 2012), which possess great potential to contribute more to the predicting power of the model. Next part, we will talk about the second-order statistical features that emphasize spatial information.

8.2.2.3 SECOND-ORDER HISTOGRAM OR TEXTURE FEATURES

Second-order feature descriptors, which are usually called texture features, provide statistical interrelationships between voxels and capture special patterns in the ROIs, which make up for the loss of information for the first-order features because of the spatial information associated with the relative positions of the gray levels. Haralick and Shanmugam (1973) introduced the idea of using textural features for image classification.

Texture pairs with matching second-order statistics cannot be distinguished by the human visual system (Julesz 1975). Second-order statistical features can provide a quantitative representation that to some degree mimic the features clinicians pay attention to and have the potential to obtain more information than are invisible to human eyes.

Once we have obtained desired isotropic interpolated voxel sizes and discretized gray levels images, features can be calculated from GLCM, gray-level run-length matrix (GLRLM), gray-level size zone matrix (GLSZM), gray-level distance zone matrix (GLDZM), neighborhood gray-tone difference matrix (NGTDM), and neighboring gray-level-dependence matrix (NGLDM) as below. Though texture feature analysis was first implemented on 2D images, we can extend it to the 3D cases.

8.2.2.3.1 Gray-level co-occurrence-based features

GLCM illustrates the distribution of the combinations of gray levels of neighboring voxels (pixels) along certain direction. There are two parameters to determine "neighboring": the distance (d) and angle (θ). If there are L gray levels in the ROI, the possible number of voxel value pairs would be $L \times L$, the count of each voxel pair will be then stored in the $L \times L$ GLCM matrix. The ranges of the parameter d is $d \in \{1,2,3,\ldots,n\}$. For θ, in the case of 3D volume, a voxel has 26 neighbors, with 13 unique direction vectors, i.e., $(0,0,1),(0,1,0),(1,0,0),(0,1,1),(1,0,1),(1,1,0),(0,1,-1),(1,0,-1),(1,-1,0),(1,1,1),(1,1,-1),(1,-1,1),(1,-1,-1)$. If we ignore the connections between slices, 2D GLCM matrix can also be used. In this case, there are eight neighbors for a voxel with four directions: $(1,0,0),(1,1,0),(0,1,0),(-1,1,0)$. Here, we want to point out that the importance of making the images to be isotropic. A lot of the times, for CT, the in-plane pixel spacing is smaller than the slice thickness, it is not reasonable to calculate 3D GLCM unless we interpolate the images to make them isotropic.

A GLCM matrix corresponds to one direction. M_θ is the GLCM matrix, with θ being the direction. Element (i,j) means the frequency of combination of neighboring voxels that have gray level of i and j along a certain direction δ and $-\delta$. $M_\theta = M_\delta + M_{-\delta} = M_\delta + M_\delta^T$. An example of GLCM calculation is shown in Figure 8.1.

Once we get the matrix, we would like to first calculate the distribution of the combination of gray levels, P_θ by dividing the matrix by the sum of all the elements. Thus, p_{ij} denotes the probability of occurrence of the neighboring voxel pair with gray levels (i,j) along certain direction. The probabilities for diagonal and cross-diagonal are worth to be expressed:

$$p_{i-j}(k) = \sum_{i=1}^{L}\sum_{j=1}^{L} p_{ij}\delta\left(k-|i-j|\right) \qquad k=0,\ldots,L-1 \tag{8.29}$$

$$p_{i+j}(k) = \sum_{i=1}^{L}\sum_{j=1}^{L} p_{ij}\delta\left(k-(i+j)\right) \qquad k=2,\ldots,2L \tag{8.30}$$

(a) Images with 4 grey levels

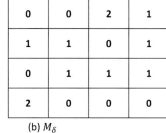

(b) M_δ

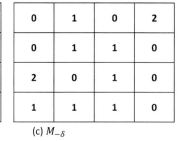

(c) $M_{-\delta}$

Figure 8.1 GLCM (a) Gray level image; (b) corresponding GLCM matrix along $\theta = 0°$; and (c) corresponding GLCM matrix along $\theta = 180°$.

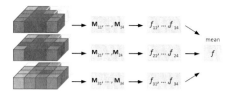
(a) by slice, without merging

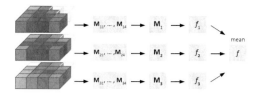
(b) by slice, with merging by slice

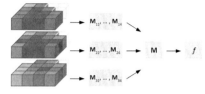
(c) by slice, with full merging

(d) as volume, without merging

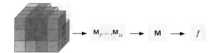
(e) as volume, with merging

Figure 8.2 (See color insert.) Methods to calculate GLCM-based features. $M_{\Delta k}$ are matrices calculated for direction Δ in slice k and $f_{\Delta k}$ is the corresponding features. (From Zwanenburg, A. et al., Image biomarker standardisation initiative. ArXiv e-prints, 1612. Available: http://adsabs.harvard.edu/abs/2016arXiv161207003Z, 2016.)

These probabilities are very useful to pick out certain combination of gray levels that we are interested in. Next step is the calculation of features from the matrix. As mentioned above, 3D and 2D matrix can be computed. Based on this idea, five methods of extraction of features have been proposed, as shown in Figure 8.2: either extract as slices (2D) (a)(b)(c) or extract as volumes (3D) (d)(e). After the extraction, there are still two options, either merge the matrices that are obtained from different directions by summing up the counts for each combination of (i, j) or not. It is still unclear which method works best right now. For each method, one or multiple matrices will be obtained, then single feature value can be computed from one merged matrix or by calculating the mean of features extracted from multiple matrices.

1. **Joint maximum:** the maximum probability corresponding to the most frequent occurred gray level combination.

$$joint\ maximum = \max\left(p_{ij}\right). \tag{8.31}$$

2. **Joint average:** gray-level-weighted sum of the joint probabilities.

$$joint\ average = \sum_{i=1}^{L}\sum_{j=1}^{L} i p_{ij}. \tag{8.32}$$

3. **Joint variance:**

$$joint\ variance = \sum_{i=1}^{L}\sum_{j=1}^{L}\left(i - \mu\right)^2 p_{ij}, \tag{8.33}$$

where μ is the joint average.

4. **Joint entropy:**

$$joint\ entropy = -\sum_{i=1}^{L}\sum_{j=1}^{L} p_{ij} log_2 p_{ij}. \tag{8.34}$$

5. **Difference average:** the average of the diagonal probabilities.

$$difference\ average = \sum_{k=0}^{L-1} k p_{i-j}(k). \tag{8.35}$$

k is the gray level difference for combination (i, j).

6. **Difference variance:** the variance for diagonal probabilities.

$$difference\ variance = \sum_{k=0}^{L-1} (k-\mu)^2 p_{i-j}(k). \tag{8.36}$$

8.2.2.3.2 Gray-level run-length-based features

GLRLM was first proposed by Galloway (1975). It is defined as the frequency of occurrence of contiguous voxels with some run length along certain direction that have the same gray level. It characterizes the distribution of combination of gray levels in different directions. The direction definition is the same as GLCM matrix. The row number (L) of the matrix stands for the number of gray level in the ROI, the column number (R) is the maximum possible run length along one direction θ. Then the element (i, j) is the count that shows how many times the continuous voxels with gray level i and run length j appear along a certain direction. An example of GLRLM matrix of a 2D image is shown in Figure 8.3. Similar with GLCM, the matrices along different directions can be merged to get a combined matrix, the methods are the same as GLCM as shown in Figure 8.2. Then features could be extracted. We denote the $L \times R$ matrix element as r_{ij}. $r_{i.} = \sum_{j}^{R} r_{ij}$ is the marginal sum over all run lengths for a certain gray level i. Similarly, $r_{.j} = \sum_{i}^{L} r_{ij}$ is the marginal sum over all gray levels for certain run length j. $S = \sum_{i}^{L}\sum_{j}^{R} r_{ij}$ is the sum of all elements in the matrix along a certain direction. Using similar method of calculation of GLCM features, some GLRLM features can be extracted are listed below.

1. **Short runs emphasis:**

$$\frac{1}{S}\sum_{j=1}^{R} \frac{r_{.j}}{j^2}. \tag{8.37}$$

1	3	4	1	5	0	0	0
2	2	4	1	4	1	0	0
3	2	1	4	4	1	0	0
2	1	3	3	3	0	0	0
(a) Images with 4 grey levels				(b) M_δ			

Figure 8.3 (a) Gray level image and (b) corresponding GLRLM matrix along $\theta = 0°$.

2. **Long runs emphasis:**

$$\frac{1}{S}\sum_{j=1}^{R}j^2 r_{.j}. \tag{8.38}$$

3. **Low gray-level run emphasis:**

$$\frac{1}{S}\sum_{i=1}^{L}\frac{r_{i.}}{i^2}. \tag{8.39}$$

4. **High gray-level run emphasis:**

$$\frac{1}{S}\sum_{i=1}^{L}i^2 r_{i.}. \tag{8.40}$$

5. **Run entropy:** If $p_{ij} = \dfrac{r_{ij}}{S}$, the entropy is then

$$-\sum_{i=1}^{L}\sum_{j=1}^{R}p_{ij}log_2 p_{ij}. \tag{8.41}$$

8.2.2.3.3 Gray-level-size zone-based features

The gray-level size-zone matrix gives the statistics of groups of voxels that are connected and have a certain gray level. It is calculated according to the run-length matrix principle: with the rows to be the gray levels L and columns the maximum size zone S. So the GLSZM is a $L \times S$ matrix with element z_{ij} showing the number of zones with gray level i and size j in the image. The total number of zones is $T = \sum_{i=1}^{L}\sum_{j=1}^{S}z_{ij}$, and the marginal sums: $z_{.j} = \sum_{i}^{L}z_{ij}$ is the number of zones with size j, and $z_{i.} = \sum_{j}^{S}z_{ij}$ is the number of zones with gray level i. The connectivity is eight directions for 2D case and 26 directions for 3D case. The matrix requires no directional computation as GLCM or GLRLM, thus, the features emphasize regional properties and describe non-periodic heterogeneous textures, while the GLCM and GLRLM may be more suited for periodic textures. Considering this, the ways we use to compute features is a bit different. Figure 8.4 shows three ways to do the feature extraction: either calculate 2D matrix or 3D matrix, for the 2D case, the features will be extracted

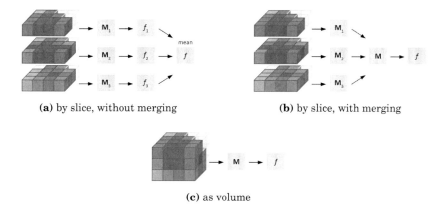

(a) by slice, without merging   **(b)** by slice, with merging

(c) as volume

Figure 8.4 (see color insert.) Methods to calculate GLSZM-based features. M_k is the texture matrices calculated for slice k, and f_k are the corresponding features. (From Zwanenburg, A. et al., Image biomarker standardisation initiative. ArXiv e-prints, 1612. Available: http://adsabs.harvard.edu/abs/2016arXiv161207003Z, 2016.)

1	3	4	1
2	2	4	1
3	2	1	4
2	1	3	3

1	0	0	1
0	0	0	1
2	1	0	0
0	0	1	0

(a) Images with 4 grey levels   (b) Grey level size zone matrix

Figure 8.5 GLSZM (a) Gray level image and (b) corresponding GLSZM matrix under eight connectedness.

with or without merging for different slices. An example of GLSZM calculation is shown in Figure 8.5. Some typical features are shown below:

1. **Small zone emphasis:**

$$\text{small zone emphasis} = \frac{1}{T}\sum_{j=1}^{S}\frac{z_{.j}}{j^2}. \tag{8.42}$$

2. **Large zone emphasis:**

$$\text{large zone emphasis} = \frac{1}{T}\sum_{j=1}^{S}j^2 z_{.j}. \tag{8.43}$$

3. **Low gray-level zone emphasis:**

$$\text{low gray level zone emphasis} = \frac{1}{T}\sum_{i=1}^{L}\frac{z_{i.}}{i^2}. \tag{8.44}$$

4. **High gray-level zone emphasis:**

$$\text{high gray level zone emphasis} = \frac{1}{T}\sum_{i=1}^{L}i^2 z_{i.} \tag{8.45}$$

Except for the GCLM-, GLRLM-, and GLSZM-based features discussed above, there are some more matrices that texture features can be extracted from, such as the GLDZM, neighborhood gray-tone difference matrix, NGTDM and neighboring gray-level-dependence matrix. We will not discuss about these matrices here.

8.2.3 DYNAMIC FEATURES

For time varying acquisition protocols, such as dynamic PET and MR, radiomic features are extracted based on kinetic analysis of the dynamic images. Compartment models are widely used for the tracer transport, its binding rates, and metabolism. For example, the FDG-PET imaging has shown great success in tumor detection and cancer staging, which uses 18F-labeled FDG as the tracer to visualize the intra-tumoral glucose metabolism. General concepts for the interpretation of PET data are four compartment model as shown in Figure 8.6 (Watabe et al. 2006), with the first compartment being the arterial blood, second the free compartment, and third and

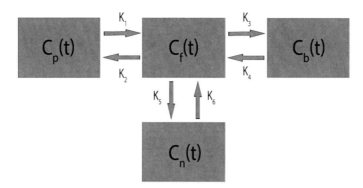

Figure 8.6 (see color insert.) General four compartment model with arterial blood, free ligand in tissue, specific and nonspecific binding and six rates (K_1~K_6). (From Watabe, H. et al., *Ann. Nucl. Med.*, 20, 583, 2006.)

fourth specific and non-specific binding. There are four radioactivity concentrations at each time point for each compartment: $C_p(t)$, $C_f(t)$, $C_b(t)$, and $C_n(t)$ and six transport and binding rates $K_1, K_2, K_3, K_4, K_5, K_6$. These rates are assumed to be linearly related to the concentrations between two compartments by simple differential equations. By fitting the model and solving these parameters, the glucose metabolic rate can be calculated, which can be used as dynamic radiomic features in outcome modeling. Since the rapid equilibrium between the non-specific-binding and free compartments for FDG tracer, three compartments model is good enough to obtain the kinetics for this example (Watabe et al. 2006). There are some literature about using the dynamic features to do the prediction. Thorwarth et al. reported that by using a compartmental model based on dynamic [18F]-Fmiso PET patient data, identification and quantification of hypoxia in human head-and-neck tumors are better than standardized uptake value alone (Thorwarth et al. 2005). Choi et al. showed that there is a good correlation between residual regional metabolic rate of glucose (MRglc) after chemoradiotherapy and the degree of pathologic tumor control in locally advanced-stage non-small cell lung cancer (NSCLC), where the MRglc is calculated by a three-compartment kinetic model (Choi et al. 2002).

8.2.4 FEATURELESS APPROACHES (CONVOLUTION NEURAL NETWORKS)

With thousands of features extracted from the ROIs, there are probably a lot of irrelevant or redundant features that may harm the performance of the outcome modeling. One way to solve the problem is to use different techniques to do feature selection, which we will discuss in the next section. Here, we want to present an alternative approach to hand-crafted features discussed above, a topic referred as "feature engineering" such as using Convolution Neural Networks (CNNs). With the development of hardware and computing power, especially the graphics processing units (GPUs) and the growing size of data, deeper and deeper NNs are possible to be applied, which is called the deep learning. Deep learning turned out to outperform conventional shallow NNs and other machine learning algorithms at discovering complicated structures in high-dimensional data (LeCun et al. 2015). Each feature map or representation in the deep network will be transformed into a representation at a higher and more abstract level, and finally higher layers of representation emphasize the features that are important for discrimination by backpropagation training or learning. CNNs are one particular type of deep, forward network. CNNs are similar to regular neural networks, but the architecture is modified to fit to the specific input of large scale images. Regular NNs receive a vector as input and transform it through a series of hidden layers that consists of neurons. Each neuron is fully connected to all neurons in the previous layer, and each connection has its own weight. When the input is an image, the number of weights explodes and become non-manageable. Inspired by the Hubel and Wiesel's work on the animal visual cortex (Hubel and Wiesel 1968), local filters are used to slide over the input space in CNNs, which not only exploit the strong local correlation in natural images, but also reduce the number of weights significantly by sharing weights for each filter. During the forward pass, we slide each filter over the width and height of the input and compute the dot products between the filter and the input in all positions, so that each filter will form a feature map.

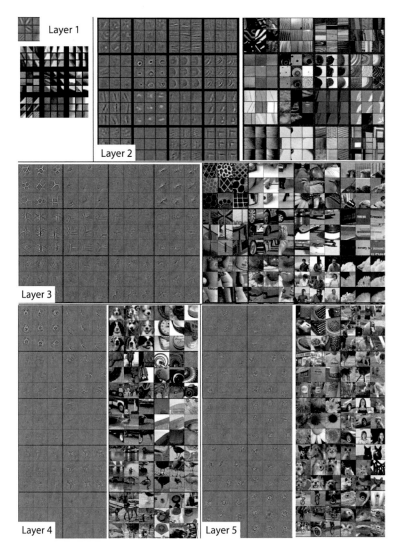

Figure 8.7 (See color insert.) Visualization of features from a fully trained model for layers 1–5 using deconvolutional network approach. (From Zeiler and Fergus, 2014a.)

Figure 8.7 is an example of visualization of different feature maps in a CNN, where low-level layers provide simple features-like edges, high-level layers give more semantic or abstract features (Zeiler and Fergus 2014a). Unlike hand-crafted features mentioned above, CNNs can engineer such features automatically by learning from the samples by aids of designing specialized filters (kernels), which act as feature identifiers (Goodfellow et al. 2016). Conventionally, the convolutional filters used in CNN are square with different sizes, and the weights will be automatically learned by the data through the network. However, this approach relies on large amounts of data. For domains with only small samples, prior domain knowledge should be exploited in the learning process to facilitate the driving of data to obtain high performance. Thus, instead of using generic square filters, carefully designed filters can be incorporated into the CNN to extract some features that we are interested in or we know that are important for the task so that domain knowledge will be added into the network. Adding designed filters to traditional filters in CNNs help us exploit the power of data as well as the privilege of domain knowledge, which holds the potential to achieve better performance in small sample size fields. There are a few works in this direction: Li et al. proposed an approach that automatically designs layers of different customized filter shapes for application in gene sequence analysis (Li et al. 2017). Xie et al. designed a donut shape filter for CNNs to solve the structured labeling problem (Xie et al. 2015).

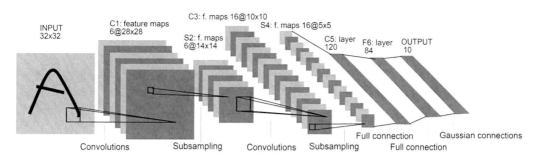

Figure 8.8 Architecture of LeNet-5. (From LeCun, Y. et al., *Proc. IEEE*, 86, 2278–2324, 1998.)

Typical architecture of CNNs contains convolution layer (Conv), pooling, and fully connected layer (FC), and the activation function is usually RELU. Conv extracts features and detects local correlation from the previous layer, while pooling merges semantically similar features into more abstract level, and RELU adds non-linearity to the system. Different combinations of these layers are stacked and form various architectures. There are several famous architectures: LeNet is the first application of CNN to read zip codes, digits, etc. by Yann LeCun in the 1990s, shown in Figure 8.8 (LeCun et al.1998), and AlexNet was used in the ImageNet challenge in 2012 and showed significantly better performance than other methods. It is similar with LeNet, but with deeper and more complex Conv structure across two GPUs (Krizhevsky et al. 2012); ZF Net is the winner of 2013 ImageNet Large Scale Visual Recognition Challenge (ILSVRC), which is a fine tuned and modified network based on the AlexNet (Zeiler and Fergus 2014b); GoogLeNet, a 22-layers CNN, winner of 2014 ILSVRC, introduced inception module in the architecture, which reduced the parameters up to 12 times more than AlexNet and found out an optimal local sparse structure (Szegedy et al. 2015); and VGGNet was introduced in 2014, and it is characterized by its simplicity with respect to the architecture, using only 3 by 3 convolutional layers stacked on each other with increasing depth (Simonyan and Zisserman 2014). There are VGG16 and VGG19, where 16 and 19 are the layer depth. This research shows that the performance increases with the depth of the architecture. However, the weights of this network are very large, and the training is quite slow; ResNet, the winner of 2015 ILSVRC, uses as many as 152 layers with extra layers set as identity, which are called the residual modules. In their paper, they showed that with the increase of the depth, the training and testing errors increase after 20 layers, which is not caused by overfitting, but the difficulty in optimization. They reformulate the layers as learning residual functions with respect to the input layer to make the networks easier to train and gain accuracy from increased depth (He et al. 2015).

The idea of CNNs has been applied to medical image processing as early as 1993, Zhang et al. (1994) used a shift-invariant artificial neural network to detect clustered microcalcifications in digital mammograms. Sahiner et al. (1996) investigated the classification of region of interest on mammograms using a CNN with spatial domains and texture images. Recently, large numbers of inspiring work applying deep CNNs to medical images analysis have been presented. For example, Shin et al. used three CNN architectures, namely, CifarNet, AlexNet, and GoogLeNet with transfer learning to computer-aided detection problems. They reported that the applications of CNN image features can be improved by either exploring the hand-crafted features or by transfer learning (Shin et al. 2016). Although CNN methods require little engineering by hand and learn features automatically, they are limited by the available data size. In the medical field, it is relatively difficult to collect comparable amount of data as in other fields. Another limitation is the lack of labeling for the data, which need the help from experienced clinicians. Even if the former two issues are solved, the features obtained from CNNs are hard to interpret, which is not satisfying for medical applications and patient care. Regarding the former two problems, transfer learning, data augmentation, semi-supervised learning, and some other techniques have been proposed. We will address this in the next section.

8.3 FEATURE RANKING AND QUALIFICATION

It is challenging to build an accurate and stable model with large number of features extracted from the images, which exceed the sample number largely. Typically, the number of variables we have would reach thousands or even higher, while the patient number is about hundreds or even less. In these situations, overfitting and high variance are the major concerns because of the curse of high dimensionality.

Feature selection is important and necessary procedure before building predictive models. By doing feature selection, we can reduce the variable space, mitigate the overfitting problem, and improve the prediction performance. Furthermore, it is less expensive for computation and storage of the data. It is also beneficial for interpreting and understanding the underlying biological mechanisms if we can capture the structure and the relationships among these features. Also, visualization of the data is easy to implement after feature selection process.

Our main goal for feature selection is to select a subset of features that can help us build a good model and eliminate irrelevant and redundant features to avoid overfitting. "Good" means not only has a good performance on training data, but also can be generalized to independent testing data. The generalization performance is essential in the sense that it is actually what we care for in practical applications. Besides, the "goodness" facilitates the process of feature selection, e.g., the evaluation of prediction power for single variable filter methods and wrapper methods, which will be discussed later, and the overall model assessment. Model order or complexity is a proper indicator to reflect the goodness of a model. We do not want the model to be too complex to harm the generalization to unseen data, nor do we want it to be too simple to fail to fit the data well. A balance between the model order and generalization ability leads to the well-known bias-variance dilemma, as shown in Figure 8.9, in which too complex model results in low bias, but high variance (overfitting) and over simple model causes high bias while low variance (under-fitting). So a good model is the optimum point that balance complexity and generalization (including stability as well). Filtering the features by removing irrelevant and redundant ones is one powerful strategy to find the balance we need. "Features are relevant if their values vary systematically with category membership" (Gennari et al. 1989). "A good feature subset is one that contains features highly correlated with the class, yet uncorrelated with each other" (Hall 1999). Consider a set of N samples (x_k, y_k) $(k = 1, \cdots, N)$ with each sample being a p dimensional vector representing p features, $x_k = \left[x_k^{(1)} \cdots x_k^{(p)} \right]$. Based on how we process the features, there are two types of feature selection technique: (1) select features from our existing feature pool $\boldsymbol{x}_k = \left[x^{(1)} \cdots x^{(p)} \right]$ or (2) create new features that are functions of original features, e.g., $\theta = \left[x^{(1)} e^{-x^{(2)}} \cdots x^{(2)} x^{(n)} \right]$. For the former case, based on the relationship between the feature selection and the model building processes, there are mainly three categories: filter methods, wrapper methods, and embedded methods. While for the latter case, varieties of machine learning algorithms exist, such as principle component analysis (PCA) and different clustering methods. One important method in this case is the CNN we mentioned in the feature type part, which automatically create features from the data.

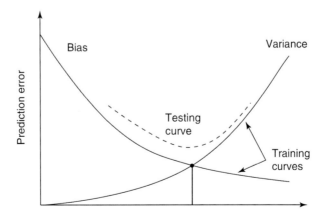

Figure 8.9 The bias-variance trade-off. (From El Naqa et al., 2015.)

Ranking and feature subset selection are two main operations that can be implemented for filter, wrapper, and embedded methods. A feature ranking based on individual importance of the features will be evaluated by these methods, then the top features that perform well for prediction task could be selected. In some cases, we first get the ranking list and do the subset selection based on the list. While in other cases, we may only get the subset features from the method. Usually, subset selection is supervised, while obtaining a ranking might be supervised or not (Roffo et al. 2015).

8.3.1 FILTER METHOD

Filter method can be regarded as a pre-processing step that explores the intrinsic properties of the features. Filter method is a ranking-based method that is independent of the classifier we choose. A ranking list of features will be provided using a relevance index. Relevance index is the metric that we can measure how relevant the feature is to the data or the outcome. There are many relevance indices that can be used.

8.3.1.1 CORRELATION CRITERIA

1. Pearson correlation coefficient is one of the simplest linear relevance index:

$$R(i) = \frac{cov(X_i, Y)}{\sqrt{var(X_i) var(Y)}}.$$ (8.46)

Pearson correlation can only detect linear dependencies between the features and target. In order to improve it to the non-linear regime, we can first pre-process the data, such as squaring, log to introduce non-linearity, and then use this correlation.

2. Relief (Kira and Rendell 1992) estimates the quality of features based on how well they can distinguish between instances that are close to each other. It searches for two nearest neighbors: one from the same class (nearest hit H), and the other from the different class (nearest miss M). The formula below shows the weight updating:

$$W_k = W_k - \left(X_i^{(k)} - H_i\right)^2 + \left(X_i^{(k)} - M_i\right)^2,$$ (8.47)

where W_k is the weight for feature X_k, i denotes the randomly sampled instance, and H_i and M_i are the nearest hit and miss for instance i. Higher the weight is for one feature, the better ranking position it will get. Kononenko modified the original algorithm by using k nearest neighbors other than the nearest one and Manhattan norm rather than Euclidean norm. However, this method is not able to remove redundant features.

8.3.1.2 CLASSICAL TEST STATISTICS

Test statistics have also been used to do feature selection, such as t-test, f-test, and chi2-test. Here, we only talk about Chi2-squared to give an example. Chi-squared is the most common statistical test that can be used to reject the null hypothesis that the data are independent. The higher score Chi-squared gets, the more correlated are the features correspondingly.

8.3.1.3 SINGLE VARIABLE PREDICTORS

The importance of individual features could be evaluated according to the goodness of fit using various classifiers. In order to implement this idea, we need to first build a classifier using a single variable and find a metric for the predictive power of individual features. There are several metrics available: area under the receiver operating characteristic (ROC) curve, error rate, etc. In essence, for a classification problem, scores are always calculated by the algorithms before the final step so that it can be used as a threshold to discriminate among classes. These scores can naturally be used as the relevance index for the filter method.

8.3.1.4 INFORMATION THEORETIC CRITERIA

1. Mutual information (cross entropy or information gain) is commonly used to measure the stochastic dependency of discrete random variables (Soofi 2000). It is based on the concept of entropy—the entropy of a random variable is a measure of the uncertainty:

$$H(y) = -\sum_y P(y)\log(P(y)). \qquad (8.48)$$

Then, the conditional entropy measures the uncertainty on y when x is known.

$$H(y|x) = H(y,x) - H(x). \qquad (8.49)$$

Thus, the mutual information is the uncertainty decrease of y when knowing x:

$$I(x,y) = H(y) - H(y|x). \qquad (8.50)$$

The overall formula for mutual information is:

$$I(i) = \sum_{x_i}\sum_y P(X = x_i, Y = y)\log\frac{P(X = x_i, Y = y)}{P(X = x_i)P(Y = y)}. \qquad (8.51)$$

The probabilities above are estimated by frequency counts. In addition, there are several variants for the mutual information, such as gain ratio $\left(GR(X,Y) = \frac{MI(X,Y)}{H(X)}\right)$, which is the normalized mutual information, that helps to reduce the bias toward features.

2. Markov blanket filter:
   However, one main problem filter method meets with is that it is univariate and ignores the interactions of features and probably leads to a suboptimal solution. In order to overcome this, several multivariate filter methods were introduced, such as the Markov blanket above, mutual information-based feature selection (MIFS) algorithm, correlation-based feature selection (CFS) (Hall 1999), and infinite feature selection recently published.

8.3.2 WRAPPER METHOD

Unlike filter method, that separates the steps of feature selection and model building, wrapper method combines these two steps together. Similar with the single variable method in the filter part, the prediction performance could be used for the feature subset selection. The difference is whether subsets of features or one single feature is fed to the classifier for wrapper and filter, respectively. Incorporating all the features enable the interactions among features to avoid redundancy. However, there are three issues to think about: (1) which classifier to use; (2) metric to measure the prediction power; and (3) the search strategy.

There is no explicit rule for choosing the classifiers. It basically depends a lot on the data underlying structure and characteristics. Some common classifiers are linear discriminant analysis (LDA), support vector machine (SVM), decision trees, and Naïve Bayes.

For the second issue, the most widely used method is cross-validation when the data is limited, which is usually the case in medical applications. K-fold cross-validation roughly split the data into K roughly equal-sized folds and each time use one-fold as validation set with other folds as training set, then average the K prediction errors to give an estimation of prediction performance of the feature subset we used in the model. The choice of K is usually 5, 10, or n (n is the number of samples). N folds is the famous "leave-one-out" cross-validation (LOO-CV or "jackknife"). In each iteration, all data are used to train the model except one data point being left out for testing. Though being a high variance and overly optimistic estimator, it gives approximately unbiased prediction results. Bootstrapping is another resampling method to assess the model accuracy. Typical way of implementation is to randomly draw datasets with replacement from original datasets with a probability of inclusion of 0.632 and obtain a "new" dataset the same size as the original. This process should be repeated typically thousands of times to make the prediction good enough. With these techniques, we can get a good estimation and select the features.

For a small feature set, exhaustive search can be used to get an optimal result. However, it is computationally very expensive or almost impossible as the number of features increases (non-deterministic polynomial-time [NP]-hard problem to solve [Amaldi and Kann 1995]). In order to overcome this search space explosion issue, two kinds of strategies, namely, deterministic search and stochastic search (Foster 2001) were introduced. In deterministic search, there are sequential forward selection (SFS) (Whitney 1971) and sequential backward elimination (SBE) (Marill and Green 2006) methods. SFS starts from empty set, sequentially incorporating feature that maximize the objective function when combined with the features that have already been selected. Oppositely, SBE starts from the full set, and sequentially eliminates the features that least reduce the objective function. These methods suffer from the "nesting effect", which means the features discarded for SBE or selected for SFS cannot be reselected or discarded later. This will lead to a suboptimal result. Michael and Lin developed Plus-l-Minus-r search to resolve the problem (Michael and Lin 1973). But it is still suboptimal and the choosing of l and r is not theoretical. Then Kittler and Pudil et al. further developed a "floating" selection algorithm (sequential floating forward/backward selection [SFFS and SFBS]) by using flexible values of l and r to approximate the optimal solution (Pudil et al. 1994). Siedlecki et al. developed a method called beam search that performs a best-first search in the feature subset space. Stochastic search is less prone to the local minimum, they also introduced the genetic algorithm (GA) for feature selection, with the chromosomes as the initial subset features together with their fitness measured by model performance (Siedlecki and Sklansky 1988, 1989). Those chromosomes with larger fitness undergo cross-over or mutation to breed into the next generation and thus gradually evolve to the better and better subsets. There are also simulated annealing, randomized hill climbing (Skalak 1994) stochastic methods for searching.

A widely used wrapper method is the SVM-recursive feature elimination (RFE) method (Guyon et al. 2002). The RFE is a kind of backward selection method: we train the classifier (in this case SVM), compute the ranking score for all features (the weights for the variables for SVM), and then remove the feature with the smallest score, repeat above steps until the set is empty. Though wrapper methods consider the feature dependencies, they are computationally intense and are prone to overfitting.

8.3.3 EMBEDDED METHODS

Embedded methods embed the feature selection step into the model construction process, so that the optimal set can be obtained while training the model. Embedded methods are more efficient without the large amount of subsets search and have the potential to perform better with larger available dataset, as data will not be split into training and validation sets. There are two families of embedded methods: estimates changes in the objective function value by making moves in the feature subset space and combine with some search engine to obtain the feature set; formalize the objective function with maximum likelihood and a regularization term, and directly optimize the objective function. The least absolute shrinkage and selection operator (LASSO) for linear regression is a typical embedded method belonging to the second family that solves the least-squared problem with an L1-penalized regularization term. The key is that the weights for the above objective function are sparse, so that the algorithm can automatically select relevant features. The main shortage of these L1 norm penalty methods is that they cannot be kernelized (i.e., mapped easily to higher dimensions), in addition they are prone to discontinuous derivative issues.

Another typical embedded method are decision trees (Breiman et al. 1984). When growing a tree, the selection of the feature is based on how much it decreases the impurity. For classification, the impurity is usually Gini index or information gain which provides the score for the feature ranking.

8.3.3.1 FEATURE EXTRACTION

Feature selection and feature extraction both belong to the feature engineering regime, aiming at reducing the dimensionality, extracting most useful information, and finally achieve the best prediction result. Comparing with feature selection, feature extraction has some advantages: (1) the unsupervised characteristic for most feature extraction methods makes them less prone to overfitting; (2) for data that are unlabeled or with few labeled samples; (3) more flexible to represent the data structure and thus have the potential to be more efficient in prediction tasks; and (4) the redundant and feature interaction issues are taken care of automatically. However, it is not as interpretable since the new set is a function of the original variables, and it is usually not reversible.

Different algorithms can be used to transform the original features to a new set that can better represent the original data with a lower dimension, such as the basic linear transforms PCA, factor analysis, and clustering and some other non-linear dimension reduction algorithms like the most recent one t-Distributed Stochastic Neighbor Embedding (t-SNE) (Maaten and Hinton 2008).

The intuition of PCA is that to fit the data with a n-dimensional ellipsoid, each axis of the ellipsoid stands for one principle component (Hotelling 1933). If the axis in one direction is small, then the corresponding variance of data along that direction is also small, we will not lose much information if we omit such component. PCA reduces dimension by finding a linear combination of the original basis that represents the data with the least information loss. Mathematically, let $X \in R^d$ be our dataset, $\mu \in R^d$ as the average of the original data, $A \in R^{d \times n}$ is the linear mapping, and $Y \in R^k$ is the new representation of the data, where k is the dimension of reduced result. We want to minimize the following objective function:

$$min_{\mu \in R^d, A, y_i \in R^k} \sum_{i=1}^{n} x_i - \mu - Ay_i^2 \text{ or } min_{P \in p_k} \sum_{i=1}^{n} x_i - Px_i^2, \tag{8.52}$$

where P is AA^T, the set of rank k, $d \times d$ projection matrix. Above optimization problem can be solved using Eckart & Young theorem which incorporates the singular value decomposition (SVD). A is given by the left singular vectors of X. Basically, PCA is solving the least-square rank-k linear approximation of the dataset. Maximizing the variance is another way of thinking about the problem, e.g., the variance of the first principle component is $Var\left(y^{(i)}\right) = \frac{1}{n}\sum_{i=1}^{n}\left(a_1^T x_i\right)^2 = a_1^T\left(\frac{1}{n}\sum x_i x_i^T\right)$, we can use generalized Rayleigh quotient to derive the PCA from the maximum variance perspective then, so that the reduced basis is just the eigenvectors of the covariance matrix XX^T. PCA is powerful, but it cannot deal with non-linear problems. Other non-linear dimension reduction techniques are now available, such as the kernelized PCA (Schölkopf et al. 1997), Isomap (Balasubramanian and Schwartz 2002), t-SNE, etc. Among which t-SNE seems to be the most powerful one recently.

Clustering is another feature extraction algorithm which aims at finding relevant features and combining them by their cluster centroid based on some similarity measure. Similar with the supervised wrapper methods mentioned before, where the prediction performance can be used as a guide for the selection of feature subsets, a criterion that evaluates the quality of the feature set generated by some clustering algorithm is necessary. Two popular clustering methods are k-means and hierarchical clustering. K-means algorithm is an iteratively expectation-maximization (EM) method, given number of clusters and a set of points in a Euclidean space, aiming to minimize the total sum of the distances of each point to the nearest cluster. Thus, the optimal k clusters that group the data will be found. Dy and Brodley have reported a feature subset selection using EM clustering (FSSEM) with two criteria—scatter separability and maximum likelihood criterion (Dy and Brodley 2000). The scatter separability used in discriminant analysis is used for the former criterion to evaluate the feature subsets based on how good they can group the data into clusters that are separable. The latter criterion is based on a Gaussian assumption of the clusters, it tells how good the model we build using certain features, which is a Gaussian mixture, fits the data. Furthermore, using their framework, different combination of clustering algorithms and criteria can be explored. Hong et al. came up with a clustering ensembles guided feature selection algorithm (CEFS) that combines the clustering ensembles and population-based incremental learning algorithm (Hong et al. 2008). Different clustering methods might give contradictory results. Clustering ensembles method leverages various clustering solutions and achieves a more robust consensus result. So CEFS searches for a subset of features that the clustering algorithm trained on this can achieve the most similar solution with the ensembled one. Dhillon et al. developed an information-theoretic feature clustering method by minimizing the mutual information measured by generalized Jensen-Shannon divergence (Dhillon et al. 2003).

Since the ranking method is mentioned, some more discussions about this would be presented here. Feature ranking shows the individual significance of features based on the correlation with outcome or prediction ability, it usually ignores the interaction of different features, which makes the consequent subset selection not optimal. Based on the investigation of Guyon and Elisseeff (2003), redundant features may still be able to reduce noise and thus lead to better class separation. Highly correlated features might provide complementary information. A feature that is useless in discriminating classes by itself could achieve a significant improvement in prediction if combined with some other features. These results clearly pointed out the limitation of ranking operation. Individually and independently ranking the features suffers from the loss of best combination of features. However, ranking methods are still in use for a lot of applications with its special benefits. Even if the feature set might not be optimal, it could give satisfactory results with the advantage of computational efficiency. Traditionally, feature selection is a NP-hard problem, the optimal combination can be obtained by exhausting all the possible sets of features which make the computation impossible to be implemented. While we only need to calculate n scores for the ranking and sort them. Another advantage is the statistical stability. However, for direct selection methods, especially for the data that contain lots of correlated and redundant features, the subset of features provided might change dramatically every time. Since sometimes one feature will be selected, other times the highly correlated features will be selected. The good news is we are able to obtain good stability by using some ranking aggregation techniques that we can get a convergent and near optimal ranking list in prepare for use, which makes the result more stable and robust. Typical aggregation methods are Borda, Kemeny, etc. (Kemeny 1959, Young and Levenglick 1978). We will not go into the details of ranking aggregation. Interested readers may want to refer to the reference papers to learn more.

8.3.4 NEURAL NETWORK FOR FEATURE SELECTION

As mentioned in the types of features part, neural networks could be used to construct features, specifically the features extracted by CNN for images. Besides, if the inputs of the networks are features, further feature extraction (transformation) and selection can be realized. Using neural networks for feature selection is not a new technique. Numerous research about feature selections using neural networks were available from the 90s. Basically, there are three types: zero-order methods using only the network parameter values; first-order methods using the first derivatives of network parameters; and second-order methods using the second derivatives of parameters. In order to take into account the feature dependencies, only one variable was suggested to be selected and retrain the network to evaluate the relevance of the remaining variables. Yacoub and Bennani exploited both the weight value and the network structure of a multi-layer perceptron (Yacoub and Bennani 2000). First-order methods evaluate the relevance by derivative of the error or the outcome with respect to that feature. Saliency-based pruning (SBP) is one example of this type. It measured the usefulness of one variable by replacing it with its empirical mean and then calculated the error change (Utans et al. 1995).

With the development of computing ability, NN becomes deeper and deeper, which makes deep feature selection possible. Ruangkanokmas et al. selected features and reduced noise by using deep belief networks (DBN) as the classifier with the feature ranking obtained by filter-based Chi-squared technique (Ruangkanokmas et al. 2016). Ibrahim et al. proposed a multi-level feature selection approach for gene selection (Ibrahim et al. 2014). They first used DBN to generate a high level representation that is able to extract feature interaction, then implemented normal feature selection (ReliefF) on the representation to select a subset of the genes, finally unsupervised active learning was applied to reduce the subset of genes based on the accuracy of classification. Zou et al. used DBN to achieve feature abstraction by minimizing the feature reconstruction error. And then selected the features by filtering off those with higher reconstruction error (Zou et al. 2015).

Here, we want to revisit the CNN-based featureless approaches. In order to address the problem of short-age of data, especially labeled data, transfer learning technique was proposed to transfer knowledge from a related task that has already been learned. This idea originates from the fact that human learners recognize and apply relevant knowledge from previous experience to new problems. Thus, in the medical image classification studies, transfer learning is a powerful tool to avoid overfitting and improve the performance. Specifically, CNN architectures, such as AlexNet, GoogLeNet, or ResNet that are pre-trained on large dataset (e.g., ImageNet) can be used as generic feature extractors, and then do the classification based on these features. We can either use other traditional machine learning algorithms such as SVM or LDA to do the classification or directly using CNN itself. For the latter case, the number of layers that will be pre-trained is a question we need to think about. Usually, it depends on how transferable the source dataset and target data-sets are, and also the amount of data we have for the new task (Yosinski et al. 2014). The more related the two datasets are, the more layers we can directly use in the new problem. However, in the case of medical application, we do not have a very similar and large dataset as for other natural image classification tasks. It would be better to use only the first few layers and then fine tune for the rest of the layers. While another issue is the labeled data we have is still not enough to fine tune on the rest of the layers, so that it is still very common to freeze all layers, and only tune the last fully connected layer using the training data, (freezing the layers means to adopt the weights obtained from training on the ImageNet directly). To increase the available data size, data augmentation is a useful technique. Specifically, images from different views, different time points, images scaled to various resolutions, images being rotated, shifted, etc. are simple ways to increase the training samples. Pezeshk et al. presented a new method to augment the data size (Pezeshk et al. 2017). They first blend a nodule to a new location to increase the diversity of the samples, then in order to break the diversity limitation that arose from inserting the same lesion to different backgrounds, they implemented a deformable transformation to create synthetic nodules and thus amplify the number of samples. The result showed a large improvement for multiple classifiers. Another powerful technique is the generative adversarial nets (GAN) that can generate new data from the existing samples (Goodfellow et al. 2014). GAN consists of two models: a generative model that captures the data distribution and a discriminative model that estimates the probability that a sample is from the training data rather than the generative model. Several research applying GAN to medical fields showed the potential of this method to provide useful new samples with little cost. Alex et al. used GAN in the detection of brain lesions (Alex et al. 2017) and Nie et al. developed a 3D GAN for estimating CT images from MRI images by taking MRI patches as input and CT patches as output (Nie et al. 2016). The shortage of GAN is that it could be sometimes relatively hard to train, and sometimes the models will not converge. And the result will be too similar to the original samples. It probably is not able to provide new information to the problem at hand.

8.4 DISCUSSION

Now we have discussed various types of features and large amount of methods that can be applied to implement feature selection and extraction. There are two more topics that we want to further discuss a bit.

8.4.1 Hand-crafted radiomic features vs CNN features

It is not clear to decide which kind of features are better when we deal with a specific task. Although quite time-consuming to obtain, hand-crafted features take the advantage of domain knowledge and tend to outperform CNN methods when the available dataset is small and noisy. In addition, we can choose feature selection and classification models that contain less fitting parameters to avoid overfitting, rather than the CNN methods that usually have thousands or even more parameters to fit. Another advantage is that engineered features are easier to interpret, since we know what features we extract and feed into the model. CNN methods are attracting more and more attention, with the advantage of avoiding tedious feature engineering and higher prediction accuracy. However, it suffers from the scarcity of data and the difficulty in interpretability. Since we do not know what is actually being learned, sometimes the "good" result might only come from bias in the data. For the worst case, it might be just garbage in, garbage out. For small dataset, transfer learning, data augmentation, and GAN have been implemented to amplify the dataset and showed promising results in image segmentation, disease detection, and endpoint prediction tasks. Another direction to think about is the combination of traditional features and CNNs directly applied to images. Features carefully designed can provide experts experience that might be hard to learn by CNN based on limited samples, while CNN can extract other important information missed by hand-crafted features. Several possible architectures could be input radiomic features into CNN, output classification results; input CNN features to different traditional machine learning classifiers; input radiomic features and images together to CNN; or input radiomic features and images to CNN to extract abstract features and use ensemble learning of machine learning algorithms.

8.4.2 FEATURE SELECTION STABILITY

As mentioned in the beginning of this section, our goal is to find a subset of features that machine learning models can be built on to give good prediction performance. "Good" here has two aspects: high accuracy and generalizability or stability. We want the features to give good performance not only on the training dataset, but also the testing set. During the implementation of these feature selection algorithms, we found that for one dataset, the features being selected show large variation for different methods and different folds of data. There are two reasons: firstly, a lot of the features might be correlated with each other and the dataset usually contains noise that will affect the feature selection algorithms; and secondly, all feature selection methods have statistical biases, they reflect different statistical aspects of the data. According to this, one solution to get robust features is the use of ensemble feature selection techniques, where multiple feature selection methods are combined to generate robust results (Saeys et al. 2008, Abeel et al. 2010). For this method, first step is to obtain a set of different feature selection methods, second is to aggregate the outputs of those methods. In order to get various ranking lists, we can use different feature selection methods, different folds of data, and different combination of features. For the second aggregation part, Borda and Kemeny are two typical aggregation methods (Sarkar et al. 2014). The ensemble method has the potential to mitigate the statistical bias issue for various feature selection methods. Then how about the correlation and redundant features issue? The ultimate solution should be revealing the underlying data structure (or relationship) in the feature space. Pavan and Pelillo developed a graph-theoretic approach to map the data to be clustered to the nodes of a weighted graph, where the nodes are features, and edge weights are similarity between two features (Pavan and Pelillo 2007). Later Zhang and Hancock further developed the graph-based feature selection method. They first used the dominant-set clustering algorithm to find the feature clusters that have the greater internal similarity inside. For each dominant set, they directly estimated the multidimensional interaction information (MII), which represent the amount of information those features contain about the output. The advantage of using MII is that it incorporates higher order feature interaction, so that the correlation and redundancy are considered when implementing feature selection (Zhang and Hancock 2011).

8.5 CONCLUSIONS

Feature selection remains an important and crucial topic for radiomics analyses. Current features can be divided into statics and dynamic. New trends driven by machine learning advances are pushing towards featureless approaches to avoid redundancy issues related to off-the-shelf challenges when dealing with these features that may not be represented by the data. However, the question of optimal feature engineering and selection of the most appropriate features to build a particular radiomics model without over-fitting is still ongoing.

REFERENCES

Abeel, T., Helleputte, T., Van De Peer, Y., Dupont, P. & Saeys, Y. 2010. Robust biomarker identification for cancer diagnosis with ensemble feature selection methods. *Bioinformatics*, 26, 392–398.

Aerts, H. J. W. L., Velazquez, E. R., Leijenaar, R. T. H. et al., 2014. Decoding tumour phenotype by noninvasive imaging using a quantitative radiomics approach. *Nat Commun*, 5, 4006.

Alex, V., Mohammed Safwan K. P., Chennamsetty, S. S. & Krishnamurthi, G. 2017. Generative adversarial networks for brain lesion detection. In Medical Imaging 2017: Image Processing (Vol. 10133, p. 101330G). *International Society for Optics and Photonics*.

Amaldi, E. & Kann, V. 1995. The complexity and approximability of finding maximum feasible subsystems of linear relations. *Theor Comput Sci*, 147, 181–210.

Avanzo, M., Stancanello, J. & El Naqa, I. 2017. Beyond imaging: The promise of radiomics. *Phys Med*, 38, 122–139.

Balasubramanian, M. & Schwartz, E. L. 2002. The Isomap algorithm and topological stability. *Science*, 295, 7–7.

Breiman, L., Friedman, J., Stone, C. J. & Olshen, R. A. 1984. *Classification and Regression Trees*, Boca Raton, FL: CRC Press.

Choi, N. C., Fischman, A. J., Niemierko, A., Ryu, J. S., Lynch, T., Wain, J., Wright, C., Fidias, P. & Mathisen, D. 2002. Dose-response relationship between probability of pathologic tumor control and glucose metabolic rate measured with FDG PET after preoperative chemoradiotherapy in locally advanced non-small-cell lung cancer. *Int J Radiat Oncol Biol Phys*, 54, 1024–1035.

Dhillon, I. S., Mallela, S. & Kumar, R. 2003. A divisive information theoretic feature clustering algorithm for text classification. *J Mach Learn Res*, 3, 1265–1287.

Doi, K. 2007. Computer-aided diagnosis in medical imaging: Historical review, current status and future potential. *Computerized Medical Imaging and Graphics: The Official Journal of the Computerized Medical Imaging Society*, 31, 198–211.

Dy, J. G. & Brodley, C. E. 2000. Feature subset selection and order identification for unsupervised learning. *Proceedings of the Seventeenth International Conference on Machine Learning.* San Francisco, CA: Morgan Kaufmann Publishers.

Echegaray, S., Nair, V., Kadoch, M., Leung, A., Rubin, D., Gevaert, O. & Napel, S. 2016. A rapid segmentation-insensitive "digital biopsy" method for radiomic feature extraction: Method and pilot study using CT images of non-small cell lung cancer. *Tomography*, 2, 283–294.

El Naqa, I. E. 2014. The role of quantitative PET in predicting cancer treatment outcomes. *Clin Transl Imaging*, 2, 305–320.

El Naqa, I., Grigsby, P., Apte, A. et al., 2009. Exploring feature-based approaches in PET images for predicting cancer treatment outcomes. *Pattern Recognit*, 42, 1162–1171.

Foster, J. A. 2001. Evolutionary computation. *Nat Rev Genet*, 2, 428–436.

Galloway, M. M. 1975. Texture analysis using gray level run lengths. *Comput Gr Image Process*, 4, 172–179.

Gennari, J. H., Langley, P. & Fisher, D. 1989. Models of incremental concept formation. *Artif Int*, 40, 11–61.

Gillies, R. J., Kinahan, P. E. & Hricak, H. 2016. Radiomics: Images are more than pictures, they are data. *Radiology*, 278, 563–577.

Gillman, R. 2002. Geometry and gerrymandering. *Math Horizons*, 10, 10–22.

Goodfellow, I., Bengio, Y. & Courville, A. 2016. *Deep Learning*, Cambridge, MA: MIT Press.

Goodfellow, I., Pouget-Abadie, J., Mirza, M., Xu, B., Warde-Farley, D., Ozair, S., Courville, A. & Bengio, Y. Generative adversarial nets. *Adv Neural Inf Process Syst*, 2014. 2672–2680.

Guyon, I. & Elisseeff, A. 2003. An introduction to variable and feature selection. *J Mach Learn Res*, 3, 1157–1182.

Guyon, I., Weston, J., Barnhill, S. & Vapnik, V. 2002. Gene selection for cancer classification using support vector machines. *Mach Learn*, 46, 389–422.

Hall, M. A. 1999. *Correlation-Based Feature Selection for Machine Learning.* Doctoral dissertation, University of Waikato, Department of Computer Science.

Haralick, R. M. & Shanmugam, K. 1973. Textural features for image classification. *IEEE Trans Syst Man Cybern*, 6, 610–621.

He, K., Zhang, X., Ren, S. & Sun, J. 2015. Deep residual learning for image recognition. *ArXiv e-prints*, 1512. Available: http://adsabs.harvard.edu/abs/2015arXiv151203385H [Accessed December 1, 2015].

Hong, Y., Kwong, S., Chang, Y. & Ren, Q. 2008. Unsupervised feature selection using clustering ensembles and population based incremental learning algorithm. *Pattern Recognit*, 41, 2742–2756.

Hotelling, H. 1933. Analysis of a complex of statistical variables into principal components. *J Educ Psychol*, 24, 417.

Hubel, D. H. & Wiesel, T. N. 1968. Receptive fields and functional architecture of monkey striate cortex. *J Physiol*, 195, 215–243.

Ibrahim, R., Yousri, N. A., Ismail, M. A. & El-Makky, N. M. 2014. Multi-level gene/MiRNA feature selection using deep belief nets and active learning. *Conf Proc IEEE Eng Med Biol Soc*, 2014, 3957–3960.

Jain, A. K. 1989. *Fundamentals of Digital Image Processing*, New York: Prentice Hall.

Julesz, B. 1975. Experiments in the visual perception of texture. *Sci Am*, 232, 34–43.

Kemeny, J. G. 1959. Mathematics without numbers. *Daedalus*, 88, 577–591.

Kickingereder, P., Burth, S., Wick, A. et al., 2016. Radiomic profiling of glioblastoma: Identifying an imaging predictor of patient survival with improved performance over established clinical and radiologic risk models. *Radiology*, 280, 880–889.

Kira, K. & Rendell, L. A. 1992. A practical approach to feature selection. *Proceedings of The Ninth International Workshop On Machine Learning.* Aberdeen, UK: Morgan Kaufmann Publishers Inc.

Krizhevsky, A., Sutskever, I. & Hinton, G. E. Imagenet classification with deep convolutional neural networks. *Adv Neural Inf Process Syst*, 2012. 1097–1105.

Kumar, V., Gu, Y., Basu, S. et al., 2012. QIN "Radiomics: The process and the challenges." *J Magn Reson Imaging*, 30, 1234–1248.

LeCun, Y., Bengio, Y. & Hinton, G. 2015. Deep learning. *Nature*, 521, 436–444.

Lecun, Y., Bottou, L., Bengio, Y. & Haffner, P. 1998. Gradient-based learning applied to document recognition. *Proceedings of the IEEE*, 86, 2278–2324.

Leijenaar, R. T., Nalbantov, G., Carvalho, S., Van Elmpt, W. J., Troost, E. G., Boellaard, R., Aerts, H. J., Gillies, R. J. & Lambin, P. 2015. The effect of SUV discretization in quantitative FDG-PET radiomics: The need for standardized methodology in tumor texture analysis. *Sci Rep*, 5, 11075.

Li, X., Li, F., Fern, X. & Raich, R. 2017. *Filter Shaping For Convolutional Neural Networks. International Conference on Learning Representations (ICLR).*

Maaten, L. V. D. & Hinton, G. 2008. Visualizing data using t-SNE. *J Mach Learn Res*, 9, 2579–2605.

Marill, T. & Green, D. 2006. On the effectiveness of receptors in recognition systems. *IEEE Trans Inf Theor*, 9, 11–17.

Michael, M. & Lin, W. C. 1973. Experimental study of information measure and inter-intra class distance ratios on feature selection and orderings. *IEEE Trans Syst Man Cybern*, 172–181.

Nie, D., Trullo, R., Petitjean, C., Ruan, S. & Shen, D. 2016. Medical image synthesis with context-aware generative adversarial networks. ArXiv e-prints, 1612. Available: http://adsabs.harvard.edu/abs/2016arXiv 161205362N (Accessed December 1, 2016).

Parekh, V. & Jacobs, M. A. 2016. Radiomics: A new application from established techniques. *Expert Rev Precis Med Drug Dev*, 1, 207–226.

Pavan, M. & Pelillo, M. 2007. Dominant sets and pairwise clustering. *IEEE Trans Pattern Anal Mach Intell*, 29, 167–172.

Pezeshk, A., Petrick, N., Chen, W. & Sahiner, B. 2017. Seamless lesion insertion for data augmentation in CAD training. *IEEE Trans Med Imaging*, 36, 1005–1015.

Pudil, P., Novovicova, J. & Kittler, J. 1994. Floating search methods in feature selection. *Pattern Recognit Lett*, 15, 1119–1125.

Roberts, L. G. 1963. *Machine Perception of Three-Dimensional Solids*. (Doctoral dissertation, Massachusetts Institute of Technology).

Roffo, G., Melzi, S. & Cristani, M. 2015. Infinite feature selection. *Proceedings of the 2015 IEEE International Conference on Computer Vision (ICCV)*. IEEE Computer Society.

Ruangkanokmas, P., Achalakul, T. & Akkarajitsakul, K. 2016. Deep belief networks with feature selection for sentiment classification. *Intelligent Systems, Modelling and Simulation (ISMS), 2016 7th International Conference on IEEE*.

Saeys, Y., Abeel, T. & Van De Peer, Y. 2008. Robust feature selection using ensemble feature selection techniques. In Daelemans, W., Goethals, B. & Morik, K. (Eds.), *Machine Learning and Knowledge Discovery in Databases: European Conference, ECML PKDD 2008*, Antwerp, Belgium, September 15–19, 2008, Proceedings, Part II. Berlin, Germany: Springer.

Sahiner, B., Chan, H. P., Petrick, N., Wei, D., Helvie, M. A., Adler, D. D. & Goodsitt, M. M. 1996. Classification of mass and normal breast tissue: A convolution neural network classifier with spatial domain and texture images. *IEEE Trans Med Imaging*, 15, 598–610.

Sarkar, C., Cooley, S. & Srivastava, J. 2014. Robust feature selection technique using rank aggregation. *Appl Artif Intell*, 28, 243–257.

Schölkopf, B., Smola, A. & Müller, K.-R. Kernel principal component analysis. *International Conference on Artificial Neural Networks*, 1997. Springer, 583–588.

Segal, E., Sirlin, C. B., Ooi, C. et al., 2007. Decoding global gene expression programs in liver cancer by noninvasive imaging. *Nat Biotech*, 25, 675–680.

Shin, H.-C., Roth, H. R., Gao, M., Lu, L., Xu, Z., Nogues, I., Yao, J., Mollura, D. & Summers, R. M. 2016. Deep convolutional neural networks for computer-aided detection: CNN architectures dataset characteristics and transfer learning. ArXiv e-prints, 1602. Available: http://adsabs.harvard.edu/abs/2016arXiv160203409S (Accessed February 1, 2016).

Siedlecki, W. & Sklansky, J. 1988. On automatic feature selection. *Intern J Pattern Recognit Artif Intell*, 02, 197–220.

Siedlecki, W. & Sklansky, J. 1989. A note on genetic algorithms for large-scale feature selection. *Pattern Recognit Lett*, 10, 335–347.

Simonyan, K. & Zisserman, A. 2014. Very deep convolutional networks for large-scale image recognition. ArXiv e-prints, 1409. Available: http://adsabs.harvard.edu/abs/2014arXiv1409.1556S (Accessed September 1, 2014).

Skalak, D. B. 1994. Prototype and feature selection by sampling and random mutation hill climbing algorithms. *Proceedings of the Eleventh International Conference on International Conference on Machine Learning*. New Brunswick, NJ: Morgan Kaufmann Publishers.

Soofi, E. S. 2000. Principal information theoretic approaches. *J Am Stat Assoc*. 95, 1349–1353.

Szegedy, C., Liu, W., Jia, Y., Sermanet, P., Reed, S., Anguelov, D., Erhan, D., Vanhoucke, V. & Rabinovich, A. 2015. Going deeper with convolutions. *Proceedings of the IEEE Conference on Computer Vision and Pattern Recognition*, 1–9.

Thibault, G., Angulo, J. & Meyer, F. 2014. Advanced statistical matrices for texture characterization: Application to cell classification. *IEEE Trans Biomed Eng*, 61, 630–637.

Thorwarth, D., Eschmann, S. M., Paulsen, F. & Alber, M. 2005. A kinetic model for dynamic [18F]-Fmiso PET data to analyse tumour hypoxia. *Phys Med Biol*, 50, 2209–2224.

Utans, J., Moody, J., Rehfuss, S. & Siegelmann, H. 1995. Input variable selection for neural networks: Application to predicting the US business cycle. *Computational Intelligence for Financial Engineering, Proceedings of the IEEE/IAFE 1995, IEEE*, 118–122.

Vaidya, M., Creach, K. M., Frye, J., Dehdashti, F., Bradley, J. D. & El Naqa, I. 2012. Combined PET/CT image characteristics for radiotherapy tumor response in lung cancer. *Radiother Oncol*, 102, 239–245.

Vallières, M., Kay-Rivest, E., Perrin, L. J. et al., 2017. Radiomics strategies for risk assessment of tumour failure in head-and-neck cancer. *Sci Rep*, 7, 10117.

Van Dijk, L. V., Brouwer, C. L., Van Der Schaaf, A., Burgerhof, J. G., Beukinga, R. J., Langendijk, J. A., Sijtsema, N. M. & Steenbakkers, R. J. 2017. CT image biomarkers to improve patient-specific prediction of radiation-induced xerostomia and sticky saliva. *Radiother Oncol*, 122, 185–191.

Van Velden, F. H., Kramer, G. M., Frings, V., Nissen, I. A., Mulder, E. R., De Langen, A. J., Hoekstra, O. S., Smit, E. F. & Boellaard, R. 2016. Repeatability of radiomic features in non-small-cell lung cancer [(18)F] FDG-PET/CT studies: Impact of reconstruction and delineation. *Mol Imaging Biol*, 18, 788–795.

Van Velden, F. H. P., Cheebsumon, P., Yaqub, M., Smit, E. F., Hoekstra, O. S., Lammertsma, A. A. & Boellaard, R. 2011. Evaluation of a cumulative SUV-volume histogram method for parameterizing heterogeneous intratumoural FDG uptake in non-small cell lung cancer PET studies. *Eur J Nucl Med Mol Imaging*, 38, 1636–1647.

Watabe, H., Ikoma, Y., Kimura, Y., Naganawa, M. & Shidahara, M. 2006. PET kinetic analysis—Compartmental model. *Ann Nucl Med*, 20, 583.

Whitney, A. W. 1971. A direct method of nonparametric measurement selection. *IEEE Trans Comput*, 20, 1100–1103.

Xie, S., Huang, X. & Tu, Z. 2015. Top-down learning for structured labeling with convolutional pseudo-prior. ArXiv e-prints, 1511. Available: http://adsabs.harvard.edu/abs/2015arXiv151107409X (Accessed November 1, 2015).

Yacoub, M. & Bennani, Y. 2000. Features selection and architecture optimization in connectionist systems. *Int J Neural Syst*, 10, 379–395.

Yip, S. S. & Aerts, H. J. 2016. Applications and limitations of radiomics. *Phys Med Biol*, 61, R150–166.

Yosinski, J., Clune, J., Bengio, Y. & Lipson, H. 2014. How transferable are features in deep neural networks? Proceedings of the 27th International Conference on Neural Information Processing Systems—Volume 2. Montreal, Canada: MIT Press.

Young, H. P. & Levenglick, A. 1978. A consistent extension of condorcet's election principle. *SIAM J Appl Math*, 35, 285–300.

Zeiler, M. D. & Fergus, R. 2014a. Visualizing and understanding convolutional networks. In FLEET, D., Pajdla, T., Schiele, B. & Tuytelaars, T. (Eds.), *Computer Vision—ECCV 2014: 13th European Conference, September 6–12, 2014, Proceedings*, Part I. Zurich, Switzerland: Springer International Publishing.

Zeiler, M. D. & Fergus, R. 2014b. Visualizing and understanding convolutional networks. *European Conference on Computer Vision*. Berlin, Germany: Springer, 818–833.

Zhang, W., Doi, K., Giger, M. L., Wu, Y., Nishikawa, R. M. & Schmidt, R. A. 1994. Computerized detection of clustered microcalcifications in digital mammograms using a shift-invariant artificial neural network. *Med Phys*, 21, 517–524.

Zhang, Z. & Hancock, E. R. 2011. A graph-based approach to feature selection. *Proceedings of the 8th International Conference on Graph-based Representations in Pattern Recognition*. Münster, Germany: Springer-Verlag.

Zou, Q., Ni, L., Zhang, T. & Wang, Q. 2015. Deep learning based feature selection for remote sensing scene classification. *IEEE Geosci Remote Sens Lett*, 12, 2321–2325.

Zwanenburg, A., Leger, S., Vallières, M., Löck, S. & Image Biomarker Standardisation Initiative, F. T. 2016. Image biomarker standardisation initiative. ArXiv e-prints, 1612. Available: http://adsabs.harvard.edu/abs/2016arXiv161207003Z (Accessed December 1, 2016).

Predictive modeling, machine learning, and statistical issues

PANAGIOTIS KORFIATIS, TIMOTHY L. KLINE, ZEYNETTIN AKKUS, KENNETH PHILBRICK, AND BRADLEY J. ERICKSON

9.1 DATA EXPLOSION AND DECISION SUPPORT SYSTEMS

9.1.1 INTRODUCTION

The electronic health record enables the routine storage of medical records in digital format. The Picture Archiving and Communications System (PACS) or other medical image archives store the image data generated in clinical care. The amount of both text and imaging data keep growing as the clinical practice keeps improving. As a physician cares for a patient, she/he must evaluate information that originates from pathology, radiology, physical examination, as well as other physicians. In the case of radiology, modern magnetic resonance (MR) and computed tomography (CT) imaging systems produce high quality imaging data that contain several images of the disease being investigated. Pathology generates high resolution 2D images that may contain more than 10^9 pixels (Wang et al. 2011). Clinicians in everyday clinical practice are expected to evaluate and integrate these data in order to design the best treatment plan for their patients. Integrating all these types of data with their knowledge of medicine is a substantial challenge that can lead to delays or errors. Lack of adequate access to data can also lead to delays or errors. In addition, the knowledge base required to properly use the data is rapidly advancing, and it is not feasible for every physician to master every type of data, in order to provide the best possible care.

Machine learning has been proposed as a technology that can be leveraged to create decision support systems (Brink et al. 2017; Hassanpour et al. 2017; Pons et al. 2016; Shin and Markey 2006), and in several reports, is highlighted as an important component in delivering personalized medicine (Cruz and Wishart 2007; Davis et al. 2008; Lee et al. 2018; Moon et al. 2007). Machine learning has been described for a range of tasks including classification between malignant and benign, prediction of molecular biomarkers, creation of quantitative measurements by segmentation of interest, or even classification of CT scans based on present of pathology to prioritize reading of those scans. Figure 9.1 captures the general schema of supervised learning.

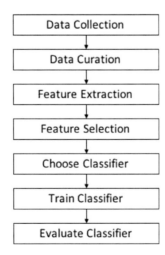

Figure 9.1 Supervised classification overview. A first step is data collection followed by curation. Next, features are extracted (computed) and the best features are selected. Those features are then provided to the classifier that has been selected for this task, the classifier is trained, and that trained system is evaluated to determine if it meets performance requirements.

9.1.2 LEARNING FROM MEDICAL IMAGING DATA

9.1.2.1 DATA COLLECTION

Every major medical institution stores imaging data in the PACS or in a separate vendor-neutral archive (VNA), nearly always using digital image communications in medicine (DICOM) (2001). The radiology information system (RIS) is used for the text component of an imaging department, such as the order, the indication for examination, the interpretation, and billing information. The PACS provides an image viewer that enables quick image curation (typically by the radiologic technologist) and image viewing and measurement for examination interpretation (typically by a radiologist).

The first step in developing a machine-learning-based tool for decision support is to identify the set of imaging examinations relevant to the question to be addressed. For instance, if one wishes to develop a tool for detecting nodules on lung CT, one must be able to select and extract the lung CT examinations from the PACS, collect the associated metadata that are needed to establish the "truth" being predicted (presence and location of nodules in this example). PACS and RIS are primarily designed for efficient collection and interpretation of every clinical examination, but are not well-suited for supporting research on imaging biomarkers or radiomics. Both RIS and PACS can identify lung CT examinations, but identifying those that were done for a research cohort (e.g., lung cancer screening) may be more challenging, and determining which have nodules, let alone the locations and contours of those nodules, can be very difficult to extract in an automated fashion.

Once the examinations of interest are identified, one may query the PACS, and transfer (using DICOM) to a local file system or a content management system (Herrick et al. 2016; Korfiatis and Erickson 2016). It is quite common to convert the DICOM data file format to other ones like NeuroImaging informatics Technology Initiative (NIfTI). Before using the imaging data for research purposes, anonymization needs to be performed since the DICOM file format includes tags that can contain protected health information (PHI) among other important information such as the acquisition parameters.

9.1.2.2 PRE-PROCESSING

Once the data (with correct labels) are collected, they are typically split into three groups: training, validation, and test. Training data are the data used for training the network: that is, the weights are updated to produce a good result with this set of data. The validation set is then used to see how general the trained network is. In many cases, training and validation data are used and then re-mixed as training progresses (this is known as cross validation). The test set is typically held out and never "shown" to the learning algorithm, but is kept purely for testing the system accuracy. If datasets are limited, it is possible to only create training and validation sets, though there is a greater risk of overfitting (described later). In some cases, the validation and test set definitions are reversed, but in this manuscript, we will maintain this nomenclature.

Before creating the training, validation, and testing sets, it is necessary to pre-process the data. Pre-processing may include image registration, bias correction, intensity standardization, artifact correction, and denoising. The selection of the pre-processing to be performed is dependent on the images. For instance, intensity values originating from CT have very specific meaning, while intensities obtained from MRI scanners often have no absolute intensity scale that has meaning. In the case of functional imaging, parametric maps must first be computed by application of mathematical modeling to the raw data such as the generation of cerebral blood volume maps from dynamic susceptibility contrast MRI imaging (Korfiatis and Erickson 2014). The importance of the pre-processing steps is often overlooked. However, the selection of the pre-processing steps and their parameters can significantly affect the performance of the classification system build based on these data (Nanni et al. 2015).

9.1.2.3 FEATURE EXTRACTION

Feature extraction is a necessary step for most classical machine learning approaches. The features capture the information needed to have the correct output. The features may include quantitative or visible image properties, but deep learning methods may also identify features that are not visible or obvious. Image features should be robust against variations in noise, intensity, and rotation angles, as these are some of the most common variations observed when working with medical imaging data (Erickson et al. 2017).

9.1.2.4 FEATURE SELECTION

A large number of features have been proposed in the literature with several tools publicly available to calculate them (van Griethuysen et al. 2017). Utilizing a large number of features can lead to overfitting rather than learning the true basis of a decision (Way et al. 2010). Furthermore, by reducing the number of features, we reduce the complexity of the models and create models that are easier to understand.

Selecting the best features to use is not always simple. The process of selecting the subset of features that should be used to make the best predictions is known as feature selection (Saeys et al. 2007). Feature selection algorithms may be categorized as filter, wrapper, or an embedded method. Filter methods utilize a statistical measure to assign a score to each feature, and the features with the highest scores are used for the classifier. Wrapper methods select a set of features based on search, where each combination is evaluated and compared to other combinations. A predictive model is used to test a combination of features, and the resulting model accuracy is used as the score. Embedded methods learn which features best contribute to the accuracy of the model while the model is being created. Regularization methods are the most common type of embedded feature selection method.

9.1.2.5 MACHINE LEARNING AND PREDICTIVE MODELING

Machine learning is the field of computer science that gives computers the ability to learn without being explicitly programmed (Samuel 1988). Machine learning explores the study and construction of algorithms that can learn from and make predictions on data (Kohavi and Provost 1998). Such algorithms are not based solely on static program instructions, but instead use data to build a model from which predictions or decisions can be made from future/new data.

In general, the machine learning algorithms can be separated into three categories: supervised, unsupervised, and semi-supervised depending on the type of input data. This chapter will focus on supervised approaches.

9.1.2.5.1 Regression algorithms

Regression algorithms are used to model the relationship between variables (Schneider et al. 2010). The most commonly utilized regression algorithms are: linear, least squares, logistic, and stepwise regression. Linear regression is considered to be the most simple form of regression since the model representation is so simple (Eq. 9.1).

$$y = B_0 + B_1 \times x. \tag{9.1}$$

Given a set of training data, the algorithm estimates the best values for the coefficients (B_0 and B_1 in Eq. 9.1) using some error metric. The least squares method fits a straight line to the training data and calculates the distance from the line to each data point, and the sum of all the squared distances is minimized.

Regularization methods can be used to train the linear regression model. These methods also aim to minimize the sum of squared errors, but in addition they minimize the complexity of model. Two examples of these methods are Lasso (Tibshirani 1994) and Ridge regression (Ng 2004). Such methods are valuable because they reduce the chance of overfitting.

9.1.2.5.2 Instance-based algorithms

Instance-based algorithms construct their hypotheses directly from the training instances themselves. Two of the most common algorithms of this category are k-nearest neighbor (kNN) (Altman 1992) and self-organizing maps (SOM) (Kohonen 1982).

With, a test case is classified based on the number of (known) neighbors that are the most similar to it. The similarity function may be the Euclidean distance in parameter space between the values of the input vector versus the values of the vector for the other examples, but other similarity metrics may work better for specific use cases. It is worth noting that normalization of the feature vector values is critical for this classifier since this metric optimized is based on distance. One disadvantage of this type of algorithm is that the complexity (and thus computation time) grows with the number of data points.

9.1.2.5.3 Kernel-based methods

Kernel methods owe their name to the use of kernel functions, which enable them to operate in a high-dimensional space. A kernel computes the dot product of two vectors X and Y in a different and usually higher-dimensional feature space. This mapping of data to some other dimension is called the "kernel trick." Kernel methods include kernel perceptrons, support vector machines (SVMs) (Cortes and Vapnik 1995), Gaussian processes, and principal components analysis (PCA) (Hotelling 1933).

Support vector machines (SVMs) (Cortes and Vapnik 1995) are one of the most popular classifiers in this category. The algorithm computes a mapping of training data to enable an optimal hyperplane that separates the training data. That mapping and hyperplane are then used for classifying new examples. SVM can be used for both classification and regression. Non-linear SVM is also possible and means that the boundary "plane" is not a flat or straight plane.

SVMs utilize kernels, and the kernels allow SVMs to solve problems which aren't linearly separable in the original space. A common example is two interleaved spirals—this cannot be separated by a linear method, but can be separated with non-linear kernels (Cortes and Vapnik 1995).

9.1.2.5.4 Decision tree algorithms

Decision trees have natural appeal because the decision process it creates is human-readable (Kingsford and Salzberg 2008). The method gets its name by creating a tree or sequence of decisions. The classification process thus consists of answering the question at each decision point about the example in question. Each node of the tree can be viewed as a question, and the next question/node that follows depends on the answer.

There are several considerations the algorithm uses to create the decision tree. A node criterion attempts to split the training data with heterogeneous class labels into subsets with nearly homogeneous labels, stratifying the data so that there is little variance. Entropy and the Gini impurity (Flach 2012) are the most common metrics of heterogeneity. Complex trees can result in overfitting, thus stopping rules are required. The minimum number of training inputs to use on each leaf or the maximum depth of the tree can be used as rules. Decision trees are robust and can handle numeric and categorical features as well as missing data.

9.1.2.5.5 Ensemble methods

Ensembles classify data by learning the optimal combination of base model predictions. Figure 9.2 depicts an ensemble architecture. The classifiers used in the ensemble are called base learners, and the algorithm used to find the best combination is called the meta learner (Flach 2012).

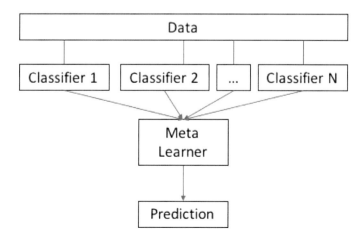

Figure 9.2 (See color insert.) A simple ensemble architecture consisting of multiple classifiers that each classify the examples. The output of those weak classifiers is then used by the meta-learner to produce the ensemble prediction, which is typically superior to any single weak classifier.

Boosting is an ensemble method that creates a strong classifier from a number of weak classifiers (Freund and Schapire 1997). A model is built utilizing the training data and then a second model is created in an effort to correct the errors from the first model. Models are added until the training set is predicted perfectly or a maximum number of models are added.

Random forests (Breiman 2001) are an ensemble of untrained decision trees referred to as weak learners. They take randomly sampled subsets of the training data to create a decision tree. This is done several times, and the resulting trees are then combined into an optimal model.

9.1.2.6 TRAINING

Training is the step that follows the data collection, curation, and feature extraction. Training for supervised classification uses a set of training cases (features and labels) to minimize a cost function.

There are two basic approaches to choosing a cost function of a classifier: the empirical risk minimization and structural risk minimization. Empirical risk minimization seeks the function that best fits the training data, while structural risk minimization includes a penalty function that controls the bias/variance trade-off (Vapnik 2000). Once the system is trained, it can be used to classify (also called "inference") an unknown example.

Cross validation (de Lacerda et al.) is a method to estimate the performance of a given machine learning method for a given problem. During cross validation, data are split into N parts (often called "folds"). The classifier is trained utilizing N-1 parts of the data and validated using the set of data left out. This is repeated N times, where a different fold is left out each time. When N is set to the number of examples, then one trains up N models, and tests on the 1 example left out in each of the N cases, and this is referred to as "leave one out cross validation." Cross validation is computationally expensive, but useful in case of smaller dataset.

9.1.2.7 EVALUATION

Evaluating a classifier is not always straightforward. Sometimes measures can suggest very good performance, and hide the fact that a classifier is performing poorly, for example, it does well on common classes, but does poorly on rare classes. This problem is quite common in cases of unbalanced datasets (i.e., where one class is much more common than other classes in the data). One way to help assure the robustness of results is to apply several different types of performance metrics. While not fool-proof, it would be less common for several metrics to all be wrong (assuming they are correctly applied).

The performance of the classifier can be summarized using a confusion matrix. In this matrix, each row corresponds to the true classes while each column corresponds to the predicted classes. Using the confusion matrix, we can calculate the most commonly utilized metrics including false positives, true positives, false negatives, true negatives, accuracy, sensitivity, specificity, error rate, precision, recall, and F1 score.

Accuracy captures the proportion of the correctly classified cases. The complementary measure is the error rate (wrongly classified cases). Both metrics are bounded between 0 and 1. True positives and true negatives correspond to correctly classified positive and negative test cases. False negatives are the incorrectly classified positive cases and false positives are the misclassified negative cases. Sensitivity and specificity are defined as the true positive and true negative fractions. Precision is the proportion of the true positives over the predicted positives (TP/(TP + FP). So given a set that contains 50 cases of class "1" and 50 cases of class "2," if our classifier predicted all 100 cases as class "1," then the precision is $50/100 = 0.5$. Recall is the fraction of true positives that are correctly predicted (TP/(TP + FN)). The F1 score is the harmonic mean of precision and recall (Eq. 9.2).

$$F1 = \left(\text{precision} - \text{recall} \right) / \left(\text{precision} + \text{recall} \right). \tag{9.2}$$

The area under the receiver operating curve (ROC), is a commonly utilized metric in case of binary problems. In contrast to overall accuracy that it is based on a specific threshold of the classifier output (usually 0.5), ROC plots the sensitivity and specificity for all the thresholds applied to the classifier output (Hajian-Tilaki 2013).

9.1.2.8 CHALLENGES

The classifier output on an example is a "score" for that example to be that class (Niculescu-Mizil and Caruana 2005). It is *not* a probability. In order to obtain probabilities, a calibration usually needs to be performed. A reliability graph evaluates how well a classifier is calibrated.

Missing data can be a challenge when dealing with machine learning and medical imaging, especially in case of data originating from multiple sites. One way to deal with missing data is to just remove all cases with missing information. Another is to interpolate the missing information based on correlated features. More sophisticated methods have been proposed to account for this issue (Belanche et al. 2014; Sovilj et al. 2016). These techniques could be applied in classical machine learning approaches. In deep learning, generative adversarial networks can be utilized to generate images for training from example images and other information, but these training examples are less independent and may not prevent overfitting.

Pre-processing images can affect image texture measures and consequently classification accuracy (Arivazhagan et al. 2006; Kolarević et al. 2017; Pietikäinen et al. 2000; Zhang and Tan 2002). Ideal texture features are intensity, resolution, noise, and rotation invariant (Brynolfsson et al. 2017), but often, pre-processing steps have to be utilized to account for features that don't present this characteristic, but this can also present challenges (Gourtsoyianni et al. 2017; Juntu et al. 2010).

Medical imaging datasets are quite often unbalanced (more examples that are one class versus another class). There are several techniques that can be used to deal with unbalanced datasets. One option is random undersampling of the class with the majority samples to produce an approximately equal number of examples of each class in the training set. Another option is to apply data augmentation (Ben-Cohen et al. 2018) to create more examples of the less common class. Finally, one may also weight datasets based on their rarity: the weights are higher for the uncommon class(es).

9.2 DEEP LEARNING

9.2.1 CONVOLUTIONAL NEURAL NETWORKS

Traditional machine learning requires extensive feature engineering. Recent advances in neural network technology have brought deep learning (DL) methods to the forefront, and it does not require feature engineering. Development of DL networks has been accelerated due to the availability of graphic processing units (GPUs) and libraries that can map the DL calculations onto. Increases in the number of GPU cores, speed, and memory have accelerated the rapid development of the field. Deep learning networks get their name because they have many layers of nodes, in some cases approaching 1000 layers.

There are several improvements in neural networks that enabled the success of DL. One specific subcategory of these algorithms, called convolutional neural network (CNN), has found extensive application in medical imaging, including segmentation (Badrinarayanan et al. 2015; Guo et al. 2016; Havaei et al. 2017; Kline et al. 2017; Korfiatis et al. 2016; Milletari et al. 2016) and classification (Esteva et al. 2017; Greenspan et al. 2016; Shin et al. 2016) tasks. Figure 9.3 describes the architecture of a simple CNN.

CNNs can be separated in two parts: the convolution part, where the features are computed, and the fully connected part where the features are combined using weights to produce a classification. In addition to convolution, the input data also undergo activation and pooling stages that improve the identification of important features.

The key concept of a neural network is that while the initial set of values are random, errant outputs due to these random weights provide feedback that allow improvement in the weights. A popular method for improving the weights in a CNN is stochastic gradient descent (SGD) (Bottou 2011). SGD is simple to implement and fast even in large datasets. An important parameter when training a CNN is the learning rate. The learning rate defines the magnitude of change to the weights. If the learning rate is too fast, the CNN may never "settle" on a good solution. If it is too low, learning will be slow, and may get caught in a local minimum. Currently, there is not an easy way to select the learning rate, so grid search and cross validation are often used for this purpose. Learning rate schedulers have also been proposed (Bengio 2012).

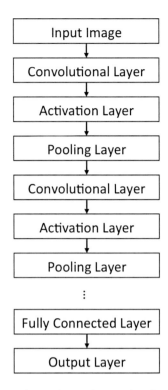

Figure 9.3 A depiction of a simple convolutional neural network architecture. The image is fed into the input layer, and the output of each layer is then passed to the next layer below. There are usually several groups of convolution/activation/pooling and the final is then fed into a fully connected network with several layers until a final output layer.

Several algorithms have been proposed to overcome these limitations. AdaGrad (Duchi et al. 2011), for instance, adapts the learning rate to the parameters, performing larger updates for infrequent and smaller updates for frequent parameters. The learning rate in some cases can shrink and eventually become infinitesimally small, at which point the algorithm stops learning. AdaDelta (Zeiler 2012) has been proposed as an extension of AdaGrad to solve the above problem. Other options for optimizers include: Adam (Kingma and Ba 2014), AdaMax, RMSprop, and Nandam (Dozat 2016; Kingma and Ba 2014).

The inputs to the CNN are usually normalized to zero mean and unit variance. During the training process, the normalization is lost and the training process slows down. Batch normalization (Ioffe and Szegedy 2015) re-establishes the range of values for every mini-batch and changes are back-propagated through the operation as well, this allows the use of higher learning rates.

CNNs can benefit from a technique called "transfer learning." Transfer learning means that a CNN trained for one task is used as a starting point in training a network for a different task. Needing to only fine tune a CNN is especially useful in cases where there is limited data for a specific task. When performing transfer learning, a factor to consider is the similarity of the original dataset to the one we want to perform classification since this can guide the fine-tuning strategy (Tajbakhsh et al. 2016). For instance, if the new dataset is of small size and the context is completely different, it might be better to train only the fully connected layer of the CNN. Additionally, the learning rates should be small since we assume that the network weights are relatively good.

9.2.2 OVERVIEW OF COMMON DEEP LEARNING ARCHITECTURES

Although CNNs started as relatively shallow architectures, they soon evolved to more complex and deeper architectures. LeNet (LeCun et al. 2015) was one of the first architectures introduced followed by

AlexNet (Krizhevsky et al. 2017) and VGG (Simonyan and Zisserman 2014). More complex architectures include GoogLeNet (22 layers deep) which introduced the inception module. With the introduction of ResNets (He et al. 2015), network architectures reached up to 1000 layers deep, and the residual connection or identity mapping was introduced to deal with the vanishing gradient problem. The most recent version of ResNet is called ResNeXt, and it combines the idea of ResNet and the inception layer (Xie et al. 2017). Architectures aimed toward mobile applications have been proposed (Iandola et al. 2016). Deep neural architectures specifically aimed at segmentation have also been proposed (Badrinarayanan et al. 2015; Long et al. 2015; Ronneberger et al. 2015).

GAN is an entirely different breed of neural network architecture, in which a neural network is used to *generate* an entirely new image which is not present is the training dataset, but which is realistic enough to be in the dataset (Creswell et al. 2017; Goodfellow et al. 2014).

9.3 APPLICATION OF DEEP LEARNING IN RADIOMICS

Machine learning and deep learning techniques are being rapidly applied to a wide range of medical imaging tasks, including: (i) reconstruction, (ii) denoising, (iii) registration, (iv) segmentation, and (v) classification tasks. For many tasks (e.g., predicting survival or predicting underlying genetic information from imaging phenotypes), image segmentation is a necessary first step as it gives the system knowledge of the important location to characterize (e.g., specific organ or tumor).

9.3.1 SEGMENTATION

Convolutional neural networks are the most commonly used architecture for medical imaging segmentation. They require a large database of images and corresponding annotations and are typically trained to predict whether a voxel is a part of the object or not. One approach is to extract image patches and learn to classify voxels in these smaller patches. However, more successful approaches have been proposed which are able to perform segmentation on the entire image (or volume) and classify each voxel individually.

Due to the fact that CNN architectures often downsample the image (e.g., strided convolutions, max pooling, etc.) in order to obtain a larger receptive field, and therefore have far lower output resolution, a number of approaches have been developed to prevent this decrease in resolution. One approach, known as "shift-and-stitch" (Long et al. 2015) applies the CNN to shifted versions of the input image and then stitches the results back together to obtain a full resolution version of the final output. A weakness of this approach is that by focusing on a small region of the image, the context and location of the patch within the image cannot be used to help make decisions.

Autoencoders are also being utilized for medical image segmentation tasks as well (Shin et al. 2013). An autoencoder is a symmetrical convolutional neural network consisting of an encoding and a decoding part (Korfiatis et al. 2016). U-Nets (Ronneberger et al. 2015), an evolution of autoencoders, combine a traditional CNN followed by an upsampling path which introduced connections to earlier layers (so-called "skip connections"). This approach has been successfully used in a number of areas of biomedical imaging to perform image segmentation (Çiçek et al. 2016; Kline et al. 2017). Figure 9.4 shows an architectural diagram of a U-Net, and one can see visually how it got its name.

The removal of user interaction to create a segmentation, which introduces another level of variability, is an important step for consistent and reproducible segmentation of the images.

9.3.2 CLASSIFICATION

Segmentation of an image is often a prerequisite for classification. Deep learning approaches outperform the majority of other, more traditional, machine learning classification methods. Traditional methods typically rely on feature engineering (i.e., identifying important features within the image), while DL automates the

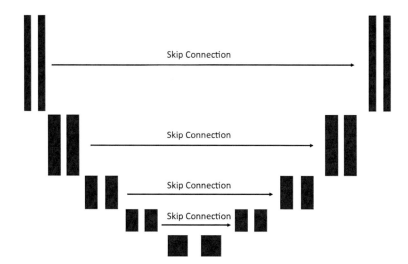

Figure 9.4 Architectural diagram of a U-Net. The input image goes through multiple layers of a CNN to identify important features. That output then goes through a series of steps that help to "restore" the resolution by skip-connections that access the original resolution pixel data.

feature computation and selection process and may also find features that were not thought to be important (and thus not included in the feature set considered).

A significant benefit of DL is that radiomic signatures can be learned without the need for investigation/computation of possible features. Deep learning has become the dominant technology for image classification tasks, such as the challenge of predicting malignancy from lung CT in the 2017 Data Science Bowl (Kuan et al. 2017). It has also been used to predict molecular markers in gliomas (Akkus et al. 2017; Korfiatis et al. 2017). Other application areas include classification of skin lesions (Kawahara and Hamarneh 2016), nodule detection in chest CT (Setio et al. 2016), and assessment of patient survival by training a CNN on MRI images of high-grade gliomas (Nie et al. 2016).

9.4 HOW MACHINE AND DEEP LEARNING CAN BE MISLEADING

9.4.1 OVERFITTING

Overfitting happens when a model learns a complex function that predicts the training data well, but performs poorly when assessed against data not in the training set. Essentially, overfitting occurs when the model learns predictive properties unique to the training examples, but not the features important to the general problem. Getting good performance on training data does not always mean that the model generalizes well.

Recall from the start of the chapter that labeled data are typically split into training, validation, and test sets. The training and validation sets are used to update the network parameters and validate that the model is properly generalizing. The difference in performance between the training data and validation data is a good indicator of overfitting. If they are very different, overfitting is probably occurring as seen in Figure 9.5, which shows training loss becomes minimal after a certain epoch, but the validation loss increases. It is common practice that after a few steps of training that the training and validation data are put back together and a new validation set is randomly (or in a stratified way) extracted for the next round of training (as in cross validation). This will invalidate the curves as an indicator of overfitting. If there is a separate test set that is never "seen," it can also be used to assess whether overfitting is occurring.

There are multiple strategies to address overfitting. The most common is increasing the amount of training data. Models trained on a large diverse dataset typically generalize better. When getting more data is not possible, regularization techniques such as L1 or L2 normalization of learnable parameters (model weights) can help.

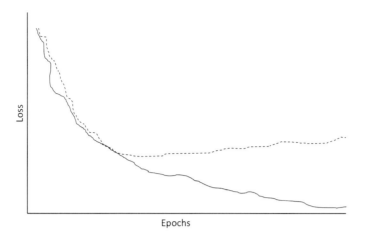

Figure 9.5 Loss curves show that while the network continues to perform better on the training data with each epoch, the performance on validation data is declining. This is likely due to overfitting. (The solid line shows the loss on the training data and the dashed line shows the loss on validation data.)

L1 and L2 regularization places constraints on the complexity of the model by forcing its learnable parameters to have only small values, additionally, in the case of L1, regularization by limiting the number of parameters included in the model (promoting spartisity).

Deep learning models can be prone to overfitting, especially in cases where not enough data are available. Dropout has been proposed as a simple technique to reduce overfitting (Srivastava et al. 2014). Dropout works by randomly removing a subset of the network nodes during each training epoch, which changes the composition of the network across training epochs. As a result, the model generates predictions which are less reliant on specific neurons in the network, which prevents the network from amplifying random noise. Other techniques to address overfitting strive to boost the size and variability of the training set and include batch normalization, data augmentation, and kernel and activity regularization. Such techniques are known as regularization techniques, and they are valuable because they can tune the model as well as controlling the model's complexity.

Multiple models can be fit to a given dataset. Increased model complexity is often not beneficial. Simple models are less likely to overfit than more complex models. A simple model is a model that has fewer learnable parameters or the one that has less entropy in the distribution of its learnable parameters. More learnable parameters means the more "memorizing" capacity and the more chance of overfitting. However, the models must have enough parameters to be able to learn about the problem to be solved. At this point in time, the design of networks is still an "art" to get the right balance of enough capacity to learn, but not enough to memorize.

9.5 WHAT DOES DEEP LEARNING SEE?

Machine learning models generate predictions by identifying predictive features identified within the input. When applied to images, features are combinations of shapes and textures present within an image that predict the desired output. Understanding the features identified and used by a machine learning model to form predictions is required if a models decision making process is to be understood.

One critique of using DL over traditional machine learning (ML) is that DL is often viewed as a "black box." However, there are a number of approaches available to gain insights into what a DL architecture has learned in terms of the important features, as well as what parts of an image the network most greatly responds to.

9.5.1 TRADITIONAL MACHINE LEARNING

Traditional machine learning methods (e.g., random forest, SVM, etc.) begin with an informed user that calculates a number of possible features, which may then be reduced to the most important subset using feature selection methods. Those features are then combined by the machine learning method to find the optimal combination for the task. As a result, the predictive power of the model is inherently limited by the set of input features quantified. Imaging features can be quantified across the entire image or, more commonly, for sub-regions of images that are believed to contain region of interest (ROI). As an example, for radiology imaging, ROIs may be defined that describe one or more slices of an organ of interest or regions of pathology to selectively quantify the images within these sub-regions. Following definition, ROI features are quantified from the identified region(s) using feature quantification libraries and/or custom engineered algorithms.

Pyradiomics (van Griethuysen et al. 2017) is a widely used radiomics feature quantification library (https://github.com/Radiomics/pyradiomics). The library has been designed to quantify imaging features based on shape and texture. This software tool is capable of computing more than 1500 textures for an ROI. Imaging features quantified using this tool have implicit definitions (e.g., mean voxel intensity, variance of voxel intensity, shericity, etc.). As a result, imaging features found to be predictive are inherently describable.

ROI texture quantifications generated using pyradiomics are a function of the ROI boundary definition and the pyradiomics texture definition. The ROI boundary describes the region across which the feature is sampled. The texture definition describes the texture features quantified. Alterations to ROI boundary definitions can substantially alter pyradiomics ROI texture quantifications. Relevant imaging features will be unidentified if the features reside outside the ROI definition or are not described within the fixed set of features quantified by the library. The reproducibility of multiple observers to extract similar measurements may be limited if there are significant differences between observer ROI annotations.

9.5.2 DEEP LEARNING

Deep learning methods learn relevant patterns in images and utilize these patterns to make predictions. Owing to their ability to self-identify patterns, deep learning classifiers can typically be built with substantially less descriptive annotations than would be required for a similar classifier built using traditional machine learning methods. As an example, a deep learning classifier designed to classify tumor imaging could be built using only the imaging and descriptive annotations that describe the type of a tumor exhibited in the imaging as training data. Given sufficient training examples, the classifier would then learn to highly weight features that correctly predict tumor label and ignore features not associated with tumor label. Notably, this method does not directly restrict the spatial location of the features identified by the classifier. Any feature in the imaging that is correlated with the desired output can be used by the classifier to formulate its prediction. As a result, determining the imaging features identified by a deep learning classifier and their relative importance on the models output are required to more fully understand a deep learning models decision making process.

9.5.2.1 VISUALIZING KERNEL ACTIVATION

CNNs learn to identify features contained in imaging data by training convolutional kernels to identify complex non-linear pattern (Krizhevsky et al. 2017). Following training, kernel weights from the first layer can be extracted and visualized to directly describe the most basic imaging features that a network has learned to detect. Convolutional kernels within the first layer, layer that receives imaging as input directly, are analogous to individual gradient-based texture features in a traditional radiomics feature set [e.g., pyradiomics (van Griethuysen et al. 2017)]. Kernel weights for the first layer typically converge to states which detect simple high frequency patterns (e.g., edges, gradients, etc.) (Krizhevsky et al. 2017).

Understanding the features detected in subsequent layer kernels is more informative, but also more difficult. To learn non-linear relationships across network layers, activation functions (e.g., ReLU, Tanh, PReLU, etc.) are applied to convolutional kernel output (Glorot et al. 2011). These functions transform the relationship between kernel input and output from linear to a non-linear. These non-linearities enable vastly different inputs, shallower network activations, to induce similar responses in deep neurons. One approach to understand imaging that activates lower layer convolutional kernels has been to track the kernel activations induced by images and then, for specific kernels of interest, return a set of images which induce the highest kernel activations (Girshick et al. 2014). For large datasets, this approach will produce examples of images which exhibit patterns that strongly activate a kernel. A clear limitation of this approach is that it does not directly indicate the features in the imaging which activated the kernel. If the kernel acts to identify a relatively common texture, the pattern identified by the kernel may not be obvious in the returned datasets. Also, and perhaps more importantly, this method is entirely dependent upon the dataset containing images which strongly activate the kernel of interest. Alternatively, generative methods have been used to create novel images which "maximize" the activation of a convolutional kernel from within a network (Erhan et al. 2009; Mahendran and Vedaldi 2015; Olah et al. 2017). In brief, these methods work by generating an image at random and then progressively altering the image based on the gradient of the image with respect to a kernel's activation (Erhan et al. 2009). At its core, this method is similar to the gradient-based optimization used to train CNN. Images generated using this approach exhibit patterns which strongly activate the kernel they are generated against. Due to the generative nature of this approach, no existing data outside of the network itself are required to create these images (Mahendran and Vedaldi 2015). This has led to the observation that CNNs retain "knowledge" of the images they were trained to predict (Mahendran and Vedaldi 2015). However, due to the effects of: random initialization, cross-layer non-linearities, and other parameters, this method can converge to multiple solutions.

Generating convolutional kernel activating images using a simple gradient descent-based approach typically results in the generation of non-informative high-frequency images which do not appear to visually reflect the set of "natural images" a network was trained against (Mordvintsev et al. 2015; Olah et al. 2017). The naive convergence to solutions which overemphasize the high frequency components of an image is likely a consequence of the sensitivity of convolutional kernels, particularly those in the uppermost levels of a network, to high frequency features in input imaging. Regardless, the representativeness of images generated to activate a CNN kernel can be substantially improved by conducting repeated minor transformations (translation, rotation, and scaling) on the image that is being generated between each epoch (Mordvintsev et al. 2015; Olah et al. 2017). Additionally, optimizations that transform the gradient computed between the kernel and activating images that promote convergence to solutions which exhibit frequency patterns that resemble those of "natural images," e.g., low pass filtering the gradient, typically substantially improve the representativeness of generated images (Mordvintsev et al. 2015; Olah et al. 2017).

9.5.2.2 INPUT FEATURE OCCLUSION

CNNs produce output based on the values in input imaging. Therefore, disruption or removal of the predictive features in input imaging should logically alter CNN predictions. Work investigating the relationship between features in images and CNN predictions have utilized this concept by masking features from input imaging, zeroing out the regions of experimental interest (Zeiler and Fergus 2014). The relative importance of the imaging features falling within masked region, then estimated by computing the difference between the classifiers output for the original image and the masked image. Clearly, removing data from images in this manner may introduce "new" artificial features into images, e.g., the boundary interface between mask and image. These artificial features could intern act to alter, increase, or decrease, the estimates of the masked features importance and thereby function to obfuscate the true importance of a masked feature.

9.5.2.3 SALIENCY MAPS

Saliency maps are visualizations which illustrate the relative importance pixels/voxels within input data have on a CNN classifiers class prediction (Simonyan et al. 2013). Saliency maps have been used to provide rough segmentations of objects based on the concept that voxels within an object typically have the greatest effect on an objects classification. For a given CNN classifier, producing N predictions, a single input image will produce N saliency maps, one for each of the N predictions. Saliency maps are computed as the derivative of a classes activation with respect to each pixel/voxel in the input. Saliency maps are perhaps most interesting when evaluated for correctly classified data against the corresponding correct class, true-positive, and for

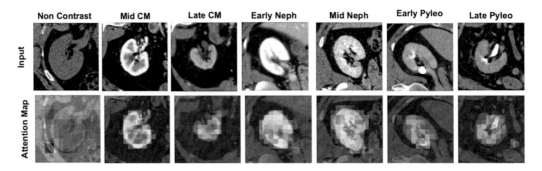

Figure 9.6 (See color insert.) Visualizations of attention maps generated to visualize the last layer of a CNN trained to identify renal scan phase from CT imaging. Attention maps illustrate that for the relative influences of regions of the image on the images classification (purple-blue = low; green = medium-low; yellow = medium; orange = medium-high; and red = high).

in-correctly classified data against the two-incorrectly classified classes, false-negative and false-positive as these maps illustrate the regions of the input data which exerted largest effect on the data classification.

9.5.2.4 ATTENTION MAPS

Attention maps illustrate the relative contribution of a CNN's output layer on its class predictions (Zhou et al. 2016). Similar to saliency maps, for a given CNN, producing N predictions, a single input image produces N attention maps, one for each of the N predictions. Attention maps were first demonstrated for a classification CNN which ended with network output undergoing a global average pooling operation which was then used as input to a single dense layer which produced the networks final classification (Zhou et al. 2016). The resulting map was then normalized, rescaled to the dimensions of the input image, and drawn transparently over it using a heat-map like projection, in which the "hottest" parts of the projection exerted the greatest effect on the classifiers output. Figure 9.6 shows a representative example of attention maps for CNN trained to identify the kidneys in CT images. Since the introduction of the attention map, more generalizable approaches, e.g., Grad-CAM map, have been developed to generate visualizations which are similar to attention maps for any layer in a network of arbitrary network complexity (Selvaraju et al. 2016). Analyzing attention maps of correctly and incorrectly classified data can provide important insights as to features in the input imaging which act to activate a classifier.

For visualizations of network output, e.g., attention maps and Grad-CAM maps, to have utility in understanding the relative importance of input features on CNNs predictions, the input data which fall under the attention map need to exert a large effect on the values in the overlaid attention map. Convolutional kernels with dimensions greater than 1×1 move information across the convolutional plane and thereby will act to disrupt the correspondence between voxels in input imaging and the attention map. For very deep networks, these effects may marginalize the utility of this type of visualization. In practice, attention maps appear to correspond well with features in input imaging. Similar to saliency maps, rough segmentations of objects have been generated from classification networks using attention maps (Zhou et al. 2016), suggesting that for many networks there is sufficient correspondence between input imaging and CNN output for attention maps to provide powerful insights into the features in an image that activated a network.

9.6 CONCLUSIONS

There has been tremendous advance of machine learning methods, and in particular, deep learning methods have now demonstrated impressive ability to identify both visible and invisible features in a very robust fashion. There is concern that Dl is a "black box" preventing us from learning the features that are important, but this seems to be a problem that is being solved. Of greater concern is that these are extremely powerful tools that can also be applied incorrectly, and detecting when overfitting has occurred is an important concern in the broader application of DL methods.

REFERENCES

Akkus, Z., Ali, I., Sedlář, J., Agrawal, J.P., Parney, I.F., Giannini, C., & Erickson, B.J. (2017). Predicting deletion of chromosomal arms 1p/19q in low-grade gliomas from MR images using machine intelligence. *J. Digit. Imaging.* **30**, 469–476.

Altman, N.S. (1992). An introduction to kernel and nearest-neighbor nonparametric regression. *Am. Stat.* **46**, 175.

Arivazhagan, S., Ganesan, L., & Padam Priyal, S. (2006). Texture classification using Gabor wavelets based rotation invariant features. *Pattern Recognit. Lett.* **27**, 1976–1982.

Badrinarayanan, V., Kendall, A., & Cipolla, R. (2015). Segnet: A deep convolutional encoder-decoder architecture for image segmentation. arXiv preprint arXiv:1511.00561.

Belanche, L.A., Kobayashi, V., & Aluja, T. (2014). Handling missing values in kernel methods with application to microbiology data. *Neurocomputing* **141**, 110–116.

Ben-Cohen, A., Klang, E., Kerpel, A., Konen, E., Amitai, M.M., & Greenspan, H. (2018). Fully convolutional network and sparsity-based dictionary learning for liver lesion detection in CT examinations. *Neurocomputing* **275**, 1585–1594.

Bengio, Y. (2012). Practical recommendations for gradient-based training of deep architectures. In *Neural Networks: Tricks of the Trade*, G. Montavon, G.B. Orr, & K.-R. Müller, eds. (Berlin, Germany: Springer), pp. 437–478.

Bottou, L. (2011). Large-scale machine learning with stochastic gradient descent. In Chapman & Hall/CRC, *Computer Science & Data Analysis*, pp. 17–25.

Breiman, L. (2001). Random forests. *Mach. Learn.* **45**, 261–277.

Brink, J.A., Arenson, R.L., Grist, T.M., Lewin, J.S., & Enzmann, D. (2017). Bits and bytes: The future of radiology lies in informatics and information technology. *Eur. Radiol.* **27**, 3647–3651.

Brynolfsson, P., Nilsson, D., Torheim, T., Asklund, T., Karlsson, C.T., Trygg, J., Nyholm, T., & Garpebring, A. (2017). Haralick texture features from apparent diffusion coefficient (ADC) MRI images depend on imaging and pre-processing parameters. *Sci. Rep.* **7**, 4041.

Çiçek, Ö., Abdulkadir, A., Lienkamp, S.S., Brox, T., & Ronneberger, O. (2016). 3D U-net: Learning dense volumetric segmentation from sparse annotation. In *Medical Image Computing and Computer-Assisted Intervention – MICCAI 2016* (Springer International Publishing), pp. 424–432.

Cortes, C., & Vapnik, V. (1995). Support-vector networks. *Mach. Learn.* **20**, 273–297.

Creswell, A., White, T., Dumoulin, V., Arulkumaran, K., Sengupta, B., & Bharath, A.A. (2017). Generative adversarial networks: An overview. *IEEE Signal Processing Magazine* **35**, 53–65.

Cruz, J.A., & Wishart, D.S. (2007). Applications of machine learning in cancer prediction and prognosis. *Cancer Inform.* **2**, 59–77.

Davis, J., Lantz, E., Page, D., Struyf, J., Peissig, P., Vidaillet, H., & Caldwell, M. (2008). Machine learning for personalized medicine: Will this drug give me a heart attack. In *The Proceedings of International Conference on Machine Learning (ICML)*.

DICOM reference guide. (2001). *Health Devices* **30**, 5–30.

Dozat, T. (2016). Incorporating Nesterov Momentum into Adam, https://openreview.net/pdf?id=OM0jvwB8jIp57ZJjtNEZ.

Duchi, J., Hazan, E., & Singer, Y. (2011). Adaptive subgradient methods for online learning and stochastic optimization. *J. Mach. Learn. Res.* **12**, 2121–2159.

Erhan, D., Bengio, Y., Courville, A., Manzagol, P.-A., & Vincent, P. (2010). Why does unsupervised pretraining help deep learning? *J. Mach. Learn. Res.* **11**, 625–660.

Erickson, B.J., Korfiatis, P., Akkus, Z., & Kline, T.L. (2017). Machine learning for medical imaging. *Radiographics* **37**, 505–515.

Esteva, A., Kuprel, B., Novoa, R.A., Ko, J., Swetter, S.M., Blau, H.M., & Thrun, S. (2017). Dermatologist-level classification of skin cancer with deep neural networks. *Nature* **542**, 115–118.

Flach, P. (2012). *Machine Learning: The Art and Science of Algorithms that Make Sense of Data* (New York: Cambridge University Press).

Freund, Y., & Schapire, R.E. (1997). A decision-theoretic generalization of on-line learning and an application to boosting. *J. Comput. System Sci.* **55**, 119–139.

Girshick, R., Donahue, J., Darrell, T., & Malik, J. (2014). Rich feature hierarchies for accurate object detection and semantic segmentation. In *Proceedings of the IEEE Conference on Computer Vision and Pattern Recognition*, pp. 580–587.

Glorot, X., Bordes, A., & Bengio, Y. (2011). Domain adaptation for large-scale sentiment classification: A deep learning approach. In *Proceedings of the 28th International Conference on Machine Learning (ICML-11)*, pp. 513–250.

Goodfellow, I.J., Pouget-Abadie, J., Mirza, M., Xu, B., Warde-Farley, D., Ozair, S., Courville, A., & Bengio, Y. (2014). Generative adversarial networks. *Adv. Neural. Inf. Process. Syst.* 2672–2680.

Gourtsoyianni, S., Doumou, G., Prezzi, D., Taylor, B., Stirling, J.J., Taylor, N.J., Siddique, M., Cook, G.J.R., Glynne-Jones, R., & Goh, V. (2017). Primary rectal cancer: Repeatability of global and local-regional MR imaging texture features. *Radiology* **284**, 552–561.

Greenspan, H., van Ginneken, B., & Summers, R.M. (2016). Guest editorial deep learning in medical imaging: Overview and future promise of an exciting new technique. *IEEE Trans. Med. Imaging* **35**, 1153–1159.

van Griethuysen, J.J.M., Fedorov, A., Parmar, C., Hosny, A., Aucoin, N., Narayan, V., Beets-Tan, R.G.H., Fillion-Robin, J.-C., Pieper, S., & Aerts, H.J.W.L. (2017). Computational radiomics system to decode the radiographic phenotype. *Cancer Res.* **77**, e104–e107.

Guo, Y., Gao, Y., & Shen, D. (2016). Deformable MR prostate segmentation via deep feature learning and sparse patch matching. *IEEE Trans. Med. Imaging* **35**, 1077–1089.

Hajian-Tilaki, K. (2013). Receiver operating characteristic (ROC) curve analysis for medical diagnostic test evaluation. *Caspian J. Intern. Med.* **4**(2), 627–635.

Hassanpour, S., Langlotz, C.P., Amrhein, T.J., Befera, N.T., & Lungren, M.P. (2017). Performance of a machine learning classifier of knee MRI reports in two large academic radiology practices: A tool to estimate diagnostic yield. *AJR Am. J. Roentgenol.* **208**, 750–753.

Havaei, M., Davy, A., Warde-Farley, D., Biard, A., Courville, A., Bengio, Y., Pal, C., Jodoin, P.-M., & Larochelle, H. (2017). Brain tumor segmentation with deep neural networks. *Med. Image Anal.* **35**, 18–31.

He, K., Zhang, X., Ren, S., & Sun, J. (2015). Deep residual learning for image recognition. *Proceedings of the IEEE Conference on Computer Vision and Pattern Recognition* 770–778.

Herrick, R., Horton, W., Olsen, T., McKay, M., Archie, K.A., & Marcus, D.S. (2016). XNAT Central: Open sourcing imaging research data. *Neuroimage* **124**, 1093–1096.

Hotelling, H. (1933). Analysis of a complex of statistical variables into principal components. *J. Educ. Psychol.* **24**, 417–441.

Iandola, F.N., Han, S., Moskewicz, M.W., Ashraf, K., Dally, W.J., & Keutzer, K. (2016). SqueezeNet: AlexNet-level accuracy with 50x fewer parameters and <0.5 MB model size. arXiv preprint arXiv:1602.07360.

Ioffe, S., & Szegedy, C. (2015). Batch normalization: Accelerating deep network training by reducing internal covariate shift. arXiv preprint arXiv:1502.03167.

Juntu, J., Sijbers, J., De Backer, S., Rajan, J., & Van Dyck, D. (2010). Machine learning study of several classifiers trained with texture analysis features to differentiate benign from malignant soft-tissue tumors in T1-MRI images. *J. Magn. Reson. Imaging* **31**, 680–689.

Kawahara, J., & Hamarneh, G. (2016). Multi-resolution-tract CNN with hybrid pretrained and skin-lesion trained layers. In *Machine Learning in Medical Imaging* (Springer International Publishing), pp. 164–171.

Kingma, D.P., & Ba, J. (2014). Adam: A method for stochastic optimization. arXiv preprint arXiv:1412.6980.

Kingsford, C., & Salzberg, S.L. (2008). What are decision trees? *Nat. Biotechnol.* **26**, 1011–1013.

Kline, T.L., Korfiatis, P., Edwards, M.E., Blais, J.D., Czerwiec, F.S., Harris, P.C., King, B.F., Torres, V.E., & Erickson, B.J. (2017). Performance of an artificial multi-observer deep neural network for fully automated segmentation of polycystic kidneys. *J. Digit. Imaging* **30**, 442–448.

Kohavi, R., & Provost, F. (1998). Applications of data mining to electronic commerce. *Data Min. Knowl. Discov.* **5**, 5–10.

Kohonen, T. (1982). Self-organized formation of topologically correct feature maps. Biol. Cybern. **43**, 59–69.

Kolarević, D., Vujasinović, T., Kanjer, K., Milovanović, J., Todorović-Raković, N., Nikolić-Vukosavljević, D., & Radulovic, M. (2017). Effects of different preprocessing algorithms on the prognostic value of breast tumour microscopic images. *J. Microsc.* doi:10.1111/jmi.12645.

Korfiatis, P., & Erickson, B. (2014). The basics of diffusion and perfusion imaging in brain tumors. *Appl. Radiol.* **43**, 22–29.

Korfiatis, P., & Erickson, B. (2016). PESSCARA: An example infrastructure for big data research. In *Big Data on Real-World Applications*, S.V. Soto, J.M. Luna, & A. Cano, eds. (London, UK: InTechOpen Limited).

Korfiatis, P., Kline, T.L., & Erickson, B.J. (2016). Automated segmentation of hyperintense regions in FLAIR MRI using deep learning. *Tomography* **2**, 334–340.

Korfiatis, P., Zhou, Z., Liang, J., & Erickson, B.J. (2017). Fully automated IDH mutation prediction in MRI utilizing deep learning. In *Proceedings of the 2nd Conference on Machine Intelligence in Medical Imaging*, B.J. Erickson & E.L. Siegel, eds., p. 23.

Krizhevsky, A., Sutskever, I., & Hinton, G.E. (2017). Imagenet classification with deep convolutional neural networks. *Commun. ACM* **60**, 84–90.

Kuan, K., Ravaut, M., Manek, G., Chen, H., Lin, J., Nazir, B., Chen, C., Howe, T.C., Zeng, Z., & Chandrasekhar, V. (2017). Deep learning for lung cancer detection: Tackling the kaggle data science bowl 2017 challenge. *arXiv [cs.CV]*. arXiv. http://arxiv.org/abs/1705.09435.

LeCun, Y., Bengio, Y., & Hinton, G. (2015). Deep learning. *Nature* **521**, 436–444.

Lee, S.-I., Celik, S., Logsdon, B.A., Lundberg, S.M., Martins, T.J., Oehler, V.G., Estey, E.H. et al. (2018). A machine learning approach to integrate big data for precision medicine in acute myeloid leukemia. *Nat. Commun.* **9**, 42.

Long, J., Shelhamer, E., & Darrell, T. (2015). Fully convolutional networks for semantic segmentation. In *2015 IEEE Conference on Computer Vision and Pattern Recognition (CVPR)*.

Mahendran, A., & Vedaldi, A. (2015). Understanding deep image representations by inverting them. In *Proceedings of the IEEE Conference on Computer Vision and Pattern Recognition*, pp. 5188–5196.

Milletari, F., Navab, N., & Ahmadi, S.A. (2016). V-Net: Fully convolutional neural networks for volumetric medical image segmentation. In *2016 Fourth International Conference on 3D Vision (3DV)*, pp. 565–571.

Moon, H., Ahn, H., Kodell, R.L., Baek, S., Lin, C.-J., & Chen, J.J. (2007). Ensemble methods for classification of patients for personalized medicine with high-dimensional data. *Artif. Intell. Med.* **41**, 197–207.

Mordvintsev, A., Olah, C., & Tyka, M. (2015). Inceptionism: Going deeper into neural networks. https://ai.google/research/pubs/pub45507.

Nanni, L., Brahnam, S., Ghidoni, S., & Menegatti, E. (2015). Improving the descriptors extracted from the co-occurrence matrix using preprocessing approaches. *Expert Syst. Appl.* **42**, 8989–9000.

Ng, A.Y. (2004). Feature selection, L1 vs. L2 regularization, and rotational invariance. In *Proceedings of the Twenty-First International Conference on Machine Learning* (New York: ACM), p. 78 doi:10.1145/1015330.1015435.

Niculescu-Mizil, A., & Caruana, R. (2005). Predicting good probabilities with supervised learning. In *Proceedings of the 22nd International Conference on Machine Learning—ICML'05*.

Nie, D., Zhang, H., Adeli, E., Liu, L., & Shen, D. (2016). 3D deep learning for multi-modal imaging-guided survival time prediction of brain tumor patients. *Medical Image Computing and Computer-Assisted Intervention: MICCAI ... International Conference on Medical Image Computing and Computer-Assisted Intervention* **9901** (October), pp. 212–220.

Olah, C., Mordvintsev, A., & Schubert, L. (2017). Feature visualization. *Distill* **2**(11), e7. https://distill.pub/2017/feature-visualization/.

Pietikäinen, M., Ojala, T., & Xu, Z. (2000). Rotation-invariant texture classification using feature distributions. Pattern Recognit. **33**, 43–52.

Pons, E., Braun, L.M.M., Hunink, M.G.M., & Kors, J.A. (2016). Natural language processing in radiology: A systematic review. *Radiology* **279**, 329–343.

Ronneberger, O., Fischer, P., & Brox, T. (2015). U-Net: Convolutional networks for biomedical image segmentation. In *Medical Image Computing and Computer-Assisted Intervention—MICCAI 2015*, N. Navab, J. Hornegger, W.M. Wells, & A.F. Frangi, eds. (Springer International Publishing), pp. 234–241.

Saeys, Y., Inza, I., & Larranaga, P. (2007). A review of feature selection techniques in bioinformatics. *Bioinformatics* **23**, 2507–2517.

Samuel, A.L. (1988). Some studies in machine learning using the game of checkers. I. In *Computer Games I* (New York: Springer-Verlag), pp. 335–365.

Schneider, A., Hommel, G., & Blettner, M. (2010). Linear regression analysis: Part 14 of a series on evaluation of scientific publications. *Dtsch. Arztebl. Int.* **107**, 776–782.

Selvaraju, R.R., Cogswell, M., Das, A., Vedantam, R., Parikh, D., & Batra, D. (2016). Grad-cam: Visual explanations from deep networks via gradient-based localization. https://arxiv.org/abs/1610.02391 v3 **7**(8). https://github.com/ramprs/grad-cam/.

Setio, A.A.A., Ciompi, F., Litjens, G., Gerke, P., Jacobs, C., van Riel, S., Wille, M.W., Naqibullah, M., Sanchez, C., & van Ginneken, B. (2016). Pulmonary nodule detection in CT images: False positive reduction using multi-view convolutional networks. *IEEE Transactions on Medical Imaging*. doi:10.1109/TMI.2016.2536809.

Shin, H., & Markey, M.K. (2006). A machine learning perspective on the development of clinical decision support systems utilizing mass spectra of blood samples. *J. Biomed. Inform.* **39**, 227–248.

Shin, H.-C., Orton, M.R., Collins, D.J., Doran, S.J., & Leach, M.O. (2013). Stacked autoencoders for unsupervised feature learning and multiple organ detection in a pilot study using 4D patient data. *IEEE Trans. Pattern Anal. Mach. Intell.* **35**(8), 1930–1943. doi:10.1109/TPAMI.2012.277.

Shin, H.-C., Roth, H.R., Gao, M., Lu, L., Xu, Z., Nogues, I., Yao, J., Mollura, D., & Summers, R.M. (2016). Deep convolutional neural networks for computer-aided detection: CNN architectures, dataset characteristics and transfer learning. *IEEE Trans. Med. Imaging* **35**, 1285–1298.

Simonyan, K., Vedaldi, A., & Zisserman, A. (2013). Deep inside convolutional networks: Visualising image classification models and saliency maps. *arXiv [cs.CV]*. arXiv. http://arxiv.org/abs/1312.6034.

Simonyan, K., & Zisserman, A. (2014). Very deep convolutional networks for large-scale image recognition. arXiv preprint arXiv:1409.1556.

Sovilj, D., Eirola, E., Miche, Y., Björk, K.-M., Nian, R., Akusok, A., & Lendasse, A. (2016). Extreme learning machine for missing data using multiple imputations. *Neurocomputing* **174**, 220–231.

Srivastava, N., Hinton, G.R., Krizhevsky, A., Sutskever, I., & Salakhutdinov, R. (2014). Dropout: A simple way to prevent neural networks from overfitting. *J. Mach. Learn. Res.* **15**, 1929–1958.

Tajbakhsh, N., Shin, J.Y., Gurudu, S.R., Todd Hurst, R., Kendall, C.B., Gotway, M.B., & Liang, J. (2016). Convolutional neural networks for medical image analysis: Full training or fine tuning? *IEEE Trans. Med. Imaging* **35**, 1299–1312.

Tibshirani, R. (1994). Regression shrinkage and selection via the lasso. *J. Royal Stat. Soc. Series B* 267–288.

Vapnik, V.N. (2000). *The Nature of Statistical Learning Theory* (Red Bank, NJ: Springer).

Wang, F., Kong, J., Cooper, L., Pan, T., Kurc, T., Chen, W., Sharma, A. et al. (2011). A data model and database for high-resolution pathology analytical image informatics. *J. Pathol. Inform.* **2**, 32.

Way, T.W., Sahiner, B., Hadjiiski, L.M., & Chan, H.-P. (2010). Effect of finite sample size on feature selection and classification: A simulation study. *Med. Phys.* **37**, 907–920.

Xie, S., Girshick, R., Dollar, P., Tu, Z., & He, K. (2017). Aggregated residual transformations for deep neural networks. In *2017 IEEE Conference on Computer Vision and Pattern Recognition* (CVPR).

Zeiler, M.D. (2012). ADADELTA: An adaptive learning rate method. arXiv preprint arXiv:1212.5701.

Zeiler, M.D., & Fergus, R. (2014). Visualizing and understanding convolutional networks. In *Computer Vision – ECCV 2014* (Springer International Publishing), pp. 818–833. doi:10.1007/978-3-319-10590-1_53.

Zhang, J., & Tan, T. (2002). Brief review of invariant texture analysis methods. *Pattern Recognit.* **35**, 735–747.

Zhou, B., Khosla, A., Lapedriza, A., Oliva, A., & Torralba, A. (2016). Learning deep features for discriminative localization. In *Proceedings of the IEEE Conference on Computer Vision and Pattern Recognition*, pp. 2921–2929. https://www.cv-foundation.org/openaccess/content_cvpr_2016/html/Zhou_Learning_Deep_Features_CVPR_2016_paper.html.

Radiogenomics
Rationale and methods

OLIVIER GEVAERT

10.1 INTRODUCTION

Vast amounts of biomedical data characterizing complex diseases are now publicly available, and new computational tools for quantitatively analyzing them are being developed routinely. No burgeoning area other than radiogenomics is a better example of the potential of quantitative analysis of biomedical data. Radiogenomics is a novel area that is centered around the idea that entities at different scales such as molecules, cells, and tissues are linked to each other and therefore can be modeled together [1–5]. The most notable examples are in the field of oncology, where radiogenomic biomedical data are routinely collected and huge opportunities exist to exploit potential synergies between molecular and imaging data.

Formally, we define radiogenomics as linking genome wide molecular data with quantitative image features extracted from radiology images. This can involve several types of molecular data with the most common being gene expression data or transcriptomic data of all known human genes. Other types of whole genome data that can be used are DNA copy number data or DNA methylation data. Moreover, all omics data can also be combined. On the imaging side, in most cases images computed tomography (CT) or magnetic resonance (MR) images are studied, as they provide the highest resolution images allowing quantitative feature extraction either using predefined features also known as (aka) radiomics [6] or learned from the data with deep learning techniques. These quantitative image features are predefined and different implementations exist [7]. New approaches for quantitatively mining are continuously being developed such that the process of extracting image features is always evolving. In any case, to be able to build a radiogenomics map, a key characteristic is that many features are needed, as in most cases there is little information on which features will be important for subsequent radiogenomics analyses.

It is important to disambiguate radiogenomics from a more narrow definition in the area of radiation therapy response prediction [8]. In the context of radiation therapy, radiogenomics has been used to refer to the study of genetic variation associated with response to radiation, aka radiation genomics. In this chapter, we will discuss radiogenomics in the broader definition as defined above.

The rationale for radiogenomics is centered on the idea to link medical imaging with molecular biology to create a better understanding of patient outcomes and treatment. Radiogenomics has shown its potential through its ability to predict clinical outcomes, e.g., prognosis, and through predicting actionable molecular properties of tumors, e.g., the activity of epidermal growth factor receptor (EGFR), a major drug target in many cancers. In addition, radiogenomics maps provide multivariate links between image features and gene expression profiles or associations between image phenotypes and potential treatment [4,5,9–12]. For example, reports have shown how quantitative image features can be mapped to metagenes for lung cancer patients [5], for defining imaging subtypes, for example, for adult brain tumors and breast cancer with implications for targeted treatment [11,13], and recent reports have shown how EGFR mutation status can be predicted from lung CT images [14–16].

One of the key questions that radiogenomics can answer is how much of the molecular make-up of complex diseases is reflected in the 3D appearance of diseases as captured by radiology images. Radiogenomics modeling thus broadly involves linking information from molecules and cells to tissues and organs. In this chapter, we will further discuss the rationale for radiogenomics and motivate why radiogenomics modeling is important and how it can contribute to precision medicine. Next, we will discuss the core methodology behind radiogenomics by describing three essential components that are needed to develop a radiogenomics model. We will conclude with challenges and future work.

10.2 RATIONALE FOR RADIOGENOMICS

10.2.1 HETEROGENEITY

One of the main motivations of radiogenomics research is the study of heterogeneity. Intra and inter individual heterogeneities are often quoted as the main challenge for studying complex diseases. These heterogeneities exist at all scales, from microscopic to macroscopic. However, the technologies used to generate data at a particular scale don't equally assess the heterogeneity. A single biopsy of a brain tissue does not reflect the known cellular and spatial heterogeneity. Whereas multi-modal radiology images provide a global picture of the heterogeneity of the complete tumor at the tissue scale. Radiogenomic data fusion can have profound contributions toward predicting diagnosis and treatment by uncovering unknown synergies and relationships. Moreover, this can reveal to novel insights in how data at different scales are linked to each other.

We will discuss the issue of heterogeneity in the context of oncology, as it is well known that there is genomic heterogeneity between tumors and also within individual tumors [17,18]. However, the only way to determine tumor genotype is through invasive surgical sampling, and given the spatial inhomogeneity of genomic expression within a tumor, information about the local in vivo microenvironment and loco regional tumor precursor cells-of-origin can be lost in biopsies (Figure 10.1). The creation of radiogenomic maps bypasses the need for surgical sampling in order to understand a tumor lesions' underlying biology and the changes that may result from treatment intervention. The application of radiogenomic maps has been demonstrated in hepatocellular carcinoma [19,20], lung cancer [1,5,9,21], and gliomas (particularly high grade gliomas such as glioblastoma [4,11,12,22–24]. In some of these reports, the investigators utilized human-based semantic descriptions of the tumor appearance on imaging such as the Visually Accessible Rembrandt Images (VASARI) features [12,24], while others utilized quantitative imaging features such as the extent of tumor-associated edema, non-enhancing, enhancement, and necrosis tumor regions from semi-automatic segmentations of tumors with imaging analysis software, and lastly, global extraction of more classical radiomics quantitative features extracted across the gross

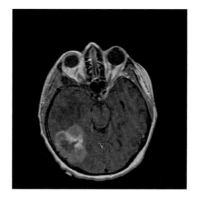

Figure 10.1 (See color insert.) Visualization of MR post contrast image of a patient from the TCGA cohort. Left, raw MR image. Right, molecular heterogeneity of a gene expression module visualized on top of the raw MR image using a radiogenomics map.

tumor volume. Importantly, there exists a significant body of literature confirming that molecular expression is variable within a single tumor lesion.

Recent studies have investigated this concept and shown the potential of radiogenomics. For example, a recent report showed that localized radiomics features that were most informative of hypoxia distinguished between short- and long-term survivors [25]. This study showed that the quantitative representations exist that capture the molecular heterogeneity of hypoxia. Secondly, in another report, multiple image-guided biopsies from glioblastoma patients were collected in multiple regions including regions of enhancement and non-enhancing parenchyma. For each biopsy, they generated DNA copy number data of key cancer driver genes for glioblastoma. They showed that high accuracy could be achieved to predict six driver genes based on local quantitative image features [26]. Taken together, initial studies in radiogenomics have provided preliminary evidence that this technique can capture molecular heterogeneity.

10.2.2 THE QUEST FOR NON-INVASIVE BIOMARKERS

One of the most compelling applications of radiogenomics is the discovery of non-invasive biomarkers. By mapping signatures capturing molecular biology of tumors to their matched quantitative images, one can discover non-invasive biomarkers that mirror the molecular properties of these tumors.

Most clinically used biomarkers, especially in the area of oncology, use a piece of tissue and apply genomics technologies to measure a molecule that is associated with the diagnostic, predictive of prognosis, or determines treatment. In many cases, targeted therapies in oncology require often the presence of a particular biomarker, "the target" to be effective. For example EGFR, is a common oncogene that is often deliberately activated in many cancers including lung cancer, and a targeted drug exist to block this gene. However, the drug is typically only used when the activated version of the gene is present, requiring to assess this activity of this biomarker. A biomarker can also be multivariate and require the measurement of multiple molecules. For example, subtyping of medulloblastoma is based on gene expression of 22 genes using technologies such as nanostring [27]. Knowing the subtype of medulloblastoma informs on the biology of the disease and what treatment is likely to help. Many more examples exist, and biomarkers are used to subtype cancers, to determine treatment, and to determine prognosis.

A main disadvantage is that surgical sampling (whether in the form of a resection or biopsy) is invasive and confers added risks to patients depending on the tumor site. Moreover, in certain cases, access to tissue is not possible due to tumors being in areas that are too sensitive or the procedure cause too much harm. Biomarkers that do not require access to the tissue thus are more easy to use and potentially also more easy to implement. It therefore is not hard to argue that quantitative imaging is a great candidate to investigate whether a single image feature or combination of image features can serve as a biomarker for the underlying molecular biology. The recent observation that a radiographic image is more than just a picture, it is mineable data [28], has provided motivation for this concept and has opened up a whole avenue of quantitative feature extraction from radiology images for biomarker discovery. Radiogenomics analysis can thus elucidate novel biomarkers by linking images with molecular biology and potential biomarkers of the disease.

10.2.3 BIOMEDICAL DECISION SUPPORT

Radiogenomics also has potential to improve the prediction of clinical outcomes. More specifically, radiogenomics modeling can aid in increasing the accuracy for predicting diagnosis, prognosis, or treatment outcomes. Biomedical decision support has been the focus of much of machine learning research in biomedicine for many decades. Due to the explosion of data that can potentially be collected before, during, and after diseases manifest, mathematical and statistical models are needed to extract patterns from the data that can help in the context of precision medicine. Yet few studies have investigated modeling of radiogenomics data for the purpose of predicting patient diagnosis and patient outcomes, while, for example, cancer studies abound with radiogenomic data.

Molecular biomarkers have already shown promise for changing treatment paradigms in many cases [29], and at the same time, recent efforts have shown that quantitative analysis of medical images is predictive of prognosis and treatment outcomes [30,31]. Combining both efforts might improve the accuracy for making predictions and uncover previously unappreciated synergy.

10.3 THE ESSENTIAL COMPONENTS TO BUILD A RADIOGENOMICS MODEL

A radiogenomics model is composed of three essential components to result in a successful model: the omics strategy, the quantitative imaging strategy, and the integration strategy. Each of these three strategies needs to be considered to develop a successful radiogenomics model.

10.3.1 THE GENOMICS STRATEGY

The genomics strategy depends on what type of molecular data are available. As mentioned in the introduction, we define radiogenomics as the linking of at least one genome-wide molecular data source with quantitative imaging. The first examples in radiogenomics made use of gene expression data [19,32] initially generated by microarray technology, but now superseded by RNA sequencing technology. Gene expression data provide the easiest and cheapest way to get a snapshot of the molecular processes ongoing in the cell. One can get an idea of the activity of all the processes that are relevant for a particular disease, from single genes such as drug targets, to pathways such as the hallmarks of cancer [33]. However, one challenge is the high dimensionality of gene expression data. In the past in the case of microarray technology [34], one would get activity values for all known genes in the human genomes. However, with the use of RNA sequencing technology, one can get activity values for known and unknown entities and not only protein coding genes. For example, long non-coding RNAs can also be investigated [35].

Taken together, this has created additional challenges to develop a successful genomics strategy as the danger of identifying spurious correlations between genomics and imaging is huge. Therefore, it is often recommended to consider using a dimensionality reduction before linking genomics data to imaging data in a radiogenomics model. A common approach is to create metagenes, clusters of highly correlated genes [5].

Besides gene expression data, also other types of molecular data have important applications in determining diagnosis or treatment of cancer patients. Examples are DNA sequencing, DNA copy number, DNA methylation analysis, and protein expression. DNA sequencing provides us with the full complement of mutations in the genome of a biological sample. For example, activation of oncogenes or inactivation of tumor suppressor genes are common drivers of many tumor types. Similarly, DNA copy number analysis allows to profile how many copies are present for each gene in the human genome. This analysis allows to identify if certain genes are amplified, e.g., oncogenes, or are deleted from the genome, e.g., tumor suppressor genes. These genes are often known as cancer driver genes and can drive tumor growth. Genome-wide DNA methylation measurements are possible thanks to microarray technology enabling reading out binary states in the genome to determine if they are methylated or not [36–40]. Finally, protein expression analysis is technologically more complicated, as profiling the human proteome is more complicated and expensive, but has recently seen an increase in use thanks to developments in mass spectrometry technology and large multi-cancer projects are being executed [41].

In the last decade, large multi-institutional projects have also been completed that have provided multi-omics data. For example, many radiogenomics investigators have taken advantage of large multi-omic databases that have been created such as The Cancer Genome Atlas (TCGA), where also radiology images are available for the same patients. These efforts at least in part have accelerated the study of radiogenomics analysis in the field of oncology and have led to several reports showing the potential of radiogenomic modeling of cancer [2–4].

A major challenge when considering a multi-omics approach is the genomics modeling strategy. It is often inadvisable to link each single gene's omics with image data, but rather first use state-of-the art bioinformatics approaches to focus on the key biological drivers of the disease context at hand, to avoid focusing on genes that are not relevant. For example, MutSigCV only considers the somatic mutations with the strongest statistical support [42], GISTIC enables focusing only on the key genes with DNA copy number changes [43], and MethylMix focuses only on genes that are differentially methylated and have an effect on gene expression [36,44,45]. In the case of a multi-omics approach, one needs to reduce the dimensions and treat groups of genes instead of single genes. For example PARADIGM combines DNA copy number and gene expression data to estimates of activities of known pathways based on decades of molecular biology work to elucidate the gene networks [46]. Similarly, AMARETTO is a data driven bioinformatics algorithm that combines gene expression, DNA methylation, and DNA copy number data, to create gene expression modules that capture the major processes [38,47,48]. Both PARADIGM and AMARETTO are ways to reduce the dimensionality of the data, and build radiogenomics maps linking image phenotypes with molecular processes instead of single genes, improve the power to identify significant statistical associations.

At the moment it is still an open problem which omics and thus which level of molecular biology is the most relevant for capturing the genomics state of the cell or even if a multi-omics strategy is appropriate. Future efforts relevant for radiogenomics are currently focusing heavily on proteogenomics, an integrated genomics and proteomics analysis in the context of radiogenomics [41].

10.3.2 QUANTITATIVE IMAGING STRATEGY

Next, one needs to consider what strategy one will use to analyze the radiology images. This strategy is also influenced by what type of radiology image is being analyzed (i.e., CT, multi-modal MR, PET, etc.), as the resolution of the image, the number of modalities, and also the number of time points whereby imaging data are available will all influence how much data are available for analysis and what can be extracted from these images.

One can observe two main strategies at the moment to analyze radiology images: extraction of pre-defined image features [6,19,22,49–51] and learning features directly from the data, for example, with the use of convolutional neural networks (CNN) [7].

In the case of pre-defined image features, current implementations of quantitative imaging pipelines evolve rapidly with increasingly more features being extracted from medical images [52,53]. Each class of features is defined by a specific characteristic that the class of features represents. Roughly, quantitative image features can be grouped in to four different areas: edge, shape, intensity, and texture [52]. Edge and shape features are more straightforward as they represent quantitative characteristics of the tumor edge, e.g., sharp versus blurry edges. Shape features represent the overall shape of the tumor including how spherical it is and how irregular the boundary is defined. For example, the compactness features are defined as follows: a value of 1.0 reflects a lesion that is completely spherical, whereas a value of 0 reflects a completely irregular lesion. Next, other features that capture shape are, for example, the radial distance signal (RDS), this is a set of features that captures the variability of the tumor shape [52]. Intensity and texture features represent first-, second-, and higher-order of statistics of the image and become increasingly harder to interpret. First-order statistics involve histogram statistics such as mean, median, skewness, and kurtosis of the intensity histogram. These features do not take into account the spatial organization of each voxel in the tumor, but rather only capture a summary of the intensity histogram of a tumor. Beyond that, higher-order statistics become increasingly less interpretable, but do capture spatial information. These are the so-called texture features and are often the most important contribution of the radiomics characterization. There is an increasingly growing list of texture features and more are being defined. Texture features were first introduced by Haralick [54]. The first texture features describe spatial relationships between voxels through a gray-level co-occurrence matrix (GLCM) defined as combinations of discretized gray levels of neighboring voxels and how they are distributed in different directions. To make GLCM features rotation invariant, these features are calculated after combining information from the different matrices [55]. Similar to GLCM, the gray-level run-length matrix (GLRLM) was introduced in 1975 [56] to capture a run length defined as the length of a consecutive sequence of pixels or voxels with the same gray level along a particular direction. Subsequently other texture features have been introduced such as the gray-level size-zone matrix (GLSZM) [57], gray-level distance-zone matrix (GLDZM) [57], and neighborhood gray-tone difference matrix (NGTDM) [58]. Taken together, there is a potential to extract 100s–1000s of texture features out of a single radiology image. However, it is well known that correlation between texture features can be high, and as such, not all features represent unique characteristics of the MR image, rather this approach is motivated by extracting any useful signal out of the image. In a second step, the radiomics research will use machine learning approaches to analyze the data and identify which features are important either using unsupervised or supervised analysis.

This quantitative image pipeline can now be used on each imaging modality and also each lesion in the image that is of interest. This means that the full set of image features can be generated multiple times to reflect each imaging modality or lesion, creating a large dataset for further analysis.

More recently, deep learning in the form of CNNs have emerged as a more powerful approach for modeling medical images [7]. The main advantage being that no decision is needed on what features to define, however, CNNs learn what features are important directly from the data. This comes at a cost, as large databases of medical images are needed to learn CNNs, and as such, the use of CNNs in radiogenomics applications is still emerging.

10.3.3 THE INTEGRATION STRATEGY

How to integrate the genomics and imaging data is the final component that defines the radiogenomic model? This component is heavily influenced depending on the application of the model. We can distinguish two broad applications of radiogenomic modeling: finding associations between image and molecular phenotypes or developing so-called radiogenomic maps, and biomedical decision support using radiogenomic data.

Radiogenomic mapping: To build radiogenomics maps for discovering associations between molecular and image phenotypes, increasingly more complicated approaches can be used: (1) univariate correlation between image features and molecular data, (2) multivariate prediction of molecular data or image features, and (3) two-way multivariate integration of image features and molecular data. In all cases, molecular data can be any genomic data such as gene expression data, DNA methylation data, or the expression of a metagene. In the case of univariate associations, it is important to consider the multiple testing problem, as many associations will be tested between molecular data and image features. This problem has been widely studied in bioinformatics, but is less known in quantitative image studies. To avoid multiple testing problems, a statistical correction using the false discovery rate (FDR) is the most common strategy [59]. Another option is to use the significance analysis of microarrays (SAM)-statistics, which is a modified t-test that also includes multiple testing corrections [60]. The univariate correlation map that results from this analysis can be used to determine any relationships between molecular features and image features. Next, multivariate modeling can be used to develop predictive models of molecular features in terms of image features, or vice versa, image features in terms of molecular features. Both have interesting applications, in the case of the former, one can develop non-invasive biomarkers of molecular features that model therapeutic properties of the tumor, e.g., a drug target. In the case of the latter, one can learn more about the molecular biology behind an image feature such as size, shape, or morphology of the lesion. In the case of multivariate modeling, any popular machine learning algorithm can be used such as decision trees, support vector machines, or regularized linear regression. Finally, one can also use more advanced statistical models such as canonical correlation analysis (CCA) to associate combinations of image features with combinations of molecular features [61]. However, in this case, it becomes harder to interpret the resulting models.

Biomedical decision support: To develop integrative models of radiogenomic data for biomedical decision support, it is important to consider what integration strategy to use. Data fusion frameworks exist for many popular machine learning algorithms including Bayesian networks [62,63], support vector machines (SVMs) [64], and regularized linear regression models. However, one needs to decide the exact integration strategy. We recommend comparing early, late, and intermediate integration of multi-scale datasets (Figure 10.2) [65]. In the vast majority of genomic analyses, the most frequently used integration strategy for combining heterogeneous datasets is concatenation, or early integration, in which all the data matrices are merged or "concatenated" by patient prior to learning the data for analysis [66] (Figure 10.2, Panel 1). Early integration is simple and intuitive to use and maintains inter-data relationships, but at the cost of losing individual properties of each dataset, like structure, and compounding the problem of high-dimensionality. The other extreme is late data fusion (Figure 10.2, Panel 2), whereby modeling is performed on each data matrix separately, and then the prediction from each model is integrated. In late integration the individual structure and properties of each dataset are preserved, which enables analysis using tools that are best suited for each dataset. However, by analyzing each dataset in isolation, possible inter-data relationships, like correlations and interactions, might be missed. Next, intermediate strategies are possible, but these will depend on what machine learning model is chosen (Figure 10.2, Panel 3).

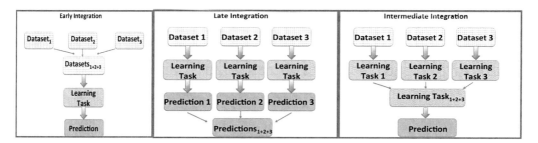

Figure 10.2 (See color insert.) Data fusion for biomedical decision support. Left panel: early integration, middle panel: late integration, and right panel, intermediate integration.

10.4 CONCLUSION

Radiogenomics has only emerged as a novel field of study about a decade ago and still much work needs to be done. The relationship between the molecular and the imaging phenotype has not been studied in depth, as previous work has only scratched the surface, while a clear rationale exists to further study the integration of radiogenomic data. Radiogenomics enables studying the relationship between molecular and imaging heterogeneity, can result in the discovery of non-invasive biomarkers, and improve biomedical decision support models.

REFERENCES

1. Zhou, M. et al., Non-small cell lung cancer radiogenomics map identifies relationships between molecular and imaging phenotypes with prognostic implications. *Radiology*, 2018. **286**(1): p. 307–315.
2. Rutman, A.M. and M.D. Kuo, Radiogenomics: Creating a link between molecular diagnostics and diagnostic imaging. *Eur J Radiol*, 2009. **70**(2): p. 232–241.
3. Karlo, C.A. et al., Radiogenomics of clear cell renal cell carcinoma: Associations between CT imaging features and mutations. *Radiology*, 2014. **270**(2): p. 464–471.
4. Gevaert, O. et al., Glioblastoma multiforme: Exploratory radiogenomic analysis by using quantitative image features. *Radiology*, 2014: p. 131731.
5. Gevaert, O. et al., Non-small cell lung cancer: Identifying prognostic imaging biomarkers by leveraging public gene expression microarray data—Methods and preliminary results. *Radiology*, 2012. **264**(2): p. 387–396.
6. Lambin, P. et al., Radiomics: Extracting more information from medical images using advanced feature analysis. *Eur J Cancer*, 2012. **48**(4): p. 441–446.
7. LeCun, Y., Y. Bengio and G. Hinton, Deep learning. *Nature*, 2015. **521**(7553): p. 436–444.
8. Kerns, S.L., H. Ostrer and B.S. Rosenstein, Radiogenomics: Using genetics to identify cancer patients at risk for development of adverse effects following radiotherapy. *Cancer Discov*, 2014. **4**(2): p. 155–165.
9. Nair, V.S. et al., Prognostic PET 18F-FDG uptake imaging features are associated with major oncogenomic alterations in patients with resected non-small cell lung cancer. *Cancer Res*, 2012. **72**(15): p. 3725–3734.
10. Echegaray, S. et al., Core samples for radiomics features that are insensitive to tumor segmentation: Method and pilot study using CT images of hepatocellular carcinoma. *J Med Imaging* (Bellingham), 2015. **2**(4): p. 041011.
11. Itakura, H. et al., Magnetic resonance image features identify glioblastoma phenotypic subtypes with distinct molecular pathway activities. *Sci Transl Med*, 2015. 7(303): p. 303ra138.
12. Nicolasjilwan, M. et al., Addition of MR imaging features and genetic biomarkers strengthens glioblastoma survival prediction in TCGA patients. *J Neuroradiol*, 2015. **42**(4): p. 212–221.
13. Wu, J. et al., Unsupervised clustering of quantitative image phenotypes reveals breast cancer subtypes with distinct prognoses and molecular pathways. *Clin Cancer Res*, 2017. **23**(13): p. 3334–3342.
14. Rios Velazquez, E. et al., Somatic mutations drive distinct imaging phenotypes in lung cancer. *Cancer Res*, 2017. **77**(14): p. 3922–3930.
15. Minamimoto, R. et al., Prediction of EGFR and KRAS mutation in non-small cell lung cancer using quantitative (18)F FDG-PET/CT metrics. *Oncotarget*, 2017. **8**(32): p. 52792–52801.
16. Gevaert, O. et al., Predictive radiogenomics modeling of EGFR mutation status in lung cancer. *Scientific Reports*, 2017. **7**(41674).
17. Sottoriva, A. et al., Intratumor heterogeneity in human glioblastoma reflects cancer evolutionary dynamics. *Proc Natl Acad Sci U S A*, 2013. **110**(10): p. 4009–4014.
18. Gerlinger, M. et al., Intratumor heterogeneity and branched evolution revealed by multiregion sequencing. *N Engl J Med*, 2012. **366**(10): p. 883–892.

19. Segal, E. et al., Decoding global gene expression programs in liver cancer by noninvasive imaging. *Nat Biotechnol*, 2007. **25**(6): p. 675–680.

20. Bakr, S. et al., Noninvasive radiomics signature based on quantitative analysis of computed tomography images as a surrogate for microvascular invasion in hepatocellular carcinoma: A pilot study. *J Med Imaging* (Bellingham), 2017. **4**(4): p. 041303.

21. Nair, V.S. et al., NF-kappaB protein expression associates with (18)F-FDG PET tumor uptake in non-small cell lung cancer: A radiogenomics validation study to understand tumor metabolism. *Lung Cancer*, 2014. **83**(2): p. 189–196.

22. Zinn, P.O. et al., Radiogenomic mapping of edema/cellular invasion MRI-phenotypes in glioblastoma multiforme. *PLoS One*, 2011. **6**(10): p. e25451.

23. Jain, R. et al., Genomic mapping and survival prediction in glioblastoma: Molecular subclassification strengthened by hemodynamic imaging biomarkers. *Radiology*, 2013. **267**(1): p. 212–220.

24. Gutman, D.A. et al., MR imaging predictors of molecular profile and survival: Multi-institutional study of the TCGA glioblastoma data set. *Radiology*, 2013. **267**(2): p. 560–569.

25. Beig, N. et al., Radiogenomic analysis of hypoxia pathway is predictive of overall survival in Glioblastoma. *Sci Rep*, 2018. **8**(1): p. 7.

26. Hu, L.S. et al., Radiogenomics to characterize regional genetic heterogeneity in glioblastoma. *Neuro Oncol*, 2017. **19**(1): p. 128–137.

27. Northcott, P.A. et al., Rapid, reliable, and reproducible molecular sub-grouping of clinical medulloblastoma samples. *Acta Neuropathol*, 2012. **123**(4): p. 615–626.

28. Gillies, R.J., P.E. Kinahan and H. Hricak, Radiomics: Images are more than pictures, they are data. *Radiology*, 2016. **278**(2): p. 563–577.

29. Gentles, A.J. et al., Integrating tumor and stromal gene expression signatures with clinical indices for survival stratification of early-stage non-small cell lung cancer. *J Natl Cancer Inst*, 2015. **107**(10).

30. Joye, I. et al., Quantitative imaging outperforms molecular markers when predicting response to chemoradiotherapy for rectal cancer. *Radiother Oncol*, 2017. 124(1).

31. Bulens, P. et al., Development and validation of an MRI-based model to predict response to chemoradiotherapy for rectal cancer. *Radiother Oncol*, 2018. **126**(3): 437–442.

32. Diehn, M. et al., Identification of noninvasive imaging surrogates for brain tumor gene-expression modules. *Proc Natl Acad Sci USA*, 2008. **105**(13): p. 5213–5218.

33. Hanahan, D. and R.A. Weinberg, Hallmarks of cancer: The next generation. *Cell*, 2011. **144**(5): p. 646–674.

34. Gevaert, O. and B. De Moor, Prediction of cancer outcome using DNA microarray technology: Past, present and future. *Expert Opin Med Diagn*, 2009. **3**(2): p. 157–165.

35. Zheng, H. et al., Benchmark of lncRNA quantification for RNA-Seq of cancer samples. *bioRxiv*, 2018:241869.

36. Gevaert, O., R. Tibshirani and S.K. Plevritis, Pancancer analysis of DNA methylation-driven genes using MethylMix. *Genome Biol*, 2015. **16**(1): p. 17.

37. Gevaert, O. and S. Plevritis, Identifying master regulators of cancer and their downstream targets by integrating genomic and epigenomic features. *Pac Symp Biocomput*, 2013: p. 123–134.

38. Champion, M. et al., Module analysis captures pancancer genetically and epigenetically deregulated cancer driver genes for smoking and antiviral response. *EBioMedicine*, 2018. **27**: p. 156–166.

39. Campbell, J.D. et al., Genomic, pathway network, and immunologic features distinguishing squamous carcinomas. *Cell Rep*, 2018. **23**(1): p. 194–212 e6.

40. Brennan, K. et al., NSD1 inactivation defines an immune cold, DNA hypomethylated subtype in squamous cell carcinoma. *Sci Rep*, 2017. **7**(1): p. 17064.

41. Zhang, B. et al., Proteogenomic characterization of human colon and rectal cancer. *Nature*, 2014. **513**(7518): p. 382–387.

42. Lawrence, M.S. et al., Mutational heterogeneity in cancer and the search for new cancer-associated genes. *Nature*, 2013. **499**(7457): p. 214–218.

43. Mermel, C.H. et al., GISTIC2.0 facilitates sensitive and confident localization of the targets of focal somatic copy-number alteration in human cancers. *Genome Biol*, 2011. **12**(4): p. R41.

44. Gevaert, O., MethylMix: An R package for identifying DNA methylation-driven genes. *Bioinformatics*, 2015. **31**(11): p. 1839–1841.

45. Cedoz, P.L. et al., MethylMix 2.0: An R package for identifying DNA methylation genes. *Bioinformatics*, 2018. **34**(17): p. 3044–3046.

46. Vaske, C.J. et al., Inference of patient-specific pathway activities from multi-dimensional cancer genomics data using PARADIGM. *Bioinformatics*, 2010. **26**(12): p. i237–i245.

47. Manolakos, A. et al., CaMoDi: A new method for cancer module discovery. *BMC Genomics*, 2014. **15**(10): p. S8.

48. Gevaert, O. et al., Identification of ovarian cancer driver genes by using module network integration of multi-omics data. *Interface Focus*, 2013. **3**(4): p. 20130013.

49. Aerts, H.J. et al., Decoding tumour phenotype by noninvasive imaging using a quantitative radiomics approach. *Nat Commun*, 2014. **5**: p. 4006.

50. Zhou, M. et al., Radiomics in brain tumor: Image assessment, quantitative feature descriptors, and machine-learning approaches. *AJNR Am J Neuroradiol*, 2018. **39**(2): p. 208–216.

51. Kumar, V. et al., Radiomics: The process and the challenges. *Magn Reson Imaging*, 2012. **30**(9): p. 1234–1248.

52. Echegaray, S. et al., Quantitative image feature engine (QIFE): An open-source, modular engine for 3D quantitative feature extraction from volumetric medical images. *J Digit Imaging*, 2018. 31(4):p. 403–414.

53. van Griethuysen, J.J.M. et al., Computational radiomics system to decode the radiographic phenotype. *Cancer Res*, 2017. **77**(21): p. e104–e107.

54. Haralick, R.M., K. Shanmugam and I. Dinstein, Textural features for image classification. *IEEE Trans Syst Man Cybern*, 1973. **Smc3**(6): p. 610–621.

55. Depeursinge, A. et al., Three-dimensional solid texture analysis in biomedical imaging: Review and opportunities. *Med Image Anal*, 2014. **18**(1): p. 176–196.

56. Galloway, M.M., Texture analysis using gray level run lengths. *Comput Gr Image Process*, 1975. **4**(2): p. 172–179.

57. Thibault, G., J. Angulo and F. Meyer, Advanced statistical matrices for texture characterization: Application to cell classification. *IEEE Trans Biomed Eng*, 2014. **61**(3): p. 630–637.

58. Amadasun, M. and R. King, Textural features corresponding to textural properties. *IEEE Trans Syst Man Cybern*, 1989. **19**(5): p. 1264–1274.

59. Benjamini, Y. and Y. Hochberg, Controlling the false discovery rate: a practical and powerful approach to multiple testing. *J R Stat Soc Series B* (Methodological), 1995: p. 289–300.

60. Tusher, V.G., R. Tibshirani and G. Chu, Significance analysis of microarrays applied to the ionizing radiation response. *Proc Natl Acad Sci USA*, 2001. **98**(9): p. 5116–5121.

61. Witten, D.M. and R.J. Tibshirani, Extensions of sparse canonical correlation analysis with applications to genomic data. *Stat Appl Genet Mol Biol.*, 2009. **8**(1): p. 1–27.

62. Gevaert, O., F. De Smet, E. Kirk et al., Predicting the outcome of pregnancies of unknown location: Bayesian networks with expert prior information compared to logistic regression. *Hum Reprod*, 2006. **21**(7): p. 1824–1831.

63. Gevaert, O., F.D. Smet, D. Timmerman, Y. Moreau and B.D. Moor, Predicting the prognosis of breast cancer by integrating clinical and microarray data with Bayesian networks. *Bioinformatics*, 2006. **22**(14): p. e184–e190.

64. Daemen, A., O. Gevaert, K. Leunen, E. Legius, I. Vergote and B. De Moor, Supervised classification of array CGH data with HMM-based feature selection. *The Pacific Symposium on Biocomputing*, 2009. p. 468–479.

65. Pavlidis, P. et al., Learning gene functional classifications from multiple data types. *J Comput Biol*, 2002. **9**(2): p. 401–411.

66. Dey, S. et al., Integration of clinical and genomic data: A methodological survey, In *Computer Science & Engineering*. 2013, Minneapolis, MN: University of Minnesota College of Science & Engineering: University of Minnesota.

Resources and datasets for radiomics

KEN CHANG, ANDREW BEERS, JAMES BROWN, AND JAYASHREE
KALPATHY-CRAMER

11.1 INTRODUCTION

Medical imaging allows for non-invasive evaluation of disease to aid with clinical decision-making. However, current imaging within typical clinical workflows is often qualitative and subject to variable interpretation. As such, there is need for more quantitative assessment of disease. Radiomics is a technique that allows for delineation of quantitative features extracted from single- or multi-modal imaging that describe disease imaging characteristics. These imaging features may include descriptors such as image intensity, shape, and texture. While the utility of individual features may be limited, the many imaging features extracted from a patient describe a disease phenotype that can be used collectively to predict clinical variables such as genomics, treatment response, and survival.[1–4] This is accomplished through the use of machine learning techniques that can extract highly predictive imaging phenotypes from high quantities of features. This approach is important in the era of "precision medicine," which is focused on a more customized approach to patient care. In this chapter, we will discuss publicly available resources for practitioners of radiomics.

This includes software, packages, and datasets. We will also briefly review best practices and lessons learned for radiomics research, developed through multi-institutional collaborative projects.

11.2 RADIOMICS PIPELINES

Radiomics is a process for extracting quantitative imaging features from region of interest (ROI) in medical images such as tumors and organs. These mathematical descriptors provide information about many aspects of the ROI such as the size, shape, intensity, and texture. These characteristics, especially in the case of tumors, have been shown to be informative in classifying underlying tumor types, predicting outcomes, and assessing response to therapy. They are thought to reflect the underlying biology through attributes such as the heterogeneity of the tumor and sub-regions or "habitats." Most radiomics approaches follow a series of steps or a "pipeline," as described below.

11.2.1 SEGMENTATION

The first step of a radiomics pipeline involves delineating the disease region of interest. To do this, an expert (such as a radiologist) annotates the ROI within the image on a voxel level using software such as 3D Slicer.[5] One challenge of using annotations from a single expert is the inherent inter- and intra-rater variability. To account for variability within segmentation, annotations from multiple experts can be obtained and combined using methods such as simultaneous truth and performance level estimation (STAPLE).[6] The added benefit of obtaining annotations from multiple experts is that it allows for assessment of stability of downstream imaging features with respect to segmentation variability.[2] The reproducibility of segmentations can be further improved using semi-automated and fully automated segmentation methods.[7–9] It is worth noting that these ROIs can be represented as label maps (when a voxel may be 1 if it is part of the ROI or zero outside), a set of contour points on axial planes, or a 3-dimensional mesh. Conversion between these different representations may be required for the next step of feature extraction. However, this process of conversion between different ROI representation is not lossless due to resolution (e.g., number of triangles in the 3D mesh), partial volume effects (e.g., different ways of dealing with voxels straddle the boundary lines), and the use of splines and other geometrical methods to describe contours.

11.2.2 FEATURE EXTRACTION

Typical feature categories include size, intensity, shape, texture, and wavelet features.[10–12] Size features include volume and maximal diameters. Examples of intensity features include energy, entropy, kurtosis, maximum, minimum, mean, median, mean absolute deviation, range, root mean square, skewness, standard deviation, uniformity, and variance.[13] Examples of shape features include compactness, maximum diameter, spherical disproportion, sphericity, surface area, surface area to volume ratio, and volume.[13] Examples of texture features include gray-level co-occurrence matrix-based features, gray-level run-length matrix-based features, gray-level size-zone matrix-based features, Neighborhood gray-tone difference matrix-based features, and Laplacian of Gaussian features.[10,14] Wavelet features are derived by first passing the region of interest through a wavelet transform (which decomposes in image into low- or high-frequency components) and then calculating intensity or texture features.[13] A comprehensive list of features and their mathematical description can be found at https://arxiv.org/pdf/1612.07003.pdf.

11.2.3 FEATURE SELECTION

Feature selection is a method of determining the most relevant features for training a machine learning classifier. Due to the high dimensionality of imaging features extracted, feature selection can improve the performance of the machine learning classifier by inclusion of the most predictive features and exclusion of the less predictive ones.

There are many different methods for feature selection including Pearson's correlation coefficient, Spearman's correlation coefficient, Fisher score, Relief, T-score, Chi-square, Wilcoxon, Gini index, random forest minimal depth, mutual information maximization, mutual information feature selection, minimum redundancy maximum relevance, conditional infomax feature extraction, joint mutual information, conditional mutual information maximization, interaction capping, double input symmetric relevance, and Cox-regression model.[15,16] In a study comparing feature selection methods for the prediction of overall survival from computed tomography (CTs) of lung cancer patients, Parmar et al. found that the Wilcoxon test had the highest prognostic performance.[15] In another study comparing feature selection methods in the prediction of overall survival in patients with head and neck squamous cell carcinoma, Leger et al. found that the use of Spearman correlation coefficient was the most effective.[16]

11.2.4 CLASSIFICATION/PREDICTION

Classification techniques allow for integration of large quantities of imaging features into a single predictive model. There are many methods for classification including bagging, bayesian classification, boosting trees, decision trees, discriminant analysis, generalized linear models, boosting gradient linear models, multiple adaptive regression splines, nearest neighbors, neural networks, partial least square regression, principal component regression, random forests, and support vector machines.[15,16] In comparing these methods, Parmar et al. found that random forests had the highest prognostic performance in the prediction of overall survival from CT scans of lung cancer patients.[15] Leger et al. found that the use of random forests and boosting trees were effective in predicting overall survival in patients with head and neck squamous cell carcinoma.

11.2.5 STATISTICS

Importantly, when assessing the predictive value of individual features, correction for multiple testing should be made.[17] Furthermore, assessment of a classification model should be performed using both cross-validation as well as validation on an independent testing set.[17]

11.3 REPEATABILITY AND REPRODUCIBILITY OF RADIOMICS

Although radiomics has been shown to have utility in a number of applications, concerns about the repeatability and reproducibility of radiomics features persist. Test-retest studies have been instrumental in assessing the repeatability of the features.[18] Multi-user studies and studies comparing the feature across software packages are important in assessing the reproducibility of the features. Most radiomic features have explicit mathematical definitions, but different choices in implementation between radiomics software packages can lead to significantly different feature values extracted.[10,19] As a result, models trained with features from one package may not be replicable without that package's original source code at the time of model training. We discuss some common options in radiomics implementations of the same feature to further elaborate on this topic.

11.3.1 IMAGE RECONSTRUCTION, INTERPOLATION, AND RESAMPLING

Images can be reconstructed from "raw" CT and MR data using a range of methods. Zhao et al. demonstrated the variability in radiomics features arising from differences in reconstruction kernels.[20] Images can also be resampled (e.g., to isotropic), again leading to variability in features. It is important to specify the methods used in publication as these choices can greatly impact radiomics features and the ability to compare across studies.

11.3.2 QUANTIZATION

Medical images are usually acquired at high dynamic ranges with bit depths of 12–16. Discretization or quantization of image intensities inside the ROI is a prerequisite of the calculation of texture features for computational and noise reasons. Examples of these include such as gray-level co-occurrence matrices (GLCM), gray-level run-length matrix (GLRLM), and neighborhood gray-tone difference matrix (NGTDM).[21,22] The quantization can either be performed using a fixed number of bins, a fixed bin size, or a clustering algorithm.[17,23] Differences in the range to which values are quantized can have significant effects on these features values.[17]

11.3.3 BOUNDARY EFFECTS AND COMPLEX SHAPES

Some radiomic features can have significantly different values depending on how voxels at the boundaries of region of interest are treated. Some software packages may mask values outside the ROI as zero, creating potentially extreme contrast gradients at the ROI boundary. Others may not compute texture features at the boundary to compensate for this potential error or automatically dilate inputted ROIs for the purpose of calculate texture at the boundary. These choices will affect the final output of the feature, particularly in ROIs, where boundary tissues are highly informative. One approach to deal with boundary effects is to utilize the concept of a virtual "core biopsy," wherein the ROI is eroded to only include the core of the tissue of interest.[24,25] Another approach is to generate three regions, the core, the boundary, and the peritumoral areas, and to calculate features from these regions separately.[26–28] ROIs, especially in cancer, can have complex shapes. The calculation of radiomics features of complex shapes can be subject to variability and differences in interpretation. For instance, in objects that have "holes" in them, some packages chose to fill the holes while others treat internal boundaries in a manner similar to external boundaries.

11.3.4 NORMALIZATION

Some features may scale with ROI volume if not properly normalized. For example, the default MATLAB implementation of gray-level contrast matrices does not normalize their output for image sizes, meaning that gray-level co-occurrence matrix values will strongly correlate with ROI volume. Images can be acquired with gaps between slices. Such information is typically available in the original Digital Imaging and Communications in Medicine (DICOM) header. However, a common first step for the feature extraction pipeline is to convert these DICOM images into a single 3D array where information about the gaps may be lost. Similarly, the DICOM data may be acquired with different resolutions in all three dimensions, information that may be lost during the conversion process. Volume calculations themselves may differ between packages, if they are not normalized to the mm^3 resolution of the data. Additionally, if the data are normalized to isotropic mm^3 resolution, there may be differences in the resulting features depending on the interpolation algorithm chosen (such as Nearest neighbor, linear, and B-spline).

11.3.5 2D VERSUS 3D FEATURES

Many features can be calculated with both 2D and 3D methods on the same 3D volume. Texture features can be calculated with 2D methods on a per-slice basis, and then averaged over all slices in a 3D region of interest. These same features can often also be calculated directly, with an increased directionality that reflects the increase from two to three dimensions. These choices in feature computation will affect the precise values of textures calculated at those points.

The differences in methods of feature extraction have made the radiomic features difficult to reproduce.[10,29–33] Furthermore, certain features are sensitive to segmentation. Kalpathy-Cramer et al. found that 32% of features had a concordance correlation coefficient of <0.75 with different underlying segmentations.[10] Dercele et al. found that the entropy imaging feature within the tumor region is not just correlated with tumor type, but also with tumor size which is a major confounding factor.[34] The authors found underestimation of entropy was greater in smaller region of interest. As such, they recommend a minimum region of interest size for accurate reproduction of entropy. Dercele et al. also found that entropy of the tumor region was correlated with entropy of reference normal tissue, which may also confound interpretation of the feature.[34] Choice of acquisition parameters and CT scanner model can also have significant effect on radiomic features.[29,30] Specifically, Berenguer et al. found that only 43.1% and 89.3% of features were reproducible when pitch factor and reconstruction kernels were varied, respectively.[31] Furthermore, when the model of CT scanner was varied, the proportion of reproducible features ranged from 15.8% to 85.3%, depending on the phantom material.[31] For example, Mackin et al. found that X-ray tube current levels had little effect on radiomic features extracted from phantoms with tissue-like textures on CT imaging.[35]

11.4 STANDARDIZATION EFFORTS

To improve reproducibility of radiomic features for more reliable integration of these methods into clinical workflow, there have been several efforts to standardize radiomic pipelines including:

1. The Image Biomarker Standardisation Initiative (IBSI)[11]—this large multi-institutional, independent, international collaboration is working toward standardizing the extraction of radiomics features. They have developed a comprehensive, community-driven document that enumerates commonly features with mathematical descriptions, a digital phantom used for calculating features, and a global effort to standardize the implementation of the feature calculations.
2. Reference Ontology for Radiomics Features[12]—The Radiomics Ontology provides a semantic framework for radiomics features. It currently focuses on first-order, shape, textural, radiomics features. In addition, it includes classes about segmentation algorithms and imaging filters. This ontology has been harmonized with the IBSI initiative.
3. DICOM for Quantitative Imaging (dcmqi)[36]—This National Institutes of Health (NIH)-funded effort is developing is a free, open-source library that implements conversion of the data stored in commonly used research formats into the standard DICOM representation. This includes support for radiomics features in DICOM as part of the DICOM SR 1500 objects.

11.4.1 DATASETS

Datasets for validating radiomics algorithms are critical for algorithm developers and to ensure that the features extracted by different software packages are interoperable. Further, it is important to have objects with known "truth" as digital reference objects as well as a range of clinical cases that capture the diversity of appearance of tumors in practice.

1. Brodatz Textures[37]—The collection of photographic images of real-life objects (such as grass, wood, cloth), originally collated in the 1960s has been a commonly used resource in the computer vision community to develop algorithms for many decades.
2. Radiomics Digital Phantom[38]—This digital phantom is provided as part of the Image Biomarker Standardisation Initiative effort and is a good resource for those seeking to validate their radiomics feature extraction code. The phantom, mathematical description of the features, as well as the expected values for the features are publicly available.
3. University of Texas MD Anderson Cancer Center Texture/Radiomics Phantom[39]—The Credence Cartridge Radiomics Phantom is a hardware phantom the consists of ten cartridges, each with a unique texture. "The first four cartridges are 3D printed ABS plastic with 20%, 30%, 40%, and 50% honeycomb fill, and they provide regular, periodic textures. The next three cartridges provide natural textures: sycamore wood, cork, and extra dense cork. A cartridge of shredded rubber particles provides textures similar to those of non-small cell lung cancer. The ninth cartridge is solid, homogeneous acrylic and provides a minimal texture control. Finally, the 10th cartridge is 3D printed plaster has the highest electron density (400 HU–600 HU) and is intended to be more similar to bone."[39] This phantom has been scanned on multiple scanners, and these scans are available in The Cancer Imaging Archive.[39]
4. Quantitative Translational Imaging in Medicine (QTIM) Phantoms Dataset—This repository hosts 3D digital reference objects (DROs) that are constructed to exacerbate differences between different GLCM texture features, directions, and lengths, and to test different morphology features. They take the form of a series of multi-slice checkerboard, static, or other simple patterns, and the form of different size and shape labels place in space over a sample brain MRI.

5. The Cancer Imaging Archive—This is a well utilized resource for cancer images and meta-data associated with the images. A number of the image collections from The Cancer Imaging Archive including the Lung Image Database Consortium (LIDC), Quantitative Imaging Network (QIN) lung nodules, The Cancer Genome Atlas-Glioblastoma (TCGA-GBM), QIN head and neck, and others have been used for radiomics research as these collections include images, ROIs, and in some cases the features and outcomes.

6. Challenges—Challenges at conference venues such as Medical Image Computing and Computer Assisted Intervention (MICCAI), Society of Photo-Optical Instrumentation Engineers (SPIE), American Association of Physicists in Medicine (AAPM), and others offer opportunities to evaluate radiomics tools. These challenges typically provide imaging data including ROIs and outcomes and can be used to validate and compare radiomics approaches. The head & neck challenge conducted at MICCAI 2017 provided such a resource as did the QIN segmentation challenge.[40]

11.5 OPEN SOURCE TOOLKITS

1. Pyradiomics[41]

   http://pyradiomics.readthedocs.io/en/latest/

   This is an open-source package for extraction of both 2D and 3D radiomic features. The package can be called either within Python, as a 3D-Slicer module, or from Docker. [2] The features include 19 first-order statistics, 16 shape-based, 23 gray-level co-occurrence matrix, 16 gray-level run-length matrix, 16 gray-level size-zone matrix, 5 neighboring gray-tone difference matrix, and 14 gray-level-dependence matrix. There are also options for filters that can be applied before feature calculation including Laplacian of Gaussian, wavelet, square, square root, logarithm, exponential, gradient, and local binary pattern. The package is operating system (OS) independent and can be used with both Python 2.7 and Python $\geq$3.4.

2. Py-rex

   https://github.com/zhenweishi/Py-rex

   This is an extension of the Pyradiomics package with specific extensions to support radiation therapy use-cases by using DICOM-RT objects to define the region of interest. Further, the tool promotes semantic inter-operability by supporting the radiomics and radiation oncology ontologies. Data can be exported as comma separated value (CSV) or resource description framework (RDF) formats.

3. Imaging Biomarker Explorer (IBEX)[42,43]

   http://bit.ly/IBEX_MDAnderson

   This radiomics package is maintained by researchers at The University of Texas MD Anderson Cancer Center. The platform allows for viewing of imaging data, data sharing between users to ensure reproducibility, an integrated development environment for pipeline modification, and a graphical user interface. There are two versions of the package available: a standalone windows program and the source code written in Matrix Laboratory (MATLAB) and c/c++. Pre-processing functions available include image smoothing, image enhancement, image deblur, change enhancement, and resampling. Features include shape, intensity, intensity histogram, gray-level co-occurrence matrix (2D and 3D), neighbor intensity difference (2D and 3D), gray-level run-length matrix (2D only), and intensity histogram Gaussian fit features.

4. MaZda[44–46]

   http://www.eletel.p.lodz.pl/programy/mazda/

   This package is designed for the extraction of radiomic features from imaging and was developed at the Technical University of Lodz. The functionality applies to both 2D and 3D medical imaging and includes methods for feature reduction, feature visualization, and classification. It is written in C++ and Delphi and compiled for Windows 9x/NT/2000/XP operating systems. The features include image histogram, gradient, co-occurrence matrix, run-length matrix, autoregressive model, and Haar wavelet. The methods for feature reduction include Fisher discriminant, classification error combined with the correlation coefficient, mutual information, nearest neighbors, principal component analysis, linear discriminant analysis, and non-linear discriminant analysis. The methods for classification include nearest neighbors, artificial neural networks, agglomerative hierarchical clustering, similarity-based clustering methods, and k-means.

5. Stanford University Quantitative Image Feature Engine (QIFE)[47]
   https://github.com/riipl/3d_qifp
   This radiomics package was developed at Stanford University and is available in both MATLAB and Docker formats. It offers both pre-processing and feature calculation. In capable machines, it offers out-of-the-box parallel capacity for rapid output calculation. Pre-processing functions include segmentation deformation, topology preservation, maximum connected volume selection, and hole filling. Features include size distribution, intensity distribution, edge sharpness, local volume invariant integral, roughness, sphericity, Haralick's texture. The input format are DICOM images along with DICOM segmentation objects.

6. Radiomics[14,48]
   https://github.com/mvallieres/radiomics
   An open-source MATLAB package created and maintained by Martin Vallières at McGill University. It includes 2D and 3D texture features, shape features, modality-specific metrics (such as SUV metrics for positron emission tomography [PET] images), and utilities for performing simple machine learning models. It also includes utilities for manipulating medical images, including the conversion of radiotherapy structure (RTStruct) DICOM segmentations to 3D masks, and tools for performing simple multivariable modeling. Links to public datasets are provided, with which one can validate previous results achieved with the package.

7. Cancer Imaging Phenomics Toolkit (CaPTk)[49]
   https://www.med.upenn.edu/cbica/captk/
   CaPTk is a medical image viewer and analysis tool provided by the University of Pennsylvania and written in C++. In addition to texture and intensity features, it provides utilities for image segmentation, registration, noise reduction, and format conversions. It also includes utilities for generating predictive models from the features extracted, and specific tools for interacting brain, breast, and lung cancer data. CaPTk provides utilities to extend both the source code of the package and create pre-packaged executable extensions for use in individual projects.

8. DeepNeuro
   https://github.com/QTIM-Lab/DeepNeuro
   DeepNeuro is a Python package used primarily for deep learning that includes a wide array of radiomics features for benchmarking purposes. In addition to modules for deep learning applications, and the ability to template such applications oneself, it includes a module for generating GLCM texture features, volumetric features, and morphology features. DeepNeuro is easily distributed via Docker containers and contains many Python level functions for medical image manipulation, input/output, and conversion, as well as built in tools for multivariate analysis.

9. PET oncology radiomics test suite (PORTS)
   https://nciphub.org/groups/ports
   PORTS is an open-source MATLAB package for extracting features from PET modalities, specifically in the case of oncology. Texture and volumetric features are included, as well as reference objects to standardize the implementation of these features.

11.5.1 Next generation: Deep learning

One challenge of radiomics is its reliance on manually formulated or "hand-crafted" features, which may not capture the full range of information contained within the imaging. In contrast, the use of neural networks does not rely on manually formulated features. Neural networks take raw images as input and apply many layers of transformations to calculate the output of interest. As images are used to train the neural network, the weights within these transforms are adjusted to increase the accuracy of prediction. The many degrees of freedom within the neural network allow the network to learn highly complex patterns.[30] Careful design and optimization of neural networks allow for robust predictive performance. Recent studies have shown the potential of deep learning in many clinical tasks, including detection of retinopathy of prematurity, diagnosis of skin lesions, and predicting mutation in glioma.[51–53]

Tool	License	Platform	URL	Features
Pyradiomics	3-clause Berkeley Software Distribution (BSD) License	Python 3D-Slicer Docker	http://pyradiomics.readthedocs.io/en/latest/	First-order statistics, shape-based, gray-level co-occurrence matrix, gray-level run-length matrix, gray-level size-zone matrix, Neighboring gray-tone difference matrix, and gray-level-dependence matrix
Py-rex	Creative commons attribution 3.0 unported license.	Python	https://github.com/zhenweishi/Py-rex	First-order statistics, shape-based, gray-level co-occurrence matrix, gray-level run-length matrix, gray-level size-zone matrix, Neighboring gray-tone difference matrix, and gray-level-dependence matrix
Imaging Biomarker Explorer	2-clause BSD license	Windows MATLAB	http://bit.ly/IBEX_MDAnderson	Shape, intensity, intensity histogram, gray-level co-occurrence matrix (2D and 3D), Neighbor intensity difference (2D and 3D), gray-level run-length matrix (2D only), and intensity histogram Gaussian fit features
MaZda	End-user license agreement	Windows	http://www.eletel.p.lodz.pl/programy/mazda/	Histogram, gradient, co-occurrence matrix, run-length matrix, autoregressive model, and Haar wavelet
Stanford University Quantitative Image Feature Engine	Open source	MATLAB Docker	https://github.com/riipl/3d_qifp	Size distribution, intensity distribution, edge sharpness, local volume invariant integral, roughness, sphericity, and Haralick's texture
Radiomics (mvallieres)	GNU General public license	MATLAB	https://github.com/mvallieres/radiomics	gray-level co-occurrence matrix, gray-level run-length matrix, gray-level size-zone matrix, Neighborhood gray-tone difference matrix, PET-SUV intensity features, volumetric features, and morphology features
Cancer Imaging Phenomics Toolkit (CaPTk)	Section of biomedical image analysis (SBIA) Software license	C++ Graphical user interface	https://github.com/CBICA/CaPTk	Intensity statistics, gray-level co-occurrence matrix, gray-level run-length matrix, gray-level size-zone matrix, Neighborhood gray-tone difference matrix, local binary patterns, volumetric features, and morphology features
DeepNeuro	Massachusetts Institute of Technology (MIT) License	Python	https://github.com/QTIM-Lab/DeepNeuro	Intensity statistics, gray-level co-occurrence matrix, volumetric features, morphology features, and CT intensity features
PET Oncology Radiomics Test Suite (PORTS)	Open source	MATLAB	https://nciphub.org/groups/ports	Intensity and volumetric features, gray-level co-occurrence matrix, Neighborhood gray-tone difference matrix, size-zone matrix

REFERENCES

1. Zhang, B. et al. Multimodal MRI features predict isocitrate dehydrogenase genotype in high-grade gliomas. *Neuro. Oncol.* **19**, 109–117 (2017).

2. Chang, K. et al. Multimodal imaging patterns predict survival in recurrent glioblastoma patients treated with bevacizumab. *Neuro. Oncol.* **18**, 1680–1687 (2016).

3. Kickingereder, P. et al. Radiomic profiling of glioblastoma: Identifying an imaging predictor of patient survival with improved performance over established clinical and radiologic risk models. *Radiology* **280**, 880–889 (2016).

4. Grossmann, P. et al. Quantitative imaging biomarkers for risk stratification of patients with recurrent glioblastoma treated with bevacizumab. *Neuro. Oncol.* 1–32 (2017). doi:10.1093/neuonc/nox092.

5. Fedorov, A. et al. 3D Slicer as an image computing platform for the quantitative imaging network. *Magn. Reson. Imaging* **30**, 1323–1341 (2012).

6. Warfield, S. & Zou, K. Simultaneous truth and performance level estimation (STAPLE): An algorithm for the validation of image segmentation. *IEEE Trans. Med. Imaging* **23**, (2004).

7. Zhu, Y. et al. Semi-automatic segmentation software for quantitative clinical brain glioblastoma evaluation. *Acad. Radiol.* **19**, 977–985 (2012).

8. Kamnitsas, K. et al. Efficient multi-scale 3D CNN with fully connected CRF for accurate brain lesion segmentation. *Med. Image Anal.* **36**, 61–78 (2017).

9. Havaei, M. et al. Brain tumor segmentation with deep neural networks. *Med. Image Anal.* **35**, 18–31 (2017).

10. Kalpathy-Cramer, J. et al. Radiomics of lung nodules: A multi-institutional study of robustness and agreement of quantitative imaging features. *Tomogr. J. imaging Res.* **2**, 430–437 (2016).

11. Zwanenburg, A., Leger, S., Vallières, M. & Löck, S. Image biomarker standardisation initiative. arXiv preprint arXiv:1612.07003 (2016).

12. Radiomics Ontology. Available at: https://bioportal.bioontology.org/ontologies/RO. (Accessed: June 19, 2018).

13. Aerts, H. J. W. L. et al. Decoding tumour phenotype by noninvasive imaging using a quantitative radiomics approach. *Nat. Commun.* **5**, 4006 (2014).

14. Vallières, M., Freeman, C. R., Skamene, S. R. & El Naqa, I. A radiomics model from joint FDG-PET and MRI texture features for the prediction of lung metastases in soft-tissue sarcomas of the extremities. *Phys. Med. Biol.* **60**, 5471–5496 (2015).

15. Parmar, C., Grossmann, P., Bussink, J., Lambin, P. & Aerts, H. J. W. L. Machine learning methods for quantitative radiomic biomarkers. *Sci. Rep.* **5**, 13087 (2015).

16. Leger, S. et al. A comparative study of machine learning methods for time-to-event survival data for radiomics risk modelling. *Sci. Rep.* **7**, (2017).

17. Hatt, M. et al. Characterization of PET/CT images using texture analysis: The past, the present... any future? *Eur. J. Nucl. Med. Mol. Imaging* **44**, 151–165 (2017).

18. Balagurunathan, Y. et al. Test-retest reproducibility analysis of lung CT image features. *J. Digit. Imaging* **27**, 805–823 (2014).

19. Emaminejad, N. et al. The effects of variations in parameters and algorithm choices on calculated radiomics feature values: Initial investigations and comparisons to feature variability across CT image acquisition conditions. in *Medical Imaging 2018: Computer-Aided Diagnosis* (eds. Mori, K. & Petrick, N.) 140 (SPIE, 2018). doi:10.1117/12.2293864.

20. Zhao, B. et al. Reproducibility of radiomics for deciphering tumor phenotype with imaging. *Sci. Rep.* **6**, 23428 (2016).

21. Patel, M. B., Rodriguez, J. J. & Gmitro, A. F. Effect of gray-level re-quantization on co-occurrence based texture analysis. In *Proceedings—International Conference on Image Processing, ICIP* 585–588 (2008). doi:10.1109/ICIP.2008.4711822.

22. Galloway, M. M. Texture analysis using gray level run lengths. *Comput. Graph. Image Process.* **1**, 172–179 (1975).

23. Cook, G. J. R., Azad, G., Owczarczyk, K., Siddique, M. & Goh, V. Challenges and promises of PET radiomics. *Int. J. Radiat. Oncol.* (2018). doi:10.1016/J.IJROBP.2017.12.268.

24. Prasanna, P., Tiwari, P. & Madabhushi, A. Co-occurrence of local anisotropic gradient orientations (CoLlAGe): A new radiomics descriptor. *Sci. Rep.* **6**, 37241 (2016).

25. Echegaray, S. et al. A rapid segmentation-insensitive "Digital biopsy" method for radiomic feature extraction: Method and pilot study using CT images of non-small cell lung cancer. *Tomogr. J. Imaging Res.* **2**, 283–294 (2016).

26. Grove, O. et al. Quantitative computed tomographic descriptors associate tumor shape complexity and intratumor heterogeneity with prognosis in lung adenocarcinoma. *PLoS One* **10**, e0118261 (2015).

27. Echegaray, S. et al. Core samples for radiomics features that are insensitive to tumor segmentation: Method and pilot study using CT images of hepatocellular carcinoma. *J. Med. Imaging (Bellingham)* **2**, 041011 (2015).

28. Tunali, I. et al. Radial gradient and radial deviation radiomic features from pre-surgical CT scans are associated with survival among lung adenocarcinoma patients. *Oncotarget* **8**, 96013–96026 (2017).

29. Kim, H. et al. Impact of reconstruction algorithms on CT radiomic features of pulmonary tumors: Analysis of intra- and inter-reader variability and inter-reconstruction algorithm variability. *PLoS One* **11**, (2016).

30. Mackin, D. et al. Measuring computed tomography scanner variability of radiomics features. *Invest. Radiol.* **50**, 757–765 (2015).

31. Berenguer, R. et al. Radiomics of CT features may be nonreproducible and redundant: Influence of CT acquisition parameters. *Radiology* 172361 (2018). doi:10.1148/radiol.2018172361.

32. Shafiq-ul-Hassan, M. et al. Intrinsic dependencies of CT radiomic features on voxel size and number of gray levels. *Med. Phys.* **44**, 1050–1062 (2017).

33. Fave, X. et al. Can radiomics features be reproducibly measured from CBCT images for patients with non-small cell lung cancer? *Med. Phys.* **42**, 6784–6797 (2015).

34. Dercle, L. et al. Limits of radiomic-based entropy as a surrogate of tumor heterogeneity: ROI-area, acquisition protocol and tissue site exert substantial influence. *Sci. Rep.* **7**, 7952 (2017).

35. MacKin, D. et al. Effect of tube current on computed tomography radiomic features. *Sci. Rep.* **8**, (2018).

36. Herz, C. et al. dcmqi: An open source library for standardized communication of quantitative image analysis results using DICOM. *Cancer Res.* **77**, e87–e90 (2017).

37. Brodatz Texture Database. Available at: http://multibandtexture.recherche.usherbrooke.ca/original_brodatz_more.html. (Accessed: June 19, 2018).

38. Radiomics Digital Phantom. Available at: https://www.cancerdata.org/resource/doi:10.17195/candat.2016.08.1 (Accessed: June 19, 2018).

39. Mackin, D. et al. Data from credence cartridge radiomics phantom CT scans. *Cancer Imaging Arch.* doi:10.7937/K9/TCIA.2017.zuzrml5b.

40. QIN Lung CT Segmentation Challenge. Available at: https://wiki.cancerimagingarchive.net/display/Public/QIN+Lung+CT+Segmentation+Challenge. (Accessed: 19th June 2018).

41. Van Griethuysen, J. J. M. et al. Computational radiomics system to decode the radiographic phenotype. *Cancer Res.* **77**, e104–e107 (2017).

42. Zhang, L. et al. Ibex: An open infrastructure software platform to facilitate collaborative work in radiomics. *Med. Phys.* **42**, (2015).

43. Ger, R. B. et al. Guidelines and experience using imaging biomarker explorer (IBEX) for radiomics. *J. Vis. Exp.* (2018). doi:10.3791/57132.

44. Strzelecki, M., Szczypinski, P., Materka, A. & Klepaczko, A. A software tool for automatic classification and segmentation of 2D/3D medical images. *Nucl. Instruments Methods Phys. Res. Sect. A Accel. Spectrometers, Detect. Assoc. Equip.* **702**, 137–140 (2013).

45. Szczypiński, P. M., Strzelecki, M., Materka, A. & Klepaczko, A. MaZda–A software package for image texture analysis. *Comput. Methods Programs Biomed.* **94**, 66–76 (2009).

46. Szczypinski, P. M., Strzelecki, M. & Materka, A. MaZda—A software for texture analysis. in *Proceedings—2007 International Symposium on Information Technology Convergence, ISITC 2007*, 245–249 (2007). doi:10.1109/ISITC.2007.66.

47. Echegaray, S., Bakr, S., Rubin, D. L. & Napel, S. Quantitative image feature engine (QIFE): An open-source, modular engine for 3D quantitative feature extraction from volumetric medical images. *J Digit Imaging* 1–12 (2017). doi:10.1007/s10278-017-0019-x.

48. Zhou, H. et al. MRI features predict survival and molecular markers in diffuse lower-grade gliomas. *Neuro. Oncol.* **19**, 862–870 (2017).

49. Davatzikos, C. et al. Cancer imaging phenomics toolkit: Quantitative imaging analytics for precision diagnostics and predictive modeling of clinical outcome. *J. Med. Imaging* **5**, 1 (2018).

50. LeCun, Y. A., Bengio, Y. & Hinton, G. E. Deep learning. *Nature* **521**, 436–444 (2015).

51. Brown, J. M. et al. Automated diagnosis of plus disease in retinopathy of prematurity using deep convolutional neural networks. *JAMA Ophthalmol.* (2018). doi:10.1001/jamaophthalmol.2018.1934.

52. Esteva, A. et al. Dermatologist-level classification of skin cancer with deep neural networks. *Nature* **542**, 115–118 (2017).

53. Chang, K. et al. Residual convolutional neural network for the determination of *IDH* status in low- and high-grade gliomas from MR imaging. *Clin. Cancer Res.* **24**, 1073–1081 (2018).

PART ③

CLINICAL APPLICATIONS

12

Pathways to radiomics-aided clinical decision-making for precision medicine

TIANYE NIU, XIAOLI SUN, PENGFEI YANG, GUOHONG CAO, KHIN K. THA, HIROKI SHIRATO, KATHLEEN HORST, AND LEI XING

12.1 INTRODUCTION

The field of radiomics and radiogenomics is a rapidly evolving field where a large number of shape and texture features are extracted from the image datasets. When transformed into quantitative features, tumor radiology properties can be linked to the underlying genetic alterations [1–3] and medical outcomes [4–6]. Radiomics bridges the gap between imaging data and clinical outcome, thus proving a powerful tool for more comprehensive characterization of tumors or lesions. Newly developed radiomics related methods, including the intratumor variation analysis [7], delta radiomics [8], deep radiomics with state-of-the-art deep learning techniques, and in combination with other omics data such as radiogenomics, further expands the horizons of radiomics and facilitates quantitative imaging and big data in the clinic. Radiomics and radiogenomics represent a new frontier of biomedicine and promise to transform the way imaging information is used in clinical practice [9].

In the previous chapters, some important technical aspects, such as image and data processing pipelines and identification and extraction of features in a variety of imaging and non-imaging data, have been described extensively. Some key tasks in omics-driven applications, such as optimization of imaging protocols, image segmentation and analysis, identification and extraction of various features in imaging and/or genomic data, and various algorithms developed to find the relationships between the extracted features and the clinical outcome, are discussed. The radiomics correlative models characterizing the disease progression and predicting the therapeutic response and outcome is of great importance for research and clinical decision support. In this chapter, we will review the issues and strategies relevant to the clinical implementation of omics-tools, such as clinical workflow, imaging protocols, use of multi-parametric morphologic and functional imaging information, efficient integration of new techniques into clinical processes, data sources and sharing mechanism, and potential roles of omics-tools in routine clinical practice.

12.2 IMAGING AND ITS ROLES IN PATIENT CARE

The discoveries of computed tomography (CT), MRI, and single-photon emission computed tomography/positron emission tomography (SPECT/PET) have significantly changed the landscape of healthcare. Clinically, CT and MRI are the two major imaging modalities in the past few decades. Other imaging modalities, such as radionuclide imaging (SPECT/PET), ultrasound, optical computed tomography, and some emerging modalities, such as photoacoustic and X-ray acoustic imaging [10,11], X-ray fluorescence and luminescence imaging [12–14], microwave imaging, and MR particle imaging [15], are also important to modern medical practice. With the advancement of imaging techniques and the picture archiving and communication system, the usage of imaging in medicine has grown dramatically. From 1997 to 2006, the annual percentage growth rate of the examinations of CT, MRI, and ultrasound is around 14%, 21%, and 5%, respectively [16]. According to the study of Mokhtar et al., CT examination is performed approximately 70 million times annually in the United States [17]. Based on the data from International Society for Magnetic Resonance in Medicine, over 150 million patients have MRI examinations to date. Every year, approximately 10 million patients undergo MRI procedures.

Medical imaging plays an increasingly important role in the screening, diagnosis, staging, treatment planning, assessment of therapeutic response, and monitoring of various diseases in today's clinical practice. To illustrate this, in Figure 12.1, we show the workflow of clinical radiation therapy. Many other clinical procedures such as image guided surgery share a similar workflow. As can be seen, imaging and image analysis are involved in every step of the patient care process, ranging from simulation imaging, patient modeling, treatment planning, patient setup and dose delivery, to patient follow-up, and therapeutic evaluation [18]. Radiomics analysis with or without dosimetric information can be well integrated in the workflow and facilitates treatment decision-making. When the isodose distribution data are included in the predictive model, which is often referred to as the dosiomics or deep dose analysis [19], the achievable accuracy of outcome prediction will likely be improved, as the outcome of radiation therapy depends critically on the spatial distribution and fractionation scheme [20]. With these tools, the image features from planning CT/MRI/PET, pre-treatment cone beam computed tomography (CBCT) [21], as well as the patient's treatment plan can readily be extracted and exploited to improve the disease management.

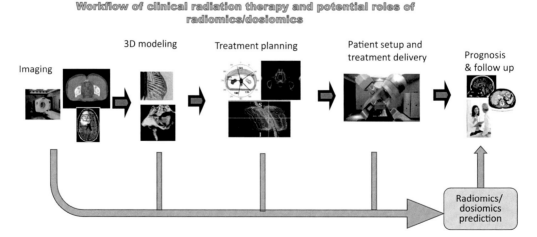

Figure 12.1 (See color insert.) The role of imaging in the workflow of radiation therapy and radiomics- and/or dosiomics-based prediction of outcome.

12.3 LEARNING AND CLINICAL DECISION-MAKING FROM IMAGING DATA

For decades, physicians have been making diagnoses and other patient management decisions based on the use of imaging information [22]. It is estimated that imaging data are involved in approximately 70% of clinical decision-making. Clinically, the interpretation of imaging data still relies heavily on visual assessment and some primitive imaging metrics. In cancer treatment, a simple, but commonly used quantitative measurement is tumor size or tumor volume [23]. For example, Response Evaluation Criteria in Solid Tumors (RECIST), which refers to a set of published rules used to assess tumor burden in order to provide an objective assessment of response to therapy, is recommended for therapeutic assessment in practice. This type of simple measurement, however, fails to reflect the complexity of tumor morphology and heterogeneity. With the advancements in digital technology and biotechnology, clinical decision-making is becoming increasingly complicated and intractable because of the increased amount of information and dimensionality of the variables that need to be taken into consideration. In reality, it is well known that integration of up to five factors to make a decision is traditionally considered as the limit of our cognitive capacity [24]. On the other hand, by year 2020, making a clinical decision for an individual patient may well depend on up to 10,000 parameters [25], which stresses the critical need for automated and quantitative methods.

Using computers to facilitate clinical diagnosis and treatment can be traced back to the early days shortly after the volumetric imaging techniques were invented. In the 1980s, various computer-aided diagnosis (CAD) schemes were developed at the University of Chicago and other intuitions. The output of CAD systems is used by radiologists as an option in making diagnostic decisions for comprehensive disease characterization, reliable prediction, or assessment of treatment response and prognosis toward the goal of precision medicine. In recent years, data-driven decision-making has gradually become the mainstay of computer-aided clinical decision-making. While data science and machine learning-based approaches such as radiomics and radiogenomics are relatively new, the concept of learning from prior knowledge and extracting value from data has been around for decades. Nevertheless, what has changed is enormous. Notably, these changes include: (i) computational power; (ii) increased awareness and accessibility of annotated data, including digital imaging data; and (iii) computer algorithms and the availability of machine learning frameworks. In addition to a variety of open source options, there are options of proprietary packages for radiomics and machine learning from commercial entities, such as HY_RADIOMICS toolkits (Huiyihuiying Medical Technology Co., Beijing, China) and TexRAD™ (Feedback plc, Cambridge, United Kingdom). Development and timely translation of data science into clinical practice will enable us to leverage the existing imaging data of previous patients to benefit new patients. A comprehensive mechanism to make full use of these data through radiomics, radiogenomics, and deep learning would thus be highly desirable.

In addition to the conventional radiomics approaches which rely on the use of hand-crafted features to build a predictive model, deep learning is being increasingly investigated for clinical decision-making [19,26]. The deep learning techniques apply a designed network to extract features rather than using the pre-designed features [27,28]. The network structure in deep learning method achieves the feature extraction of externally input data from low to advanced levels and the interpretation of external data. As a branch of machine learning, deep learning can be divided into supervised and unsupervised learning [29]. Supervised learning includes multi-layer perceptron, convolutional neural networks (CNN), etc. Unsupervised learning includes deep belief nets, auto encoders, denoising autoencoders, and sparse coding. There are several popular CNN-based architectures including AlexNet, GoogLeNet, VGGNet, and ResNet [30]. These architectures have shown superior accuracy in natural object classification on the ImageNet Large Scale Visual Recognition Challenge (ILSVRC) competitions and have shown great potential in medical image analysis [31–34].

Deep learning methods have been applied to disease detection, staging, and outcome predictions of multiple tumors. Esteva et al. [35] developed a CNN-based skin cancer diagnostic framework. Hongyoon et al. [36] used a CNN to learn from PET images and screen out potential populations of Alzheimer's disease. Yasaka et al. [37]

developed a deep CNN for liver fibrosis staging using the hepatobiliary phase in the contrast-enhanced MR images. Kawahara et al. [38] proposed a BrainNetCNN for predicting clinical neurodevelopmental outcomes. Zhang et al. [39] developed a diagnostic tool based on a deep learning framework for screening generally treatable blinding retinopathy. In addition, the study also proves that the deep learning framework has universal applicability in the diagnosis of pneumonia in children with chest X-ray images. Ibragimov et al. [19] investigated the feasibility of using CNN for individualized hepatobiliary toxicity prediction after liver stereotactic body radiotherapy (SBRT).

12.4 PRACTICAL ISSUES AND CRITICAL PATHS IN TRANSLATING OMICS-TOOLS INTO CLINICAL PRACTICE

Successful application of radiomics in clinical practice entails consideration of many technical and operational details, and a seamless integration of the tools into clinical workflow is significant for patients to truly benefit from the state-of-the-art technology. Here, we highlight some key issues relevant to these aspects.

12.4.1 DATA ACQUISITION AND QUANTIFICATION FOR DATA-DRIVEN ANALYTICS TOOLS

Standard data format is important for accessing and sharing images Digital Imaging and Communications in Medicine (DICOM) format is commonly used in storage and exchange of medical images. A DICOM file contains both image data and metadata of the patient. The metadata includes information about the image acquisition parameters, scanner, and other patient and hospital related information. The definition of region of interest (ROI), such as tumor volume, is useful for data analysis. There are several commonly used ROI file formats including DICOM-radiotherapy structure set (RTSTRUCT), Analyze, Neuroimaging Informatics Technology Initiative (NifTI), etc. These formats are applied in the ROI delineation platforms in the radiation treatment planning software or the open-source software including the 3D Slicer [40], the Insight Toolkit-SNAP (ITK-SNAP) [41], etc.

For radiomics feature extraction, which is used to convert the imaging data into quantitative parameters, the Haralick texture features [42] are most commonly used. This type of feature is modeled to capture the gray-level co-occurrence patterns in the images. The Gabor features [42] are another type of radiomics feature which are designed to mimic the way human visual system deciphers object appearances. Histogram of gradient orientations features are also useful radiomics features which yield a global patch-based signature by computing histogram distribution of orientations [43]. Recently, Lee et al. [44] developed a novel quantitative feature called the "cell orientation entropy" to capture the differences in orientation of nuclei with respect to the neighboring nuclei across different pathologies. They demonstrated the differential behavior of the nuclear orientations across aggressive and benign conditions in the context of prostate cancer. Prasanna et al. [45] presented a new radiomic descriptor, the co-occurrence of local anisotropic gradient orientations, to capture the anisotropic tensor gradient differences across similar appearing pathologies in an image.

A prerequisite in establishing a robust radiomics model is the availability of a large amount of annotated data. As all said, data are the new oil for biomedical data science. For beginners, it is useful to practice on some high quality datasets available in the public domain. For example, for deep learning applications, a list of openly available datasets has been curated (www.analyticsvidhya.com). The use of synthetic data has proved to be a valuable approach for certain types of problems [46]. The use of publicly shared medical datasets provides a source for gaining experiences and insights into some important clinical problems. Sun et al. [47] developed an image biomarker using radiomics method to predict the responses of tumor immune phenotype in patients. Four retrospectively independent cohorts were used. The first dataset is the Molecular Screening for Cancer Treatment Optimization (MOSCAT), the immunotherapy-treated and immune phenotype dataset, for the model training. The other datasets are used to validate the developed model, which are collected from the open source databases of The Cancer Genome Atlas (TCGA) and The Cancer Imaging Archive (TCIA). Li et al. [48] developed an individualized immune signature to estimate the prognosis in patients with early stage non-squamous non-small cell lung cancer (NSCLC). This study retrospectively analyzes the gene expression profiles of frozen tumor tissue samples from 19 public NSCLC cohorts, including 18 microarray datasets and an RNA Sequencing (RNA-Seq) dataset for TCGA lung adenocarcinoma cohort. The independent multi-cohort validation confirms the general usefulness of the developed model. Vallières et al. [49] developed a joint FDG-PET and MRI texture-based model for early evaluation of lung metastasis risk in soft-tissue sarcomas. The dataset is comprised of 51 patients with pre-treatment fluorodeoxyglucose (FDG)-PET images and the T1- and T2-weighted MRI images. The data are uploaded to TCIA website for sharing. National Biomedical Image Archive (NBIA) is one of the largest online CT image repositories hosted by the National Cancer Institute (NCI). The Cancer Imaging Archive is another open-access database of medical images for cancer research funded by the NCI Cancer Imaging Program [50]. There are many other open-access data sharing platforms, such as NCI's Genomic Data Commons, the National Library of Medicine, and the Swiss Institute for Computer Assisted Surgery (SICAS) medical image repository.

Training of a predictive model requires a large number of annotated datasets, and this often presents a bottleneck problem in the construction of a radiomics or radiogenomics method, especially when the difference between phenotypes is subtle or if a large heterogeneity exists within the population. Because of the avoidance of data exchange in the process, distributed learning may prove to be a powerful technique to alleviate the problem of limited size of training datasets. Several approaches have been proposed to distributed training, such as model averaging and asynchronous stochastic gradient descent. The deep learning platforms such as TensorFlow are capable of asynchronous learning and decision-making based on temporally and/or geographically distributed data. Chang et al. [51] have simulated the distribution of deep learning models across institutions using various non-parallel training heuristics and compared the results with a model trained on centrally hosted patient data. Distribution of learning models across laboratories or institutions may help to overcome the practical hurdles seen with sharing patient data.

12.4.2 CASE STUDIES

In this subsection, we will illustrate the implementation of radiomics tools by highlighting a few recent studies from the literature. The first study is a radiomics prognosis evaluation on osteosarcoma. Osteosarcoma is the primary bone malignancy in the world. Nearly 90% of osteosarcoma patients are classified as high-grade osteosarcoma at the time of diagnosis. Although the incidence rate of high-grade osteosarcoma is not high, the 5-year overall survival rate is only about 50% world wide. Aggressive treatment methods, such as multi-cycle treatments and adjuvant chemotherapies, are valuable for patients who have high risk disease. On the contrary, these treatments might produce adverse effects to the patients who are likely to have a favorable prognosis. If the prognosis could be predicted pre-operatively, then personalized treatment strategies could benefit patients with osteosarcoma. Wu et al. developed a radiomics nomogram for pre-operative survival evaluation (Figure 12.2) [52]. Image features are extracted based on the pre-treatment diagnostic CT images. The radiomics nomogram is constructed by combining radiomics features and clinical risk features. The radiomics nomogram shows expected accuracy of area under the curve (AUC) of 0.86 for the training cohort and 0.84 for the independent validation cohort. The Kaplan-Meier survival curves shows significant differences between the nomogram-predicted survival and non-survival groups.

The second example is to use radiomics for rectal cancer management. For locally advanced rectal cancer (LARC), pre-operative concurrent chemoradiation treatment (CRT) followed by total mesorectal excision is the standard treatment strategy. After CRT, approximately 20% of patients show a pathologic complete response (pCR). For these patients with pCR, a "wait-and-see" policy has been proposed since the total mesorectal excision is associated with significant morbidity and functional complications. It is important to provide physicians with accurate information using non-invasive approaches to identify patients who are likely to achieve a pCR. Nie et al. [53] applied the artificial neural network method to predict pathologic response after pre-operative CRT for LARC patients. To obtain comprehensive information of the tumor, they analyzed the image features based on the multi-parametric MRI images of anatomical T1/T2 images, diffusion-weighted images, and dynamic contrast-enhanced MRI images. A total number of 103 image features were obtained for each patient. Their result suggested combined quantitative image features may lead to better prognostic value for allowing earlier treatment alternation and more accurate non-invasive surveillance. Figure 12.3 shows the MR images and receiver operating characteristic curves (ROCs) in this study.

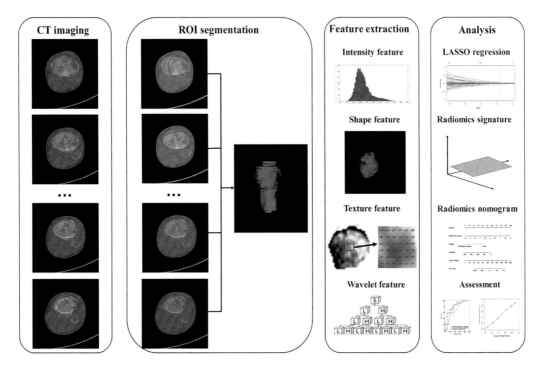

Figure 12.2 (See color insert.) The image quantitation method by using radiomics in the osteosarcoma study. The features include the features of the intensity, the shape, the texture, and the wavelet.

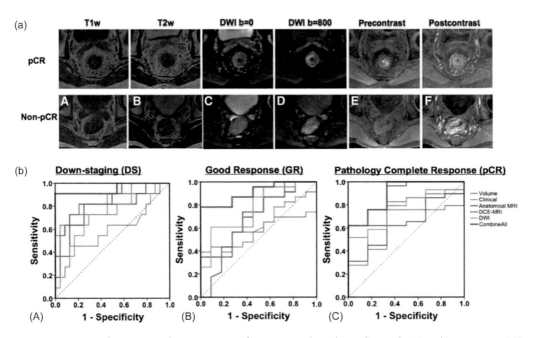

Figure 12.3 (See color insert.) The MR images of patients and results in the study. (a) multi-parametric MR images of two male patients, both at 60 years old with mid-rectum cancer; (b) the model performance for down stage, good response and pathology complete response.

Numerous radiomics studies have been carried out to deal with various clinical problems, and some of these studies will be summarized in the chapters to follow. We note that the use of deep learning-based method is on the horizon with promising results as compared to the conventional omics approaches [54,55]. Deep neural network has recently attracted much attention for its unprecedented ability to learn complex relationships and incorporate existing knowledge into the inference model through feature extraction and representation learning. The high dimensionality of hierarchical neural network transformations enables the algorithms to learn complex patterns with a high level of abstraction. Recent studies have shown the potential of deep learning in classifying dermatological lesions, identifying and classifying abnormalities in colonoscopy images, detecting diabetic retinopathy, grading prostate cancer, predicting mutations in glioma, suggesting optimal interventions for stroke patients, and assessing medical records. Due to the infant nature of these works, we will skip the subject and focus on more widely used radiomics techniques.

12.5 SUMMARY

The volume of medical image data in patient care is rapidly growing. It is estimated that data for a single patient would amount to about 7 GB [24]. In addition to the image data, clinical features, treatment features, and molecular features are also important and should be considered in clinical decision-making [56]. The computer-aided decision-making tools aiming to learn from imaging and other clinical data to guide the diagnosis, treatment, and outcome prediction will play more important roles in future precision medicine. The connection between the radiology characteristics and the tumor/lesions as well as treatment outcome has the potential to generate a "radiomics"-based workflow to improve medical care. The establishment of the omics-tools will benefit greatly from the development of advanced machine-learning methods and quantitative imaging [57]. Noteworthy, deep learning technology has shown great application prospects in medical image analysis and clinical assistant decision-making. To move forward, establishing data-sharing mechanisms, promoting quantitative imaging, optimizing imaging protocols, and adopting new imaging techniques are critical aspects to drive radiomics to clinical practice. Finally, it is important to emphasize that the discipline of radiomics and radiogenomics is interdisciplinary in nature, and timely translation of the novel technical developments requires close collaboration among medical physicists, clinicians, engineers, and computer scientists.

REFERENCES

1. Yamamoto, S. et al., Radiogenomic analysis of breast cancer using MRI: A preliminary study to define the landscape. *Am J Roentgenol*, 2012. **199**(3): 654–663.
2. Kickingereder, P. et al., Radiogenomics of glioblastoma: Machine learning-based classification of molecular characteristics by using multiparametric and multiregional MR imaging features. *Radiology*, 2016. **281**(3): 907–918.
3. Brisse, H.J. et al., Radiogenomics of neuroblastomas: Relationships between imaging phenotypes, tumor genomic profile and survival. *PLoS One*, 2017. **12**(9): e0185190.
4. Gillies, R.J., P.E. Kinahan, and H. Hricak, Radiomics: images are more than pictures, they are data. *Radiology*, 2016. **278**(2): 563–577.
5. Lambin, P. et al., Radiomics: The bridge between medical imaging and personalized medicine. *Nat Rev Clin Oncol*, 2017. **14**(12): 749–762.
6. Lambin, P. et al., Radiomics: Extracting more information from medical images using advanced feature analysis. *Eur J Cancer*, 2012. **48**(4): 441–446.
7. Wu, J. et al., Robust intratumor partitioning to identify high-risk subregions in lung cancer: A pilot study *Int J Radiat Oncol Biol Phys*, 2016. **95**(5): 1504–1512.
8. Fave, X. et al., Delta-radiomics features for the prediction of patient outcomes in non-small cell lung cancer. *Sci Rep*, 2017. **7**(1): 588.

9. Xing, L., E.A. Krupinski, and J. Cai, Artificial intelligence will soon change the landscape of medical physics research and practice. *Med Phys*, 2018. **45**(5): 1791–1793.

10. Li, L. et al., Single-impulse panoramic photoacoustic computed tomography of small-animal whole-body dynamics at high spatiotemporal resolution. *Nat Biomed Eng*, 2017. **1**(5).

11. Xiang, L. et al., X-ray photoacoustic tomography with pulsed X-ray beam from a medical linear accelerator. *Med. Phys.* (Letters), 2013. **40**: 10701–10705.

12. Pratx, G. et al., X-ray luminescence computed tomography via selective excitation: A feasibility study. *IEEE Trans Med Imaging*, 2010. **29**(12): 1992–1999.

13. Pratx, G. et al., Tomographic molecular imaging of x-ray-excitable nanoparticles. *Opt Lett*, 2010. **35**(20): 3345–3347.

14. Bazalova, M. et al., Investigation of X-ray fluorescence computed tomography (XFCT) and K-edge imaging. *IEEE Trans Med Imaging*, 2012. **31**(8): 1620–1627.

15. Lu, K. et al., Multi-channel acquisition for isotropic resolution in magnetic particle imaging. *IEEE Trans Med Imaging*, 2018. **37**(9): 1989–1998.

16. Smith-Bindman, R., D.L. Miglioretti, and E.B. Larson, Rising use of diagnostic medical imaging in a large integrated health system. *Health Aff* (Millwood), 2008. **27**(6): 1491–1502.

17. Smith-Bindman, R. et al., Radiation dose associated with common computed tomography examinations and the associated lifetime attributable risk of cancer. *Arch Intern Med*, 2009. **169**(22): 2078–2086.

18. Peeken, J.C., F. Nusslin, and S.E. Combs, "Radio-oncomics": The potential of radiomics in radiation oncology. *Strahlenther Onkol*, 2017. **193**(10): 767–779.

19. Ibragimov, B. et al., Development of deep neural network for individualized hepatobiliary toxicity prediction after liver SBRT. *Med Phys*, 2018. **45**: 4763–4774.

20. Yang, Y. and L. Xing, Optimization of radiation dose-time-fractionation scheme with consideration of tumor specific biology. *Med Phys*, 2005. **32**(12): 3666–3677.

21. Xing, L. et al., Overview of image-guided radiation therapy. *Med Dosim*, 2006. **31**(2): 91–112.

22. Timmerman, R. and L. Xing, 2009. *Image Guided and Adaptive Radiation Therapy*. Baltimore, MD: Baltimore Lippincott Williams & Wilkins.

23. Gatenby, R.A., O. Grove, and R.J. Gillies, Quantitative imaging in cancer evolution and ecology. *Radiology*, 2013. **269**(1): 8–15.

24. Bibault, J.E., P. Giraud, and A. Burgun, Big data and machine learning in radiation oncology: State of the art and future prospects. *Cancer Lett*, 2016. **382**(1): 110–117.

25. Abernethy, A.P. et al., Rapid-learning system for cancer care. *J Clin Oncol*, 2010. **28**(27): 4268–4274.

26. Choi, J.Y., Radiomics and deep learning in clinical imaging: What should we do? *Nucl Med Mol Imaging*, 2018. **52**(2): 89–90.

27. Afshar, P. et al., From hand-crafted to deep learning-based cancer radiomics: Challenges and opportunities. arXiv preprint arXiv:1808.07954, 2018.

28. LeCun, Y., Y. Bengio, and G. Hinton, Deep learning. *Nature*, 2015. **521**: 436.

29. Litjens, G. et al., A survey on deep learning in medical image analysis. *Med Image Anal*, 2017. **42**: 60–88.

30. Canziani, A., A. Paszke, and E. Culurciello, An analysis of deep neural network models for practical applications. arXiv preprint arXiv:1605.07678, 2016.

31. Russakovsky, O. et al., Imagenet large scale visual recognition challenge. *Int J Comput Vis*, 2015. **115**(3): 211–252.

32. Ibragimov, B. and L. Xing, Segmentation of organs-at-risks in head and neck CT images using convolutional neural networks. *Med Phys*, 2017. **44**(2): 547–557.

33. Qin, W. et al., Superpixel-based and boundary-sensitive convolutional neural network for automated liver segmentation. *Phys Med Biol*, 2018. **63**(9): 095017.

34. Liu, H. et al., Learning deconvolutional deep neural network for high resolution (HR) medical image reconstruction. *Inf Sci*, 2018. **468**: 142–154.

35. Esteva, A. et al., Dermatologist-level classification of skin cancer with deep neural networks. *Nature*, 2017. **542**(7639): 115–118.

36. Choi, H. and K.H. Jin, Predicting cognitive decline with deep learning of brain metabolism and amyloid imaging. *Behav Brain Res*, 2018. **344**: 103–109.

37. Yasaka, K. et al., Liver Fibrosis: Deep convolutional neural network for staging by using gadoxetic acid-enhanced hepatobiliary phase MR images. *Radiology*, 2018. **287**(1): 146–155.

38. Kawahara, J. et al., BrainNetCNN: Convolutional neural networks for brain networks; towards predicting neurodevelopment. *Neuroimage*, 2017. **146**: 1038–1049.

39. Kermany, D.S. et al., Identifying medical diagnoses and treatable diseases by image-based deep learning. *Cell*, 2018. **172**(5): 1122–1131.e9.

40. Fedorov, A. et al., 3D Slicer as an image computing platform for the quantitative imaging network. *Magn Reson Imaging*, 2012. **30**(9): 1323–1341.

41. Yushkevich, P.A., G. Yang, and G. Gerig, ITK-SNAP: An interactive tool for semi-automatic segmentation of multi-modality biomedical images. *Conf Proc IEEE Eng Med Biol Soc*, 2016. **2016**: 3342–3345.

42. Wibmer, A. et al., Haralick texture analysis of prostate MRI: Utility for differentiating non-cancerous prostate from prostate cancer and differentiating prostate cancers with different Gleason scores. *Eur Radiol*, 2015. **25**(10): 2840–2850.

43. Pallavi, T. et al., Texture descriptors to distinguish radiation necrosis from recurrent brain tumors on multi-parametric MRI. *Proc SPIE Int Soc Opt Eng*, 2014. **9035**: 90352b.

44. Lee, G. et al., Cell orientation entropy (COrE): Predicting biochemical recurrence from prostate cancer tissue microarrays. *Med Image Comput Assist Interv*, 2013. **16**(Pt 3): 396–403.

45. Prasanna, P., P. Tiwari, and A. Madabhushi, Co-occurrence of local anisotropic gradient orientations (CoLlAGe): A new radiomics descriptor. *Sci Rep*, 2016. **6**: 37241.

46. Zhao, W. et al., Visualizing the invisible in prostate radiation therapy: Markerless prostate target localization via a deep learning model and monoscopic kV projection X-ray image in 2018 Annual Meeting of ASTRO. *Int J Radiat Oncol Biol Phys*, 2018. **10**(3): S128–S129.

47. Sun, R. et al., A radiomics approach to assess tumour-infiltrating CD8 cells and response to anti-PD-1 or anti-PD-L1 immunotherapy: An imaging biomarker, retrospective multicohort study. *Lancet Oncol*, 2018. **19**(9): 1180–1191.

48. Li, B. et al., Development and validation of an individualized immune prognostic signature in early-stage nonsquamous non-small cell lung cancer. *JAMA Oncol*, 2017. **3**(11): 1529–1537.

49. Vallieres, M. et al., A radiomics model from joint FDG-PET and MRI texture features for the prediction of lung metastases in soft-tissue sarcomas of the extremities. *Phys Med Biol*, 2015. **60**(14): 5471–5496.

50. Clark, K. et al., The cancer imaging archive (TCIA): Maintaining and operating a public information repository. *J Digit Imaging*, 2013. **26**(6): 1045–1057.

51. Chang, K. et al., Distributed deep learning networks among institutions for medical imaging. *J Am Med Inform Assoc*, 2018. **25**(8): 945–954.

52. Wu, Y. et al., Survival prediction in high-grade osteosarcoma using radiomics of diagnostic computed tomography. *EBioMedicine*, 2018. **34**: 27–34.

53. Nie, K. et al., Rectal cancer: Assessment of neoadjuvant chemoradiation outcome based on radiomics of multiparametric MRI. *Clin Cancer Res*, 2016. **22**(21): 5256–5264.

54. Yuan, Y. et al., Densely connected neural network with unbalanced discriminate and category sensitive constraint for polyp recognition. *IEEE Trans Bio-Med Eng*, 2018: submitted.

55. Yuan, Y. et al., Prostate cancer classification with multiparametric MRI transfer learning model. *Med Phys*, 2019. **46**(2): 756–765.

56. Lambin, P. et al., Predicting outcomes in radiation oncology—Multifactorial decision support systems. *Nat Rev Clin Oncol*, 2013. **10**(1): 27–40.

57. Wu, Y. and L. Xing, Deciphering the inherent relationship between MR relaxometry properties: A simultaneous acquisition of qualitative and quantitative MRI (Q^2MRI) from a single relaxometry weighted MR image using deep learning. *IEEE T Bio-Med Eng*, 2018: submitted.

13

Brain cancer

WILLIAM D. DUNN JR. AND RIVKA R. COLEN

13.1 INTRODUCTION

Glioblastoma (GBM) is the most aggressive primary brain tumor. Despite decades of research, glioblastoma has a very dismal post-diagnosis survival time of slightly over one year.[1] Different factors contribute to its aggressive nature and poor prognosis, including heterogeneous genomic and epigenetic events and adapting multiple signaling pathways that cause resistance to therapies.[2] Promising research in the last few years has focused on the subtle molecular differences between patients (inter-individual) and within a single tumor (intra-tumoral) heterogeneity with the hope of discovering more effective therapeutics tailored to each person's characteristics of each tumor rather than a "one-treatment fits all" standard of care approach.[3] This new approach is considered the basis of personalized medicine.

MRI has made enormous strides since its humble beginnings in the 1970s and is currently regarded as the gold standard imaging modality for brain tumors. Progress in MRI scanners, sequences, protocols, etc. from simple T1-weighted imaging (T1WI) to complex dynamic advanced protocols not only has increased its diagnostic and prognostic power, but has also opened new frontiers in understanding the complex biological nature of glioblastoma.[4] Accordingly, MRI's potential role in identification of non-invasive biomarkers and therapeutic targets through radiogenomics and radiomic correlations signifies its weight as a cornerstone in the personalized treatment approach.[5,6]

In this article, we will outline the role of imaging in glioblastoma care, offer a landscape on the imaging and genomic biomarkers that are associated with glioblastoma, and discuss the role of MRI through the emerging fields of radiogenomics and radiomics. Finally, we end with some perspectives into the future of radiogenomics and radiomics research as well as innovative uses of MRI currently being investigated that could potentially change the way we treat glioblastoma and ultimately improve patient care.

13.2 MR IMAGING AS A CLINICAL TOOL

MRI is the imaging modality of choice for diagnosis, surgical planning, and monitoring of therapeutic response in brain tumor patients due to its high signal-to-noise ratio that provides exquisite anatomical and functional detail, superseding that provided by other imaging modalities.[7] Moreover, MRI provides a comprehensive 3D assessment of the entire tumor and tissue that can be leveraged to overcome current limitations due to inherent sampling errors in data obtained through routine biopsies that only represent a portion of the tumor.[8]

Conventional MRI sequences, such as T1-weighted, T2-weighted (T2WI), gadolinium-enhanced T1WI, and fluid-attenuated inversion recovery (FLAIR) images are the most commonly obtained sequences in neuroradiology clinical practice. Visualized on gadolinium-enhanced T1WI, glioblastoma typically appears as a heterogeneously ring enhancing lesion, representing the active proliferating portion of the tumor with rapid angiogenesis and blood brain barrier disruption, surrounding non-enhancing hypointense regions that represent necrosis (Figures 13.1 and 13.2).[9] Beyond the area of enhancement is the peritumoral T2-weighted/FLAIR hyperintensity, reflecting a mixture of edema and tumor infiltration in the surrounding tissue.[10]

In addition to conventional MRI sequences that evaluate the morphological/anatomical structures, more advanced imaging techniques such as diffusion-weighted imaging (DWI), perfusion-weighted imaging (PWI), MR spectroscopy (MRS), and susceptibility-weighted imaging (SWI) provide insight into tumor physiology.[11]

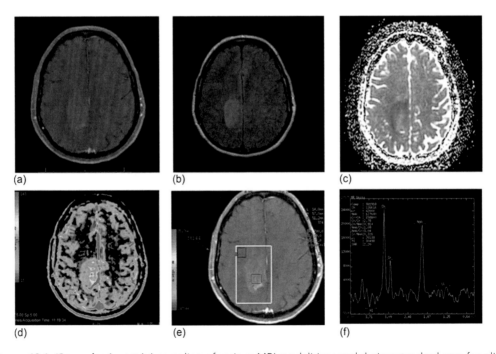

(a) (b) (c)

(d) (e) (f)

Figure 13.1 (See color insert.) A sampling of various MRI modalities used during standard care for glioblastoma treatment. Images come from a 49-year-old male recently diagnosed with GBM. Conventional imaging methods gadolinium-enhanced T1-weighted (a) and T2-weighted FLAIR (b) as well as more advanced imaging methods diffusion-weighted imaging (c), perfusion-weighted imaging (d), and magnetic resonance spectroscopy (e,f) are illustrated here.

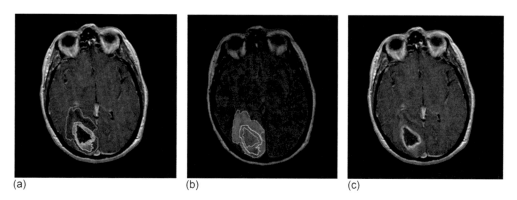

Figure 13.2 (See color insert.) Overview of image segmentation methodology based on gadolinium-enhanced T1-weighted (a) and fluid-attenuated inversion recovery (FLAIR) imaging (b) from a 59-year-old male with enhancing right parieto-occipital lobe GBM. Segmentation of areas of hypointense necrotic core (red) and marginal enhancement (yellow) are visualized in panel A and segmentation of edema/tumor invasion (blue) are visualized in panel B. Panel C illustrates co-registration of panels A and B. Segmentation was performed using 3D Slicer 4.4.1.

For example, through visualizing the restriction of Brownian motion of water molecules in tumors, DWI is able to provide more specific information about tumor density and cell compactness, beyond what is evaluable with traditional techniques.[12] It has been shown that changes in DWI can precede structural/morphological changes when assessing treatment response, suggesting that identification of changes in DWI could lead to swifter therapeutic interventions.[13] PWI (including dynamic susceptibility contrast MRI [DSC] and dynamic contrast-enhanced MRI [DCE]) captures information regarding the amount of blood within the tumor and the integrity of the blood brain barrier, respectively, through pixel-by-pixel analysis based on MR signal intensity changes associated with contrast agents.[14] PWI can be helpful in distinguishing tumor progression/recurrence and post-treatment changes (and pseudoprogression).[15,16] It is also helpful in evaluating the effectiveness of anti-angiogenic drugs such as bevacizumab by comparing post-treatment parametric response maps.[17] MRS, another advanced MRI technique, interrogates metabolites found in tissue or tumor.[18] In highly mitotic cancers, such as glioblastoma, choline (Cho) typically increases as it is associated with cell membrane constituents and proliferation, while N-acetylaspartate (NAA), a marker for neuronal and axonal integrity, often decreases, leading to the characteristic increase in the Cho/NAA ratio and Cho/Creatine ratio.[18,19] Creatine, typically present in stable quantities in healthy brain tissue, decreases in tumor regions due to increased metabolic demands of the tumor.[19] Other MRS characteristics of gliomas include accumulation of lactate resulting from anaerobic glycolysis from increasing cell density and ischemia as well as a buildup of lipids resulting from the breakdown of myelin sheaths associated with necrosis.[20] In addition, in tumors such as gliomas with isocitrate dehydrogenase 1 (*IDH1*) mutations, where alpha-ketoglutarate (α-KG) to 2-hydroxyglutarate (2HG) conversion is increased, MRS has been shown to detect 2HG in *IDH* mutant gliomas compared to non-mutated tumors.[21] Because of the advantages related to imaging of distinct functional phenomena when compared to conventional T1WI and T2WI series, advanced MRI techniques are finding an increased presence in the clinic.[22]

MRI is also important as a monitoring modality. Several response criteria based on MR imaging features of glioblastoma have been developed to assess response to therapy in clinical trials.[23,24] In brain cancer, imaging assessment criteria used in clinical trials includes four therapeutic responses based on changes of size measured by two-dimensional measures of enhancing tumor burden: complete response, partial response, stable disease, and disease progression.[23] In addition to post-contrast T1WI, changes on the T2WI and FLAIR images are included in the more recent Response Assessment in Neuro-Oncology (RANO) criteria, developed in 2010.[24] RANO includes additional measures to help distinguish true progression from pseudo-progression and attempts to define pseudo-response.[24,25]

13.2.1 KEY GLIOBLASTOMA GENOMIC BIOMARKERS

Glioblastoma is characterized by a significant level of heterogeneity in its genetics and molecular constitution, tumor morphologies, proliferation rates and behavior, and therapeutic response.[26] This heterogeneity is believed to arise either spontaneously through clonal evolution or cancer stem cells (CSCs) that generate diverse progenies.[27,28] The outcome of these mechanisms is a unique molecular makeup of each tumor that leads to significant differences between patients.[29] The manifestations of this diversity at the genetic and epigenetic level have already led to the identification of important genomic markers and differing targeted treatment strategies.[30–32] Biomarkers based on genomic analyses (for example, methylation, mutations, or gene expression levels in tissue and blood) are beginning to be used in clinical settings for a variety of cancers for predictive purposes to more precisely match therapies and patients, as well as for prognostic purposes to shed light on response or survival outcome.[33,34] Recent investigations, discussed in detail below, have uncovered many such biomarkers, because they offer valuable information, their identification and interpretation may serve as a pathway to tailored treatment of individual patients.[35–44]

Some of the important genomic markers related to glioblastoma are O-6-methylguanine-DNA methyltransferase (*MGMT*) promoter methylation status, isocitrate dehydrogenase (*IDH1*) mutation, *1p/19q* codeletion, and epidermal growth factor receptor (*EGFR*) mutation. Promoter methylation status of *MGMT* is an important prognostic and predictive biomarker for glioblastoma.[36,37,45] MGMT is an excision repair enzyme coded by *MGMT* gene located on chromosome *10q26* that removes alkyl groups from O6-alkylguanine adducts on DNA.[46] Alkylating agents, such as temozolomide (TMZ), exert their actions mainly through alkylating the O6-methylguanine position, among other actions, causing it to preferentially pair with thymine during replication, and activating cycles of ineffective mismatch repair (MMR) pathways that lead to double-strand breaks and eventually apoptosis of the cancer cell.[47,48] Since the MGMT enzyme is irreversibly inactivated upon binding alkyl groups, it must be continually regenerated to continue its repair activity against the influx of alkylating agents. Therefore, when *MGMT* is inactivated, through epigenetic silencing by hypermethylation of its promoter region, alkylating agents are better able to exert their anti-tumor activity. As a result, *MGMT* promoter methylation, seen in roughly 40%–50% of glioblastomas, is used clinically as an indication to suggest a favorable response to alkylating chemotherapeutic agents.[36,37,45,48]

Somatic mutations in *IDH1* and *IDH2* are found in approximately 5% of primary glioblastomas and 83% of secondary glioblastomas. Such mutations are frequently reported in low grade astrocytomas, oligoastrocytomas, and oligodendrogliomas (73% in grade II astrocytomas, 65% in grade III astrocytomas, 87% in grade II oligodendrogliomas, 74% grade III oligodendrogliomas, 83% in grade II oligoastrocytomas, 72% in grade III oligoastrocytomas).[49] *IDH1* is a homodimer in peroxisomes that catalyzes oxidative decarboxylation of isocitrate to alpha-ketoglutarate, a reaction that generates NADPH (reduced nicotinamide adenine dinucleotide phosphate) from NADP+ (nicotinamide adenine dinucleotide phosphate) leading in turn to the generation of reduced glutathione which is involved in protecting the cells from oxidative damage.[50–52] Accordingly, cells with impaired IDH1 function have increased susceptibility to oxidative stress and DNA damage.[51] In addition, as detailed above, *IDH1* mutation leads to the reduction of alpha ketoglutarate to 2-hydroxygluterate ((R)-2HG), an onco-metabolite which can be detected by MRS. This eventually activates the sub-unit of the transcription factor hypoxia-inducible factor (HIF) in response to a hypoxic cellular environment, facilitating angiogenesis and, hence, tumorigenesis.[53,54] *IDH1* mutation in gliomas is also associated with prolonged progression-free and overall survival.[38–40,55] Hartmann et al. identified *IDH1* as the single most important prognostic factor in glioma, suggesting that the prognostic significance of age, largely considered a prognostic factor in itself, could be mainly explained by higher *IDH1* mutation rates in younger patients.[38] The typical arginine to histidine/serine/cysteine point mutation at codon 132 is likely an early event in the development and progression of most astrocytic, oligodendroglial, and secondary glioblastomas and has been proposed to be a predictive biomarker of progression of low-grade gliomas to secondary glioblastomas.[50,56]

Loss of heterozygosity (LOH) or co-deletion of chromosome *1p/19q* is also considered an important genomic marker in gliomas.[57] In this aberration, there is an unbalanced translocation between chromosomes *1p* and *19q* and a loss of heterozygosity that leads to a loss of genetic material.[35] While considered a favorable prognostic marker for lower grade gliomas in general, it is still unclear the prognostic significance of this aberration in glioblastoma patients.[35]

EGFR mutation is another genetic aberration found in glioblastoma.[41] *EGFR* is located on the short arm of chromosome 7 at position 11.2 and contains 28 exons coding for a transmembrane glycoprotein belonging to the protein kinase family. The most frequent genetic alteration reported in glioblastoma is *EGFR* gene amplification that refers to overexpression of *EGFR* transmembrane tyrosine kinase receptors.[41] *EGFR* amplification can be found to occur concurrently with an *EGFR* mutation, *EGFRvIII*, where exons 2–7 are deleted resulting in a frame deletion variant characterized by truncation of an extracellular domain and lack of ligand binding ability.[58,59] When mutated, the receptor tyrosine kinase is constitutively phosphorylated regardless of the presence of ligand and subsequent intracellular signaling and cell proliferation pathways are permanently engaged.[60] The *EGFRvIII* variant occurs in 24%–67% of glioblastomas.[61]

Other markers include interleukins (ILs) and phosphatase and tensin homolog (*PTEN*). ILs play significant roles in a variety of cancers by activating proliferation or signaling pathways that involve cell migration and invasion, ultimately leading to resistance to chemotherapy.[62] The expression of specific ILs such as IL-6 have been associated with lower survival in glioblastoma patients.[63] *PTEN* is a tumor suppressor found mutated or silenced in approximately 40% of glioblastomas and causes disruption in cell cycle, apoptosis, and cell migration pathways.[64] Heterogeneous *PTEN* expression was found to lead to poorer prognosis in glioblastoma patients.[42]

MicroRNAs are small molecules of non-coding RNA that function in the post-transcriptional regulation of gene expression by binding to a specific sequence of a target gene.[65] One interesting biomarker is the microRNA-21 and its gene target stemness regulator Sox2 axis.[43] Based on the classification of glioblastoma into high miR-21/low Sox2 or low miR-21/high Sox2 sub-types, microRNA-21 and its target gene have been shown to significantly differentiate patients based on molecular, radiological, and survival characteristics.[43]

Elucidation of these important genomic markers have led to changes in glioblastoma classification systems.[44,66,67] The 2016 WHO classification of central nervous tumors provides a significant update to the 2007 version and redefines central nervous system tumors by their molecular characteristics.[66] In particular, significant changes in classification are seen in *IDH*-wildtype and *IDH*-mutant glioblastomas, medulloblastomas, diffuse gliomas, and embryonal tumors. Glioblastomas are grouped into three classes based on *IDH* mutation status: *IDH*-wildtype (90% of cases), *IDH*-mutant, and NOS (not otherwise specified) glioblastomas. *IDH*-wildtype tumors typically correspond to primary tumors, while *IDH*-mutant tumors correspond more with secondary glioblastomas that progress from a lower grade diffuse glioma. New glioblastoma variants include epithelioid glioblastoma, often harboring a *BRAF* V600E mutation, glioblastoma with primitive neuronal component, as well as small cell glioblastoma/astrocytoma and granular cell glioblastoma/astrocytoma.

In addition to the WHO molecular-based classification of glioma tumors, Verhaak et al. described a glioblastoma molecular classification using more than 200 samples to identify gene clusters classifying glioblastoma into classical, mesenchymal, and proneural sub-types based on aberrations and gene expression of several genes, most notably *EGFR, NF1,* and *PDGFRA/IDH1,* respectively.[44] A neural sub-type was also defined based on patterns with close similarity to normal brain cells. The molecular patterns that characterize each sub-type offer insight into growth and behavior. A follow-up study of more than 500 glioblastoma tumors found that the proneural sub-type could be divided into G-CIMP and non-G-CIMP based on CpG island methylation.[68] G-CIMP proneural tumors offered a survival advantage while non-G-CIMP proneural and non-mesenchymal tumors performed least favorably in regard to 12-month outcomes.

13.3 RADIOGENOMICS

An exciting field of research that bridges imaging data and corresponding underlying genomic data is termed "radiogenomics" or "imaging genomics" (Figure 13.3).[9,69-72] Radiogenomics is defined as the linkage of imaging information with genomic data. Bidirectional correlations and predictions between

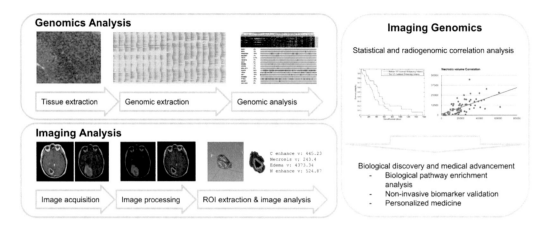

Figure 13.3 (See color insert.) General description of imaging genomics. In early stages, genetic data and corresponding imaging data are collected, pre-processed, and analyzed. In later stages, genomic and imaging data are analyzed together to discover correlations that provide insight to the underlying biological mechanisms involved in neoplasm development.

imaging features and genomics can be performed and have been shown to non-invasively identify genomic events and explore potential biological or molecular-targetable events by analyzing widely available imaging data.[9,31,69,73–75] By detecting these characteristics prior to therapy, it is hoped that these diagnostic, predictive, and prognostic markers can be used to identify optimal steps that can be taken to combat the tumor at an earlier stage, leading to a possibly safer alternative that avoids the side effects that may result from sub-optimal therapies.[74] This idea is supported by early work showing that genetic and biologic features of glioblastoma significantly influence the macroscopic properties that we see in MR imaging.[76] Radiogenomics research has become more and more feasible in the last several years through the advent of multi-institutional imaging and genomic repositories of cancer patients such as The Cancer Genome Atlas (TCGA) program, Repository of Molecular Brain Neoplasia Data (Rembrandt), and the Cancer Imaging Archive (TCIA).[77–79] Initial research in this field has been promising, with various imaging features correlated with a wide variety of molecular events such as mutation, expression, and epigenetic regulation among other events.[80–82] Radiogenomics studies using conventional as well as advanced imaging methods are discussed in the following sections.

13.3.1 QUALITATIVE RADIOGENOMICS USING CONVENTIONAL MR IMAGING

Studies in radiogenomics have largely involved correlations between genomics and imaging features derived from conventional MRI.[31,80,81,83,84] Early glioblastoma radiogenomics were based on imaging features derived from the presence or absence of characteristics such as cysts or multifocality or from visual estimations of features using general categorical ranges (i.e., contrast enhancement) or Visually AcesSAble Rembrandt Images (VASARI) standard.[23,31,85]

Several studies have found significant associations between imaging features derived from conventional imaging techniques and gene expression.[5,76,80,86] In an early 2004 study, Raza et al. identified 26 genes (using microarray gene expression profiling) negatively and positively correlated to the amount of necrosis.[83] Further analyses indicated that these genes were associated with mechanisms such as tumor necrosis factor alpha, apoptosis, or hypoxia.[83] Another early qualitative imaging genomic study by Diehn et al. used conventional MR images from 22 glioblastoma patients and found a significant correlation between ten binary imaging traits (i.e., contrast enhancement, mass effect, cortical involvement, etc.) and cDNA microarray gene expression profiles.[87] It was found that tumor contrast enhancement and mass effect were significantly correlated to expression clusters of genes involved in hypoxia and proliferation, respectively. They also found that glioblastomas with a more invasive phenotype (identified by higher grayness on the FLAIR sequence by visual inspection) harbored different expression of genes compared to those glioblastomas which harbored a large edematous non-enhancing glioblastoma component, they found that glioblastomas with high invasion had higher expression levels of genes involved in invasion. Pope et al. compared the genetic differences in 52 completely ring enhancing versus incompletely ring enhancing glioblastomas.[80] Results indicated that completely enhancing tumors expressed significantly higher levels of *VEGF*, neuronal pentraxin, interleukin-8, neuritin-1, and other factors related to hypoxia-induced angiogenesis pathways.

Carrillo et al. investigated associations between genetic mutations and imaging features.[81] In their retrospective study using conventional MRI sequences from over 200 patients, they found that predominantly non-contrast enhancing tumors, large tumor size, and presence of cysts and satellite lesions were associated with *IDH1* mutation. Gutman et al. looked at associations between mutation status of several genes (*TP53, PTEN, EGFR, NF1,* and *IDH1*) and VASARI-defined imaging features in 88 glioblastoma cases.[31] While no significant associations were found between the distribution of mutation status and specific imaging features using Fisher exact test, tumors measured on T2-FLAIR images were significantly larger in *EGFR*-mutated and significantly smaller in *TP53*-mutated glioblastomas compared to wild type. In general, measuring correlations between mutations and imaging features remains a challenge, especially in studies with small data sets, due to the heterogeneous nature of glioblastoma mutations and small number of tumors manifesting specific gene mutations or abberations.[31] This study also found lower contrast enhancement in glioblastomas of the proneural sub-types ($p = 0.02$), whereas mesenchymal glioblastomas showed less non-enhanced tumor content ($p < 0.01$).[31] In a study by Colen et al., 104 glioblastoma tumors were divided into invasive or non-invasive groups based on qualitative VASARI features related to invasion such as deep white matter tract involvement, ependymal enhancement, and pial enhancement.[88] Subsequent analyses of differing gene expression between the two groups followed by pathway analysis found that tumors showing invasive features significantly correlated to canonical pathways such as mitochondrial dysfunction, epithelial adherens junction remodeling, and potentially to *MYC* and peroxisome proliferator-activated receptor alpha (*PPARA*) transcription factor. Transcription factor analysis revealed NF-KB inhibitor-alpha (*NFKBIA*) to be the top transcription factor inhibited in the invasive group, which demonstrated poor survival when compared to the non-invasive group. This can be possibly attributed to the inhibition of *NFKBIA* in the invasive group. *NFKBIA* inhibits signaling from *EGFR* and NF-kB pathways and is considered to be a tumor suppressor.[89,90]

Apart from mutations and gene expression, imaging features have also been correlated to epigenetic events. As discussed above, *MGMT* methylation has strong prognostic implications, and while *MGMT* promoter methylation status has traditionally been difficult to predict using imaging features, several significant associations have nonetheless been described.[81,82] For example, in a study based on 86 glioblastoma patients (72 primary, 14 secondary), *MGMT* promoter methylated tumors tended to be found in the parietal and occipital lobes, whereas non-methylated *MGMT* promoter tumors were more commonly found in the temporal lobes.[91] It was also noted that mixed-nodular enhancement occurred more often in *MGMT* promoter methylated and secondary glioblastomas, whereas ring enhancement was more frequent in non-methylated *MGMT* glioblastomas ($p < 0.005$).[91] A recent retrospective study based on TCGA data of 86 treatment-naive glioblastoma patients looked at different feature selection and machine learning classification techniques to predict *MGMT* promoter methylation status using a total of 24 qualitative variables based on the VASARI featureset and ten quantitative [two-dimensional (2D) and volumetric] metrics.[92] On multivariate analysis, *MGMT* promoter methylation status could be predicted with an accuracy of 73.6% using quantitative image features (significant variables including edema/necrosis volume ratio, tumor/necrosis volume ratio, and edema volume), whereas, qualitative variables such as tumor location were reported to have lower prediction accuracy (70.2%).[92]

13.3.2 QUANTITATIVE RADIOGENOMICS USING CONVENTIONAL MR IMAGING

While imaging features have been traditionally measured using qualitative criteria, quantitative techniques, where features are digitally analyzed, are becoming more and more popular. Though still limited by time and computational challenges, volumetric measurements have a strong potential in the future to improve consistency and accuracy of imaging and hence radiogenomic analysis.[93] Using 2D measurements, a tumor is typically assumed to take on a regular ellipsoid shape and the "volume" is inferred based on the cross-section with the greatest area.[93] However, using segmentation of the entire tumor, one captures the entire 3D tumor volume, which provides a more accurate, robust, and representative measurement than 2D measurements, especially in tumors with an irregular shape or tumors with satellite lesions.[93] In addition, numerous studies have demonstrated an increased reproducibility of volumetric measurements compared to 2D measurements, potentially increasing the power of clinical trials and research studies.[94] It can be anticipated that radiogenomic studies using information captured from the entire 3D tumor volume can advance our understanding of tumor behavior.

Zinn et al. published the first comprehensive quantitative radiogenomic analysis correlating quantitative MRI volumetrics with gene and microRNA expression profiles.[5] In this study, 78 newly diagnosed treatment-naive glioblastoma patients were randomly segregated into discovery and validation sets, both sets were further sub-grouped into high, medium, and low volume groups based on FLAIR volumes (edema/cellular invasion) obtained after image segmentation and volumetric analysis. Subsequently, high and low volume groups were selected for analysis, the FLAIR radiophenotype was then correlated to gene and miRNA expression data. Periostin *(POSTN)* was found to be the most upregulated gene in both sets (4-fold in discovery and 11-fold in validation) and was associated with poor overall survival and shorter time for disease progression.[5] miRNA-219 was identified as the most downregulated microRNA across both sets and was bioinformatically predicted to be a regulator of *POSTN*, diminishing POSTN protein levels.[5] In another such study, Colen et al. investigated the molecular profiles and quantitative necrosis volumes in 99 (69 males, 30 females) glioblastoma patients.[95] Based on tumor necrosis volumes extracted from MR imaging, both males and females were sub-stratified into high and low necrosis volume groups. Females were found to have lower necrosis volumes in comparison to their male counterparts ($p = 0.03$), females with high volumes of necrosis additionally showed much shorter survival ($p = 0.01$, 6.5 months versus 14.5 months). Transcription factor analysis demonstrated that the mechanism of cell death appears to be different in females versus males. In fact, this study found a distinct gender-specific molecular mechanism of cell death (defined as necrosis on MRI), with cell death being driven by the *TP53* (apoptotic) pathway versus *MYC* (necrotic) pathway in males versus females, respectively. These data are concordant with the gender survival difference in males versus females found in the study.[95]

Gutman et al. delineated the volumes of various phenotypes such as contrast-enhancement, edema, and necrosis in 76 glioblastoma TCGA patients and assessed the association of these volumetric features/ratios with mutation status of nine genes (*TP53, RB1, NF1, EGFR, IDH1, PIK3R1, RB1, PIK3CA,* and *PDGFRA).*[96,97] Results showed correlations between imaging and mutations involving *TP53, RB1, EGFR, PDGFRA,* and *NF1* genes. For example, compared to wildtype tumors, *TP53*-mutated tumors were characterized by significantly smaller contrast enhancement ($p = 0.012$) and necrosis volumes ($p = 0.017)$, *RB1*-mutated tumors by significantly smaller edema volumes ($p = 0.015$), and *EGFR*-mutated tumors by notably higher necrosis/contrast enhancing ratio ($p = 0.05$). Another study using Gene Set Enrichment Analysis (GSEA) to calculate enrichment scores to measure the association between volumetric features and gene sets found 64 biological processes to be correlated with one or more volumetric features (such as necrosis, contrast enhancement, edema, tumor bulk, and total tumor volume) or with their ratios.[98] Necrosis and tumor bulk were associated with apoptosis and immune response pathways, and contrast enhancement was correlated with protein folding and signal transduction pathways (Figure 13.4). In another study of 92 glioblastoma patients, Rao et al. used hierarchical clustering to identify two imaging clusters based on volume-class, T1/FLAIR ratio, and hemorrhage.[99] Patients with low values in the latter three variables harbored a better prognosis and also had differential expression of 384 mRNA and 23 miRNA compared to patients with higher values. These differential expressions corresponded to distinct biological processes such as cell-to-cell signaling and interaction, cellular assembly and organization, and oxidative phosphorylation.[99]

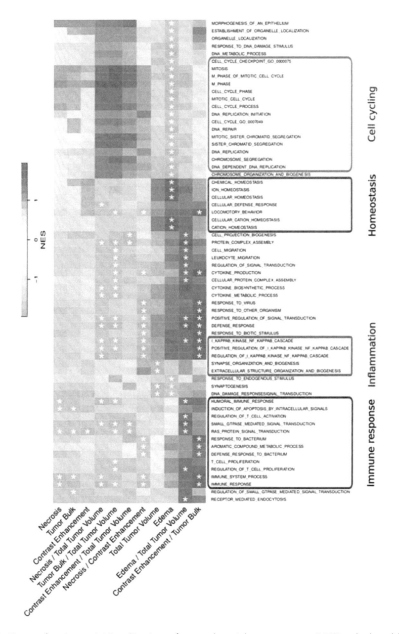

Figure 13.4 (See color insert.) Visualization of normal enrichment scores (NES) calculated by GSEA to highlight the association between volumetric features and expert-curated gene ontology gene sets. A total of 64 biological processes were found to be correlated with one or more volumetric features or their ratios. For example, necrosis and tumor bulk were markedly enriched with apoptosis and immune response pathways, and contrast enhancement was negatively correlated with protein folding and signal transduction pathways. (Adapted from Grossmann, P. et al., *BMC Cancer*, 16, 611, 2016.)

13.3.3 Advanced radiogenomics

In addition, genomic correlations with advanced imaging techniques have also been performed.[100–104] For example, a study by Gupta et al. derived three metrics [maximum relative cerebral blood volume (rCBV), percent signal recovery (PSR), and relative peak height (rPH)] from DSC PW-MR imaging and measured their association with *EGFR* amplification and *EGFRvIII* mutation status.[100] Univariate analyses showed that

EGFR amplification was associated with higher median rCBV and lower PSR and *EGFRvIII* mutation was associated with a higher rPH levels. Preliminary work in the TCGA imaging working group also identified specific genes and microRNA expression patterns associated with a specific radiophenotype when comparing tumors with low and high levels of CBV values in various tumor phenotypes (i.e. contrast enhancement, necrosis, edema).[105] Another study using the TCGA dataset divided 50 treatment-naive patients into low and high rCBVmax groups based on the median across the dataset and analyzed differences in mRNA, miRNA, and protein (reverse-phase protein array [RPPA]) between the two classes.[101] A total of 326 genes, 76 miRNAs, and 8 proteins were found to be differentially expressed between these two groups, in addition, these differences were associated with differential pathways related to cellular development, cellular proliferation, and inflammatory response.[101]

Several studies have correlated DW-MRI-derived apparent diffusion coefficient (ADC) measurements to *MGMT* promoter methylation status.[106,107] For example, Sunwoo et al. analyzed the association between methylation status and ADC values in a study involving glioblastomas from 26 patients.[106] In addition to demonstrating significant correlations between mean ADC and *MGMT* promoter methylation ratio ($p = 0.015$) and between mean ADC and methylation status ($p = 0.011$), the authors also noted significant correlations between mean ADC and progression-free survival as well as between fifth percentile ADC values and the Ki-67 labeling index. Romano et al. performed a similar study with 47 patients and found that *MGMT* promotor methylation status was associated with higher minimum ADC values.[107] In a study by Pope et al. involving 38 glioblastoma patients obtained from a bevacizumab-treated cohort, the contrast-enhanced portion of the tumor was segmented and ADC histogram values were obtained from this region, patients were dichotomized into low versus high ADC groups based on the mean.[102] Thirteen genes were found to be overexpressed in the high ADC group, six of which coding for collagen isoforms, collagen-binding proteins, and extracellular matrix. In another study, Zinn et al. probed the relationship ADC maps and gene and miRNA expression profiles.[103] Using 37 treatment-naïve TCGA glioblastoma patients with available imaging and genomic annotations, the ADC maps were registered to the segmented FLAIR and post-contrast T1WI sequences, and region of interest (ROIs) were obtained in the regions of most restricted diffusion in the FLAIR region. Patients were dichotomized into low versus high ADC groups based on the median ADC value and compared. This study demonstrated differential upregulated (tumor cell invasion, cell movement, and migration) and downregulated (apoptosis, growth failure, and tissue hypoplasia) oncogenic functions/pathways in restricted versus facilitated diffusion phenotype groups.[103] The inhibitor of kappa light polypeptide gene enhancer in B-cells, kinase gamma (IKBKG), an activator of NF-κB signaling, was predicted to be the top upstream regulator driving the restricted diffusion phenotype.

IDH1 mutations have also been correlated to imaging.[108,109] A recent retrospective study by Yamashita et al. on 66 glioblastoma patients (55 *IDH1* wild type, 11 *IDH1* mutant) correlated imaging features obtained from DWI, arterial spin labeling (ASL) and conventional MR sequences to *IDH* mutation status.[108] Differences in minimum ADC and mean ADC between the wildtype and mutant *IDH1* groups were not significant. However, it was found that absolute and relative tumor blood flow, largest cross-sectional necrosis area, and the percentage of non-enhancing area inside the largest cross-sectional enhancing lesion variables were markedly higher in *IDH1* wildtype when compared to *IDH1* mutant tumors. The ability of MRS to detect 2-HG as a surrogate of *IDH1* mutation can be helpful for non-invasively diagnosing *IDH1* mutation status and monitoring response to treatment such as with *IDH1* inhibitors.[109–112]

As mentioned earlier, *1p/19q* chromosomal co-deletion has been shown to predict overall survival and treatment response.[113] A [^{18}F] fluoroethyltyrosine positron emission tomography (FET-PET) study surveying 144 patients found that *1p/19q* co-deleted tumors had significantly higher FET uptake compared to patients without the abnormality.[114]

Importantly, radiogenomic studies such as those indicated above implicate the possibility of using imaging to enhance and advance personalized medicine.[9,70,96,115] Apart from furthering our understanding of cancer, this type of research will have important clinical implications, potentially opening the door for monitoring programs and personalized therapy of brain tumors by describing the tumor at the molecular level before any invasive intervention is performed. It can be hoped that radiogenomic studies such as these are able to capture information from the entire tumor volume and advance our understanding of the molecular characterization.

13.4 LOOKING INTO THE FUTURE OF RADIOGENOMIC

As it becomes more feasible and less expensive to collect, store, and analyze heavy genomic and imaging datasets, future radiogenomic correlations and predictions will likely lead to interesting advances in clinical care.[116] Glioblastoma imaging analysis has come a long way from preliminary methods which used scans based on a visual assessment and from one representative slice.[117–119] Currently, more advanced techniques are being implemented including measuring not only overall volumes, but also volume sub-compartments that can help increase robustness and detail.[120] Also, features of interest will expand from gross observances and measurements such as "enhancement" or "necrosis" or "size" to more complex texture properties that cannot be visualized or appreciated by the naked eye.[121]

The field of radiogenomics is still in its initial stages, and the current standards are continuously being changed and updated. For example, a limitation of the current TCGA is that it was based on surgical samples obtained from a biopsy obtained without spatial localization. Thus, not only is the location of biopsy unknown, but the biopsies themselves are only small samples that are not representative of the entire heterogeneous glioblastoma lesion.[122] Indeed, recent investigations on a wide range of cancers and including glioblastoma have shown rich heterogeneity between samples taken from the same tumor during surgical resection and suggest that regional differences in signal intensities in a tumor provide valuable information on tissue structure and the local environment.[123–127] As routine sampling techniques become more standardized and targeted, the ability to detect meaningful radiogenomic correlations can be expected to increase. However, image-guided biopsies would be important to match the spatial location on imaging with genomic spatial heterogeneity. With advances in microscale imaging such as with radiomics, as detailed below, spatial encoding of radiogenomic and genomic mapping will provide a more accurate and robust picture of the totality of the tumor in each individual patient.

13.4.1 RADIOMICS

Radiomics is defined as the automated extraction of high, multi-dimensional quantitative imaging features from standard medical images.[128] Radiomics is able to extract microscale imaging information not otherwise visualized by the naked eye.[128] It can be anticipated that such microscale quantitative high-dimensional imaging characterization methods can more accurately investigate the complex heterogeneity inherent in solid tumors.[129] Given that one imaging series could contain millions of voxels giving rise to numerous size, texture, and shape features in each sub-region, radiomics offers an enormous amount of imaging features that could be leveraged to facilitate cancer diagnosis, disease monitoring, treatment response prediction, and ultimately more precise personalized treatment.[129] Advances in high-performance computing, computational power, and computational methods have helped facilitate the groundwork for the successes of radiomics and quantitative radiogenomics.[130] In addition, advances in segmentation software and tools that provide automated, semi-automated, and manual segmentation have provided the mechanism needed for contouring and selecting the tumor or tissue of interest that will then undergo automated radiomic analysis. Of note, to date, no fully automated segmentation tool has been developed to provide sufficient accuracy in contouring tumor or tissue borders in a precise manner, and thus semi-automated and manual segmentations remain the best path for accurate segmentation.[131,132] With segmentation alone, tumors and their sub-regions can be quantified based on volumes, however, there remains various limitations such as the inability to capture subtle differences and heterogeneity within the tumor not visible to the naked human eye.[133] Thus, radiomics is an emerging field that can help fill in these gaps through the high-throughput extraction and analysis of a wide range of quantitative microscale imaging features.[128]

The radiomic workflow, as in all image analysis, begins with image acquisition. Although it has been thought that images produced using similar equipment and parameters should be preferred, a recent study, slated to become the landmark study on radiomics and radiogenomics (due to its establishment of radiogenomic causality, see below), found that through implementation of novel preprocessing methods, radiomics can be scanner, protocol, and institution agnostic through robust clinical and pre-clinical validation methods.[134]

The novel robust pipeline described by Zinn et al. in the latter paper consists of multiple steps involving segmentation, preprocessing analytics prior to radiomic-feature extraction, feature extraction, feature normalization/selection, and predictive-modeling (Figure 13.5). In brief, MRI images are segmented, in the case of glioblastoma, this study segmented distinct imaging phenotypes such as edema/invasion, active enhancing tumor, and necrosis using a semi-automated method with expert radiologist review.[9,134] Patient-specific normalization to account for differences in image acquisition tools between different institutions using patient-specific contralateral hemisphere normalization of the extracted features from the contralateral normal appearing white matter was performed in this study.[134] Subsequently, multiple preprocessing techniques were employed such as skull-stripping using Functional Magnetic Resonance Imaging of the Brain (FMRIB) Brain Extraction Tool (http://fsl.fmrib.ox.ac.uk/fsl/fslwiki/BET) and intensity of all images was normalized to a reference image set randomly selected from the study cohort. In subsequent stages, various location, shape, size, intensity, and texture-based features can be extracted from the various MR imaging sequences such as T1WI, T2WI, FLAIR, DWI, and PWI. Extracted features fall into two categories, semantic and agnostic.[129] Semantic features are commonly used and described by radiologists such as shape, vascularity, or necrosis, whereas in radiomics, a focus is on agnostic, mathematically extracted features that are independent of visual assessments. Agnostic features can be grouped into first-, second-, and third-order statistics features, first order features such as histogram analysis (based on single voxel), such as mean, variance, kurtosis, etc., are derived from the regional intensity distribution, while second-order features (based on joint probability

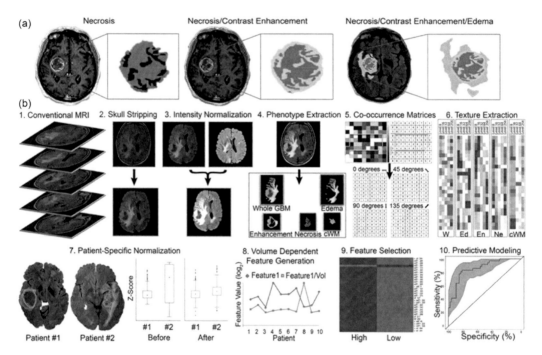

Figure 13.5 (See color insert.) Radiomic sequencing pipeline for solid tumor (glioblastoma) and tissue characterization described by Zinn et al.[134] "(a) Semi-automated segmentation of the three imaging phenotypes: necrosis (left), post-contrast active enhancing tumor (middle), and FLAIR peritumoral edema/invasion, representing edematous tumor as well as sites of cellular invasion into brain tissue (right). (b) Automated segmentation-based radiomic feature extraction is followed by patient specific normalization and feature selection, which consists of the normalization of contralateral normal-appearing white matter as an internal control normalizer to account for various potential intra- and inter-institutional biases, a crucial step that ensures comparability. The next step is volume-dependent feature generation, which uses the necrosis, post-contrast enhancement, and FLAIR volumes for the corresponding radiomic feature sets acquired from the respective sequences, thus doubling the amount of radiomic features and creating a set of tumor volume-independent and -dependent radiomic features. Finally, this homogenous dataset can then enter feature selection and predictive modeling for any GBM trait of interest."

distributions of pairs of voxels) include metrics such as entropy, homogeneity, etc.[135] Finally, higher-order features can be obtained by assessing the interrelationship of three or more voxels using methods such as run-length matrix or intensity size-zone matrix.[136] The final step in the radiomic pipeline is biostatistical/bioinformatic analysis. While this step includes the entire spectrum from simple descriptive statistics to sophisticated multivariate analyses and machine learning, the sheer amount of diverse imaging and radiogenomic data collected in today's growing technological landscape is best analyzed with big data machine learning analytics and predictive modeling techniques.[137] Machine learning can help predict unseen data based on training models carefully developed following best practice guidelines for confronting common issues such as overfitting and model calibration problems.[138] Feature selection step is an important step in big data analysis, including radiomics and radiogenomics and is done prior to predictive modeling/machine learning classification methods.[134,139] Multiple feature selection techniques such as maximum relevance minimum redundancy (MRMR) and least absolute shrinkage and selection operator (LASSO) regularization (L1 regularization) can be used to limit the extracted variables to those significantly associated with the outcome of interest in the concerned study.[134,140,141] Finally, these selected features are used to build predictive models.

As indicated by its historical name "glioblastoma multiforme," glioblastoma is an incredibly heterogeneous lesion composed of a dynamically changing cellular population with distinct phenotypes and proliferation characteristics both within a single tumor (intra-tumoral heterogeneity) and between tumors of different patients (inter-individual heterogeneity).[26,142] As such, glioblastoma is a particularly suited solid tumor for studying tumor heterogeneity using radiomic analysis and various applications have been discussed in recent studies.[95,115,143–147] For example, radiomics using traditional or advanced MRI methods is used to more in depth characterize certain imaging phenotypes such as the edema/invasion portion of the tumor or spatial heterogeneity, these can then be correlated to genomic information with the goal of better understanding underlying mechanisms.[126] A recent study conducted by Hu et al. on 82 image-guided biopsy samples collected from enhancing tumor core and surrounding peri-enhancing tumor regions from 18 glioblastoma patients measured the ability of MRI texture features to predict spatial histologic heterogeneity.[143] Texture features were derived from various pre-operative MRI sequences (contrast-enhanced T1, diffusion tensor imaging [DTI], and DSC MRI) and spatial histologic heterogeneity was graded categorically (high versus low tumor content, $\geq$80% or $\leq$80% tumor nuclei, respectively). Over 250 features were extracted from eight different MRI contrasts (post-contrast T1, T2WI, fractional anisotrophy [FA], isotropic [p], anisotropic [q], mean diffusivity [MD], rCBV, gradient-echo echo-planar contrast-enhanced [EPI+C]). Subsequent principal component analyses and classification algorithms identified three texture features (rCBV, gray-level co-occurrence matrix [EPI+C-GLCM], and local binary patterns [T1+C-LBP]) that could distinguish high versus low tumor content with an accuracy of 85% and 81.8% in training ($n = 11$) and validation ($n = 7$) sets, respectively. Due to its infiltrative nature, particularly at the tumor margin, it is difficult to completely surgically remove glioblastoma.[148] In the future, techniques like these might help identify areas of high tumor density that could be useful in guiding neurosurgeons toward removing highly infiltrative tumor regions.

In addition to glioblastoma, recent studies have shown the potential application of radiomics in other brain tumors such as brain metastasis.[149,150] In a recent study by Ortiz-Ramon et al., 2D and 3D radiomic texture features were analyzed to determine if radiomics could classify brain metastases from lung, melanoma, and breast cancer based on their primary site of origin.[149] They found that 3D features were more discriminative than 2D features in distinguishing between lung cancer brain metastasis from breast cancer brain metastases (area under the ROC curve [AUC] = 0.963) and melanoma brain metastasis (AUC = 94), however, classification was unsatisfactory for distinguishing breast cancer and melanoma brain metastases (AUC = 0.607). In another study, it was shown that radiomics could differentiate lung brain metastasis from melanoma brain metastasis (AUC > 0.90).[150]

Another promising application of radiomics relates to post-treatment monitoring of tumors.[145,151–153] These studies have important clinical implications in patient management. An interesting phenomenon that has surfaced over the past decade has been distinguishing between pseudoprogression and true progression. Pseudoprogression, as its name suggests "to mimic true progression," occurs when there is an increase in contrast enhancement on post-contrast T1W1 images or development of new lesion that subsequently decreases in size or stabilizes.[154] Pseudoprogression can strongly resemble true progression and can alter a physician's judgement about recurrent disease. The misdiagnosis of true progression can lead to premature discontinuation of effective therapy, whereas the misdiagnosis of pseudoprogression can lead to unnecessary risk of toxicity from the continuation of ineffective treatment. Thus, being able to differentiate the two can give the physician a more accurate judgment of the disease, leading to more appropriate treatment decisions.[153] In a recent study to differentiate radiation necrosis from true tumor recurrence, Haralick, Law, and Laplacian features were extracted from post-contrast T1WI, T2WI, and FLAIR images of 58 patients with histological evidence of various grades of primary and metastatic brain tumors.[145] Subsequent feature selection using minimum redundancy and maximum relevance combined with support vector machine (SVM) classifier was able to classify tissue as radionecrosis versus tumor recurrence with an overall accuracy of nearly 80%.[145] In a similar study, Zhang et al. extracted 285 radiomic features from standard MR sequences to classify necrosis from tumor progression in patients with brain metastasis after gamma knife surgery.[152] The temporal change in radiomic features longitudinally was used to build a model classifying lesions having radionecrosis versus progressive disease with an overall accuracy of 73.2%.[152] Similarly, our group investigated radiomic texture features' ability to differentiate progressive disease from pseudoprogression in over 300 post-treatment glioblastoma patients.[153] The MRMR feature selection method was used to build a SVM model, leave-one-out cross-validation (LOOCV) analyses showed 97% sensitivity, 72% specificity, and 90% accuracy to distinguish pseudoprogression from true progression. Lohmann et al. investigated the potential of PET radiomic features to distinguish radiation necrosis from tumor.[151] In their study, 62 radiomic texture features were extracted from 18-F-FET-PET scans from 54 patients, it was found that radiomic features increased the diagnostic accuracy of F-FET-PET variables alone to differentiate brain metastasis recurrence and radiation necrosis.[151]

Currently, histopathological analysis is the gold standard for brain tumor grading and diagnosis.[155] However, emerging studies show that non-invasive radiomics correlate with tumor grade and diagnosis.[156,157] In a recent study, four histogram features (global statistics features) and 14 textural features (local statistics features) applied to conventional MRI predicted high versus low grade gliomas with accuracies of 76%, 83%, and 88% based on global, local, and combination of both global and local features, respectively.[156] Similar results using more advanced imaging techniques also showed the potential diagnostic value in glioblastoma grading.[157] Ryu et al. extracted radiomic features from ADC maps and found that entropy was higher in high grade gliomas compared to low grade gliomas ($p = 0.006$).[157] Interestingly, although, fifth percentiles of the ADC histogram were also significantly different between high and low-grade gliomas, ADC entropy remained a more accurate predictor.[157]

Radiomics has been shown to help predict patient prognosis.[99,158–161] For example, in a recent study using perfusion DSC MR data of 24 glioblastoma patients, rCBV parametric maps were co-registered with contrast-enhanced T1 and T2/FLAIR images, and ROI were manually segmented.[162] Textural feature ratios and kinetic textural features based on the rCBV values in these contrast-enhancing and non-enhancing regions were then extracted and used in further analyses to measure association with survival.[162] Contrast-enhancing-derived features describing homogeneity, angular second moment, inverse difference moment, and entropy significantly distinguished survival groups based on receiver operating characteristic (ROC)-optimized cutoffs.[162] In a recent study by Liu et al. using 117 glioblastoma patients, 46 quantitative PWI features based on voxel values in the enhancing tumor region were extracted and underwent hierarchical consensus clustering with agglomerative average linkage to identify two robust imaging groups.[163] Subsequent analyses showed that the cluster associated with high intra-tumor PWI features (cluster 2) was characterized by 13 gene sets involved in biological processes such as angiogenesis-signaling, vascular development, and hypoxic response as well as by a relative survival benefit when patients were administered antiangiogenic treatment.[163] In another study by Kickingereder et al., nearly 5000 features were extracted from MR images from 172 patients and analyzed to predict response to anti-angiogenic therapy response in terms of progression-free survival (PFS) and overall survival (OS).[164] Features underwent principal component analysis to create a model that was found to be able to stratify patients into high and low risk groups for PFS and OS in both discovery (HR = 1.60; $p = 0.017$, HR = 2.14; $p < 0.001$, respectively) and validation sets (HR = 1.85; $p = 0.030$, HR = 2.60; $p = 0.001$, respectively).[164]

Finally, radiomic features have been linked to genomic information to further advance radiogenomics.[44,115,144,147] In a recent study by Zinn et al., radiomic features extracted from conventional MRI sequences were able to differentiate between *TP53*, *PTEN*, and *EGFR* mutated versus wildtype glioblastomas (Figure 13.6).[147] Interesting, volumes obtained from the three standard phenotypes of necrosis, edema/invasion, and enhancement did not show statistically significant difference between the mutated versus wildtype tumors, except for a statistically

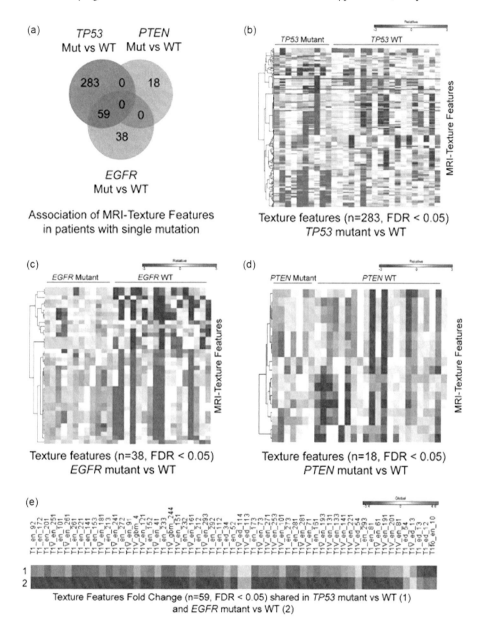

Figure 13.6 (See color insert.) MRI texture feature signatures are uniquely associated with patients with *TP53*, *PTEN*, and *EGFR* gene mutations. (a) Venn diagram depicting numbers of MRI texture features, their unique and overlapping association with GBM patients with specific gene mutation. (b–d) Heatmap generated using uniquely associated and significant texture features (false discovery rate [FDR] < 0.05) in *TP53* mutant versus WT (n = 283); (b) *EGFR* mutant versus wild type (WT) (n = 38); (c) and *PTEN* mutant versus WT (n = 18) (d). Color scale bar above heatmap shows range of texture feature values across patients. (e) Heatmap generated using fold change values of shared and significant texture features (n = 59; FDR < 0.05) in *TP53* mutant versus WT and *EGFR* mutant versus WT. Color scale above heatmap show range of fold change values for specific texture feature." (Adapted from Zinn, P.O. et al., *Neurosurgery*, 64, 203–210, 2017.)

significant difference in enhancing tumor volumes in patients with *EGFR* wildtype compared to *EGFR* mutant (*EGFR* mutant glioblastoma had more enhancing tumor volume).[147] Geveart et al. found that radiomic features correlated with the four glioblastoma molecular sub-types described by Verhaak et. al. (classic, neural, proneural, and mesenchymal).[44,144] They found that the mesenchymal sub-type was correlated with an edema-related feature, and the classic sub-type was correlated with necrosis and edema imaging features related to edge sharpness and intensity. In addition, the study also found interesting correlations between quantitative imaging features and gene expression modules.[165] Itakura et al. sought to identify sub-types based on imaging.[115] Using consensus clustering, they found three robust clusters (pre-multifocal, spherical, and rim-enhancing clusters) which significantly differed from one another in terms of survival and canonical pathways. It was proposed that the pre-multifocal cluster, being associated with upregulation of c-Kit pathway may be more sensitive to tyrosine kinase inhibitors, whereas the rim-enhancing cluster, being associated with vascular endothelial growth factor receptor (VEGFR) would be more sensitive to a VEGF inhibitor such as bevacizumab.[115]

In regard to the correlation between radiomics and epigenetics, a study by Korfiatis et al. identified four radiomic features (correlation, energy, entropy, and local intensity) that differentiated between *MGMT* promoter methylated and unmethylated glioblastomas.[160] Features were extracted based on run-length matrix (RLM) and GLCM. SVM predicted *MGMT* promoter methylation status with a sensitivity of 80% and specificity of 81%.[160] In a similar recent study by Yi et al., more than 1600 radiomic features were extracted from 48 methylated and 50 unmethylated glioblastomas.[166] LASSO methods were used to reduce the quantity of these features, which were then used to build a SVM model that was shown to predict methylation status with an accuracy of 86.5% when features were combined from T2WI and pre- and post-contrast T1WI images.[166] Finally, in a recent study, our group investigated the causal relationship between imaging and genomic information.[158] In this study, an automated pipeline was applied to extract 4800 MRI-derived texture features from glioblastoma tumors, and it was found that MGMT methylation/expression, EGFR amplification, survival, and molecular sub-types could all be significantly predicted by radiomic features (AUC 92%, $p = 0.001$; AUC 86%, $p < 0.0001$; AUC 90%, $p < 0.001$; AUC 88%, $p = 0.001$, respectively).

13.4.2 Looking into the future of radiomics

Radiomics is making significant strides in advancing studies related to the understanding and prediction of prognosis, genomics, epigenomics, and other -omics data and is likely to contribute to the establishment of personalized medicine.[147,167,168] However, robust validation both using independent non-overlapping patients and, if possible, prospective patient cohorts is needed. In addition, equally or even more importantly, the robust pre-clinical and clinical validation (to include scalability and generalizability) of radiomic/imaging software and clinical/translation research is needed in order for translation of radiomics and radiogenomics to clinical practice.[129,134,169]

13.5 CONCLUSION

In this work, we have outlined several important topics that focus on the current and future use of radiogenomics and radiomics. Genomic and imaging data are being collected at an incredibly fast pace, as data analysis techniques and storage and processing technology adapt, the clinical landscape can be anticipated to significantly change. Imaging, due to its versatility and ability to capture detailed data, can be expected to play a pivotal role in these advances as the field of radiomics and radiogenomics matures in the coming years.

ACKNOWLEDGMENTS

We would like to thank all members of the Colen Laboratory at the University of Texas MD Anderson Cancer Center who provided assistance in the preparation of this chapter, especially Srishti Abrol MD and Mohamed G. Elbanan MD for their contribution and input that greatly improved the quality of this work.

REFERENCES

1. Wen PY, Kesari S. Malignant gliomas in adults. *N Engl J Med* 2008; **359**: 492–507.
2. Wang H, Xu T, Jiang Y, Xu H, Yan Y, Fu D et al. The challenges and the promise of molecular targeted therapy in malignant gliomas. *Neoplasia* 2015; **17**: 239–255.
3. Stupp R, Hegi ME, Gilbert MR, Chakravarti A. Chemoradiotherapy in malignant glioma: Standard of care and future directions. *J Clin Oncol* 2007; **25**: 4127–4136.
4. Chung C, Metser U, Ménard C. Advances in magnetic resonance imaging and positron emission tomography imaging for grading and molecular characterization of glioma. *Semin Radiat Oncol* 2015; **25**: 164–171.
5. Zinn PO, Mahajan B, Majadan B, Sathyan P, Singh SK, Majumder S et al. Radiogenomic mapping of edema/cellular invasion MRI-phenotypes in glioblastoma multiforme. *PLoS One* 2011; **6**: e25451.
6. Anil R, Colen RR. Imaging genomics in glioblastoma multiforme: A predictive tool for patients prognosis, survival, and outcome. *Magn Reson Imaging Clin N Am* 2016; **24**: 731–740.
7. Puttick S, Bell C, Dowson N, Rose S, Fay M. PET, MRI, and simultaneous PET/MRI in the development of diagnostic and therapeutic strategies for glioma. *Drug Discov Today* 2014. doi:10.1016/j.drudis.2014.10.016.
8. Rutman AM, Kuo MD. Radiogenomics: Creating a link between molecular diagnostics and diagnostic imaging. *Eur J Radiol* 2009; **70**: 232–241.
9. Zinn PO, Colen RR. Imaging genomic mapping in glioblastoma. *Neurosurgery* 2013; **60**(Suppl 1): 126–130.
10. Pavlisa G, Rados M, Pavlisa G, Pavic L, Potocki K, Mayer D. The differences of water diffusion between brain tissue infiltrated by tumor and peritumoral vasogenic edema. *Clin Imaging* 2009; **33**: 96–101.
11. Kao H-W, Chiang S-W, Chung H-W, Tsai FY, Chen C-Y. Advanced MR imaging of gliomas: An update. *Biomed Res Int* 2013; **2013**: 970586.
12. Schmainda KM. Diffusion-weighted MRI as a biomarker for treatment response in glioma. *CNS Oncol* 2012; **1**: 169–180.
13. Gupta A, Young RJ, Karimi S, Sood S, Zhang Z, Mo Q et al. Isolated diffusion restriction precedes the development of enhancing tumor in a subset of patients with glioblastoma. *AJNR Am J Neuroradiol* 2011; **32**: 1301–1306.
14. Ferré JC, Shiroishi MS, Law M. Advanced techniques using contrast media in neuroimaging. *Magn Reson Imaging Clin N Am* 2012; **20**: 699–713.
15. Hu LS, Baxter LC, Smith KA, Feuerstein BG, Karis JP, Eschbacher JM et al. Relative cerebral blood volume values to differentiate high-grade glioma recurrence from posttreatment radiation effect: Direct correlation between image-guided tissue histopathology and localized dynamic susceptibility-weighted contrast-enhanced perfusion MR imaging measurements. *AJNR Am J Neuroradiol* 2009; **30**: 552–558.
16. Baek HJ, Kım HS, Kim N, Choi YJ, Kim YJ. Percent change of perfusion skewness and kurtosis: A potential imaging biomarker for early treatment response in patients with newly diagnosed glioblastomas. *Radiology* 2012; **264**: 834–843.
17. Aquino D, Di Stefano AL, Scotti A, Cuppini L, Anghileri E, Finocchiaro G et al. Parametric response maps of perfusion MRI may identify recurrent glioblastomas responsive to bevacizumab and irinotecan. *PLoS One* 2014; **9**: e90535.
18. Horská A, Barker PB. Imaging of brain tumors: MR spectroscopy and metabolic imaging. *Neuroimaging Clin N Am* 2010/8; **20**: 293–310.
19. Ranjith G, Parvathy R, Vikas V, Chandrasekharan K, Nair S. Machine learning methods for the classification of gliomas: Initial results using features extracted from MR spectroscopy. *Neuroradiol J* 2015; **28**: 106–111.
20. Bulik M, Jancalek R, Vanicek J, Skoch A, Mechl M. Potential of MR spectroscopy for assessment of glioma grading. *Clin Neurol Neurosurg* 2013; **115**: 146–153.

21. Nagashima H, Tanaka K, Sasayama T, Irino Y, Sato N, Takeuchi Y et al. Diagnostic value of glutamate with 2-hydroxyglutarate in magnetic resonance spectroscopy for IDH1 mutant glioma. *Neuro Oncol* 2016; **18**: 1559–1568.

22. Van Meir EG, Hadjipanayis CG, Norden AD, Shu H-K, Wen PY, Olson JJ. Exciting new advances in neuro-oncology: The avenue to a cure for malignant glioma. *CA Cancer J Clin* 2010; **60**: 166–193.

23. Macdonald DR, Cascino TL, Schold SC Jr, Cairncross JG. Response criteria for phase II studies of supratentorial malignant glioma. *J Clin Oncol* 1990; **8**: 1277–1280.

24. Wen PY, Macdonald DR, Reardon DA, Cloughesy TF, Sorensen AG, Galanis E et al. Updated response assessment criteria for high-grade gliomas: Response assessment in neuro-oncology working group. *J Clin Oncol* 2010; **28**: 1963–1972.

25. Agarwal A, Kumar S, Narang J, Schultz L, Mikkelsen T, Wang S et al. Morphologic MRI features, diffusion tensor imaging and radiation dosimetric analysis to differentiate pseudo-progression from early tumor progression. *J Neurooncol* 2013; **112**: 413–420.

26. Bonavia R, Inda M-M, Cavenee WK, Furnari FB. Heterogeneity maintenance in glioblastoma: A social network. *Cancer Res* 2011; **71**: 4055–4060.

27. Inda M-D-M, Bonavia R, Seoane J. Glioblastoma multiforme: A look inside its heterogeneous nature. *Cancers* 2014; **6**: 226–239.

28. Cho D-Y, Lin S-Z, Yang W-K, Lee H-C, Hsu D-M, Lin H-L et al. Targeting cancer stem cells for treatment of glioblastoma multiforme. *Cell Transplant* 2013; **22**: 731–739.

29. Lee J, Narang S, Martinez J, Rao G, Rao A. Spatial habitat features derived from multiparametric magnetic resonance imaging data are associated with molecular subtype and 12-month survival status in glioblastoma multiforme. *PLoS One* 2015; **10**: e0136557.

30. Lemasson B, Chenevert TL, Lawrence TS, Tsien C, Sundgren PC, Meyer CR et al. Impact of perfusion map analysis on early survival prediction accuracy in glioma patients. *Transl Oncol* 2013; **6**: 766–774.

31. Gutman DA, Cooper LAD, Hwang SN, Holder CA, Gao J, Aurora TD et al. MR imaging predictors of molecular profile and survival: Multi-institutional study of the TCGA glioblastoma data set. *Radiology* 2013; **267**: 560–569.

32. Ellingson BM, Cloughesy TF, Lai A, Mischel PS, Nghiemphu PL, Lalezari S et al. Graded functional diffusion map-defined characteristics of apparent diffusion coefficients predict overall survival in recurrent glioblastoma treated with bevacizumab. *Neuro Oncol* 2011; **13**: 1151–1161.

33. Nalejska E, Mączyńska E, Lewandowska MA. Prognostic and predictive biomarkers: Tools in personalized oncology. *Mol Diagn Ther* 2014; **18**: 273–284.

34. Roychowdhury S, Chinnaiyan AM. Translating cancer genomes and transcriptomes for precision oncology. *CA Cancer J Clin* 2016; **66**: 75–88.

35. Boots-Sprenger SHE, Sijben A, Rijntjes J, Tops BBJ, Idema AJ, Rivera AL et al. Significance of complete 1p/19q co-deletion, IDH1 mutation and MGMT promoter methylation in gliomas: Use with caution. *Mod Pathol* 2013; **26**: 922–929.

36. Hegi ME, Diserens A-C, Gorlia T, Hamou M-F, de Tribolet N, Weller M et al. MGMT gene silencing and benefit from temozolomide in glioblastoma. *N Engl J Med* 2005; **352**: 997–1003.

37. Paus C, Murat A, Stupp R, Regli L, Hegi M. Role of MGMT and clinical applications in brain tumours. *Bull Cancer* 2007; **94**: 769–773.

38. Hartmann C, Hentschel B, Wick W, Capper D, Felsberg J, Simon M et al. Patients with IDH1 wild type anaplastic astrocytomas exhibit worse prognosis than IDH1-mutated glioblastomas, and IDH1 mutation status accounts for the unfavorable prognostic effect of higher age: Implications for classification of gliomas. *Acta Neuropathol* 2010; **120**: 707–718.

39. Sanson M, Marie Y, Paris S, Idbaih A, Laffaire J, Ducray F et al. Isocitrate dehydrogenase 1 codon 132 mutation is an important prognostic biomarker in gliomas. *J Clin Oncol* 2009; **27**: 4150–4154.

40. Weller M, Felsberg J, Hartmann C, Berger H, Steinbach JP, Schramm J et al. Molecular predictors of progression-free and overall survival in patients with newly diagnosed glioblastoma: A prospective translational study of the German Glioma Network. *J Clin Oncol* 2009; **27**: 5743–5750.

41. Ekstrand AJ, James CD, Cavenee WK, Seliger B, Pettersson RF, Collins VP. Genes for epidermal growth factor receptor, transforming growth factor alpha, and epidermal growth factor and their expression in human gliomas in vivo. *Cancer Res* 1991; **51**: 2164–2172.

42. Idoate MA, Echeveste J, Diez-Valle R, Lozano MD, Aristu J. Biological and clinical significance of the intratumour heterogeneity of PTEN protein expression and the corresponding molecular abnormalities of the PTEN gene in glioblastomas. *Neuropathol Appl Neurobiol* 2014; **40**: 736–746.

43. Sathyan P, Zinn PO, Marisetty AL, Liu B, Kamal MM, Singh SK et al. Mir-21-Sox2 axis delineates glioblastoma subtypes with prognostic impact. *J Neurosci* 2015; **35**: 15097–15112.

44. Verhaak RGW, Hoadley KA, Purdom E, Wang V, Qi Y, Wilkerson MD et al. Integrated genomic analysis identifies clinically relevant subtypes of glioblastoma characterized by abnormalities in PDGFRA, IDH1, EGFR, and NF1. *Cancer Cell* 2010; **17**: 98–110.

45. Esteller M, Hamilton SR, Burger PC, Baylin SB, Herman JG. Inactivation of the DNA repair gene O6-methylguanine-DNA methyltransferase by promoter hypermethylation is a common event in primary human neoplasia. *Cancer Res* 1999; **59**: 793–797.

46. Pegg AE, Dolan ME, Moschel RC. Structure, function, and inhibition of O6-alkylguanine-DNA alkyltransferase. *Prog Nucleic Acid Res Mol Biol* 1995; **51**: 167–223.

47. Quintavalle C, Mangani D, Roscigno G, Romano G, Diaz-Lagares A, Iaboni M et al. MiR-221/222 target the DNA methyltransferase MGMT in glioma cells. *PLoS One* 2013; **8**: e74466.

48. Hegi ME, Liu L, Herman JG, Stupp R, Wick W, Weller M et al. Correlation of O6-methylguanine methyltransferase (MGMT) promoter methylation with clinical outcomes in glioblastoma and clinical strategies to modulate MGMT activity. *J Clin Oncol* 2008; **26**: 4189–4199.

49. Schaap FG, French PJ, Bovée JVMG. Mutations in the isocitrate dehydrogenase genes IDH1 and IDH2 in tumors. *Adv Anat Pathol* 2013; **20**: 32–38.

50. Ichimura K, Pearson DM, Kocialkowski S, Bäcklund LM, Chan R, Jones DTW et al. IDH1 mutations are present in the majority of common adult gliomas but rare in primary glioblastomas. *Neuro Oncol* 2009; **11**: 341–347.

51. Lee SM, Koh H-J, Park D-C, Song BJ, Huh T-L, Park J-W. Cytosolic NADP(+)-dependent isocitrate dehydrogenase status modulates oxidative damage to cells. *Free Radic Biol Med* 2002; **32**: 1185–1196.

52. Kim SY, Lee SM, Tak JK, Choi KS, Kwon TK, Park J-W. Regulation of singlet oxygen-induced apoptosis by cytosolic NADP+-dependent isocitrate dehydrogenase. *Mol Cell Biochem* 2007; **302**: 27–34.

53. Zhao S, Lin Y, Xu W, Jiang W, Zha Z, Wang P et al. Glioma-derived mutations in IDH1 dominantly inhibit IDH1 catalytic activity and induce HIF-1alpha. *Science* 2009; **324**: 261–265.

54. Dang L, White DW, Gross S, Bennett BD, Bittinger MA, Driggers EM et al. Cancer-associated IDH1 mutations produce 2-hydroxyglutarate. *Nature* 2010; **465**: 966.

55. Yan H, Parsons DW, Jin G, McLendon R, Rasheed BA, Yuan W et al. IDH1 and IDH2 mutations in gliomas. *N Engl J Med* 2009; **360**: 765–773.

56. Lai A, Kharbanda S, Pope WB, Tran A, Solis OE, Peale F et al. Evidence for sequenced molecular evolution of IDH1 mutant glioblastoma from a distinct cell of origin. *J Clin Oncol* 2011; **29**: 4482–4490.

57. Jenkins RB, Blair H, Ballman KV, Giannini C, Arusell RM, Law M et al. A t (1; 19)(q10; p10) mediates the combined deletions of 1p and 19q and predicts a better prognosis of patients with oligodendroglioma. *Cancer Res* 2006; **66**: 9852–9861.

58. Wikstrand CJ, McLendon RE, Friedman AH, Bigner DD. Cell surface localization and density of the tumor-associated variant of the epidermal growth factor receptor, EGFRvIII. *Cancer Res* 1997; **57**: 4130–4140.

59. Ekstrand AJ, Longo N, Hamid ML, Olson JJ, Liu L, Collins VP et al. Functional characterization of an EGF receptor with a truncated extracellular domain expressed in glioblastomas with EGFR gene amplification. *Oncogene* 1994; **9**: 2313–2320.

60. Nishikawa R, Ji XD, Harmon RC, Lazar CS, Gill GN, Cavenee WK et al. A mutant epidermal growth factor receptor common in human glioma confers enhanced tumorigenicity. *Proc Natl Acad Sci USA* 1994; **91**: 7727–7731.

61. Heimberger AB, Suki D, Yang D, Shi W, Aldape K. The natural history of EGFR and EGFRvIII in glioblastoma patients. *J Transl Med* 2005; **3**: 38.

62. Yeung YT, McDonald KL, Grewal T, Munoz L. Interleukins in glioblastoma pathophysiology: Implications for therapy. *Br J Pharmacol* 2013; **168**: 591–606.

63. Tchirkov A, Khalil T, Chautard E, Mokhtari K, Véronèse L, Irthum B et al. Interleukin-6 gene amplification and shortened survival in glioblastoma patients. *Br J Cancer* 2007; **96**: 474–476.

64. Karsy M, Neil JA, Guan J, Mark MA, Colman H, Jensen RL. A practical review of prognostic correlations of molecular biomarkers in glioblastoma. *Neurosurg Focus* 2015; **38**: E4.

65. Bartel DP. MicroRNAs: Target recognition and regulatory functions. *Cell* 2009; **136**: 215–233.

66. Louis DN, Perry A, Reifenberger G, von Deimling A, Figarella-Branger D, Cavenee WK et al. The 2016 World Health Organization classification of tumors of the central nervous system: A summary. *Acta Neuropathol* 2016; **131**: 803–820.

67. Phillips HS, Kharbanda S, Chen R, Forrest WF, Soriano RH, Wu TD et al. Molecular subclasses of high-grade glioma predict prognosis, delineate a pattern of disease progression, and resemble stages in neurogenesis. *Cancer Cell* 2006; **9**: 157–173.

68. Brennan CW, Verhaak RGW, McKenna A, Campos B, Noushmehr H, Salama SR et al. The somatic genomic landscape of glioblastoma. *Cell* 2013; **155**: 462–477.

69. ElBanan MG, Amer AM, Zinn PO, Colen RR. Imaging genomics of glioblastoma: State of the art bridge between genomics and neuroradiology. *Neuroimaging Clin N Am* 2015; **25**: 141–153.

70. Zinn PO, Mahmood Z, Elbanan MG, Colen RR. Imaging genomics in gliomas. *Cancer J* 2015; **21**: 225–234.

71. Kim MM, Parolia A, Dunphy MP, Venneti S. Non-invasive metabolic imaging of brain tumours in the era of precision medicine. *Nat Rev Clin Oncol* 2016; **13**: 725–739.

72. Kuo MD, Jamshidi N. Behind the numbers: Decoding molecular phenotypes with radiogenomics—Guiding principles and technical considerations. *Radiology* 2014; **270**: 320–325.

73. Belden CJ, Valdes PA, Ran C, Pastel DA, Harris BT, Fadul CE et al. Genetics of glioblastoma: A window into its imaging and histopathologic variability. *Radiographics* 2011; **31**: 1717–1740.

74. Moton S, Elbanan M, Zinn PO, Colen RR. Imaging genomics of glioblastoma: Biology, biomarkers, and breakthroughs. *Top Magn Reson Imaging* 2015; **24**: 155–163.

75. Jansen RW, van Amstel P, Martens RM, Kooi IE, Wesseling P, de Langen AJ et al. Non-invasive tumor genotyping using radiogenomic biomarkers, a systematic review and oncology-wide pathway analysis. *Oncotarget* 2018; **9**: 20134–20155.

76. Barajas RF Jr, Hodgson JG, Chang JS, Vandenberg SR, Yeh R-F, Parsa AT et al. Glioblastoma multiforme regional genetic and cellular expression patterns: Influence on anatomic and physiologic MR imaging. *Radiology* 2010; **254**: 564–576.

77. Madhavan S, Zenklusen JC, Kotliarov Y, Sahni H, Fine HA, Buetow K. Rembrandt: Helping personalized medicine become a reality through integrative translational research. *Mol Cancer Res* 2009; **7**: 157–167.

78. Tomczak K, Czerwińska P, Wiznerowicz M. The Cancer Genome Atlas (TCGA): An immeasurable source of knowledge. *Contemp Oncol* 2015; **19**: A68–A77.

79. Cancer Genome Atlas Research Network. Comprehensive genomic characterization defines human glioblastoma genes and core pathways. *Nature* 2008; **455**: 1061–1068.

80. Pope WB, Chen JH, Dong J, Carlson MRJ, Perlina A, Cloughesy TF et al. Relationship between gene expression and enhancement in glioblastoma multiforme: Exploratory DNA microarray analysis. *Radiology* 2008; **249**: 268–277.

81. Carrillo JA, Lai A, Nghiemphu PL, Kim HJ, Phillips HS, Kharbanda S et al. Relationship between tumor enhancement, edema, IDH1 mutational status, MGMT promoter methylation, and survival in glioblastoma. *AJNR Am J Neuroradiol* 2012; **33**: 1349–1355.

82. Drabycz S, Roldán G, de Robles P, Adler D, McIntyre JB, Magliocco AM et al. An analysis of image texture, tumor location, and MGMT promoter methylation in glioblastoma using magnetic resonance imaging. *Neuroimage* 2010; **49**: 1398–1405.

83. Raza SM, Fuller GN, Rhee CH, Huang S, Hess K, Zhang W et al. Identification of necrosis-associated genes in glioblastoma by cDNA microarray analysis. *Clin Cancer Res* 2004; **10**: 212–221.

84. Eoli M, Menghi F, Bruzzone MG, De Simone T, Valletta L, Pollo B et al. Methylation of O6-methylguanine DNA methyltransferase and loss of heterozygosity on 19q and/or 17p are over-lapping features of secondary glioblastomas with prolonged survival. *Clin Cancer Res* 2007; **13**: 2606–2613.

85. Brandes AA, Franceschi E, Tosoni A, Blatt V, Pession A, Tallini G et al. MGMT promoter methylation status can predict the incidence and outcome of pseudoprogression after concomitant radiochemo-therapy in newly diagnosed glioblastoma patients. *J Clin Oncol* 2008; **26**: 2192–2197.

86. Van Meter T, Dumur C, Hafez N, Garrett C, Fillmore H, Broaddus WC. Microarray analysis of MRI-defined tissue samples in glioblastoma reveals differences in regional expression of therapeutic targets. *Diagn Mol Pathol* 2006; **15**: 195–205.

87. Diehn M, Nardini C, Wang DS, McGovern S, Jayaraman M, Liang Y et al. Identification of noninva-sive imaging surrogates for brain tumor gene-expression modules. *Proc Natl Acad Sci USA* 2008; **105**: 5213–5218.

88. Colen RR, Vangel M, Wang J, Gutman DA, Hwang SN, Wintermark M et al. Imaging genomic mapping of an invasive MRI phenotype predicts patient outcome and metabolic dysfunction: A TCGA glioma phenotype research group project. *BMC Med Genomics* 2014; **7**: 30.

89. Bredel M, Bredel C, Juric D, Duran GE, Yu RX, Harsh GR et al. Tumor necrosis factor-alpha-induced protein 3 as a putative regulator of nuclear factor-kappaB-mediated resistance to O6-alkylating agents in human glioblastomas. *J Clin Oncol* 2006; **24**: 274–287.

90. Bredel M, Scholtens DM, Yadav AK, Alvarez AA, Renfrow JJ, Chandler JP et al. NFKBIA deletion in glioblastomas. *N Engl J Med* 2011; **364**: 627–637.

91. Eoli M, Menghi F, Bruzzone MG, De Simone T, Valletta L, Pollo B et al. Methylation of O6-methylguanine DNA methyltransferase and loss of heterozygosity on 19q and/or 17p are overlapping features of secondary glioblastomas with prolonged survival. *Clin Cancer Res* 2007; **13**: 2606–2613.

92. Kanas VG, Zacharaki EI, Thomas GA, Zinn PO, Megalooikonomou V, Colen RR. Learning MRI-based classification models for MGMT methylation status prediction in glioblastoma. *Comput Methods Programs Biomed* 2017; **140**: 249–257.

93. Sorensen AG, Batchelor TT, Wen PY, Zhang W-T, Jain RK. Response criteria for glioma. *Nat Clin Pract Oncol* 2008; **5**: 634–644.

94. Kanaly CW, Mehta AI, Ding D, Hoang JK, Kranz PG, Herndon JE 2nd et al. A novel, reproducible, and objective method for volumetric magnetic resonance imaging assessment of enhancing glioblastoma. *J Neurosurg* 2014; **121**: 536–542.

95. Colen RR, Wang J, Singh SK, Gutman DA, Zinn PO. Glioblastoma: Imaging genomic mapping reveals sex-specific oncogenic associations of cell death. *Radiology* 2015; **275**: 215–227.

96. Gutman DA, Dunn WD Jr, Grossmann P, Cooper LAD, Holder CA, Ligon KL et al. Somatic mutations associated with MRI-derived volumetric features in glioblastoma. *Neuroradiology* 2015; **12**: 1227–1237. doi:10.1007/s00234-015-1576-7.

97. Jenkinson M, Beckmann CF, Behrens TEJ, Woolrich MW, Smith SM. FSL. *Neuroimage* 2012; **62**: 782–790.

98. Grossmann P, Gutman DA, Dunn WD Jr, Holder CA, Aerts HJWL. Imaging-genomics reveals driv-ing pathways of MRI derived volumetric tumor phenotype features in Glioblastoma. *BMC Cancer* 2016; **16**: 611.

99. Rao A, Rao G, Gutman DA, Flanders AE, Hwang SN, Rubin DL et al. A combinatorial radiographic phenotype may stratify patient survival and be associated with invasion and proliferation characteristics in glioblastoma. *J Neurosurg* 2016; **124**: 1008–1017.

100. Gupta A, Young RJ, Shah AD, Schweitzer AD, Graber JJ, Shi W et al. Pretreatment dynamic susceptibility contrast MRI perfusion in glioblastoma: Prediction of EGFR gene amplification. *Clin Neuroradiol* 2014. doi:10.1007/s00062-014-0289-3.

101. Rao A, Manyam G, Rao G, Jain R. Integrative analysis of mRNA, microRNA, and protein correlates of relative cerebral blood volume values in GBM reveals the role for modulators of angiogenesis and tumor proliferation. *Cancer Inform* 2016; **15**: 29–33.

102. Pope WB, Mirsadraei L, Lai A, Eskin A, Qiao J, Kim HJ et al. Differential gene expression in glioblastoma defined by ADC histogram analysis: Relationship to extracellular matrix molecules and survival. *AJNR Am J Neuroradiol* 2012; **33**: 1059–1064.

103. Zinn PO, Hatami M, Youssef E, Thomas GA, Luedi MM, Singh SK et al. Diffusion weighted magnetic resonance imaging radiophenotypes and associated molecular pathways in glioblastoma. *Neurosurgery* 2016; **63** Suppl 1: 127–135.

104. Jansen NL, Schwartz C, Graute V, Eigenbrod S, Lutz J, Egensperger R et al. Prediction of oligodendroglial histology and LOH 1p/19q using dynamic [(18)F]FET-PET imaging in intracranial WHO grade II and III gliomas. *Neuro Oncol* 2012; **14**: 1473–1480.

105. Colen R, Zinn P. Perfusion imaging genomic mapping uncovers potential genomic targets involved in angiogenesis and invasion. In: *Neuro-oncology*. Oxford University Press, Cary, NC, 2013, pp. 193–193.

106. Sunwoo L, Choi SH, Park C-K, Kim JW, Yi KS, Lee WJ et al. Correlation of apparent diffusion coefficient values measured by diffusion MRI and MGMT promoter methylation semiquantitatively analyzed with MS-MLPA in patients with glioblastoma multiforme. *J Magn Reson Imaging* 2013; **37**: 351–358.

107. Romano A, Calabria LF, Tavanti F, Minniti G, Rossi-Espagnet MC, Coppola V et al. Apparent diffusion coefficient obtained by magnetic resonance imaging as a prognostic marker in glioblastomas: Correlation with MGMT promoter methylation status. *Eur Radiol* 2013; **23**: 513–520.

108. Yamashita K, Hiwatashi A, Togao O, Kikuchi K, Hatae R, Yoshimoto K et al. MR Imaging-based analysis of glioblastoma multiforme: Estimation of IDH1 mutation status. *AJNR Am J Neuroradiol* 2016; **37**: 58–65.

109. Andronesi OC, Rapalino O, Gerstner E, Chi A, Batchelor TT, Cahill DP et al. Detection of oncogenic IDH1 mutations using magnetic resonance spectroscopy of 2-hydroxyglutarate. *J Clin Invest* 2013; **123**: 3659–3663.

110. Choi C, Ganji SK, DeBerardinis RJ, Hatanpaa KJ, Rakheja D, Kovacs Z et al. 2-hydroxyglutarate detection by magnetic resonance spectroscopy in IDH-mutated patients with gliomas. *Nat Med* 2012; **18**: 624–629.

111. Heo H, Kim S, Lee HH, Cho HR, Xu WJ, Lee S-H et al. On the utility of short echo time (TE) single voxel 1H-MRS in non-invasive detection of 2-hydroxyglutarate (2HG); Challenges and potential improvement illustrated with animal models using MRUI and LCModel. *PLoS One* 2016; **11**: e0147794.

112. Pope WB, Prins RM, Albert Thomas M, Nagarajan R, Yen KE, Bittinger MA et al. Non-invasive detection of 2-hydroxyglutarate and other metabolites in IDH1 mutant glioma patients using magnetic resonance spectroscopy. *J Neurooncol* 2012; **107**: 197–205.

113. Weller M, Stupp R, Hegi ME, van den Bent M, Tonn JC, Sanson M et al. Personalized care in neuro-oncology coming of age: Why we need MGMT and 1p/19q testing for malignant glioma patients in clinical practice. *Neuro Oncol* 2012; **14** Suppl 4: iv100–iv108.

114. Jansen NL, Schwartz C, Graute V, Eigenbrod S, Lutz J, Egensperger R et al. Prediction of oligodendroglial histology and LOH 1p/19q using dynamic [18F] FET-PET imaging in intracranial WHO grade II and III gliomas. *Neuro Oncol* 2012; nos259.

115. Itakura H, Achrol AS, Mitchell LA, Loya JJ, Liu T, Westbroek EM et al. Magnetic resonance image features identify glioblastoma phenotypic subtypes with distinct molecular pathway activities. *Sci Transl Med* 2015; **7**: 303ra138.

116. Cooper LA, Kong J, Gutman DA, Dunn WD, Nalisnik M, Brat DJ. Novel genotype-phenotype associations in human cancers enabled by advanced molecular platforms and computational analysis of whole slide images. *Lab Invest* 2015. doi:10.1038/labinvest.2014.153.

117. Murovic J, Turowski K, Wilson CB, Hoshino T, Levin V. Computerized tomography in the prognosis of malignant cerebral gliomas. *J Neurosurg* 1986; **65**: 799–806.

118. Andreou J, George AE, Wise A, de Leon M, Kricheff II, Ransohoff J et al. CT prognostic criteria of survival after malignant glioma surgery. *AJNR Am J Neuroradiol* 1983; **4**: 488–490.

119. Reeves GI, Marks JE. Prognostic significance of lesion size for glioblastoma multiforme. *Radiology* 1979; **132**: 469–471.

120. Egger J, Kapur T, Fedorov A, Pieper S, Miller JV, Veeraraghavan H et al. GBM volumetry using the 3D Slicer medical image computing platform. *Sci Rep* 2013; **3**: 1364.

121. Chaddad A, Zinn PO, Colen RR. Quantitative texture analysis for glioblastoma phenotypes discrimination. In *2014 International Conference on Control, Decision and Information Technologies (CoDIT)*. 2014, pp 605–608.

122. Li SC, Tachiki LML, Kabeer MH, Dethlefs BA, Anthony MJ, Loudon WG. Cancer genomic research at the crossroads: Realizing the changing genetic landscape as intratumoral spatial and temporal heterogeneity becomes a confounding factor. *Cancer Cell Int* 2014; **14**: 115.

123. Grove O, Berglund AE, Schabath MB, Aerts HJWL, Dekker A, Wang H et al. Quantitative computed tomographic descriptors associate tumor shape complexity and intratumor heterogeneity with prognosis in lung adenocarcinoma. *PLoS One* 2015; **10**: e0118261.

124. Bianchi G, Ghobrial IM. Biological and clinical implications of clonal heterogeneity and clonal evolution in multiple myeloma. *Curr Cancer Ther Rev* 2014; **10**: 70–79.

125. Lee J-Y, Yoon J-K, Kim B, Kim S, Kim MA, Lim H et al. Tumor evolution and intratumor heterogeneity of an epithelial ovarian cancer investigated using next-generation sequencing. *BMC Cancer* 2015; **15**: 1077.

126. Hu LS, Ning S, Eschbacher JM, Baxter LC, Gaw N, Ranjbar S et al. Radiogenomics to characterize regional genetic heterogeneity in glioblastoma. *Neuro Oncol* 2016. doi:10.1093/neuonc/now135.

127. Ross JL, Cooper LAD, Kong J, Gutman D, Williams M, Tucker-Burden C et al. 5-Aminolevulinic acid guided sampling of glioblastoma microenvironments identifies pro-survival signaling at infiltrative margins. *Sci Rep* 2017; **7**: 15593.

128. Lambin P, Rios-Velazquez E, Leijenaar R, Carvalho S, van Stiphout RGPM, Granton P et al. Radiomics: Extracting more information from medical images using advanced feature analysis. *Eur J Cancer* 2012; **48**: 441–446.

129. Gillies RJ, Kinahan PE, Hricak H. Radiomics: Images are more than pictures, they are data. *Radiology* 2016; **278**: 563–577.

130. Parmar C, Grossmann P, Bussink J, Lambin P, Aerts HJWL. Machine learning methods for quantitative radiomic biomarkers. *Sci Rep* 2015; **5**: 13087.

131. Fyllingen EH, Stensjøen AL, Berntsen EM, Solheim O, Reinertsen I. Glioblastoma segmentation: Comparison of three different software packages. *PLoS One* 2016; **11**: e0164891.

132. Gordillo N, Montseny E, Sobrevilla P. State of the art survey on MRI brain tumor segmentation. *Magn Reson Imaging* 2013; **31**: 1426–1438.

133. Abrol S, Kotrotsou A, Salem A, Zinn PO, Colen RR. Radiomic phenotyping in brain cancer to unravel hidden information in medical images. *Top Magn Reson Imaging* 2017; **26**: 43–53.

134. Zinn PO, Singh SK, Kotrotsou A, Hassan I, Thomas G, Luedi MM et al. A co-clinical radiogenomic validation study—Conserved magnetic resonance radiomic appearance of periostin expressing glioblastoma in patients and xenograft models. *Clin Cancer Res* 2018. doi:10.1158/1078-0432.CCR-17-3420.

135. Haralick RM, Shanmugam K, Dinstein I. Textural features for image classification. *IEEE Trans Syst Man Cybern* 1973; **SMC-3**: 610–621.

136. Lam SWC. Texture feature extraction using gray level gradient based co-occurence matrices. *In 1996 IEEE International Conference on Systems, Man and Cybernetics. Information Intelligence and Systems (Cat. No.96CH35929)*. 1996, pp. 267–271, vol. 1.

137. Parekh V, Jacobs MA. Radiomics: A new application from established techniques. *Expert Rev Precis Med Drug Dev* 2016; **1**: 207–226.

138. Luo W, Phung D, Tran T, Gupta S, Rana S, Karmakar C et al. Guidelines for developing and reporting machine learning predictive models in biomedical research: A multidisciplinary view. *J Med Internet Res* 2016; **18**: e323.

139. Wu J, Tha KK, Xing L, Li R. Radiomics and radiogenomics for precision radiotherapy. *J Radiat Res* 2018; **59**: i25–i31.

140. Peng H, Long F, Ding C. Feature selection based on mutual information: Criteria of max-dependency, max-relevance, and min-redundancy. *IEEE Trans Pattern Anal Mach Intell* 2005; **27**: 1226–1238.

141. Tibshirani R. Regression shrinkage and selection via the lasso. *J R Stat Soc Series B Stat Methodol* 1996; **58**: 267–288.

142. Soeda A, Hara A, Kunisada T, Yoshimura S-I, Iwama T, Park DM. CORRIGENDUM: The evidence of glioblastoma heterogeneity. *Sci Rep* 2015; **5**: 9630.

143. Hu LS, Ning S, Eschbacher JM, Gaw N, Dueck AC, Smith KA et al. Multi-parametric MRI and texture analysis to visualize spatial histologic heterogeneity and tumor extent in glioblastoma. *PLoS One* 2015; **10**: e0141506.

144. Gevaert O, Mitchell LA, Achrol AS, Xu J, Echegaray S, Steinberg GK et al. Glioblastoma multiforme: Exploratory radiogenomic analysis by using quantitative image features. *Radiology* 2014; **273**: 168–174.

145. Tiwari P, Prasanna P, Wolansky L, Pinho M, Cohen M, Nayate AP et al. Computer-extracted texture features to distinguish cerebral radionecrosis from recurrent brain tumors on multiparametric MRI: A feasibility study. *AJNR Am J Neuroradiol* 2016; **37**: 2231–2236.

146. Jain R, Poisson L, Narang J, Gutman D, Scarpace L, Hwang SN et al. Genomic mapping and survival prediction in glioblastoma: Molecular subclassification strengthened by hemodynamic imaging biomarkers. *Radiology* 2013; **267**: 212–220.

147. Zinn PO, Singh SK, Kotrotsou A, Abrol S, Thomas G, Mosley J et al. Distinct radiomic phenotypes define glioblastoma TP53-PTEN-EGFR mutational landscape. *Neurosurgery* 2017; **64**: 203–210.

148. Montana V. Glioma: The mechanisms of infiltrative growth. *Opera Medica et Physiologica* 2016. http://cyberleninka.ru/article/n/glioma-the-mechanisms-of-infiltrative-growth (accessed 25 March 2017).

149. Ortiz-Ramón R, Larroza A, Ruiz-España S, Arana E, Moratal D. Classifying brain metastases by their primary site of origin using a radiomics approach based on texture analysis: A feasibility study. *Eur Radiol* 2018. doi:10.1007/s00330-018-5463-6.

150. Ortiz-Ramon R, Larroza A, Arana E, Moratal D. A radiomics evaluation of 2D and 3D MRI texture features to classify brain metastases from lung cancer and melanoma. *Conf Proc IEEE Eng Med Biol Soc* 2017; **2017**: 493–496.

151. Lohmann P, Stoffels G, Ceccon G, Rapp M, Sabel M, Filss CP et al. Radiation injury vs. recurrent brain metastasis: Combining textural feature radiomics analysis and standard parameters may increase 18F-FET PET accuracy without dynamic scans. *Eur Radiol* 2017; **27**: 2916–2927.

152. Zhang Z, Yang J, Ho A, Jiang W, Logan J, Wang X et al. A predictive model for distinguishing radiation necrosis from tumour progression after gamma knife radiosurgery based on radiomic features from MR images. *Eur Radiol* 2018; **28**: 2255–2263.

153. Abrol S, Kotrotsou A, Hassan A, Elshafeey N, Idris T, Manohar N et al. Abstract 3040: Radiomics discriminates pseudo-progression from true progression in glioblastoma patients: A large-scale multi-institutional study. *Cancer Res* 2018; **78**: 3040–3040.

154. Ellingson BM, Chung C, Pope WB, Boxerman JL, Kaufmann TJ. Pseudoprogression, radionecrosis, inflammation or true tumor progression? Challenges associated with glioblastoma response assessment in an evolving therapeutic landscape. *J Neurooncol* 2017; **134**: 495–504.

155. Louis DN, Ohgaki H, Wiestler OD, Cavenee WK, Burger PC, Jouvet A et al. The 2007 WHO classification of tumours of the central nervous system. *Acta Neuropathol* 2007; **114**: 97–109.

156. Hsieh KL-C, Lo C-M, Hsiao C-J. Computer-aided grading of gliomas based on local and global MRI features. *Comput Methods Programs Biomed* 2017; **139**: 31–38.

157. Ryu YJ, Choi SH, Park SJ, Yun TJ, Kim J-H, Sohn C-H. Glioma: Application of whole-tumor texture analysis of diffusion-weighted imaging for the evaluation of tumor heterogeneity. *PLoS One* 2014; **9**: e108335.

158. Zinn PO, Singh SK, Kotrotsou A, Zandi F, Thomas G, Hatami M et al. 139 clinically applicable and biologically validated MRI radiomic test method predicts glioblastoma genomic landscape and survival. *Neurosurgery* 2016; **63**(Suppl 1): 156–157.

159. Nicolasjilwan M, Hu Y, Yan C, Meerzaman D, Holder CA, Gutman D et al. Addition of MR imaging features and genetic biomarkers strengthens glioblastoma survival prediction in TCGA patients. *J Neuroradiol* 2014. doi:10.1016/j.neurad.2014.02.006.

160. Korfiatis P, Kline TL, Coufalova L, Lachance DH, Parney IF, Carter RE et al. MRI texture features as biomarkers to predict MGMT methylation status in glioblastomas. *Med Phys* 2016; **43**: 2835.

161. McGarry SD, Hurrell SL, Kaczmarowski AL, Cochran EJ, Connelly J, Rand SD et al. Magnetic resonance imaging-based radiomic profiles predict patient prognosis in newly diagnosed glioblastoma before therapy. *Tomography* 2016; **2**: 223–228.

162. Lee J, Jain R, Khalil K, Griffith B, Bosca R, Rao G et al. Texture feature ratios from relative CBV maps of perfusion MRI are associated with patient survival in glioblastoma. *AJNR Am J Neuroradiol* 2016; **37**: 37–43.

163. Liu TT, Achrol AS, Mitchell LA, Rodriguez SA, Feroze A, Iv M et al. Magnetic resonance perfusion image features uncover an angiogenic subgroup of glioblastoma patients with poor survival and better response to antiangiogenic treatment. *Neuro Oncol* 2017; **19**: 997–1007.

164. Kickingereder P, Götz M, Muschelli J, Wick A, Neuberger U, Shinohara RT et al. Large-scale radiomic profiling of recurrent glioblastoma identifies an imaging predictor for stratifying anti-angiogenic treatment response. *Clin Cancer Res* 2016; **22**: 5765–5771.

165. Gevaert O, Villalobos V, Sikic BI, Plevritis SK. Identification of ovarian cancer driver genes by using module network integration of multi-omics data. *Interface Focus* 2013; **3**: 20130013.

166. Xi Y-B, Guo F, Xu Z-L, Li C, Wei W, Tian P et al. Radiomics signature: A potential biomarker for the prediction of MGMT promoter methylation in glioblastoma. *J Magn Reson Imaging* 2018; **47**: 1380–1387.

167. Hassan I, Kotrotsou A, Bakhtiari AS, Thomas GA, Weinberg JS, Kumar AJ et al. Radiomic texture analysis mapping predicts areas of true functional MRI activity. *Sci Rep* 2016; **6**: 25295.

168. Kotrotsou A, Zinn PO, Colen RR. Radiomics in brain tumors: An emerging technique for characterization of tumor environment. *Magn Reson Imaging Clin N Am* 2016; **24**: 719–729.

169. Kickingereder P, Burth S, Wick A, Götz M, Eidel O, Schlemmer H-P et al. Radiomic profiling of glioblastoma: Identifying an imaging predictor of patient survival with improved performance over established clinical and radiologic risk models. *Radiology* 2016; **280**: 880–889.

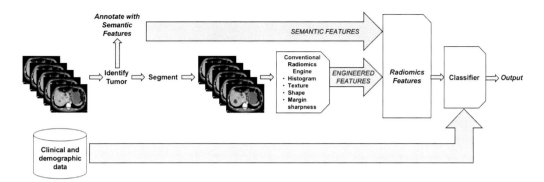

Figure 1.1 Conventional radiomics workflow combining semantic and human-engineered computational features. These radiomics features can then be combined with clinical and demographic data, before the final classification stage, which generates an output such as benign/malignant, responder/non-responder, probability of 5-year survival, etc.

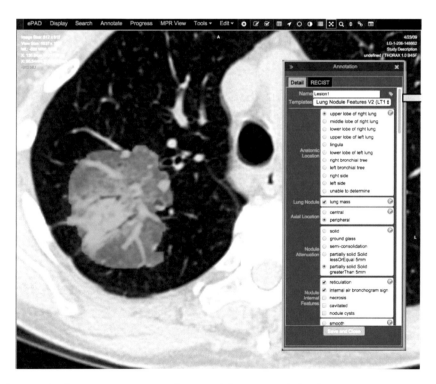

Figure 1.2 Example of semantic annotation of a part solid part ground glass lung tumor using ePAD. Following tumor segmentation (green), either manually within ePAD or created via other means, an observer selects annotations using a custom template (built, e.g., using the AIM Template Builder [96]) or one of several available. The example in this figure shows a subset of the annotation topics that are required to complete this annotation.

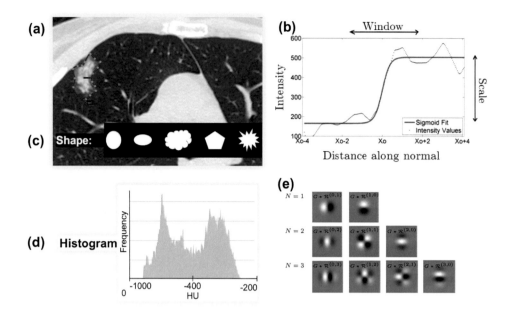

Figure 1.3 Classes of conventional radiomics features. (a) CT cross-section of a chest containing a VOI of a lung tumor (outlined). (b) Edge sharpness features (window and scale) calculated by fitting intensities along normal (red line in [a]) to a sigmoid function. (c) Shape features include measures of sphericity, lobulation, spiculation, roughness, etc. (d) Gray value histogram features include statistics such as mean, maximum, standard deviation, kurtosis, etc. (e) Texture features measure statistics of spatial frequencies represented in the VOI.

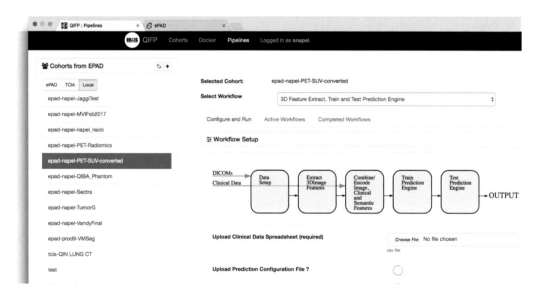

Figure 1.4 Workflow of the Quantitative Image Feature Engine, which allows cohorts of subjects with imaging studies and segmented VOIs to be processed, producing a vector of imaging features for each subject, and subsequently combine these with clinical data to derive associations and to build predictive models.

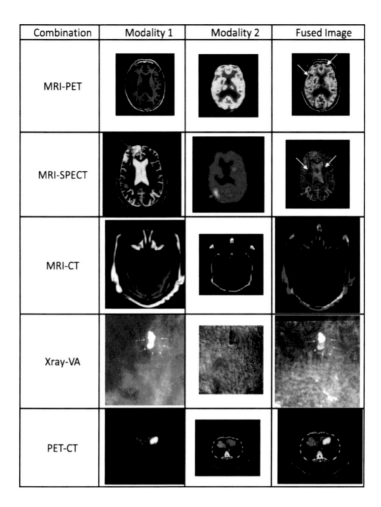

Combination	Modality 1	Modality 2	Fused Image
MRI-PET			
MRI-SPECT			
MRI-CT			
Xray-VA			
PET-CT			

Figure 2.1 Examples 6444 of multimodal medical image fusion. The combination of modality 1 with modality 2 using specific image fusion techniques results in improved feature visibility for medical diagnostics and assessments as shown in the fused image column.

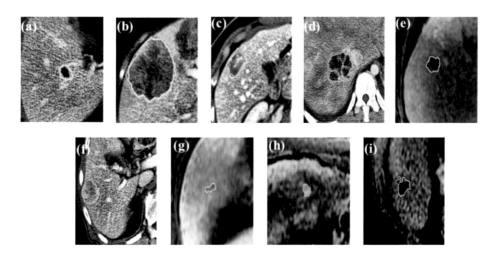

Figure 2.2 Automatic segmentations (yellow) of liver lesions in CT (a–d, f) and MRI (e, g–i) images obtained using the method that is presented in [127] with the piecewise constant model (PC) (a,h) different lesion sizes (Dice of 0.88, 0.91, respectively), (b–d) heterogeneous lesions (Dice of 0.91, 0.92, and 0.89, respectively), (e) homogeneous lesion (Dice of 0.97), (f,g) low contrast lesions (Dice of 0.92, and 0.96, respectively), (h,i) noisy background (Dice of 0.93, 0.86 respectively). Green contours represent the manual annotations.

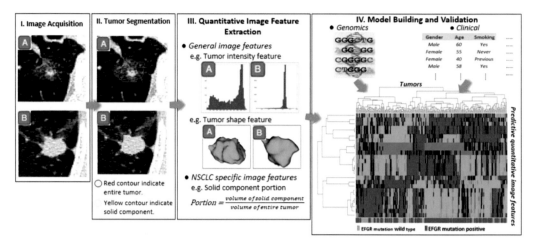

Figure 3.1 A radiomic workflow to predict EGFR mutation status in lung cancer patients.

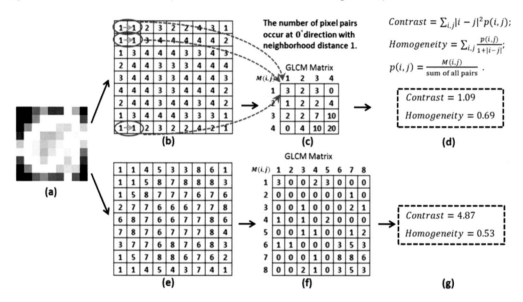

Figure 3.2 Computing the GLCM features of contrast and homogeneity using different bin levels. (a) Original image. (b) Normalized image using the bin level of four. (c) GLCM matrix derived from (b). (d) Contrast and homogeneity computed from GLCM in (c). (e) Normalized image using the bin level of eight. (f) GLCM matrix derived from (e). (g) Contrast and homogeneity computed from GLCM in (f).

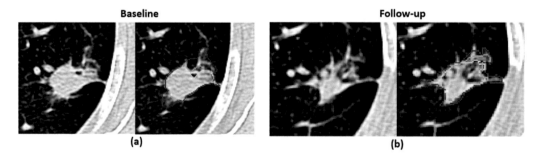

Figure 3.4 An example of one tumor in the same patient, (a) one baseline, and (b) one follow-up, segmented by three radiologists independently. The segmentations of the different radiologists are indicated by contours of different color.

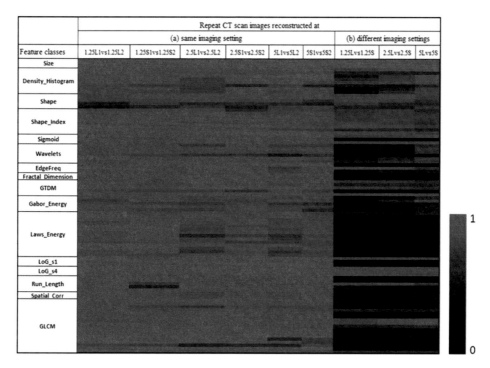

Figure 3.5 CCC heat map of radiomic features. The CCCs (0 to 1) of the studied radiomic features were computed from repeat CT images reconstructed at (a) six identical imaging settings or (b) three different imaging settings. There were 89 quantitative features grouped into 15 feature classes. The brighter the red color, the higher the CCC value (i.e., the more reproducible) of a feature computed for the repeat scans. The label of "1.25L1 versus 1.25L2" means both first and second scans were reconstructed at 1.25 mm slice thickness using the lung algorithm. "2.5L versus 2.5S" means both scans were reconstructed at 2.5 mm slice thickness, but using different algorithms (i.e., lung versus standard algorithms). (From Zhao, B. et al., *Sci. Rep.*, 6, 23428, 2016.)

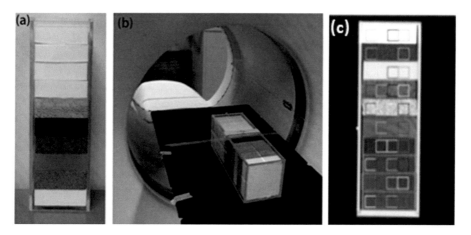

Figure 3.6 Credence cartridge radiomics (CCR) phantom. (a) CCR phantom with 10 cartridges. (b) CCR phantom set up for scanning. (c) Coronal views showing each of the 10 layers of the phantom (regions of interest for analysis shown as colored squares). (From Mackin, D. et al., *Invest. Radiol.*, 50, 757–765, 2015. Acquired reuse permission from Wolters Kluwer Health, Inc.)

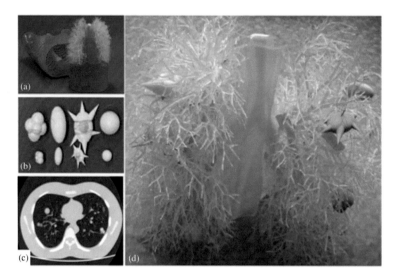

Figure 3.7 The anthropomorphic thorax phantom. (a) The phantom, (b) phantom lesions of different shapes and sizes, (c) an example of a phantom CT image, and (d) an example of phantom lesions attached to vasculature. (Reprinted from *Transl. Oncol.*, 7, Zhao, B. et al., Exploring variability in CT characterization of tumors: A preliminary phantom study, 88–93, Copyright 2014, with permission from Elsevier.)

Settings QIF groups	1.25S vs 2.5S	1.25L vs 2.5L	2.5S vs 5S	2.5L vs 5L	1.25S vs 5S	2.5S vs 5L	1.25S vs 5L	1.25L vs 5L	5L vs 5S	1.25S vs 2.5L	2.5L vs 2.5S	1.25L vs 2.5S	1.25L vs 1.25S	2.5L vs 5S	1.25L vs 5S	Average CCC of QIF groups
1	0.880	0.994	0.971	0.983	0.985	0.848	0.808	0.869	0.827	0.905	0.927	0.883	0.910	0.894	0.881	0.914
2	0.980	0.880	0.985	0.912	0.954	0.970	0.864	0.839	0.966	0.842	0.895	0.884	0.825	0.843	0.764	0.913
3	0.880	0.986	0.949	0.938	0.910	0.909	0.900	0.893	0.882	0.907	0.848	0.902	0.878	0.787	0.758	0.899
4	0.904	0.939	0.842	0.898	0.931	0.967	0.948	0.968	0.910	0.923	0.823	0.844	0.819	0.713		0.895
5	0.934	0.945	0.949	0.909	0.943	0.820	0.825	0.867	0.876	0.817	0.865	0.855	0.903	0.875	0.864	0.884
6	0.938	0.946	0.918	0.896	0.919	0.877	0.894	0.902	0.824	0.892	0.858	0.883	0.853	0.806	0.822	0.882
7	0.978	0.974	0.936	0.946	0.928	0.842	0.851	0.941	0.790	0.837	0.822	0.833	0.825	0.757	0.756	0.868
8	0.887	0.763	0.827	0.758	0.855	0.888	0.888	0.825	0.883	0.809	0.781	0.851	0.911	0.739	0.844	0.834
9	0.898	0.853	0.900	0.892	0.800	0.691	0.656	0.893	0.637	0.707	0.731	0.647	0.694	0.704	0.636	0.763
10	0.804	0.783	0.671	0.729	0.673	0.750	0.723	0.812	0.775	0.730	0.750	0.830	0.771	0.624	0.664	0.739
11	0.915	0.761	0.840	0.706	0.691	0.788	0.639	0.699	0.847	0.665	0.713	0.546	0.618	0.636	0.575	0.709
12	0.913	0.925	0.922	0.944	0.791	0.609	0.635	0.803	0.602	0.636	0.582	0.562	0.512	0.538	0.433	0.694
13	0.635	0.735	0.675	0.755	0.301	0.848	0.438	0.415	0.867	0.748	0.893	0.901	0.576	0.560	0.285	0.642
14	0.913	0.832	0.813	0.741	0.850	0.761	0.754	0.557	0.673	0.488	0.491	0.374	0.371	0.409	0.289	0.621
15	0.906	0.772	0.824	0.658	0.807	0.772	0.692	0.426	0.654	0.385	0.478	0.279	0.324	0.339	0.195	0.567
16	0.941	0.790	0.929	0.781	0.852	0.496	0.553	0.527	0.495	0.426	0.388	0.281	0.259	0.373	0.246	0.556
17	0.857	0.835	0.826	0.766	0.696	0.464	0.577	0.566	0.410	0.428	0.321	0.341	0.264	0.275	0.209	0.522
18	0.965	0.631	0.976	0.660	0.822	0.560	0.578	0.355	0.568	0.278	0.252	0.101	0.073	0.264	0.063	0.483
19	0.892	0.674	0.933	0.709	0.787	0.264	0.372	0.350	0.226	0.215	0.161	0.110	0.084	0.135	0.068	0.399
20	0.637	0.856	0.471	0.761	0.277	0.373	0.598	0.574	0.127	0.405	0.239	0.256	0.159	0.077	0.046	0.390
21	0.777	0.560	0.525	0.460	0.289	0.634	0.672	0.182	0.322	0.354	0.245	0.164	0.116	0.112	0.055	0.364
22	0.611	0.534	0.292	0.339	0.116	0.523	0.466	0.155	0.181	0.369	0.184	0.180	0.088	0.044	0.016	0.273
23	0.801	0.711	0.712	0.563	0.489	0.059	0.097	0.297	0.034	0.039	0.025	0.021	0.014	0.015	0.008	0.259
Average CCC of setting pairs	0.875	0.820	0.815	0.768	0.725	0.684	0.674	0.636	0.627	0.604	0.583	0.543	0.515	0.504	0.441	

Group (a) Fixing reconstruction algorithm while changing slice thickness
Group (b) Fixing slice thickness while changing reconstruction algorithm
Group (c) Smooth reconstruction algorithm (S) plus thin slice thickness versus sharp reconstruction algorithm (L) plus thick slice thickness
Group (d) Sharp reconstruction algorithm (L) plus thin slice thickness versus smooth reconstruction algorithm (S) plus thick slice thickness

1   0.5   0

Figure 3.8 The CCCs of non-redundant quantitative imaging feature (QIF) groups under the 15 inter-setting comparisons. Columns are arranged in descending order according to the average CCC of the inter-setting comparisons. Rows are arranged in descending order according to average CCCs of Non-redundant QIF Groups. (From Lu, L. et al. *PLoS One*, 11, e0166550, 2016.)

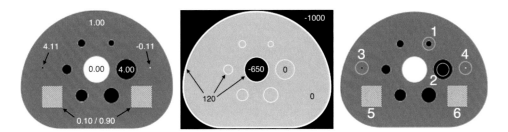

Figure 4.1 A digital reference object (DRO) for quantitative PET/CT data analysis and reporting. Known PET SUV, CT HU, and ROI allow for quality assurance testing of commercial image analysis software and improved reproducibility in multi-center imaging trials. (Reproduced from Pierce, L.A. et al., *Radiology*, 277, 538–545, 2015. With permission.)

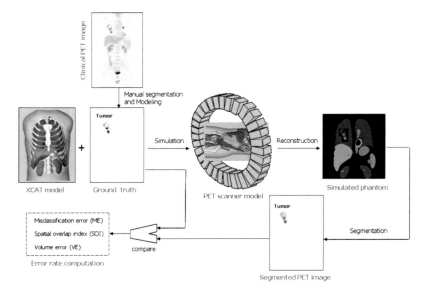

Figure 4.3 Quantitative PET/CT segmentation QA workflow. Realistic patient tumors are converted to ground truth contours and activity distributions that can be used to simulate PET images and evaluate the performance of different automatic PET segmentation algorithms. (Reproduced from Zaidi, H. and El Naqa, I., *Eur. J. Nucl. Med. Mol. Imag.*, 37, 2165–2187, 2010. With permission.)

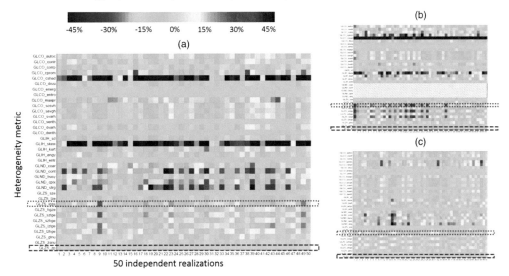

Figure 4.4 Radiomics array of quantitative PET metric variability due to stochastic image noise. (a) A subset of co-occurrence matrix, (b) intensity histogram features, and (c) display variability about the mean in excess of 45%. (Reproduced from Nyflot, M.J. et al., *J. Med. Imaging (Bellingham)*, 2, 041002, 2015. With permission.)

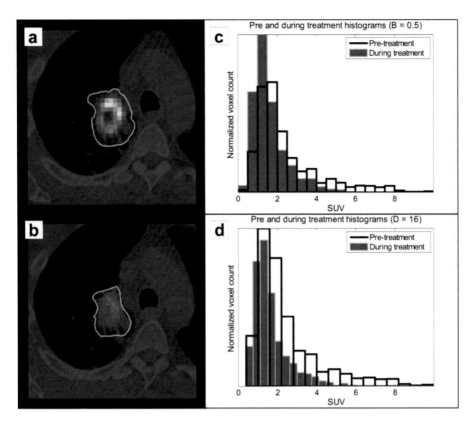

Figure 4.5 Effect of SUV discretization on response assessment. Fixed SUV bin size (0.5 SUV) quantized image (a) and histogram (c) is compared against fixed SUV bin number (16 bins) quantized image (b) and histogram (d). (Reproduced from Leijenaar, R.T. et al., *Sci. Rep.* 5, 11075, 2015. With permission.)

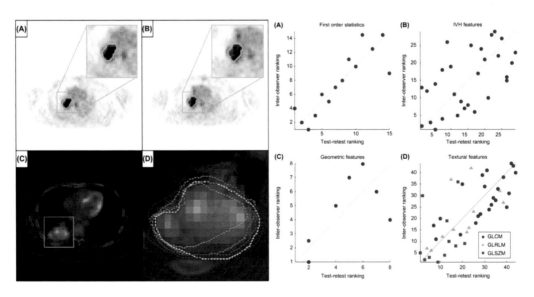

Figure 4.6 Test-retest stability and inter-observer variability of PET radiomics. PET SUV summary statistics and zone-size matrix (GLSZM) features (a and c) achieved high ranks in both test-retest and inter-observer variability, while intensity histogram (IH) and co-occurrence matrix (GLCM) features (b and d) had unfavorable trade-offs in variability. (Reproduced from Leijenaar, R.T. et al., *Acta Oncol.*, 52, 1391–1397, 2013. With permission.)

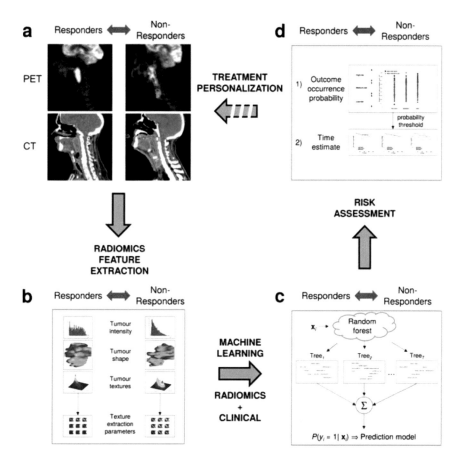

Figure 4.7 Workflow for combining PET/CT (a) radiomics feature extraction (b) and machine learning analytics (c) for predicting risk and clinical outcome (d) in head-and-neck cancer patients. (Reproduced from Vallieres, M. et al., *Sci. Rep.*, 7, 10117, 2017. With permission.)

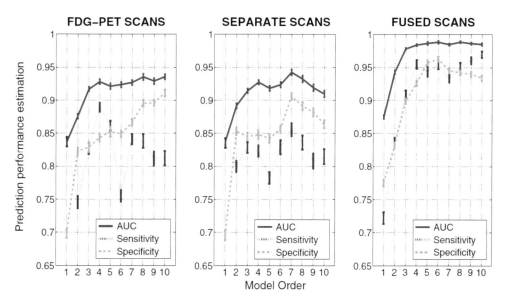

Figure 4.8 Joint FDG-PET/MR radiomics model for predicting lung metastatic incidence from primary sarcoma. Prediction performance improves when operating jointly on fused PET/MR images (right) relative to individual PET images (left) or separate PET and MR images (middle). (Reproduced from Vallieres, M. et al., *Phys. Med. Biol.*, 60, 5471–5496, 2015. With permission.)

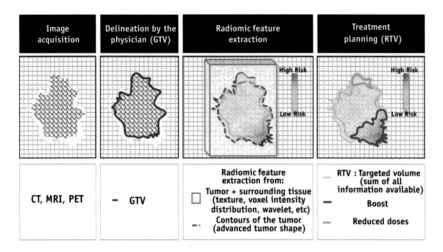

Image acquisition	Delineation by the physician (GTV)	Radiomic feature extraction	Treatment planning (RTV)
CT, MRI, PET	— GTV	☐ Radiomic feature extraction from: Tumor + surrounding tissue (texture, voxel intensity distribution, wavelet, etc) -· Contours of the tumor (advanced tumor shape)	— RTV : Targeted volume (sum of all information available) — Boost — Reduced doses

Figure 4.9 Concept of the radiomic target volume (RTV) based on feature extraction and parametric maps derived from CT, MRI, and PET. Spatially distinct risk regions with the RTV can guide dose escalation or dose de-escalation strategies. (Reproduced from Sun, R. et al., *Int. J. Radiat. Oncol. Biol. Phys.*, 95, 1544–1545, 2016. With permission.)

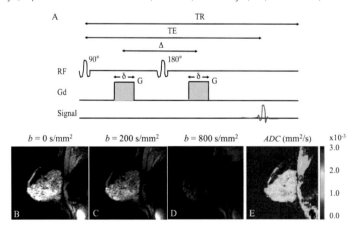

Figure 4.10 Deep learning algorithm for PET/CT feature extraction combined with conventional support vector machine for pulmonary lesion detection and classification. The convolution neural network (CNN) is made more transparent by replacing the prediction model with a simple SVM classifier, allowing greater user control and interpretability. (Reproduced from Teramoto, A. et al., *Med. Phys.*, 43, 2821–2827, 2016. With permission.)

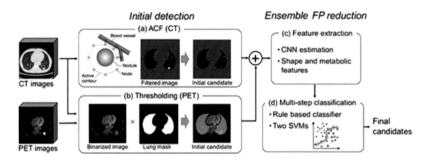

Figure 5.1 An example of DWI acquisition and analysis. (A) Representative schematic of a typical diffusion-weighted spin-echo sequence where diffusion sensitizing gradients with amplitude G and duration δ separated by the diffusion time Δ, are applied symmetrically around the 180° refocusing pulse. Panel B shows an image of the normal breast without diffusion weighting (b-value = 0), whereas panels C and D show diffusion-weighted images with b-values equal to 200 s/mm² and 800 s/mm², respectively. Image voxels from panels B, C, and D are fit exponentially to the diffusion-weighted spin-echo signal equation (Eq. [5.2]) to generate an ADC parametric map (E).

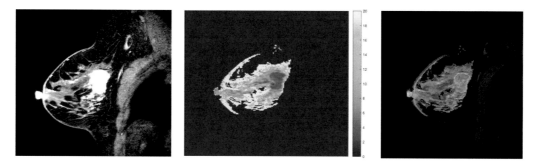

Figure 5.3 Quantitative magnetization transfer image of an estrogen and progesterone receptor positive human breast tumor (3.7 cm diameter) reveals a low pool size ratio in the tumor compared with the surrounding fibroglandular tissue. Shown are the contrast-enhanced anatomical image (left), pool size ratio map (middle), and pool size ratio map overlaid with the anatomical image (right).

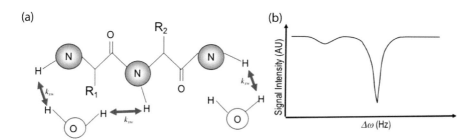

Figure 5.4 Panel (a) displays an example macromolecular nitrogen (N) backbone with exchangeable hydrogens (H) demonstrating direct chemical exchange with the water protons as denoted by the red arrows. Panel (b) presents an example z-spectrum with the signal intensity shown as a function of saturation offset frequency, $\Delta\omega$, demonstrating a slight signal decrease with the CEST effect and nearly complete saturation around the water frequency.

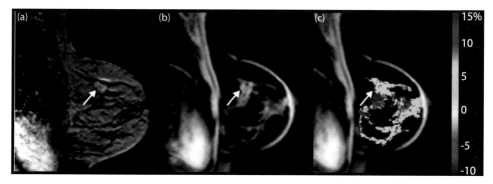

Figure 5.5 CEST imaging of breast cancer patient at 3 Tesla. (a) Contrast-enhanced image with arrow indicating lesion location, (b) S_0 image where $\Delta\omega - 3.5$ ppm, (c) APTasym map demonstrating a higher APT within the lesion relative to surrounding fibroglandular tissue indicating increased concentration of proteins/peptides or tissue pH. The heterogeneity seen in Panel C is due to the low SNR that is inherent to the CEST technique.

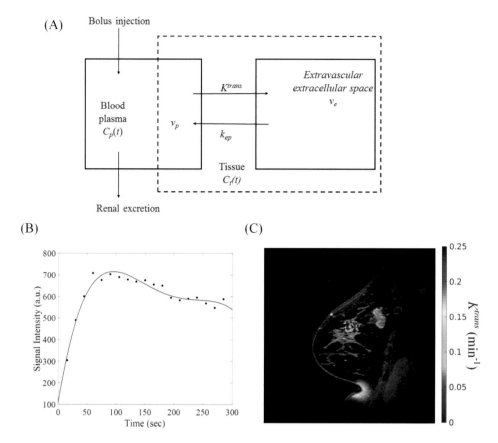

Figure 5.6 DCE-MRI can be quantified using the two-compartment Tofts-Kety model (panel A) evaluating the measured voxel-based time-intensity curve resulting from sequential T_1-weighted images acquired before, during, and after contrast injection (panel B). Panel C displays an example of a parametric map of the DCE-MRI pharmacokinetic parameter K^{trans} is shown for a malignant breast lesion overlaid on a cross-sectional high-resolution MRI.

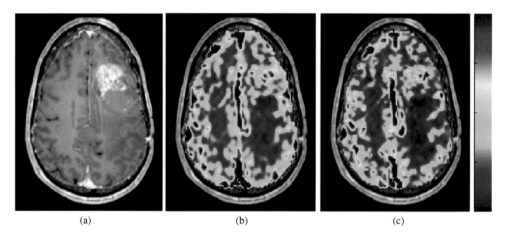

Figure 5.7 An in vivo example of a DSC-MRI experiment in a patient with a glioblastoma multiforme (GBM). (a) Post-contrast T_1-weighted image exhibits the characteristic enhancement accompanying a disrupted blood brain barrier. DSC-MRI derived cerebral blood volume (CBV) and cerebral blood flow (CBF) maps (b and c, respectively) reveal high vascular density and perfusion within the enhancing tumor, with values substantially higher than those observed in gray or white matter. Also note that normal appearing gray matter CBV and CBF values are roughly 1.5–2.5 times higher than those found in white matter.

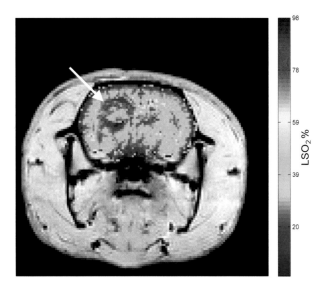

Figure 5.8 An example LSO$_2$ map obtained with the multiparametric qBOLD protocol in a C6 rat glioma model. The LSO$_2$ values in tumor (indicated by the arrow) are much lower than those found in the surrounding normal-appearing tissue.

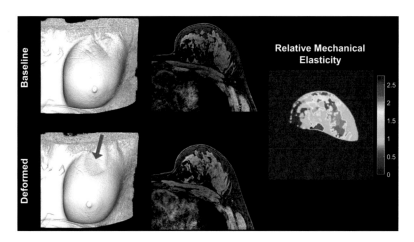

Figure 5.9 An example of quasi-static MRE in the breast. Conventional anatomical MR image volumes are acquired in both a baseline and a deformed state and biomechanical model-based reconstruction techniques are used to generate a map of mechanical elasticity within the tissue. Note that the red arrow denotes the location of the external deformation source.

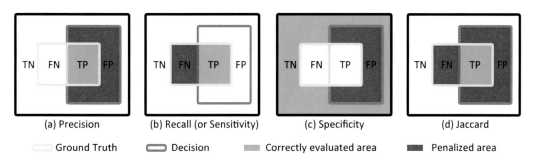

Figure 6.1 Visual representation of the performance evaluation metrics of (a) Precision, (b) Recall, (c) Specificity, and (d) Jaccard.

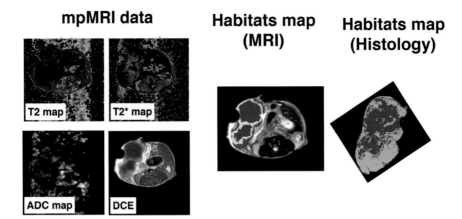

Figure 7.1 Multiparametric MRI data were clustered to create a habitats map, which is corroborated by histology in a pre-clinical model of breast cancer ($4T_1$). Histological images (H&E and immunohistochemical sections stained with a hypoxia marker pimonidazole) were segmented to generate a habitat map at histological level. Blue represents necrotic areas and green represents viable tumor cells; yellow represents a perinecrotic region stained for hypoxia, while pink represents regions of hypoxic viable tumor cells.

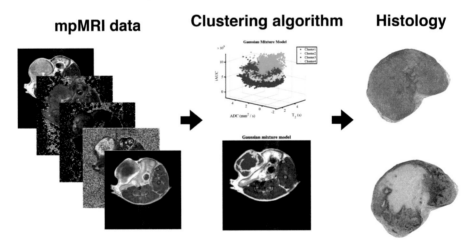

Figure 7.2 Multiparametric MRI was used as raw data of different clustering algorithms to create habitats maps, which each cluster is represented by different colors. Evaluation of the corresponding histological slices is crucial to confirm the cellular phenotype of the habitats and determinate the best image segmentation algorithm. Example of a murine breast cancer ($4T_1$) allograft in a mouse model. In the Gaussian mixture model, cluster green represents regions of viable tumor cells; clusters yellow and magenta represent regions with moderate enhancement in DCE-MRI and are confirmed as hypoxic in hypoxia (positive for pimonidazole); cluster blue represents region with non-viable tumor cells.

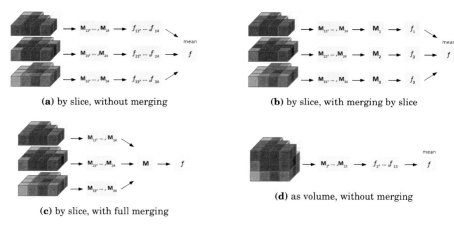

(a) by slice, without merging

(b) by slice, with merging by slice

(c) by slice, with full merging

(d) as volume, without merging

(e) as volume, with merging

Figure 8.2 Methods to calculate GLCM-based features. $M_{\Delta k}$ are matrices calculated for direction Δ in slice k and $f_{\Delta k}$ is the corresponding features. (From Zwanenburg, A. et al., Image biomarker standardisation initiative. ArXiv e-prints, 1612. Available: http://adsabs.harvard.edu/abs/2016arXiv161207003Z, 2016.)

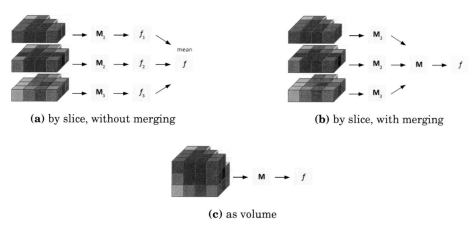

(a) by slice, without merging

(b) by slice, with merging

(c) as volume

Figure 8.4 Methods to calculate GLSZM-based features. M_k is the texture matrices calculated for slice k, and f_k are the corresponding features. (From Zwanenburg, A. et al., Image biomarker standardisation initiative. ArXiv e-prints, 1612. Available: http://adsabs.harvard.edu/abs/2016arXiv161207003Z, 2016.)

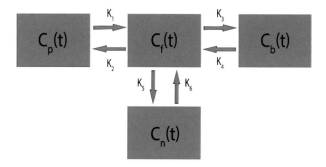

Figure 8.6 General four compartment model with arterial blood, free ligand in tissue, specific and nonspecific binding and six rates ($K_1 \sim K_6$). (From Watabe, H. et al., *Ann. Nucl. Med.*, 20, 583, 2006.)

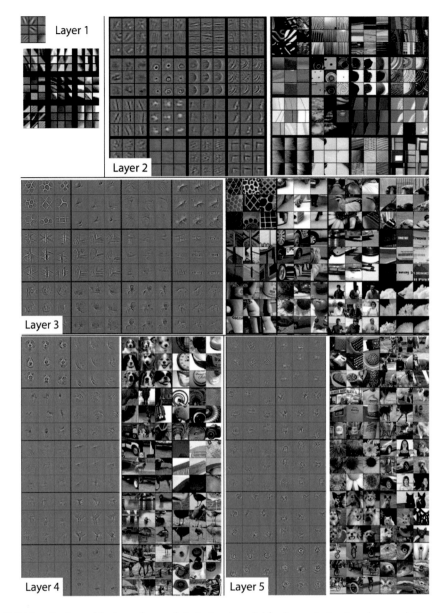

Figure 8.7 Visualization of features from a fully trained model for layers 1–5 using deconvolutional network approach. (From Zeiler and Fergus, 2014a.)

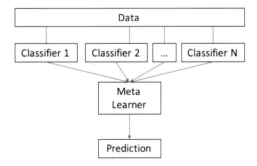

Figure 9.2 A simple ensemble architecture consisting of multiple classifiers that each classify the examples. The output of those weak classifiers is then used by the meta-learner to produce the ensemble prediction, which is typically superior to any single weak classifier.

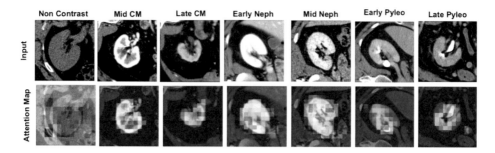

Figure 9.6 Visualizations of attention maps generated to visualize the last layer of a CNN trained to identify renal scan phase from CT imaging. Attention maps illustrate that for the relative influences of regions of the image on the images classification (purple-blue = low; green = medium-low; yellow = medium; orange = medium-high; and red = high).

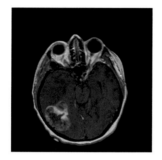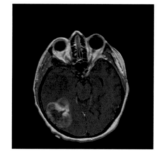

Figure 10.1 Visualization of MR post contrast image of a patient from the TCGA cohort. Left, raw MR image. Right, molecular heterogeneity of a gene expression module visualized on top of the raw MR image using a radiogenomics map.

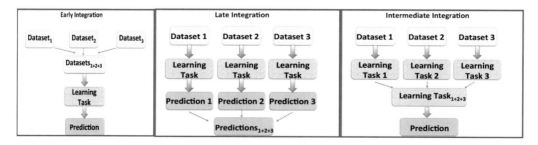

Figure 10.2 Data fusion for biomedical decision support. Left panel: early integration, middle panel: late integration, and right panel, intermediate integration.

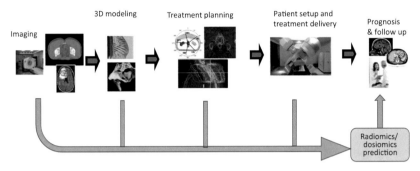

Figure 12.1 The role of imaging in the workflow of radiation therapy and radiomics- and/or dosiomics-based prediction of outcome.

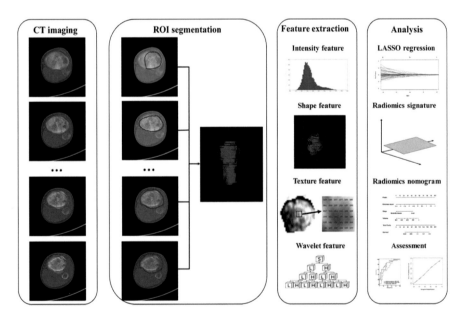

Figure 12.2 The image quantitation method by using radiomics in the osteosarcoma study. The features include the features of the intensity, the shape, the texture, and the wavelet.

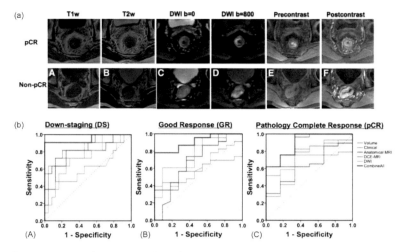

Figure 12.3 The MR images of patients and results in the study. (a) multi-parametric MR images of two male patients, both at 60 years old with mid-rectum cancer; (b) the model performance for down stage, good response and pathology complete response.

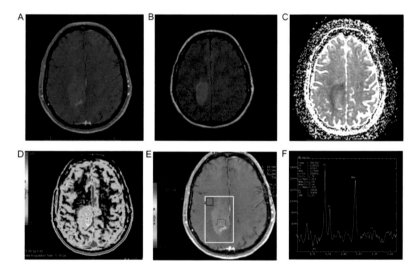

Figure 13.1 A sampling of various MRI modalities used during standard care for glioblastoma treatment. Images come from a 49-year-old male recently diagnosed with GBM. Conventional imaging methods gadolinium-enhanced T_1-weighted (a) and T_2-weighted FLAIR (b) as well as more advanced imaging methods diffusion-weighted imaging (c), perfusion-weighted imaging (d), and magnetic resonance spectroscopy (e,f) are illustrated here.

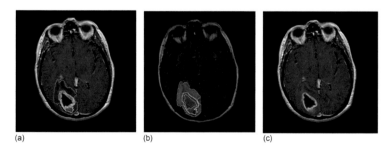

Figure 13.2 Overview of image segmentation methodology based on gadolinium-enhanced T_1-weighted (a) and fluid-attenuated inversion recovery (FLAIR) imaging (b) from a 59-year-old male with enhancing right parieto-occipital lobe GBM. Segmentation of areas of hypointense necrotic core (red) and marginal enhancement (yellow) are visualized in panel A and segmentation of edema/tumor invasion (blue) are visualized in panel B. Panel C illustrates co-registration of panels A and B. Segmentation was performed using 3D slicer 4.4.1.

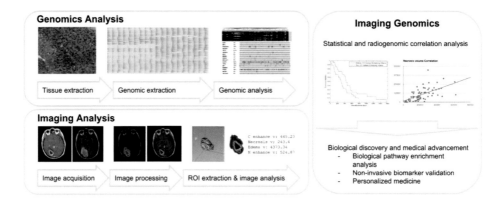

Figure 13.3 General description of imaging genomics. In early stages, genetic data and corresponding imaging data are collected, pre-processed, and analyzed. In later stages, genomic and imaging data are analyzed together to discover correlations that provide insight to the underlying biological mechanisms involved in neoplasm development.

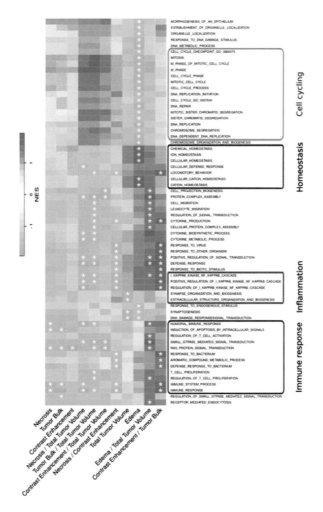

Figure 13.4 Visualization of normal enrichment scores (NES) calculated by GSEA to highlight the association between volumetric features and expert-curated gene ontology gene sets. A total of 64 biological processes were found to be correlated with one or more volumetric features or their ratios. For example, necrosis and tumor bulk were markedly enriched with apoptosis and immune response pathways, and contrast enhancement was negatively correlated with protein folding and signal transduction pathways. (Adapted from Grossmann, P. et al., *BMC Cancer*, 16, 611, 2016.)

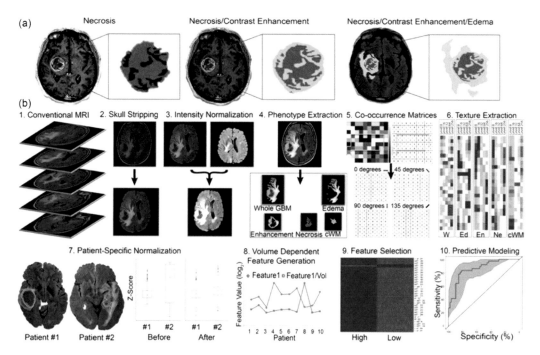

Figure 13.5 Radiomic sequencing pipeline for solid tumor (glioblastoma) and tissue characterization described by Zinn et al.[134] "(a) Semi-automated segmentation of the three imaging phenotypes: necrosis (left), post-contrast active enhancing tumor (middle), and FLAIR peritumoral edema/invasion, representing edematous tumor as well as sites of cellular invasion into brain tissue (right). (b) Automated segmentation-based radiomic feature extraction is followed by patient specific normalization and feature selection, which consists of the normalization of contralateral normal-appearing white matter as an internal control normalizer to account for various potential intra- and inter-institutional biases, a crucial step that ensures comparability. The next step is volume-dependent feature generation, which uses the necrosis, post-contrast enhancement, and FLAIR volumes for the corresponding radiomic feature sets acquired from the respective sequences, thus doubling the amount of radiomic features and creating a set of tumor volume-independent and -dependent radiomic features. Finally, this homogenous dataset can then enter feature selection and predictive modeling for any GBM trait of interest."

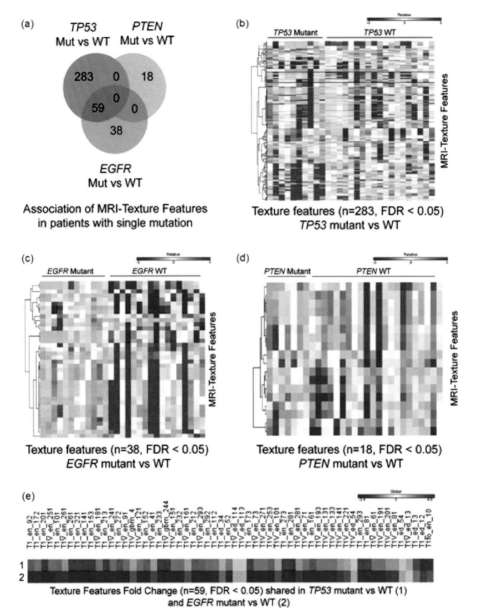

Figure 13.6 MRI texture feature signatures are uniquely associated with patients with *TP53, PTEN,* and *EGFR* gene mutations. (a) Venn diagram depicting numbers of MRI texture features, their unique and overlapping association with GBM patients with specific gene mutation. (b–d) Heatmap generated using uniquely associated and significant texture features (false discovery rate [FDR] < 0.05) in *TP53* mutant versus WT (*n* = 283); (b) *EGFR* mutant versus wild type (WT) (*n* = 38); (c) and *PTEN* mutant versus WT (*n* = 18) (d). Color scale bar above heatmap shows range of texture feature values across patients. (e) Heatmap generated using fold change values of shared and significant texture features (*n* = 59; FDR < 0.05) in *TP53* mutant versus WT and *EGFR* mutant versus WT. Color scale above heatmap show range of fold change values for specific texture feature." (Adapted from Zinn, P.O. et al., *Neurosurgery,* 64, 203–210, 2017.)

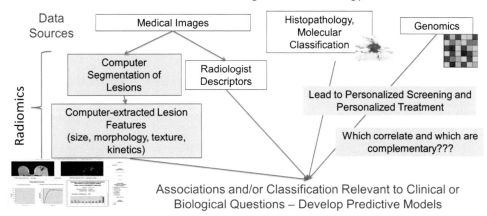

Figure 14.1 A logic diagram of an example of the field of radiogenomics for breast cancer using digital mammography and DCE MRI. (Reprinted from Giger, M., *Decoding Breast Cancer with Quantitative Radiomics & Radiogenomics: Imaging Phenotypes in Breast Cancer Risk Assessment, Diagnosis, Prognosis, and Response to Therapy*, TCGA 4th Scientific Symposium, NIH, Bethesda, MD, 2015; Colen, R. et al., *Transl. Oncol.*, 7, 556–569, 2014. With permission.)

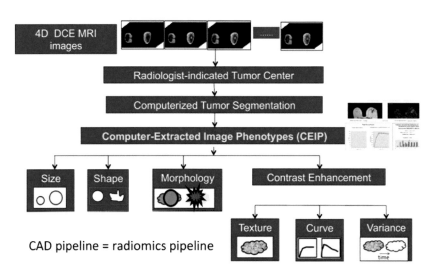

Figure 14.2 Schematic flowchart of a computerized tumor phenotyping system for breast cancers on DCE-MRI. CAD radiomics pipeline includes computer segmentation of the tumor from the local parenchyma and computer-extraction of "hand-crafted" radiomic features covering six phenotypic categories: (1) size (measuring tumor dimensions), (2) shape (quantifying the 3-D geometry), (3) morphology (characterizing tumor margin), (4) enhancement texture (describing the heterogeneity within the texture of the contrast uptake in the tumor on the first postcontrast MRIs), (5) kinetic curve assessment (describing the shape of the kinetic curve and assessing the physiologic process of the uptake and washout of the contrast agent in the tumor during the dynamic imaging series, and (6) enhancement-variance kinetics (characterizing the time course of the spatial variance of the enhancement within the tumor) CAD = computer-aided diagnosis; DCE-MRI = dynamic contrast-enhanced MRI. (Reprinted from Giger, M., *J. Am. Coll. Radiol.*, 15, 512–520, 2018. With permission.)

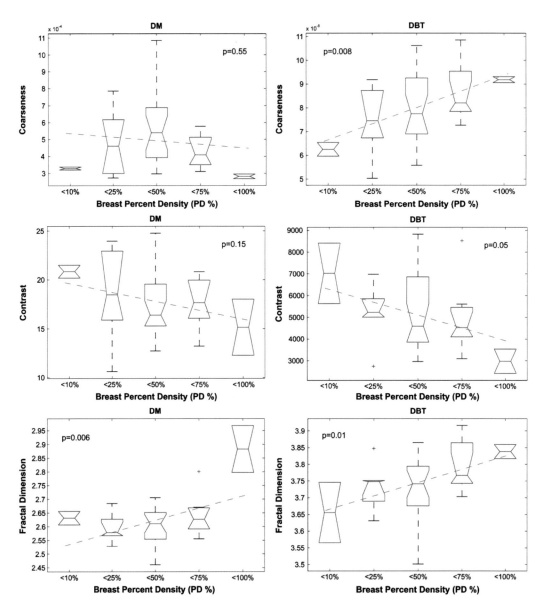

Figure 14.3 Box-plots with fitted regression lines and associated *p*-values for digital mammography (DM) and digital breast tomosynthesis (DBT) coarseness, contrast, and fractal dimension texture features versus the five groups of increasing breast percent density (PD): <10%, 10%≤…<25%, 25%≤…<50%, 50%≤…<75%, and 75%≤…<100%. (Reprinted from Kontos, D. et al., *Acad. Radiol.*, 16, 283–298, 2009. With permission.)

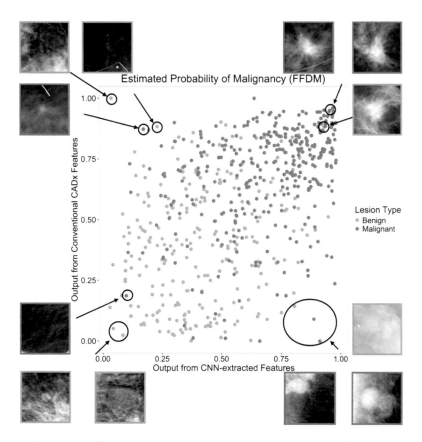

Figure 14.6 A diagonal classifier agreement plot between the CNN-based classifier and the conventional CADx classifier for FFDM. The x-axis denotes the output from the CNN-based classifier, and the y-axis denotes the output from the conventional CADx classifier. Each point represents an ROI for which predictions were made. Points near or along the diagonal from bottom left to top right indicate high classifier agreement; points far from the diagonal indicate low agreement. ROI pictures of extreme examples of agreement/disagreement are included. (Reprinted from Antropova, N. et al., *Med. Phys.*, 44, 5162–5171, 2017. With permission.)

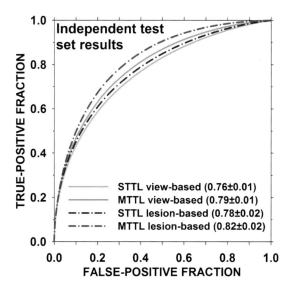

Figure 14.7 ROC curves for the independent test set using the two selected C_1 transfer networks: STTL: single-task transfer learning, MTTL: multi-task transfer learning. The difference in the AUCs was statistically significant between the two lesion-based curves (*p*-value = 0.007) and between the two view-based curves (*p*-value = 0.008). (Reprinted from Samala, R. et al., *Phys. Med. Biol.*, 62, 8894–8908, 2017. With permission.)

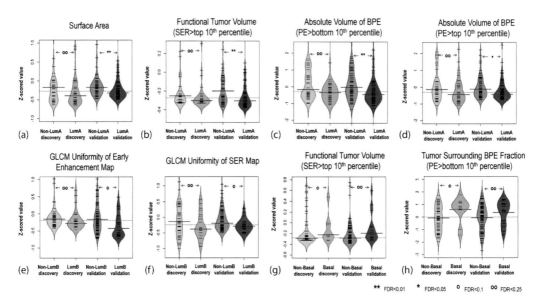

Figure 14.10 Imaging features significantly associated with molecular subtypes (after correction for multiple testing) in both discovery and validation cohorts, (a–d) four features for distinguishing luminal A versus nonluminal A; (e, f) two features for distinguishing luminal B versus nonluminal B; and (g, h) two features for distinguishing basal-like versus nonbasal-like. Wilcoxon rank sum test was implemented to investigate pairwise difference. Also, the FDR adjusted for multiple testing is reported. (Reprinted from Wu, J. et al., *J. Magn. Reson. Imaging*, 46, 1017–1027, 2017. With permission.)

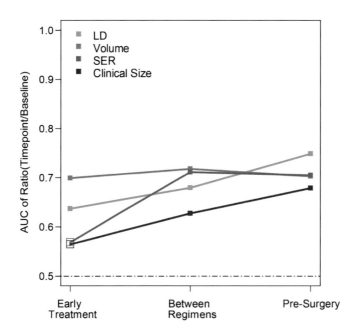

Figure 14.11 Graph shows AUCs for prediction of pCR for the four predictor variables at the early treatment, between regimens, and pre-surgery time points. Predictors are expressed as the ratio of value at each time point to baseline value for tumor longest diameter (LD) (green), volume (orange), signal enhancement ratio (SER) (blue), and clinical size (red). Solid squares = $p \leq 0.05$. (Reprinted from Hylton, N.M. et al., *Radiology*, 263, 663–672, 2012. With permission.)

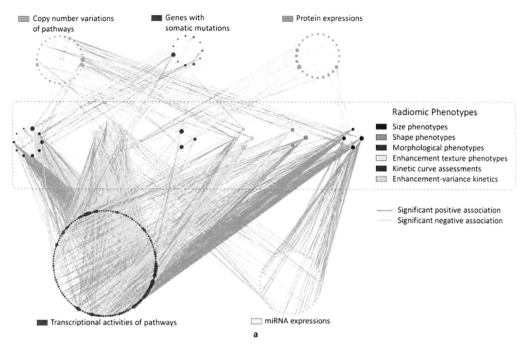

	Size phenotypes	Shape phenotypes	Morphological phenotypes	Enhancement texture phenotypes	Kinetic curve assessments	Enhancement-variance kinetics
Transcriptional activities of pathways	173	109	49	374	268	130
Copy number variations of pathways	24	7	7	14	15	21
Mutated genes	3	1	1	15	22	3
miRNA expressions	73	0	0	58	1	0
Protein expressions	10	0	9	17	0	0

b

Figure 14.12 Overview of all identified statistically significant associations. (a) In the figure, each node is a genomic feature or a radiomic phenotype. Each line is an identified statistically significant association. Genomic features without statistically significant association are not shown. Genomic features are organized into circles by data platform and indicated by different node colors. Radiomic phenotypes are divided into six categories also indicated by different node colors. The node size is proportional to its connectivity relatively to other nodes in the category. Associations are deemed as statistically significant if the adjusted p-values ≤ 0.05. The only exception is for the associations involving somatically mutated genes, for which the statistical significance criteria are: (1) p-value ≤ 0.05 and (2) the gene mutated in at least five patients. (b) A table showing the numbers of statistically significant associations between genomic features of different platforms and radiomic phenotypes of different categories. *(Continued)*

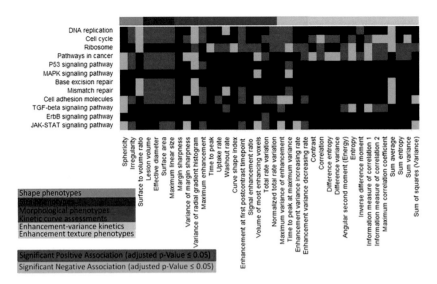

Figure 14.12 (Continued) Overview of all identified statistically significant associations. (c) Heatmap representation of statistically significant associations between radiomic phenotypes and transcriptional activities of some cancer-related genetic pathways. In the heatmap, genetic pathways are rows and radiomic phenotypes are columns. (Reprinted from Zhu, Y. et al., *Sci. Rep.*, 5, 17787, 2015. With permission.)

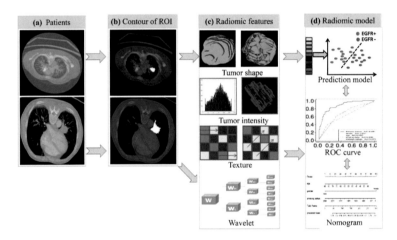

Figure 15.1 Radiomics analysis pipeline on medical images [2]. (a) CT images of lung cancer. (b) Tumor detection and segmentation in CT images. (c) Feature extraction and quantification. Features are extracted from the segmented tumor ROI to quantify the tumor intensity, shape, texture, and wavelet information. (d) Diagnostic and prognostic model building. The radiomic features and clinical outcomes are associated by machine learning algorithms. (From Zhang, L.W. et al., *Transl. Oncol.*, 11, 94–101, 2017.)

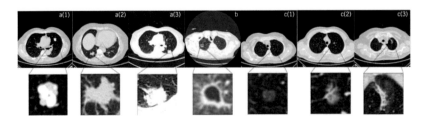

Figure 15.2 Different types of lung nodules. a(1): isolate and solid lung nodule, a(2): juxtavascular lung nodule, a(3): juxta-pleural lung nodule, (b): cavity lung nodule, c(1): isolate GGO, c(2): juxtavascular GGO, c(3): juxta-pleural GGO. (From Song, J. et al., *IEEE Trans. Med. Imaging*, 35, 337–353, 2016.)

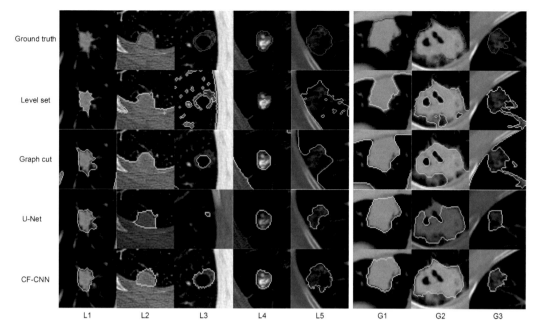

Figure 15.4 Segmentation results of various types of lung nodule. From top to bottom: ground truth, level set result, graph cut result, U-Net result, and CF-CNN result. L means images from the LIDC dataset. G means images from Guangdong General Hospital. (From Wang, S. et al., *Med. Image Anal.*, 40, 172–183, 2017.)

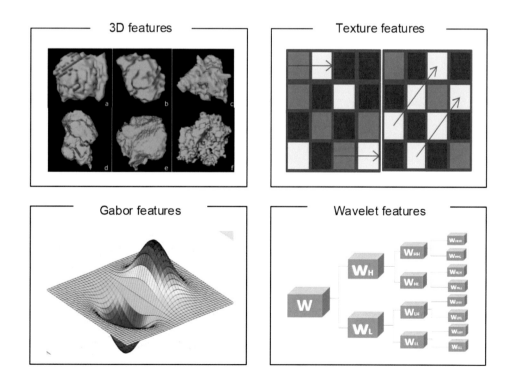

Figure 15.5 Groups of radiomic features. Normally, the radiomic features include shape features, Gabor features, texture features, and wavelet features. (From Song, J. et al., *IEEE Trans Med Imaging*, 35, 337–353, 2016; Zhang, L.W. et al., *Transl. Oncol.*,11, 94–101, 2017.)

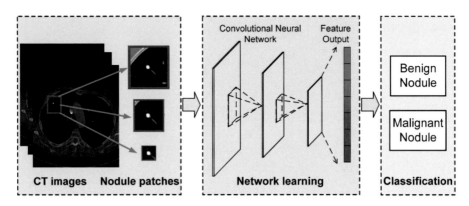

Figure 15.6 Multi-scale CNN for lung nodule classification. A rectangle bounding box covering the whole lung nodule is manually selected as the input to the CNN. Afterwards, the CNN learned discriminative features from nodule images automatically. Finally, a classifier is used to classify the input nodule as malignant or benign [27]. (From Shen, W. et al., Multi-scale convolutional neural networks for lung nodule classification, in *IPMI 2015: Information Processing in Medical Imaging*, pp. 588–599, 2015.)

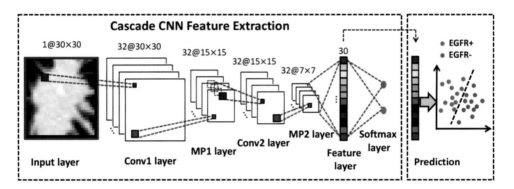

Figure 15.7 Structure of the proposed deep learning model for EGFR-mutation prediction. (From Yu, D. et al., Convolutional neural networks for predicting molecular profiles of non-small cell lung cancer biomedical imaging, in *2017 IEEE 14th International Symposium on Biomedical Imaging (ISBI 2017)*, Melbourne, Australia, pp. 569–572, 2017.)

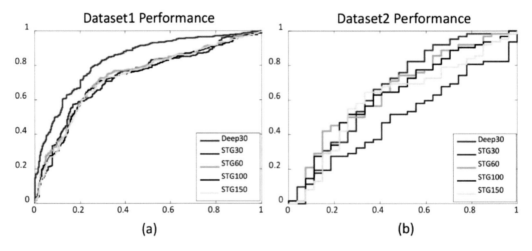

Figure 15.8 ROC curves of EGFR mutation prediction based on feature sets with different number of features in Dataset1 (a) and Dataset2 (b). (From Yu, D. et al., Convolutional neural networks for predicting molecular profiles of non-small cell lung cancer biomedical imaging, in *2017 IEEE 14th International Symposium on Biomedical Imaging, (ISBI 2017)*, Melbourne, Australia, pp. 569–572, 2017.)

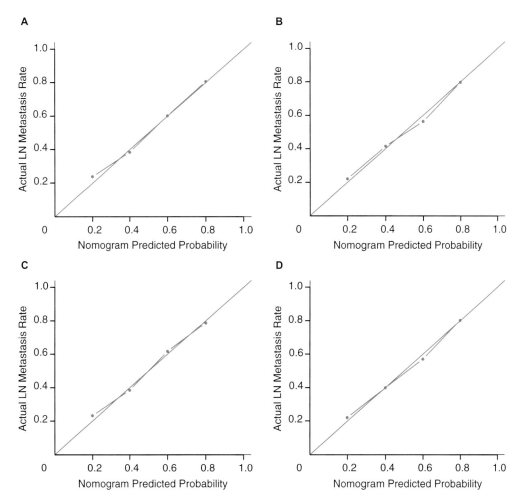

Figure 17.2 Calibration curves of the radiomics nomogram and the model with the addition of histologic grade prediction in each cohort. (A) Calibration curve of the radiomics nomogram in the primary cohort. (B) Calibration curve of the radiomics nomogram in the validation cohort. (C) Calibration curve of the model with addition of histologic grade in the primary cohort. (D) Calibration curve of the model with addition of histologic grade in the validation cohort. Calibration curves depict the calibration of each model in terms of the agreement between the predicted risks of LN metastasis and observed outcomes of LN metastasis. The y-axis represents the actual LN metastasis rate. The x-axis represents the predicted LN metastasis risk. The diagonal dotted line represents a perfect prediction by an ideal model. The pink solid line represents the performance of the nomogram, of which a closer fit to the diagonal dotted line represents a better prediction. The calibration curve was drawn by plotting $\hat{P}$ on the x-axis and $P_C = [1 + exp -(y_0 + y_1 L)]^{-1}$ on the y-axis, where P_C is the actual probability, $L = $ logit $(\hat{P})$, $\hat{P}$ is the predicted probability, y_0 is the corrected intercept, and y_1 is the slope estimates.

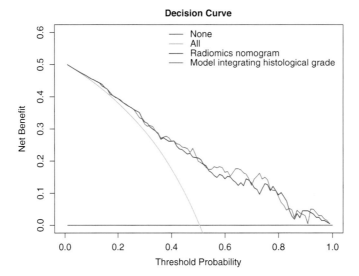

Figure 17.3 Decision curve analysis for the radiomics nomogram and the model with addition of histological grade. The Y-axis measures the net benefit. The pink line: radiomics nomogram. The blue line: the model with addition of histological grade. The green line: assume all patients have LN metastases. Thin black line: assume no patients have LN metastases. The net benefit was calculated by subtracting the proportion of all patients who are false-positive from the proportion who are true-positive, weighting by the relative harm of forgoing treatment compared with the negative consequences of an unnecessary treatment. In here, the relative harm was calculated by $\left(\frac{P_t}{1-P_t}\right)$. "$p_t$" (threshold probability) is where the expected benefit of treatment is equal to the expected benefit of avoiding treatment, at which a patient will opt for treatment is informative of how a patient weighs the relative harms of false-positive and false-negative results $((a-c)/(b-d) = (1-p_t)/p_t)$. a–c is the harm from a false-negative result; b–d is the harm from a false-positive result. a, b, c, and d give, respectively, the value of true positive, false positive, false negative, and true negative. The decision curve showed that if patient's or doctor's threshold probability is higher than 10%, then using the radiomics nomogram in the current study to predict LN metastases adds more benefit than treat all patients, or treat none scheme. For example, if a patient's personal threshold probability is 60% (i.e., the patient would opt for treatment if his probability of cancer was greater than 60%), then the net benefit is 0.145 when using the radiomics nomogram to make the decision to undergo treatment, with added benefit than treat all scheme, or treat none scheme. Besides, the net benefit was comparable (with several overlaps) basing on the radiomics nomogram and the model integrating histological grade.

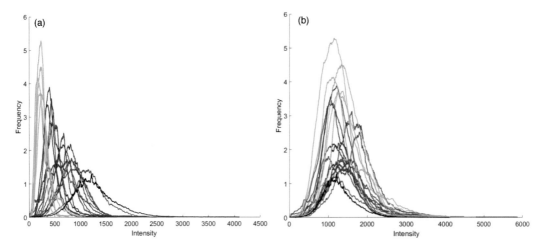

Figure 18.2 T2w signal intensity distributions (a) before standardization, and (b) after standardization for five representative MRI datasets from each site. Different colors correspond to each of the four different sites, and black corresponds to the template to which all the datasets were standardized. Note the improved alignment between intensity distributions in (b), implying tissue specific-meaning across sites, scanners, and patients as a result of standardization.

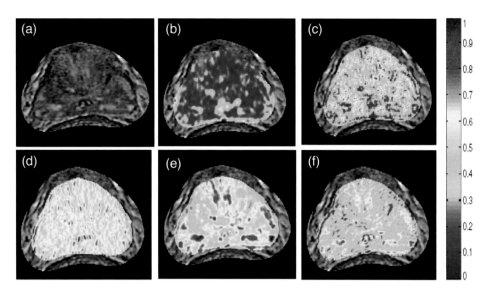

Figure 18.3 Representative radiomic feature heatmaps corresponding to (a) T2w MR image: (b) 1st-order statistics, (c) 2nd-order statistics, (d) gradient, (e) Laws, and (f) Gabor.

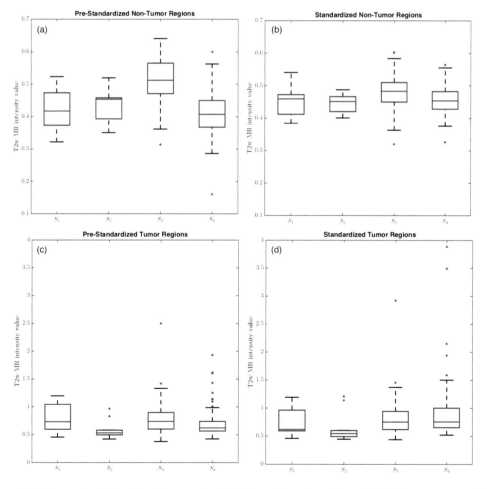

Figure 18.4 Box-and-whisker plots for normalized T2w signal intensity values within (a,b) non-tumor regions, and (c,d) tumor regions. Each box-and-whisker plot corresponds to data from a single site. T2w MR intensities appear less consistent before standardization (left column) than after standardization (right column).

Non-tumor Regions

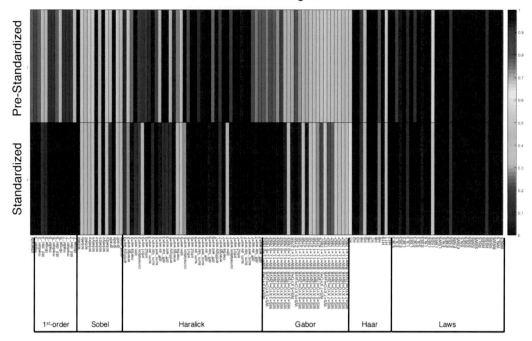

Figure 18.5 Cross-site instability score heatmap for non-tumor regions, where blue indicates IS score = 0 (i.e., highly reproducible), while red indicates IS score = 1 (i.e., highly variable). A majority of features (e.g., Haralick, Gabor) appear less reproducible pre-standardization (top row, predominantly green), compared to after standardization (bottom row, same cells are now deep blue).

Tumor Regions

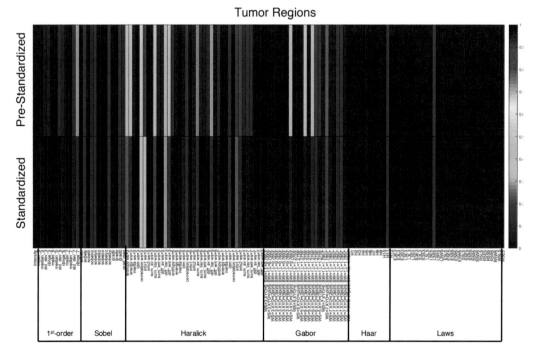

Figure 18.6 Cross-site instability score heatmap for tumor regions, where blue indicates IS score = 0 (i.e., highly reproducible), while red IS score = 1 (i.e., highly variable). A majority of features are reproducible pre-standardization (top row comprises largely blue cells) and are even more reproducible after standardization (bottom row, deeper shade of blue for most cells).

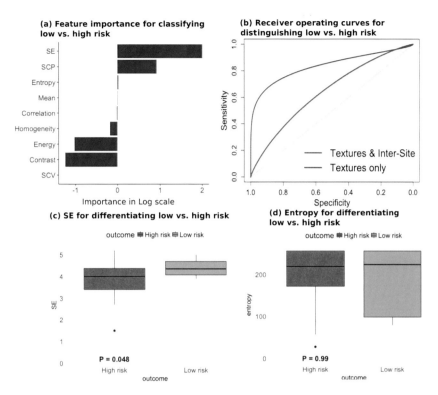

Figure 19.1 Differentiating between high-risk from low-risk HGSOC patients using single-site and inter-site texture measures. The relative importance of features when using both single-site and inter-site measures in a model trained using leave-one-out cross-validation using elastic net regression identified the inter-site measures as most relevant (a), comparison of receiver operating characteristic curves show better performance of the combined model (single-site textures and inter-site textures) compared to single-site texture (b), and comparison in differentiating high-risk from low-risk patients using inter-site entropy (SE) (c), and single-site entropy (d).

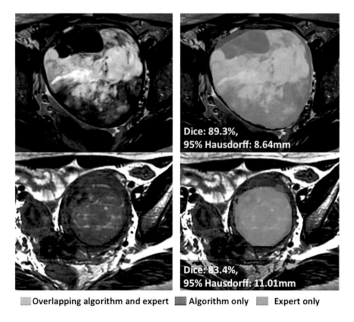

Figure 19.2 Example segmentations generated using 3D Slicer (grow-cut method) for uterine leiomyosarcomas from T2w MRI. The overlapping portion between the algorithm and the expert delineated manual segmentation (blue), the algorithm only (in brown), and the expert only portion (in green) are also shown along with the segmentation concordance computed as Dice overlap score and 95% Hausdorff metric between the algorithm and manual delineation.

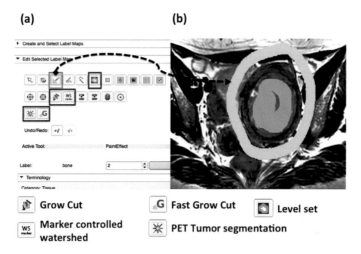

Grow Cut

Marker controlled watershed

Fast Grow Cut

PET Tumor segmentation

Level set

Figure 19.3 Example segmentation methods available through the Editor effect in 3D Slicer. (a) Editor panel in 3D Slicer and (b) user inputs produced using paint effect.

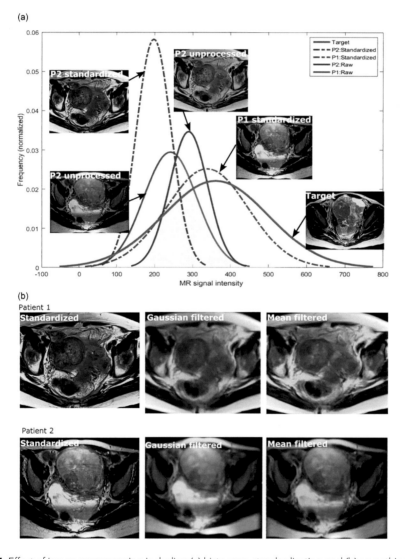

Figure 19.4 Effect of image preprocessing including (a) histogram standardization and (b) smoothing.

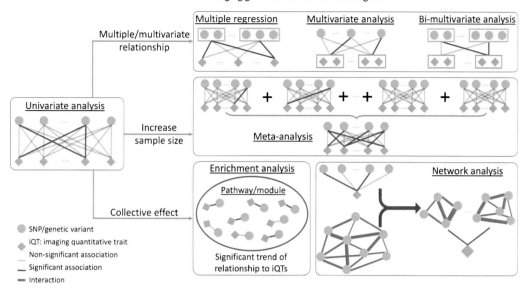

Brain imaging genomics association strategies

Figure 20.1 Schematic summary of analytic strategies for association discovery in brain imaging genomics.

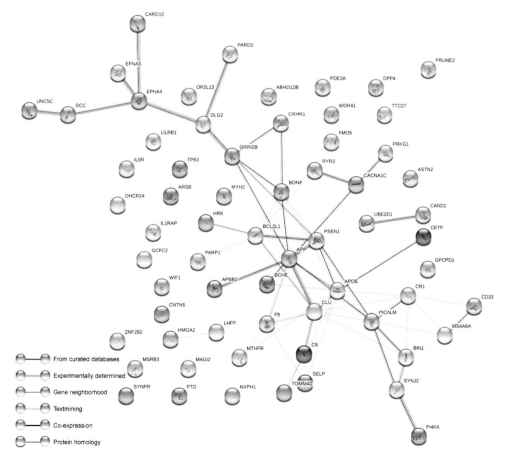

Figure 20.2 Interaction network among imaging genomic findings in the study of Alzheimer's disease. Functional interactions are shown using the STRING database. Nodes represent genes, and edges denote the interactions. Content inside node shows known or predicted 3D structure of corresponding protein, and color of the edges indicates the source of interaction evidence as shown in legend.

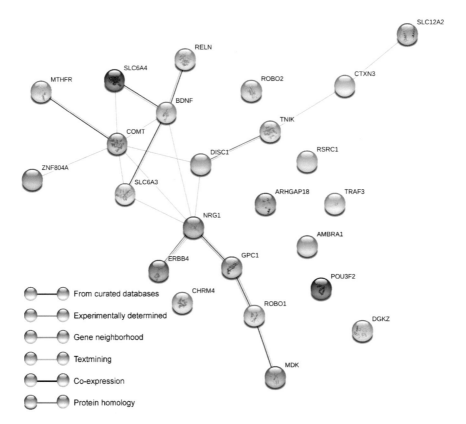

Figure 20.3 Interaction network among imaging genomic findings in the study of Schizophrenia. Functional interactions are shown using the STRING database. Nodes represent genes, and edges denote the interactions. Content inside node shows known or predicted 3D structure of corresponding protein, and color of the edges indicates the source of interaction evidence as shown in legend.

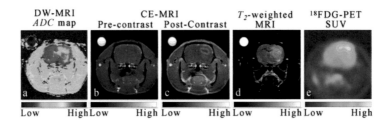

Figure 21.1 Example images are shown for a rat injected with a C6 glioma. Panel a shows an example *ADC* map. Panels b and c show a CE-MRI image before and after the injection of a gadolinium-based contrast agent. Hyper-intense signal is observed at the periphery of the tumor volume. Panel d shows a T_2-weighted MRI. Panel e shows ¹⁸FDG-PET SUV indicating high uptake in the tumor region.

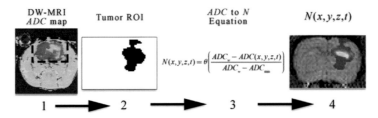

Figure 21.2 An example flow chart for a rat inject with a C6 glioma is shown demonstrating the conversion of *ADC* to tumor cell number, $N(x,y,z,t)$. First, the *ADC* map is cropped to show only the brain. Second, the tumor region of interest (ROI) is determined from the CE-MRI data. Third, $N(x,y,z,t)$ is calculated from *ADC*. Fourth, the calculated $N(x,y,z,t)$ is displayed over a grayscale anatomical image.

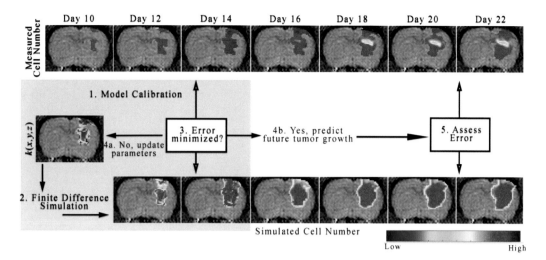

Figure 21.3 An example optimization and model prediction framework is shown for a representative rat injected with a C6 glioma. The top row shows the measured cell number from day 10 to 22, while the bottom row shows the simulated cell number from day 12 to 22. First, the model is calibrated using data from days 10–14. An initial guess of the model parameters (k, D) is assigned. Second, a finite difference simulation of the reaction-diffusion model is performed. Third, the error is evaluated between the simulated and measured tumor cell number. Fourth, if the error is not minimized the model parameters are updated (4a), otherwise the current model parameters are used to predict future tumor growth (4b). Fifth, the prediction error is assessed between the simulated and measured tumor cell number.

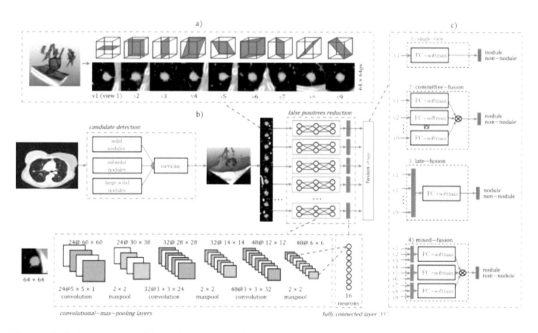

Figure 22.3 An overview of the pulmonary nodule detection study. (a) An example of extracted 2-D patches from nine symmetrical planes of a cube. The candidate is located at the center of the patch with a bounding box of 50 × 50 mm and 64 × 64 px. (b) Candidates are detected by merging the outputs of detectors specifically designed for solid, subsolid, and large nodules. The false positive reduction stage is implemented as a combination of multiple ConvNets. Each of the ConvNets stream processes 2-D patches extracted from a specific view. (c) Different methods for fusing the output of each ConvNet stream. Gray and orange boxes represent concatenated neurons from the first fully connected layers and the nodule classification output. Neurons are combined using fully connected layers with softmax or a fixed combiner (product-rule). (a) Extracted 2-D patches using nine views of a volumetric object. (b) Schematic of the proposed system. (c) Fusion methods.

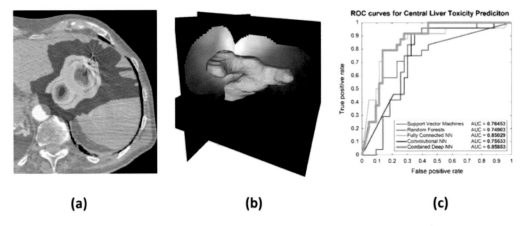

(a)   **(b)**   **(c)**

Figure 22.4 Prediction of radiation toxicity to liver tissue (portal vein). (a) The portal vein (green) on the CT image with superimposed radiation dose map. (b) A 3D dose map delivered to the central hepatobiliary tract (15 mm expansion of the portal vein), which was analyzed using deep learning for hepatobiliary toxicity prediction. (c) Receiver operating characteristic (ROC) curves for the proposed deep dose analysis (combined deep neural network) and alternative toxicity predictors (support vector machine, random forests, fully connected neural network, and convolutional neural network).

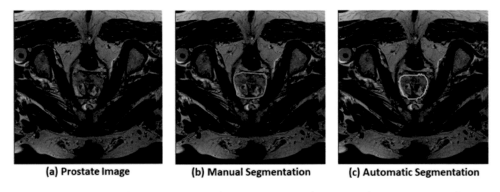

(a) Prostate Image   **(b) Manual Segmentation**   **(c) Automatic Segmentation**

Figure 22.5 Result obtained using VHDR-CNN for segmentation, where manual segmentation and automatic segmentation are compared for a single case study. (a) prostate image (b) manual segmentation and (c) automatic segmentation. The dice similarity coefficient for prostate delineation was 87% for the entire test set.

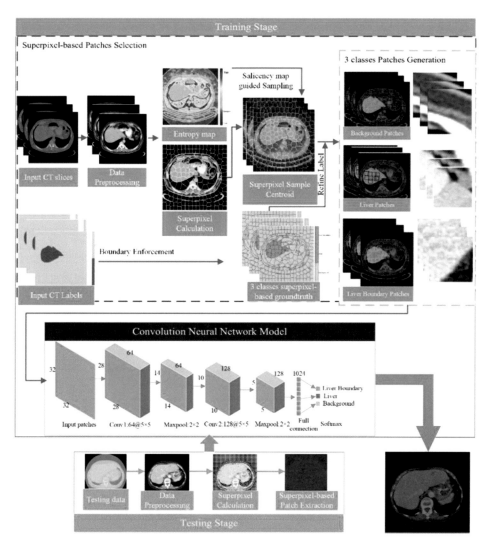

Figure 22.6 Illustration of sensitive convolutional neural network framework for segmentation. Superpixel patches were extracted (under the guidance of saliency maps) and assigned class labels (interior liver, liver boundary, and nonliver background). A CNN network was trained with given image patches and their labels. The class labels of testing data were predicted, where the image patches that cover all superpixels from individual patients were analyzed and assigned the probability for three classes, and the probability maps were integrated into the final segmentation result.

Breast cancer

HUI LI AND MARYELLEN L. GIGER

Breast image interpretation is an important task of radiologists in both breast cancer screening and diagnostic work-up to ensure optimal patient management. In addition, breast imaging and associated analyses have the potential for use in risk assessment, prognosis, response to therapy, and risk of recurrence (1,2). For a medical imaging exam to be beneficial, both good physical image quality and accurate image interpretation are necessary.

Over the past decades, advancements in image acquisition systems, computational resources, and algorithmic designs have raised the potential use of artificial intelligence in various medical imaging tasks, such as risk assessment, detection, diagnosis, prognosis, and therapy response, as well as in multi-omics disease discovery. It is of interest note, however, that as early as 1966, computer-aided detection methods had been proposed, especially in breast and lung imaging, and ultimately translated to clinical practice (1,3–5).

This chapter touches briefly on breast image acquisition systems used currently in routine clinical practice and then mainly on the various areas of rapid development in breast image analysis due to the continuous rise in machine learning, including deep learning.

14.1 OVERVIEW—CLINICAL IMAGING OF BREAST CANCER: IMAGE ACQUISITION AND DECISION-MAKING TASKS

14.1.1 MAMMOGRAPHY AND DIGITAL BREAST TOMOSYNTHESIS

Full-field digital mammography (FFDM) is used in both screening programs and diagnostic workups to detect early cancers and better diagnose/determine patient management, respectively (6). FFDM is a 2D single-projection radiographic image that allows for image formation due to the differential X-ray attenuation of the components within the breast—e.g., parenchyma stroma tissue, tumors, microcalcifications. In mammography, the breast is compressed, and attenuated X-rays are subsequently passed through a grid to reduce scatter radiation and then recorded by a digital detector system. Digital mammography systems have a large dynamic range, as compared to prior screen/film mammography, which had a nonlinear response with a limited exposure range (7). Since FFDM is a single projection image, a disadvantage is the presence of overlapping tissue from the 3D breast volume to the 2D planar image.

Investigators have explored various acquisition protocols to extend the benefit of FFDM. Such systems include dual-energy mammography (8–10), contrast-enhanced mammography (11,12), and three-component mammography (13,14). These various methods exploit attenuation changes while maintaining the high spatial resolution of FFDM.

Digital breast tomosynthesis (DBT) is similar to FFDM in patient set up (breast compression although with slightly less applied pressure and immobilization), however, the X-ray tube is not fixed, but rather moves along a trajectory (an arc), and exposes and images the breast at various time intervals and projection angle (15). The series of single-projection images is then subjected to a reconstruction algorithm to yield images of breast sections parallel to the detector. Images of these breast sections allow for the viewing of abnormalities with minimal overlapping tissue, basically, providing pseudo 3D images (or 2.5 D). Use of breast tomosynthesis in breast cancer screening is rapidly increasing (16).

Dedicated breast computed tomography (CT) systems incorporate 300–500 low-dose projections acquired in a circular trajectory around the breast, followed by tomographic image reconstruction to yield an actual 3D breast volume (17). Dose with breast CT is higher than a typical two-view screening mammogram (craniocaudal [CC] and mediolateral oblique [MLO]). However, with breast CT, 3D images of the internal structures of the breast can be visualized eliminating overlap between lesions and normal structure and allowing for assessment of extent of disease.

14.1.2 BREAST MRI

Tumors on breast magnetic resonance imaging (MRI) are characterized by the morphology of the lesion, perfusion (kinetics) from contrast-enhancement, and signal intensity on T1- and T2-weighted MR images (6,18–20).

Breast MRI uses magnetic fields and radio waves to create an image of the breast. During an MRI scan, the breast is positioned through openings in the scanning table that contain a dedicated multi-channel breast coil that detects the magnetic signal from protons. A static strong magnetic field and pulses of radio waves at an appropriate resonant in the radio frequency range frequency are applied. This changes the orientation of the nuclear spins. As the realigned nuclei emit energy, these signals are analyzed by computer and converted to breast MR image. There are two forms of relaxation. One is delay of the amplitude along the direction of main magnetic field, i.e., z axis, called T1 relaxation. The other is the increasing of the amplitude in the x–y plane, also called T2 relaxation. Different tissues have different T1 and T2 relaxation times. When the difference in T1 for the different tissues is maximized, the acquired MRI scan is a T1-weighted scan. When the difference in T2 for the different tissues is maximized, the acquired MRI scan is a T2-weighted scan.

Dynamic contrast-enhanced MRI (DCE-MRI) requires intravenous contrast agent injection to enhance the visibility of breast lesions on T1 images. A series of scans is performed before and after contrast agent injection, usually in approximately 60-second intervals, to demonstrate the uptake and washout of the contrast agent, characterizing the role of angiogenesis. Often, DCE-MRI acquisition is coupled with a fat-suppression technique (6,18,19).

Diffusion-weighted magnetic resonance imaging (DWI) is another advanced MRI technique to measure the mobility of the water molecules diffusing in the tissue (21). It measures the Brownian motion of water molecules which indirectly reflects the tissue microstructure via the apparent diffusion coefficient (ADC).

More recently, ultrafast DCE-MRI protocols are being developed for breast imaging in order to characterize the high-temporal-resolution of the contrast enhancement (22–24). In these protocols, the imaging interval after post-contrast agent injection is approximately 3–7 seconds.

Abbreviated MRI is also being proposed for use in breast cancer screening. With abbreviated MRI, only the pre-contrast and one post-contrast image, along with the maximum-intensity projection image, is used to assess morphology and kinetics (25).

14.1.3 BREAST ULTRASOUND

Ultrasound imaging (sonography) of the breast uses sound waves to produce pictures of the internal structures of the breast. Breast ultrasound is primarily used for working up suspicious lesions that may have been found during a physical exam, from screening mammograms, or other imaging modalities. Thus, ultrasound is typically used to characterize a suspicious lesion (6).

However, more recently, ultrasound has been evaluated for its potential role in breast cancer screening. Whole-breast ultrasound (WBUS) scans the entire breast and is being used as an adjunct to FFDM for breast cancer screening. There are two ways to perform WBUS, hand-held whole-breast ultrasound, and automated whole-breast ultrasound (ABUS). Hand-held whole-breast ultrasound is performed by a physician or ultrasound technologist on the entire breast using a hand-held ultrasound probe. ABUS is performed by a machine with an ultrasound probe through an automated process on the entire breast, thus yielding more consistent pressure (26).

14.1.4 CLINICAL DECISION-MAKING TASKS

Breast imaging is used throughout the screening and work-up stages for the breast cancer patient. In screening, the clinical task is to locate (detect) lesions and regions within the breast that are suspect of being cancerous. Usual signals are masses, clustered microcalcifications, asymmetries, and architectural distortions (6). While in the past, screening programs utilized the same protocol for all, nowadays women at high risk have more options. Breast density has been shown to be a risk factor for future breast cancer as well as a factor in masking "hidden" cancers (27). Beyond breast density, the pattern of the parenchyma may also contribute to assessing future risk of breast cancer. Roughly half of women undergoing screening will have breasts that are approximately 50% dense and thus require additional screening imaging such as breast MRI or ultrasound.

Once a suspicious lesion is found, the patient may have additional mammograms, breast MRI, or breast ultrasound in order for the radiologist to consider the likelihood that the lesion is cancerous and recommend specific patient management. Such decision making is a classification task.

Once cancer is diagnosed, the patient will benefit from biopsy results indicating cancer subtype and genotypes. During treatment, breast imaging can also be employed in assessing response. Although not used clinically yet, breast imaging is also being investigated to assess prognosis, risk of recurrence, and within imaging-genomic association studies (1).

14.2 RATIONALE FOR RADIOMICS AND RADIOGENOMICS OF BREAST CANCER

Image interpretation by humans is limited by multiple factors, some of which can be mitigated by incorporating computer vision and machine learning into the interpretation process. Computer reads can help handle the human limitations of incomplete visual search patterns, radiologist fatigue and distractions, large amounts of image data, and some loss of physical image quality. However, the presence of structure noise (camouflaging normal anatomical background) and the presentation of subtle and/or complex disease states can also make computer reads difficult.

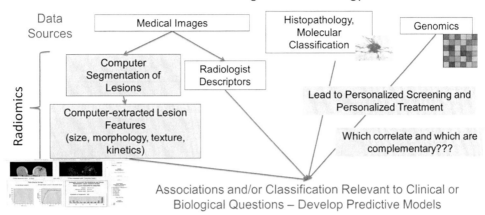

Figure 14.1 (See color insert.) A logic diagram of an example of the field of radiogenomics for breast cancer using digital mammography and DCE MRI. (Reprinted from Giger, M., *Decoding Breast Cancer with Quantitative Radiomics & Radiogenomics: Imaging Phenotypes in Breast Cancer Risk Assessment, Diagnosis, Prognosis, and Response to Therapy*, TCGA 4th Scientific Symposium, NIH, Bethesda, MD, 2015; Colen, R. et al., *Transl. Oncol.*, 7, 556–569, 2014. With permission.)

In addition, integration of information from multiple patient tests such as clinical, molecular, imaging, and genetic testing is expected to yield improved cancer diagnosis and treatment management. Thus, with regards to breast imaging, there are multiple efforts to convert image data to quantitative (numeric) data. Such efforts have existed over the past few decades, through the development of computer-aided detection (CADe) and computer-aided diagnosis (CADx) for aiding radiologists in the detection and classification of breast lesions, which incorporated both radiomics and deep learning, although such specific terminology was not used (1,28–31). Radiomics, an expansion of computer-aided diagnosis, involves computerized image analyses to convert images to minable data. In radiogenomics (or imaging-genomics), these quantitative image data are related to other "-omic" data such as clinical, pathologic, and genomic data (32,33). Radiomic phenotypes that are highly correlated with important clinical, molecular, or genomic biomarkers can potentially serve as diagnostic or prognostic tools for patient monitoring and assessing therapeutic response and thus augment the utility of medical imaging as a non-invasive technology for cancer care, like a "virtual digital biopsy" (Figure 14.1) (34,35).

14.3 RADIOMIC FEATURES/PHENOTYPES OF BREAST CANCER

14.3.1 HAND-CRAFTED RADIOMIC PHENOTYPES OF BREAST CANCER

Quantitative radiomic features are mathematical descriptors that correspond to characteristics of the tumor or tissue. Hand-crafted phenotypes (human-engineered features) correspond to those that correspond to characteristics that can be directly described by some computer algorithm and often correspond to visually discernible characteristics, such as size, shape, morphology, texture, and kinetics. Often such hand-crafted features require a prior segmentation of the tumor from the parenchyma background. Such segmentation can be automatic or semi-manual.

Such extraction of features has been conducted on mammography, tomosynthesis, ultrasound, and MRI (1). For example, various hand-crafted features can be automatically extracted from DCE-MRI on a workstation that automatically segments the tumor from the surrounding parenchymal background within

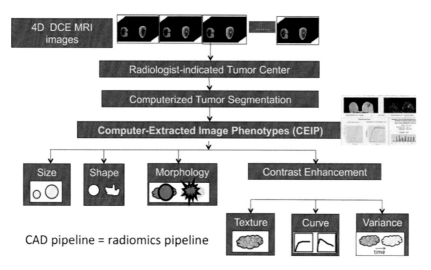

Figure 14.2 (See color insert.) Schematic flowchart of a computerized tumor phenotyping system for breast cancers on DCE-MRI. CAD radiomics pipeline includes computer segmentation of the tumor from the local parenchyma and computer-extraction of "hand-crafted" radiomic features covering six phenotypic categories: (1) size (measuring tumor dimensions), (2) shape (quantifying the 3-D geometry), (3) morphology (characterizing tumor margin), (4) enhancement texture (describing the heterogeneity within the texture of the contrast uptake in the tumor on the first postcontrast MRIs), (5) kinetic curve assessment (describing the shape of the kinetic curve and assessing the physiologic process of the uptake and washout of the contrast agent in the tumor during the dynamic imaging series, and (6) enhancement-variance kinetics (characterizing the time course of the spatial variance of the enhancement within the tumor) CAD = computer-aided diagnosis; DCE-MRI = dynamic contrast-enhanced MRI. (Reprinted from Giger, M., *J. Am. Coll. Radiol.*, 15, 512–520, 2018. With permission.)

the DCE-MR images and extracts lesion characteristics in phenotypic categories such as: (i) size (measuring tumor dimensions), (ii) shape (quantifying the 3D geometry), (iii) morphology (margin characteristics), (iv) enhancement texture (describing the heterogeneity within the texture of the contrast uptake in the tumor on the first post-contrast MRIs), (v) kinetic curve assessment (describing the shape of the kinetic curve and assessing the physiologic process of the uptake and washout of the contrast agent in the tumor during the dynamic imaging series), and (vi) enhancement-variance kinetics (characterizing the time course of the spatial variance of the enhancement within the tumor) (Figure 14.2) (36–41). Note that the particular features useful for characterization will depend on the specific clinical task as well as imaging modality.

A multitude of review papers have been written on using quantitative hand-crafted radiomics in characterizing breast lesions (1,28).

14.3.2 DEEP LEARNING YIELDING RADIOMIC FEATURES CHARACTERIZING BREAST CANCER

Deep learning algorithms can also contribute to the extraction of features from breast images. Often for such feature extraction, pre-trained convolutional neural networks (CNNs) are used with extraction from various layers yielding features (42). An advantage is that there is no need to segment the tumor or program the mathematical descriptors of features. Disadvantages of such a method include having features that may not be intuitive and a tendency to have too many features for use in classification as compared to the number of available images for training. Examples using deep learning on breast images for characterization are given in multiple papers (42–55). As with hand-crafted features, the selected features will depend on the task and modality.

14.4 DATABASES

Various databases of breast images are part of the collected de-identified datasets from The Cancer Genome Atlas (TCGA) and The Cancer Imaging Archive (TCIA)—cancer research resources supported by the National Cancer Institute of the U.S. National Institutes of Health (56). The TCIA also includes datasets from clinical trials, such as those images from the ACRIN Investigation of Serial Studies to Predict Your Therapeutic Response with Imaging And moLecular Analysis (I-SPY TRIAL) study, which consists of breast MRI images acquired over treatment intervals.

Datasets have also been made available through various challenges such as the Dialogue for Reverse Engineering Assessments and Methods (DREAM) challenge for FFDM in detecting breast cancer (dream-challenges.org), The International Society for Optics and Photonics (SPIE)-American Association of Physicists in Medicine (AAPM)-NCI Challenges (LungX and ProstateX) (57,58), and those listed at Challenges.org.

14.5 RADIOMICS AND VIRTUAL DIGITAL BIOPSIES

The computer-extraction of radiomic features that characterize a tumor or normal parenchyma can be viewed as a "digital virtual biopsy" (34). A virtual biopsy would characterize the entire tumor (or tissue), be essentially non-invasive, and be repeatable (such as over time during treatments). The extracted radiomic features can be used in association studies for discovery with other -omic data, and then in predictive models for diagnosis, prognosis, risk assessment, and response to treatment. It is important to note that such virtual biopsies are basically CADx systems, whose goals were to increase sensitivity and reduce the number of unnecessary actual biopsies. Note that the role of a virtual biopsy, in other situations, is not to replace an actual biopsy, but to be used when an actual biopsy is not practical.

14.6 RADIOMICS AND RISK OF FUTURE BREAST CANCER

Breast density and parenchymal patterns have been showed to have a role in estimating breast cancer risk (59). Breast density indicates the amount of fibroglandular tissue in the breast relative to the amount of fatty tissue. Breast density values are area-based from 2D images (i.e., FFDM), as well as volumetric-based from 3D images (i.e., MRI). Note that with mammographic density, increased density serves as a breast cancer risk factor as well as causing a masking effect that obscures lesions (27).

Besides density, investigators are characterizing the spatial distribution (i.e., pattern) of the stroma and relating it to breast cancer risk. Texture analysis is typically used to extract these radiomic parenchymal features. For FFDM, Byng et al. calculated a skewness index from the analysis of mammograms to describe the density variation (60). Huo et al. and Li et al. have used radiomic texture analysis to discriminate women at high risk (e.g., BRCA1/BRCA2 gene-mutation carriers) from women at normal risk for breast cancer and demonstrated that women at high risk for breast cancer have dense breasts with parenchymal patterns that are coarse and low in contrast (61,62). The parenchyma pattern on breast tomosynthesis images is also being examined to characterize the parenchyma pattern for ultimate use in breast cancer risk estimation (Figure 14.3) (63).

The breast parenchyma is also characterized by the BPE, i.e., background parenchymal enhancement, when dynamic contrast-enhanced MRI is employed (64). BPE has been shown by Wu et al. to be associated with the presence of breast cancer (65).

As with radiomics in general, deep learning can be incorporated into the computer vision and artificial intelligence (AI) tasks of parenchymal characterization and risk assessment (66–68). Li et al. examined the

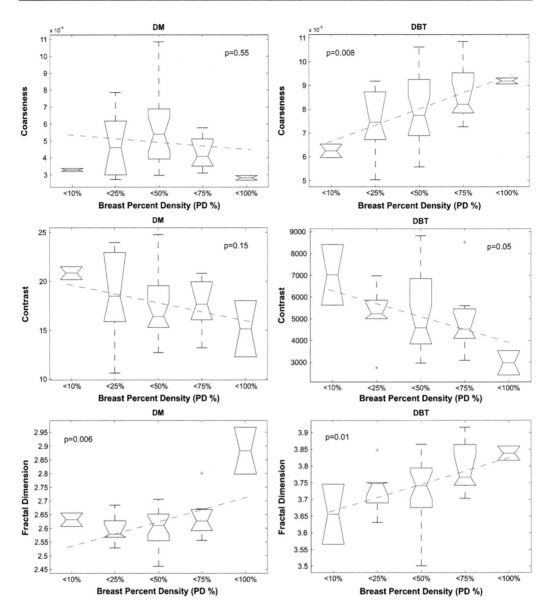

Figure 14.3 (See color insert.) Box-plots with fitted regression lines and associated *p*-values for digital mammography (DM) and digital breast tomosynthesis (DBT) coarseness, contrast, and fractal dimension texture features versus the five groups of increasing breast percent density (PD): <10%, 10%≤…<25%, 25%≤…<50%, 50%≤…<75%, and 75%≤…<100%. (Reprinted from Kontos, D. et al., *Acad. Radiol.*, 16, 283–298, 2009. With permission.)

performance of transfer learning in the distinction between women at normal risk of breast cancer and those at high risk based on their BRCA1/2 status (66). Gastounioti et al. used deep learning to merge parenchymal complexity measurements generated by texture analysis into discriminative meta-features relevant for breast cancer risk prediction (67). Wanders et al. combined breast density and a deep-learning-based texture score for breast cancer risk assessment (68).

14.7 RADIOMICS AND BREAST CANCER DIAGNOSIS

Once a suspect lesion is found (e.g., via screening), further imaging is conducted to estimate the likelihood that the lesion is cancerous to aid the radiologist in determining patient management. During this workup stage of imaging, often an actual biopsy is conducted from which histopathology and genetic information can be obtained. Investigators are examining the relationship between the computer-extracted features and the histopathology and genomics findings. Thus, the results from radiomics CADx could potentially increase the sensitivity of classifying a lesion as well as decrease the number of false positives (i.e., benign lesions sent to biopsy). In addition, investigators are developing radiomic biomarkers, e.g., on breast MRI, to help predict pathologic stage and potentially augment patient management and appropriate treatment, such as neoadjuvant chemotherapy, surgery, and/or radiation therapy.

CADx methods have been developed for mammography and digital breast tomosynthesis, ultrasound, and MRI, and have been extensively reviewed already in the literature (1,28,69). While computer-aided detection for screening mammography has been FDA approved since 1998, CADx for breast image analysis was not cleared until 2017 (70).

Various lesion characterization methods describing the kinetics of uptake in DCE-MRI have been investigated for the interpretation of clinical breast images (71–73). Platel et al. (24) performed breast lesion characterization using ultrafast DCE-MRI and found that the classification performance was significantly higher with kinetics derived from ultrafast DCE-MRI than that from regular DCE-MRI (Figure 14.4).

Deep learning has also been advancing CADx algorithms (42,47–51,74). When deep learning is being used to characterize a lesion, often the region around the lesion is input to the CNN with the output being related to the likelihood of cancer. Currently, investigators researching deep learning for breast cancer diagnosis have used transfer learning for fine tuning or feature extraction. And given the advances with hand-crafted radiomic CADx, output from such CADx methods are being merged with output from CNN-based CADx to yielded statistically significant improvements in CADx performance (Figures 14.5 and 14.6) (42,47).

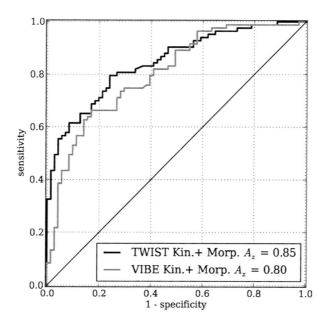

Figure 14.4 Receiver operating characteristic (ROC)-curve for both the TWIST and VIBE acquisition SVM classifications, for combined kinetic and morphological features. (Reprinted from Platel, B. et al., *IEEE Trans. Med. Imaging*, 33, 225–232, 2013. With permission.)

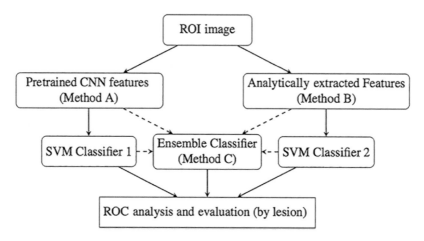

Figure 14.5 An overview of the various methods used in the task of distinguishing between benign and malignant tumors. (Reprinted from Huynh, B.Q. et al., *J. Med. Imaging*, 3, 034501, 2016. With permission.)

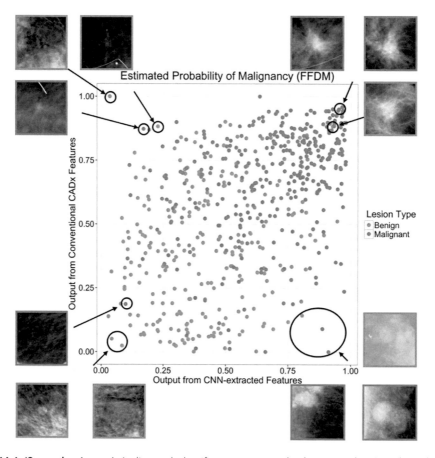

Figure 14.6 (See color insert.) A diagonal classifier agreement plot between the CNN-based classifier and the conventional CADx classifier for FFDM. The x-axis denotes the output from the CNN-based classifier, and the y-axis denotes the output from the conventional CADx classifier. Each point represents an ROI for which predictions were made. Points near or along the diagonal from bottom left to top right indicate high classifier agreement; points far from the diagonal indicate low agreement. ROI pictures of extreme examples of agreement/disagreement are included. (Reprinted from Antropova, N. et al., *Med. Phys.*, 44, 5162–5171, 2017. With permission.)

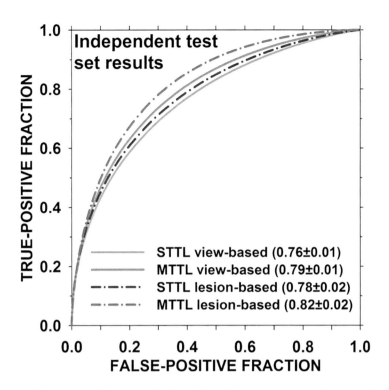

Figure 14.7 (See color insert.) ROC curves for the independent test set using the two selected C_1 transfer networks: STTL: single-task transfer learning, MTTL: multi-task transfer learning. The difference in the *AUCs* was statistically significant between the two lesion-based curves (*p*-value = 0.007) and between the two view-based curves (*p*-value = 0.008). (Reprinted from Samala, R. et al., *Phys. Med. Biol.*, 62, 8894–8908, 2017. With permission.)

In addition to single-task transfer learning, multi-task learning has also been applied to the classification of breast tumors (49,51). Samala et al. proposed a multi-task transfer learning CNN with the aim of translating the knowledge learned from nonmedical images to medical diagnostic tasks, with resulting indicating that multi-task supervised learning achieved better generalization to unknown cases than single-task learning (Figure 14.7).

Fine-tuning in the characterization of lesions enables customization of pre-trained networks for the particular breast cancer decision making task, such as density assessment (44) or classification of MRI lesions (50). In fine tuning, investigators can (i) "freeze" the early layers of a pre-trained CNN and train the later layers or (ii) train on one modality (i.e., digitized screen/film mammography) for use on a related modality (i.e., as FFDM or tomosynthesis). Samala et al. demonstrated benefit using such fine tuning on FFDMs in the training for CNN-based detection of mass lesions (75).

For deep learning assessment of breast tumors, performance will vary depending on the image input to the CNN. For example, Antropova et al. investigated various image presentations from breast DCE-MRI and found that when the maximum-intensity projection image from the 4D DCE-MRI sequences was input to the CNN, improved discrimination performance was observed (Figure 14.8) (50).

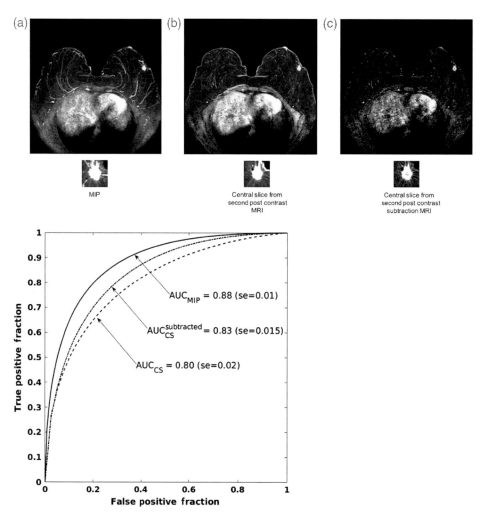

Figure 14.8 (top) Example of a malignant lesion. Full MRI images and ROIs for (a) the maximum intensity projection (MIP) image of the second post-contrast subtraction MRI, (b) the center slice of the second post-contrast MRI, and (c) the central slice of the second post-contrast subtraction MRI. (bottom) ROC curves showing the performance of three classifiers. The classifiers were trained on CNN features extracted from ROIs selected on: (a) the MIP images of second post-contrast subtraction MRIs, AUC_{MIP}, (b) the central slices of the second post-contrast MRIs, AUC_{CS}, and (c) the central slices of second post-contrast subtraction MRIs, $AUC_{CS}^{Subtracted}$. (Reprinted from Antropova, N. et al., *J. Med. Imaging.*, 5, 014503, 2018. With permission.)

14.8 RADIOMICS IN BREAST CANCER CHARACTERIZATION FOR ASSESSING PROGNOSIS

Radiomics have been investigated relative to pathologic stage and to surgically verified lymph node status (36), with classifiers designed to predict tumor pathologic stage and lymph node involvement. Tumor size was found to be a key predictor of pathologic stage, but radiomic features that captured biologic behavior also emerged as predictive [e.g., stage I and II versus stage III yielded an area under the ROC curves (AUC) of 0.83]. Notable is that no size measure was successful in the prediction of positive lymph nodes, but when a heterogeneity feature was included, a significantly improved discrimination (AUC = 0.62; p = 0.003) compared with chance was demonstrated.

Loiselle et al. (76) performed radiomic analysis on breast cancer patients with positive sentinel biopsies. The total persistent enhancement and volume adjusted peak enhancement features extracted from DCE-MRI showed promising performance in predicting lymph node burden (≥4 axillary nodes) with an AUC of 0.79 from the DCE-MRI multivariate model. Mussurakis et al. (77) reported a strong relationship between DCE-MRI radiomic features and tumor grades with AUC values ranging from 0.78 to 0.80 for classifying tumor grades for invasive ductal carcinoma (IDC) lesions. Schacht et al. (78) used quantitative radiomic analysis to classify axillary lymph nodes on breast MRI. Thus, computer-extracted MRI phenotypes have promise for predicting breast cancer pathologic stage and lymph node status.

Prognosis can also be assessed based on receptor status [estrogen receptor (ER), progesterone receptor (PR), and human epidermal growth factor receptor 2 (HER2)], indicating different breast cancer subtypes, e.g., normal-like, luminal A, luminal B, HER2-enriched, and basal-like. Different cancer subtypes could respond differently to different therapies. Thus, of interest is the correlation between the tumor radiomic features and various cancer subtypes (37). It was shown that MRI-based tumor features were able to distinguish between molecular prognostic indicators yielding performances in terms of AUC of 0.89, 0.69, 0.65, and 0.67 in the tasks of distinguishing between ER+ versus ER–, PR+ versus PR–, HER2+ versus HER2–, and triple-negative cancers versus all others, respectively (37). Authors found statistically significant associations between MRI-based tumor features and receptor status with more aggressive cancers being more likely to have larger size with contrast enhancement texture being more heterogeneity (Figure 14.9).

Wu et al. (79) studied the relationship between breast cancer molecular subtypes and DCE-MRI radiomic features of both tumor and background parenchymal enhancement, with AUC values between 0.66 and 0.79 obtained in distinguishing different molecular subtypes (i.e., luminal A/B or basal) of breast cancer (Figure 14.10).

Grimm et al. (80) investigated associations between semi-automatically extracted MRI radiomic features and breast cancer molecular subtypes. They found statistically significant associations between luminal A and luminal B cancers with MRI radiomic features.

Also for assessing prognosis, Liang et al. (81) reported on a T2-weighted MRI-based radiomics classifier for predicting Ki-67 status of breast cancers.

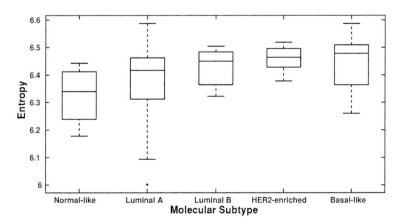

Figure 14.9 Relationship between the MRI phenotypes of enhancement texture (entropy) and the molecular subtypes. The enhancement texture is calculated at the first-post-contrast MR image thus quantitatively characterizing the heterogeneous uptake of contrast within the tumor. Shown is a statistically significant trend between entropy and molecular subtype (p-value of 0.006 from the Kendall test). HER2, human epidermal growth factor receptor 2; MRI, magnetic resonance imaging. (Reprinted from Li, H. et al., *NPJ Breast Cancer*, 2, 16012, 2016. With permission.)

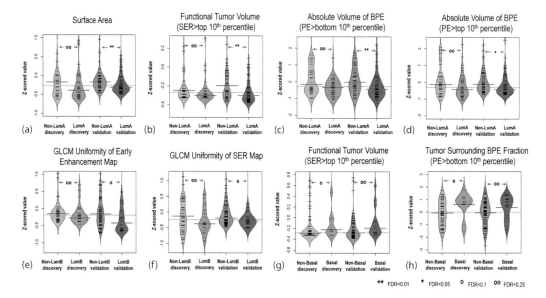

Figure 14.10 (See color insert.) Imaging features significantly associated with molecular subtypes (after correction for multiple testing) in both discovery and validation cohorts, (a–d) four features for distinguishing luminal A versus nonluminal A; (e, f) two features for distinguishing luminal B versus nonluminal B; and (g, h) two features for distinguishing basal-like versus nonbasal-like. Wilcoxon rank sum test was implemented to investigate pairwise difference. Also, the FDR adjusted for multiple testing is reported. (Reprinted from Wu, J. et al., *J. Magn. Reson. Imaging*, 46, 1017–1027, 2017. With permission.)

As artificial intelligence methods for prognosis evolve, computer-extracted image-based phenotypes may ultimately be used clinically in assisting the discrimination of breast cancer subtypes, yielding quantitative predictive signatures for assessing prognosis.

14.9 RADIOMICS AND ASSESSING RESPONSE TO BREAST CANCER THERAPY AND ASSESSING RISK OF RECURRENCE

While there is a large variation in the clinical presentation of breast cancer in women, it has been shown that features of the primary tumor can correlate with outcome and risk of cancer recurrence (82–89). Thus, development of radiomic features to assess outcome as early and as accurately as possible will be beneficial to the development of targeted and personalized breast cancer therapies.

The 21-gene Oncotype DX assay, the 50-gene PAM50 assay, and the 70-gene MammaPrint microarray assay were developed as multi-gene assays to relate breast cancer expression profiles to risk of cancer recurrence (90–96). Thus, to understand the role of radiomics in assessing risk of cancer recurrence, investigators have examined the relationships between quantitative radiomic features and risk of breast cancer recurrence (38). Results demonstrated significant associations between MRI radiomic signatures including tumor size and enhancement heterogeneity and multi-gene assay recurrence scores. Thus, radiomics may have potential for use in assessing the risk of cancer recurrence.

As breast cancer patients undergo treatment, radiomic approaches to predict response and outcome become of practical clinical use. Clinical trials, such as the American College of Radiology Imaging Network (ACRIN) trial 6657, a multi-center study of contrast-enhanced MR imaging, are being conducted to understand the role of image-based biomarkers in predicting recurrence and association with recurrence-free survival using MR images of breast tumors to assess breast tumor response to neoadjuvant chemotherapy (97). Hylton et al. (71) showed that breast MRI with a semi-automated tumor analysis method was more strongly associated with pathologic response after neoadjuvant chemotherapy than clinical examination, with the main advantage measured early in treatment by using the volumetric measurement of tumor response, which was predictive of

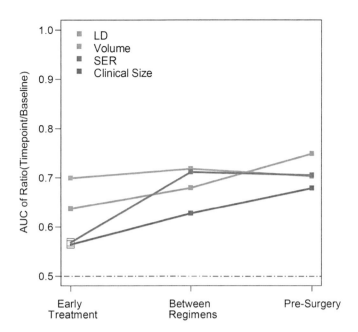

Figure 14.11 (See color insert.) Graph shows AUCs for prediction of pCR for the four predictor variables at the early treatment, between regimens, and pre-surgery time points. Predictors are expressed as the ratio of value at each time point to baseline value for tumor longest diameter (LD) (green), volume (orange), signal enhancement ratio (SER) (blue), and clinical size (red). Solid squares = $p \leq 0.05$. (Reprinted from Hylton, N.M. et al., *Radiology*, 263, 663–672, 2012. With permission.)

recurrence-free survival (Figure 14.11). Drukker et al. (98) showed that an automatic and quantitative radiomic method for the determination of the most enhancing tumor volume was also successful in the task of predicting recurrence and association with recurrence-free survival. Thus, noting the potential of automated radiomics as a role in predicting response to therapy.

Wu et al. (99) investigated intra-tumoral heterogeneity measure extracted from DCE-MRI as a predictor of recurrence-free survival (RFS) in breast cancer. The results showed that multi-regional radiomic features extracted from baseline DCE-MRI can be used to characterize intra-tumoral heterogeneity and predict aggressiveness of breast cancer.

Ashraf et al. (100,101) studied the role of kinetic heterogeneity features extracted from breast DCE-MRI in predicting response. Their findings suggested that these kinetic radiomic features may be used as image-based biomarker to improve candidate patient selection for neoadjuvant treatment.

14.10 RADIOGENOMICS OF BREAST CANCER: DISCOVERY AND PREDICTIVE MODELING

Radiogenomics, or Imaging-Genomics, seeks to discover relationships between image-based characteristics and other -omics characteristics. The potential information gleaned from such associations could indirectly bring additional information into the screening or surveillance step of breast cancer management.

The TCGA Breast Phenotype Group of the National Cancer Institute (https://wiki.cancerimagingarchive. net/display/Public/TCGA+Breast+Phenotype+Research+Group) investigated relationships between computer-extracted quantitative radiomic MRI lesion features and various clinical, molecular, and genomics markers of prognosis and risk of recurrence, including gene expression profiles using datasets from the TCGA and TCIA.

Mapping association studies were performed between radiomic features and genomic features downloaded from the TCGA (including DNA mutation, miRNA expression, protein expression, pathway gene expression, and copy number variation) in order to map the genetic mechanisms that regulate the imaging presentation of specific MRI-based tumor phenotypes (102,103). Mappings were obtained using gene-set enrichment analysis (GSEA) and linear regression analysis on the radiomic and genomic features, yielding relationships with transcriptional activities of pathways and miRNA expressions. Associations were discovered between pathway transcriptional activities and various image-based phenotypes, indicating that they may be regulating various aspects of the MRI-based characteristics (phenotypes) (Figure 14.12). Ultimately, it may be possible in the future to clinically use radiomic phenotypes to predict miRNA activities, augmenting the medical practice of tumor biopsy and miRNA profiling. Such "virtual digital biopsies" will have the benefit of assessing the entire tumor to assess heterogeneity, being basically noninvasive, and being repeatable over time, such as in the case of monitoring treatment.

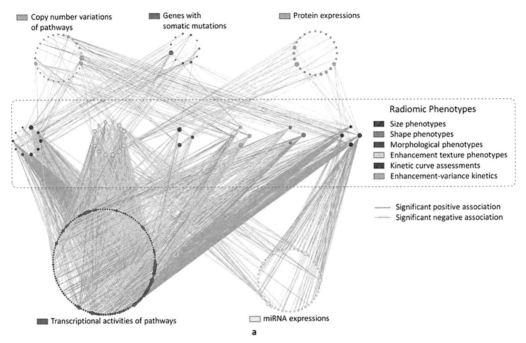

a

	Size phenotypes	Shape phenotypes	Morphological phenotypes	Enhancement texture phenotypes	Kinetic curve assessments	Enhancement-variance kinetics
Transcriptional activities of pathways	173	109	49	374	268	130
Copy number variations of pathways	24	7	7	14	15	21
Mutated genes	3	1	1	15	22	3
miRNA expressions	73	0	0	58	1	0
Protein expressions	10	0	9	17	0	0

b

Figure 14.12 (See color insert.) Overview of all identified statistically significant associations. (a) In the figure, each node is a genomic feature or a radiomic phenotype. Each line is an identified statistically significant association. Genomic features without statistically significant association are not shown. Genomic features are organized into circles by data platform and indicated by different node colors. Radiomic phenotypes are divided into six categories also indicated by different node colors. The node size is proportional to its connectivity relatively to other nodes in the category. Associations are deemed as statistically significant if the adjusted p-values ≤ 0.05. The only exception is for the associations involving somatically mutated genes, for which the statistical significance criteria are: (1) p-value ≤ 0.05 and (2) the gene mutated in at least five patients. (b) A table showing the numbers of statistically significant associations between genomic features of different platforms and radiomic phenotypes of different categories. *(Continued)*

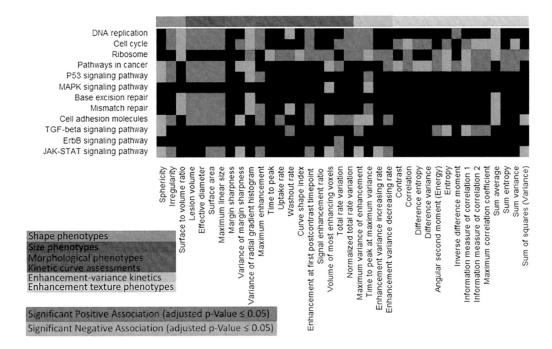

Figure 14.12 (See color insert.) (Continued) Overview of all identified statistically significant associations. (c) Heatmap representation of statistically significant associations between radiomic phenotypes and transcriptional activities of some cancer-related genetic pathways. In the heatmap, genetic pathways are rows and radiomic phenotypes are columns. (Reprinted from Zhu, Y. et al., *Sci. Rep.*, 5, 17787, 2015. With permission.)

14.11 SUMMARY

The role of machine learning, including both human-engineered radiomics and deep learning, in breast cancer analyses continues to expand for multiple decision-making tasks in the clinical interpretation of breast images as well as in cancer discovery through association with other -omics. The clinical decision-making tasks include risk assessment, detection, diagnosis, prognosis, response to therapy, and risk of recurrence. This chapter has illustrated the various applications and avenues of research, most of which have yet to enter routine clinical practice.

One should draw from this chapter, the various tasks and their clinical potential, while noting that any given performance metric (such as AUC) cannot be considered representative of the population. Algorithms cannot simply be compared as most have been evaluated on different datasets/populations as well as with different statistical validation methods. In addition, authors have used various terminology, and thus, caution is required when comparing the radiomic features and training/testing methods.

REFERENCES

1. Giger ML, Karssemeijer N, Schnabel JA. Breast image analysis for risk assessment, detection, diagnosis, and treatment of cancer. *Annu Revi Biomed Eng.* 2013;15:327–357.
2. Giger M. Machine learning in medical imaging. *J Am Coll Radiol.* 2018;15:512–520.
3. Lodwick G. Computer-aided diagnosis in radiology. A research plan. *Invest Radiol.* 1966;1:72–80.
4. Giger M. Computerized image analysis in breast cancer detection and diagnosis. *Semin Breast Dis.* 2002;5:199–210.
5. Freer T, Ulissey M. Screening mammography with computer-aided detection: Prospective study of 12,860 patients in a comunity breast center. *Radiology.* 2001;220:781–786.

6. Ikeda D, Miyake K. *Breast Imaging: The Requisites*. St. Louis, MO: Elsevier, 2016.

7. Mahesh M. Digital mammpgraphy: An overview. *Radiographics*. 2004;24(6):1747–1760.

8. Johns P, Yaffe M. Theoretical optimization of dual-energy x-ray imaging with application to mammography. *Med Phys*. 1985;12(3):289–296.

9. Johns P, Drost D, Yaffe M, Fenster A. Dual-energy mammography: Initial experimental results. *Med Phys*. 1985;12(3):297–304.

10. Boone J, Shaber G, Tecotzky M. Dual-energy mammography: A detector analysis. *Med Phys*. 1990;17(4):665–675.

11. Skarpathiotakis M, Yaffe M, Bloomquist A, Rice D, Muller S, Rick A et al. Development of contrast digital mammography. *Med Phys*. 2002;29(10):2419–2426.

12. Jong R, Yaffe M, Skarpathiotakis M, Shumak R, Danjoux N, Gunesekara A et al. Contrast-enhanced digital mammography: Initial clinical experience. *Radiology*. 2003;228(3):842–850.

13. Drukker K, Duewer F, Giger ML, Malkov S, Flowers CI, Joe B et al. Mammographic quantitative image analysis and biologic image composition for breast lesion characterization and classification. *Med Phys*. 2014;41(3).

14. Laidevant A, Malkov S, Flowers C, Kerlikowske K, Shepherd J. Compositional breast imaging using a dual-energy mammography protocol. *Med Phys*. 2010;37(1):164–174.

15. Peppard H, Nicholson B, Rochman C, Merchant J, Mayo R. Digital breast tomosynthesis in the diagnostic setting: Indications and clinical applications. *Radiographics*. 2015;35(4):975–990.

16. Durand M, Haas B, Yao X, Geisel J, Raghu M, Hooley R et al. Early clinical experience with digital breast tomosynthesis for screening mammography. *Radiology*. 2015;274(1):85–92.

17. Hendrick R. Radiation doses and cancer risks from breast imaging studies. *Radiology*. 2010;257(1):246–253.

18. Hendrick R. *Breast MRI*. New York, Springer, 2008.

19. DeMartini W, Lehman C, Partridge S. Breast MRI for cancer detection and characterization: A review of evidence-based clinical applications. *Acad Radiol*. 2007;15:408–416.

20. Marino M, Helbich T, Baltzer P, Pinker-Domenig K. Multiparametric MRI of the breast: A review. *J Magn Res Imaging*. 2017;47:301–315.

21. Partridge S, McDonald E. Diffusion weighted MRI of the breast: Protocol optimization, guidelines for interpretation, and potential clinical applications. *Magn Reson Imaging Clin N Am*. 2013;21(3):601–624.

22. Abe H, Mori N, Tsuchiya K, Schacht D, Pineda F, Jiang Y et al. Kinetic analysis of benign and malignant breast lesions with ultrafast dynamic contrast-enhanced MRI: Comparison with standard kinetic assessment. *AJR*. 2016;207:1–8.

23. Mori N, Pineda F, Tsuchiya K, Mugikura S, Takahashi S, Karczmar G et al. Fast temporal resolution dynamic contrast-enhanced MRI: Histogram analysis versus visual analysis for differentiating benign and malignant breast lesions. *AJR*. 2018;211:1–7.

24. Platel B, Mus R, Welte T, Karssemeijer N, Mann R. Automated characterization of breast lesions imaged with an ultrafast DCE-MR protocol. *IEEE Trans Med Imaging*. 2013;33(2):225–232.

25. Kuhl C, Schrading S, Stobel K, Schild H, Hilgers R-D, Bieling H. Abbreviated breast magnetic resonance imaging (MRI): First postcontrast subtracted images and maximum-intensity projection—A novel approach to breast cancer screening with MRI. *J Clin Oncol*. 2014;32(22):2304–2310.

26. Brem R, Tabar L, Duffy S, Inciardi M, Guingrich JA, Hashimoto BE et al. Assessing improvement in detection of breast cancer with three-dimensional automated breast US in women with dense breast tissue: The somoInsight study. *Radiology*. 2014;274(3):663–673.

27. Harvey J, Bovbjerg V. Quantitative assessment of mammographic breast density: Relationship with breast cancer risk. *Radiology*. 2004;230(1):29–41.

28. Giger ML, Chan HP, Boone J. Anniversary paper: History and status of CAD and quantitative image analysis: The role of medical physics and AAPM. *Med Phys*. 2008;35(12):5799–5820.

29. Zhang W, Doi K, Giger M, Wu Y, Nishikawa R, Schmidt R. Shift-invariant artificial neural network (SIANN) for CADe in mammography. *Med Phys*. 1994;21:517–524.

30. Zhang W, Doi K, Giger ML, Nishikawa RM, Schmidt RA. An improved shift-invariant artificial neural network for computerized detection of clustered microcalcifications in digital mammograms. *Med Phys.* 1996;23(4):595–601.

31. Sahiner B, Chan H-P, Petrick N, Wei D, Helvie MA, Adler DD et al. Classification of mass and normal breast tissue: A convolution neural network classifier with spatial domain and texture images. *IEEE Trans Med Imaging.* 1996;15(5):598–610.

32. Lambin P, Rios-Velazquez E, Leijenaar R, Carvalho S, van Stiphout RG, Granton P et al. Radiomics: Extracting more information from medical images using advanced feature analysis. *Eur J Cancer.* 2012;48(4):441–446.

33. Gillies R, Kinahan P, Hricak H. Radiomics: Images are more than pictures, they are data. *Radiology.* 2016;278(2):563–577.

34. Giger M. Decoding breast cancer with quantitative radiomics & radiogenomics: Imaging phenotypes in breast cancer risk assessment, diagnosis, prognosis, and response to therapy. *TCGA 4th Scientific Symposium, NIH*, Bethesda, MD, 2015.

35. Colen R, Foster I, Gatenby R, Giger ME, Gillies R, Gutman D et al. NCI workshop report: Clinical and computational requirements for correlating imaging phenotypes with genomics signatures. *Transl Oncol.* 2014;7(5):556–569.

36. Burnside ES, Drukker K, Li H, Bonaccio E, Zuley M, Ganott M et al. Using computer-extracted image phenotypes from tumors on breast magnetic resonance imaging to predict breast cancer pathologic stage. *Cancer.* 2016;122(5):748–757.

37. Li H, Zhu Y, Burnside ES, Huang E, Drukker K, Hoadley KA et al. Quantitative MRI radiomics in the prediction of molecular classifications of breast cancer subtypes in the TCGA/TCIA data set. *NPJ Breast Cancer.* 2016;2:16012.

38. Li H, Zhu Y, Burnside ES, Drukker K, Hoadley KA, Fan C et al. MR imaging radiomics signatures for predicting the risk of breast cancer recurrence as given by research versions of MammaPrint, Oncotype DX, and PAM50 gene assays. *Radiology.* 2016;281(2):382–391.

39. Chen W, Giger ML, Bick U, Newstead GM. Automatic identification and classification of characteristic kinetic curves of breast lesions on DCE-MRI. *Med Phys.* 2006;33(8):2878–2887.

40. Chen W, Giger ML, Li H, Bick U, Newstead GM. Volumetric texture analysis of breast lesions on contrast-enhanced magnetic resonance images. *Magn Reson Med.* 2007;58(3):562–571.

41. Mazurowski MA, Zhang J, Grimm LJ, Yoon SC, Silber JI. Radiogenomic analysis of breast cancer: Luminal B molecular subtype is associated with enhancement dynamics at MR imaging. *Radiology.* 2014;273(2):365–372.

42. Huynh BQ, Li H, Giger ML. Digital mammographic tumor classification using transfer learning from deep convolutional neural networks. *J Med Imaging.* 2016;3(3):034501.

43. Li H, Giger ML, Huynh B, Antropova N. Deep learning in breast cancer risk assessment: Evaluation of convolutional neural networks on a clinical dataset of FFDMs. *J Med Imaging.* 2017;4(4):041304.

44. Mohamed A, Berg Q, Peng H, Luo Y, Jankowitz R, Wu S. A deep learning method for classifying mammographic breast dnesity categories. *Med Phys.* 2018;45:314–321.

45. Li S, Wei J, Chan W, Helvie M, Roubidoux M, Lu Y et al. Computer-aided assessment of breast density: Comparison of supervised deep learning and feature-based statistical learning. *Phys Med Biol.* 2018;63:025005.

46. Lee J, Nishikawa R. Automated mammographic breast density estimation using a fully convolutional netowrk. *Med Phys.* 2018;45:1178–1190.

47. Antropova N, Huynh BQ, Giger ML. A deep feature fusion methodology for breast cancer diagnosis demonstrated on three imaging modality datasets. *Med Phys.* 2017;44(10):5162–5171.

48. Samala R, Chan H, Hadjiiski L, Helvie M, Rickter C, Cha K. Evolutionary pruning of transfer learned deep convolutional neural network for breast cancer diagnosis in digital breast tomosynthesis. *Phys Med Biol.* 2018;63:095005.

49. Samala R, Chan H, Hadjiiski L, Helvie M, Cha K, Rickter C. Multi task transfer learning deep convolutional neural network: Application to computer-aided idagnosis of breast cancer on mammograms. *Phys Med Biol.* 2017;62:8894–8908.

50. Antropova N, Abe H, Giger ML. Use of clinical MRI maximum intensity projections for improved breast lesion classification with deep CNNs. *J Med Imaging.* 2018;5(1):014503.

51. Antropova N, Huynh B, Giger M. Multi-task learning in the computerized diagnosis of breast cancer on DCE-MRIs. arXiv preprint arXiv:170103882. 2017.

52. Kooi T, Ginneken Bv, Karssemeijer N, Heeten Ad. Discriminating solitary cysts from soft tissue lesions in mammography using a pretrained deep convolutional neural network. *Med Phys.* 2017;44:1017–1027.

53. Shi B, Grimm L, Mazurowski M, Baker J, Marks J, King L et al. Prediction of occult invasive disease in ductal carcinoma in situ using deep learning features. *J Am Coll Radiol.* 2018;15:527–534.

54. Vandenberghe M, Scott M, Scorer P, Soderberg M, Balcerzak D, Barker C. Relevance of deep learning to facilitate the diagnosis of HER2 status in breast cancer. *Sci Rep.* 2017;7:45938.

55. Romo-Busheli D, Janowczyk A, Gilmore H, Romero E, Madabhushi A. A deep learning based strategy for identifying and associating mitotic activity with gene expression derived risk categories in estrogen receptor positive breast cancers. *Cytometry Part A.* 2017;91A:566–573.

56. Clark K, Vendt B, Smith K, Freymann J, Kirby J, Koppel P et al. The Cancer Imaging Archive (TCIA): Maintaining and operating a public information repository. *J Digital Imaging.* 2013;26(6):1045–1057.

57. Armato S, Hadjiiski L, Rourassi G, Drukker K, Giger M, Li F et al. LUNGx challenge for computerized lung nodule classification: Reflections and lessons learned. *J Med Imaging.* 2015;2(2).

58. Armato S, Petrick N, Drukker K. PROSTATEx: Prostate MR classification challenge. *SPIE Proceedings* 2017;10134.

59. Boyd N, Martin L, Yaffe J, Minkin S. Mammographic density and breast cancer risk: Current understanding and future prospects. *Breast Cancer Res.* 2011;13:223.

60. Byng J, Yaffe M, Jong R, Shumak R, Lockwood G, Math M et al. Analysis of mammographi density and breast cancer risk from digitized mammograms. *Radiographics.* 1998;18:1587–1598.

61. Huo Z, Giger ML, Olopade OI, Wolverton DE, Weber BL, Metz CE, et al. Computerized analysis of digitized mammograms of BRCA1 and BRCA2 gene mutation carriers. *Radiology.* 2002;225(2):519–526.

62. Li H, Giger ML, Lan L, Bancroft Brown J, MacMahon A, Mussman M et al. Computerized analysis of mammographic parenchymal patterns on a large clinical dataset of full-field digital mammograms: Robustness study with two high-risk datasets. *J Digital Imaging.* 2012;25(5):591–598.

63. Kontos D, Bakic P, Carton A-K, Troxel A, Conant E, Maidment A. Parenchymal texture analysis in digital breast tomosynthesis for breast cancer risk estimation: A preliminary study. *Acad Radiol.* 2009;16:283–298.

64. Giess C, Yeh E, Raa S, Birdwell R. Background parenchymal enhancement at breast MR imaging: Normal patterns, diagnostic challenges, and potential for false-positive and false-negative interpretation. *Radiographics.* 2014;34:234–247.

65. Wu S, Zuley M, Berg W, Kurland B, Jankowitz R, Sumkin J et al. DCE-MRI background parenchymal enhancement quantified from an early versus delayed post-contrast sequence: Association with breast cancer presence. *Sci Rep.* 2017;7:2115.

66. Li H, Giger M, Huynh B, Antropova N. Deep learning in breast cancer risk assessment: Evaluation of convolutional neural netowrkson a clinical dataset of full-field digital mammograms. *J Med Imaging* (Bellingham). 2017;4:041304.

67. Gastounioti A, Oustimov A, Hsieh M, Pantalone L, Conant E, Kontos D. Using convolutional neural networks forenhanced capture of breast parenchymal complexity patterns associated. With breast cancer risk. *Acad Radiol.* 2018;25(8):977–984.

68. Wanders J, Gils Cv, Karssemeijer N, Holland K, Kallenberg M, Peeters P et al. The combined effect of mammographic texture and density on breast cancer risk: A cohort study. *Breast Cancer Res.* 2018;20(1):36.

69. Samala R, Chan H, Hadjiiski L, Helvie M. Analysis of computer-aided detection techniques and signal characteristics for clustered microcalcifications on digital mammography and digital breast tomosynthesis. *Phys Med Biol.* 2016;61:7092–7112.

70. U.S. Food & Drug Administration. 510(k) premarket notification. Available from: https://www.accessdata.fda.gov/scripts/cdrh/cfdocs/cfpmn/pmn.cfm?ID=K170195.

71. Hylton NM, Blume JD, Bernreuter WK, Pisano ED, Rosen MA, Morris EA et al. Locally advanced breast cancer: MR imaging for prediction of response to neoadjuvant chemotherapy—Results from ACRIN 6657/I-SPY TRIAL. *Radiology*. 2012;263(3):663–672.

72. Mahrooghy M, Ashraf AB, Daye D, Mies C, Feldman M, Rosen M et al., editors. Heterogeneity wavelet kinetics from DCE-MRI for classifying gene expression based breast cancer recurrence risk. In: Mori K., Sakuma I., Sato Y., Barillot C., Navab N. (eds.), *Medical Image Computing and Computer-Assisted Intervention: MICCAI International Conference on Medical Image Computing and Computer-Assisted Intervention*, Springer, Berlin, Germany 2013.

73. Dalmis M, Gubern-Merida A, Vreemann S, Karssemeijer N, Mann R, Platel B. A computer-aided diagnosis system for breast DCE-MRI at high spatiotemporal resolution. *Med Phys*. 2016;43(1):84.

74. Antropova N, Huynh B, Giger M, editors. Long short-term memory networks for efficient breast DCE-MRI classification. NIPS: Neural Information Processing Systems, *Medical Imaging Meets NIPS*; 2017.

75. Samala R, Chan H-P, Hadjiiski L, Helvie M, Wei J, Cha K. Mass detection in digital breast tomosynthesis: Deep convolutional neural network with transfer learning from mammography. *Med Phys*. 2016;43(12):6654–6666.

76. Loiselle C, Eby P, Kim J, Calhoun K, Allison K, Gadi V et al. Preoperative MRI improves prediction of extensive occult axillary lymph node metastases in breast cancer patents with a positive sentinel lymph node biopsy. *Acad Radiol*. 2014;21(1):92–98.

77. Mussurakis S, Buckley D, Horsman A. Prediciton of axillary lymph node stastus in invasive breast cancer with dynamic contrast-enhanced MR imaging. *Radiology*. 1997;203(2):317–321.

78. Schacht DV, Drukker K, Pak I, Abe H, Giger ML. Using quantitative image analysis to classify axillary lymph nodes on breast MRI: A new application for the Z 0011 Era. *Eur J Radiol*. 2015;84(3):392–397.

79. Wu J, Sun X, Want J, Cui Y, Kato F, Shirato H et al. Identifying relations between imaging phenotypes and molecular subtypes of breast cancer: Model discovery and external validation. *J Magn Reson Imaging*. 2017;46(4):1017–1027.

80. Grimm LJ, Zhang J, Mazurowski MA. Computational approach to radiogenomics of breast cancer: Luminal A and luminal B molecular subtypes are associated with imaging features on routine breast MRI extracted using computer vision algorithms. *J Magn Reson Imaging*. 2015;42(4):902–907.

81. Liang C, Cheng Z, Huang Y, He L, Chen X, Ma Z et al. An MRI-based radiomics classifier for preoperative prediction of Ki-67 status in breast cancer. *Acad Radiol*. 2018;25:1111–1117.

82. Adamo B, Rita Ricciardi GR, Ieni A, Franchina T, Fazzari C, Sanò MV et al. The prognostic significance of combined androgen receptor, E-Cadherin, Ki67 and CK5/6 expression in patients with triple negative breast cancer. *Oncotarget*. 2017;8(44):76974–76986.

83. Criscitiello C, Bagnardi V, Pruneri G, Vingiani A, Esposito A, Rotmensz N et al. Prognostic value of tumour-infiltrating lymphocytes in small HER2-positive breast cancer. *Eur J Cancer*. 2017;87:164–171.

84. Gupta I, Ouhtit A, Al-Ajmi A, Rizvi SGA, Al-Riyami H, Al-Riyami M et al. BRIP1 overexpression is correlated with clinical features and survival outcome of luminal breast cancer subtypes. *Endocr Connect*. 2017;7:65–77.

85. Kjaer IM, Bechmann T, Brandslund I, Madsen JS. Prognostic and predictive value of EGFR and EGFR-ligands in blood of breast cancer patients: A systematic review. *Clin Chem Lab Med*. 2017;56:688–701.

86. Niméus E, Folkesson E, Nodin B, Hartman L, Klintman M. Androgen receptor in stage I-II primary breast cancer -prognostic value and distribution in subgroups. *Anticancer Res*. 2017;37(12):6845–6853.

87. Mani S, Chen Y, Li X, Arlinghaus L, Chakravarthy A, Abramson V et al. Machine learning for predicting the response of breast cancer to neoadjuvant chemotherapy. *J Am Med Inform Assoc*. 2018;20:688–695.

88. Shin S, Cho N, Lee H-B, Kim S-Y, Yi A, Kim S-Y et al. Neoadjuvant chemotherapy and surgery for breast cancer: Preoperative MRI features associated with local recurrence. *Radiology*. 2018;289:30–38.

89. Ashraf A, Daye D, Gavenonis S, Mies C, Feldman M, Rosen M et al. Identification of intrinsic imaging phenotypes for breast cancer trumors: Preliminary associations with gene expression profiles. *Radiology*. 2014;272(2):374–384.

90. Veer Lvt, Dai H, Vijver Mvd, He YD, Hart AA, Mao M et al. Gene expression profiling predicts clinical outcome of breast cancer. *Nature*. 2002;415(6871):530–536.

91. Van De Vijver MJ, He YD, van't Veer LJ, Dai H, Hart AA, Voskuil DW et al. A gene-expression signature as a predictor of survival in breast cancer. *N Engl J Med*. 2002;347(25):1999–2009.

92. Paik S, Shak S, Tang G, Kim C, Baker J, Cronin M et al. A multigene assay to predict recurrence of tamoxifen-treated, node-negative breast cancer. *N Eng J Med*. 2004;351(27):2817–2826.

93. Paik S, Tang G, Shak S, Kim C, Baker J, Kim W et al. Gene expression and benefit of chemtherapy in women with node-negative, estrogen receptor-positive breast cancer. *J Clin Oncol*. 2006;24(23):3726–3734.

94. Cronin M, Sangli C, Liu M-L, Pho M, Dutta D, Nguyen A et al. Analytical validation of the Oncotype DX genomic diagnostic test for recurrence prognosis and therapeutic response prediction in node-negative, estrogen receptor-positive breast cancer. *Clin Chem*. 2007;53(6):1084–1091.

95. Parker J, Mullins M, Cheany M, Leung S, Voduc D, Vickery T et al. Supervised risk predictor of breast cancer based on intrinsic subtypes. *J Clin Oncol*. 2009;27(8):1160–1167.

96. Prat A, Parker J, Fan C, Perou C. PAM50 assay and the three-gene model for identifying the major and clinically relevant molecular subtypes of breast cancer. *Breast Cancer Re Treat*. 2012;135(1):301–306.

97. Esserman LJ, Berry DA, Cheang MC, Yau C, Perou CM, Carey L et al. Chemotherapy response and recurrence-free survival in neoadjuvant breast cancer depends on biomarker profiles: Results from the I-SPY 1 TRIAL (CALGB 150007/150012; ACRIN 6657). *Breast Cancer Res Treat*. 2012;132(3):1049–1062.

98. Drukker K, Li H, Antropova N, Edwards A, Papaioannou J, Giger ML. MRI-based prediction of recurrence-free survival in breast cancer patients early on in neoadjuvant chemotherapy. Am Assoc Phys Med. 2017;44(6):2752.

99. Wu J, Cao G, Sun X, Lee J, Rubin D, Napel S et al. MR imaging predicts recurrence-free survival in locally advanced breast cancer treated with neoadjuvant chemotherapy. *Radiology*. 2018;288:26–35.

100. Ashraf A, Gavenonis S, Daye D, Mies C, Rosen M, Kontos D. A multichannel Markov random field framework for tumor segmentation with an application to classification of gene expression-based breast cancer recurrence risk. *IEEE Trans Med Imaging*. 2013;32(4):637–648.

101. Ashraf A, Gaonkar B, Mies C, DeMichele A, Rosen M, Davatzikos C et al. Breast DCE-MRI kinetic heterogeneity tumor markers: Preliminary associations with neoadjuvant chemotherapy response. *Transl Oncol*. 2015;8(3):154–162.

102. Zhu Y, Li H, Guo W, Drukker K, Lan L, Giger ML et al. Deciphering genomic underpinnings of quantitative MRI-based radiomic phenotypes of invasive breast carcinoma. *Sci Rep*. 2015;5:17787.

103. Guo W, Li H, Zhu Y, Lan L, Yang S, Drukker K et al. Prediction of clinical phenotypes in invasive breast carcinomas from the integration of radiomics and genomics data. *J Med Imaging*. 2015;2(4):041007.

Radiomics for lung cancer

JIE TIAN, DI DONG, AND SHUO WANG

15.1 RADIOMICS APPROACH FOR LUNG CANCER

Radiomics analysis quantifies the relationship between medical images and clinical outcomes of lung cancer [1]. As described in Figure 15.1, a typical radiomics pipeline includes three steps: (1) Tumor detection and segmentation. Since radiomics mainly analyzes tumor imaging information, segmentation of the tumor region of interest (ROI) is necessary. (2) Feature extraction and quantification. Quantitative radiomic features are extracted inside the ROI to quantify tumor phenotype. (3) Diagnostic and prognostic model building. Machine learning methods are used to establish prediction models for clinical outcomes using the radiomic features [2].

15.1.1 TUMOR DETECTION AND SEGMENTATION

Accurate tumor segmentation on CT image is necessary for radiomics analysis in lung cancer. Manual delineation of tumor contour is accurate but requires much effort and is very time-consuming. Automatic lung tumor or lung nodule segmentation methods are therefore proposed by many researches [3]. As shown in Figure 15.2, due to the various appearances of nodules, the automatic segmentation is challenging. Isolate and solid nodule is relatively easy for automatic segmentation (Figure 15.2a(1)). However, when the nodule is attached to vessel (Figure 15.2a(2)) or pleura (Figure 15.2a(3)), many methods tend to include the vessels and lung wall tissues into nodule due to the very close CT intensity between nodule and these tissues. For nodules containing cavity, segmentation methods may treat the cavity as normal lung tissue, and consequently cause under-segmentation. On the other hand, the intensity of nodule can be very close to that of normal lung tissue, such as ground-glass opacity (GGO) nodule. When vessels or pleural tissues are attached to GGO, the segmentation is more challenging (Figure 15.2c(1)–c(3)).

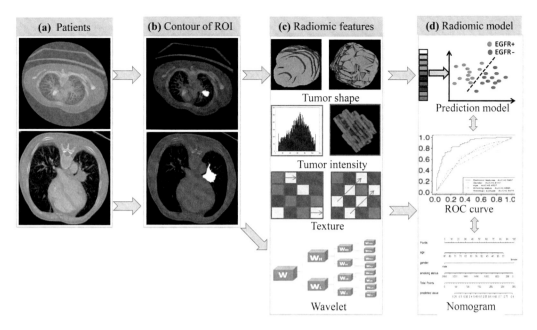

Figure 15.1 (See color insert.) Radiomics analysis pipeline on medical images [2]. (a) CT images of lung cancer. (b) Tumor detection and segmentation in CT images. (c) Feature extraction and quantification. Features are extracted from the segmented tumor ROI to quantify the tumor intensity, shape, texture, and wavelet information. (d) Diagnostic and prognostic model building. The radiomic features and clinical outcomes are associated by machine learning algorithms. (From Zhang, L.W. et al., *Transl. Oncol.*, 11, 94–101, 2017.)

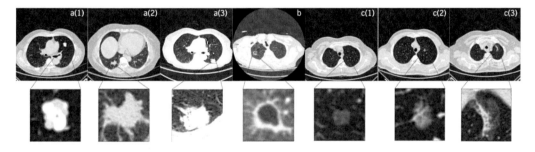

Figure 15.2 (See color insert.) Different types of lung nodules. a(1): isolate and solid lung nodule, a(2): juxta-vascular lung nodule, a(3): juxta-pleural lung nodule, (b): cavity lung nodule, c(1): isolate GGO, c(2): juxta-vascular GGO, c(3): juxta-pleural GGO. (From Song, J. et al., *IEEE Trans. Med. Imaging*, 35, 337–353, 2016.)

To deal with the diversity of lung nodules, many detection or segmentation algorithms have been proposed for lung nodule [4–13]. D. M. Campos et al. [4] used a two-step supervised method to address the lung nodule segmentation task: (1) they used two region growing methods based on volumetric shape index (VSI) and convergence index filter, and a machine learning method based on k-Nearest Neighbor (k-NN) classifier to generate three independent results. In the first region growing method, a VSI that indicated the possibility of each voxel being part of the nodule was estimated. In the second region growing method, a convergence index filter was used on the image to generate a probability map for each voxel. This filter is useful in segmenting low contrast nodules such as GGO. In the machine learning method, a k-NN classifier was trained to classify each voxel into nodule or healthy tissue. Therefore, the three segmentation methods generated three probability maps where each voxel was assigned a probability of being nodule. (2) They trained an artificial neural network (ANN) to generate the final segmentation result. Since three preliminary segmentation results were acquired in the first step, the final segmentation result needed to be generated by combining these preliminary results. The authors trained an ANN classifier which used the three preliminary probability maps as inputs and generated a final segmentation result. By testing on CT images from 20 lung nodules, this method reached a relative volume error of 12% and a Dice score of 0.9. However, the testing dataset was relatively small to include a diversity of nodules. For example, juxta-pleural nodules may be difficult for segmentation using region growing methods. S. Diciotti et al. [5] proposed a 3D geodesic distance map representation of local shape analysis to particularly deal with juxta-vascular nodules. This method required only a simple intensity threshold to acquire initial nodule shapes. In their study, 88.5% of the 157 nodules from the Lung Image Database Consortium image collection (LIDC-IDRI) dataset (https://wiki.cancerim-agingarchive.net/display/Public/LIDC-IDRI) were correctly segmented. In [8], the researchers proposed a segmentation approach based on the level set algorithm. Defining a signed distance function based on the implicit spaces, they proposed a general shape model. Finally, more than 94.3% lung nodules were correctly segmented. To concentrate on segmenting juxta-vascular nodules, another study that used flow entropy and geodesic distance was proposed [9]. Based on the prior that vessel pixels were always far from the nodule center in geodesic distance, they combined k-means clustering, geodesic distance, and flow entropy to detach vessels from nodules. Recently, a lung tumor delineation method requiring only a single user-selected seed point was proposed [10]. They first segmented lung boundaries to detach nodule from pleura. Afterwards, a region growing algorithm was performed starting from a user-specified seed point. In a dataset containing 129 patients, the method achieved similarity index of more than 0.78 when compared with two manual segmentation results. In [13], the authors proposed a new segmentation method to detect all abnormal imaging patterns including consolidations, GGO, and nodules. By combining fuzzy connectedness and rib cage, this method could delineate many abnormal patterns or structures in lung.

To realize fully automatic lung lesion segmentation, it is necessary to develop an automatic lung nodule detection method that can initialize many lung nodule segmentation methods. In [3], a region growing-based method was proposed for automatic nodule detection and segmentation. Their method achieved good results especially in segmenting GGO nodules, as shown in Figure 15.3. Methods for 3D lung volume segmentation were also provided in [14] and [15], where the methods achieved less than one-pixel error compared to human segmentation results. In other similar studies, lung nodule detection performance achieved an accuracy of 89%, and the lung volume segmentation accuracy achieved 90% [16–20], where intensity and texture features were both used. In [17], a deep learning method was proposed for convenient and fast lung nodule detection. By incorporating a deep learning model named faster-RCNN, the average Free-response Receiver Operating Characteristic (FROC) value was 0.89 in public LUNA dataset (https://luna16.grand-challenge.org/). In [19],

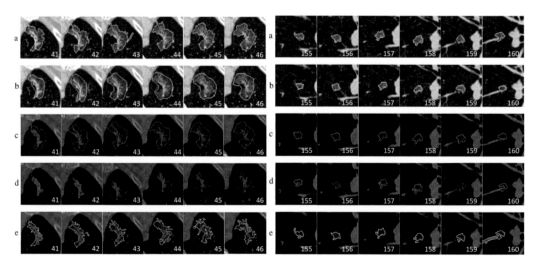

Figure 15.3 Segmentation results of two GGOs (left and right) from two patients: (a) result from method in [3], (b) manual segmentation by one radiologist, (c) manual segmentation by another radiologist, (d) level set result, and (e) skeleton graph cut result. (From Song, J. et al., *IEEE Trans. Med. Imaging*, 35, 337–353, 2016.)

candidate nodules were extracted by 15 different logic opening operations. Afterwards, phenotypical features were used for reducing false-positive candidate nodules. However, the segmentation accuracy was only 63% in this study, which was relatively poor.

Recently, S. Wang et al. [18] proposed a model, termed the Central Focused Convolutional Neural Networks (CF-CNN). This data-driven model could segment lung nodules from heterogeneous CT images. This model had three characteristics. First, they proposed a CF-CNN model to segment a variety of lung nodules. When dealing with juxta-pleural nodules, this method did not require nodule shape hypothesis or complicated parameter settings by users. Second, the CF-CNN model included two CNN branches to extract both multi-scale 2D features and 3D features for the tumor. Third, they developed a well-designed central pooling layer to consider the space information for each voxel in CT image. The central pooling layer reserved many features close to image center, and compressed features close to image edges. Several representative segmentation results of this method are shown in Figure 15.4.

15.1.2 FEATURE EXTRACTION AND QUANTIFICATION

Extraction and quantification of tumor phenotypes are the key to construct a classification model and to implement clinical analysis. The informative and discriminative features play a crucial role in building an effective model. Given that feature extraction is always followed by feature selection or dimension reduction in radiomics workflow, it is required to quantify the characteristics of tumor phenotype from multiple aspects to obtain as complete a feature set as possible. In a previous study, the authors focused on analyzing the quantitative phenotype characteristics associated with the overall survival in Non-Small Cell Lung Cancer (NSCLC) [21]. They used a series of feature extraction methods to mine the representative information from CT images. As introduced in their study, the radiomic features could be divided into the four groups: shape, texture, wavelet, and Gabor. The illustration of these groups is shown in Figure 15.5, which is similar to [1]. The four groups of features are described in detail below.

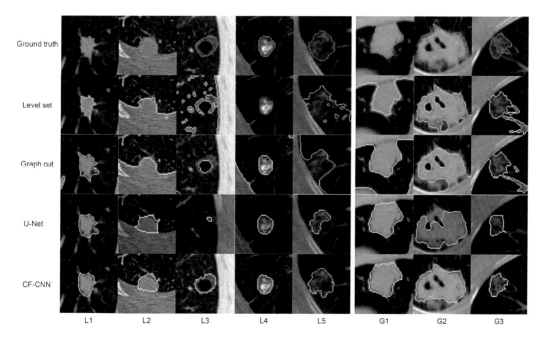

Figure 15.4 (See color insert.) Segmentation results of various types of lung nodules. From top to bottom: ground truth, level set result, graph cut result, U-Net result, and CF-CNN result. L means images from the LIDC dataset. G means images from Guangdong General Hospital. (From Wang, S. et al., *Med. Image Anal.*, 40, 172–183, 2017.)

3D features 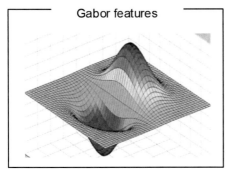	Texture features 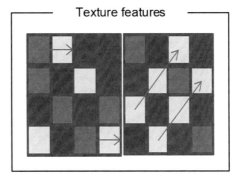
Gabor features 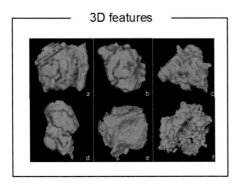	Wavelet features 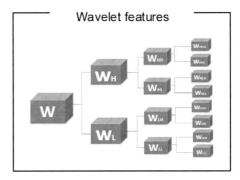

Figure 15.5 (See color insert.) Groups of radiomic features. Normally, the radiomic features include shape features, Gabor features, texture features, and wavelet features. (From Song, J. et al., *IEEE Trans Med Imaging*, 35, 337–353, 2016; Zhang, L.W. et al., *Transl. Oncol.*,11, 94–101, 2017.)

Group 1: Texture

1. **Gray-level run-length matrix (GLRLM)**

   The GLRLM could describe the distribution of the local gray-level continuity in the tumor region. The run-length is defined as the length of the consecutive pixels that has the same gray-level value. The GLRLM features are generated from the GLRLM using a series of statistical functions. The GLRLM $M(i, j|\theta)$ is a high-order statistical texture matrix, where the value of the (i, j)th element is the number of the consecutive pixels with i gray-level and j run-length in the θ direction.

2. **Gray-level co-occurrence matrix (GLCM)**

   The GLCM measures the textural characteristics of the image by quantifying the gray-level uniformity and variation of the co-occurrence pairs, i.e. the pairs of the adjacent pixels. Same with GLRLM features, the GLCM features are also calculated by evaluating the matrix via several algorithms. The GLCM $M(i, j; D, \theta)$ with a size $N(g) \times N(g)$ is a second-order statistical texture matrix, where the value of the (i, j)th element is the number of the pixel pairs separated by D pixels distance in the θ direction and with i and j gray-level respectively.

Group 2: Gabor

For Gabor feature extraction, the images should firstly be filtered by a group of Gabor filters. This kind of filters is defined to detect the gray edge in some scales and directions. The Gabor features are usually integrated into the automatic systems for object recognition and object detection. By enhancing the specific discriminative phenotypes, such as the tissues' and organs' edges, they are thought to represent some valuable information of the lung lesion.

Group 3: Shape

The shape features are generated based on the delineated tumor ROIs and could describe their morphological properties. It reflects and describes the three-dimensional size and detailed appearance of the tumor ROIs, such as surface area, volume, compactness, sphericity and so on.

Group 4: Wavelet

The wavelet transform efficaciously decouples the texture information through decomposing the tumor in both low and high frequencies, which is similar to Fourier analysis.

The comprehensive phenotype features are dominative in prognosis analysis, which is supported in [22]. In this study, a feature set of 329 radiomic features including 219 three-dimensional and 110 two-dimensional features were extracted. To select the reproducible and non-redundant features, statistical analysis was applied. Meanwhile, the authors used a repeated experiment to investigate whether the stability of features extraction and selection was enough to resist the influence of different readers. In the end, 29 key features were proved to be significant and reserved for further study.

Furthermore, the tumor phenotype keeps changing during the radiation therapy process. This variation is associated with therapeutic response and prognosis of the patients and could be measured using the delta radiomic features [23]. By introducing a time component and comprising different time-dependent feature groups, delta radiomics could provide the information on the evolution of feature values, and therefore facilitate the image-based monitor on the patients [24]. Considering the tumor status changes at all stages of treatment, the delta radiomics will play an important role in clinical individualized management of lung cancer.

15.1.3 DIAGNOSTIC AND PROGNOSTIC MODEL BUILDING

After extracting radiomic features, a diagnostic or prognostic model will be constructed to build the relationship between image and clinical outcomes. Depending on clinical applications, different machine learning methods can be used to predict the clinical outcomes. For example, using the support vector machine (SVM) method to identify malignant lung nodules from the benign lesions by radiomic features; using a neural network model to predict the prognosis of lung cancer; and using logistic regression to predict the distant metastasis of lung cancer.

In [25], researchers used radiomic features to predict the emergence of cancer on National Lung Cancer Screening Trial (NLST) dataset (https://www.cancer.gov/types/lung/research/nlst/). A machine learning method named random forest was used to predict the probability of the nodule to become cancerous in one and two years. The corresponding accuracies were about 79%. D. Kumar et al. proposed a method to identify benign and malignant lesions from 93 patients in the LIDC-IDRI dataset [26]. Compared with the clinical-proved ground truth, their method achieved an average accuracy of 77.52% in cross-validation, and the sensitivity was 79.06% and the specificity was 76.11%. A recent study found that through a self-learning strategy, convolutional neural network (CNN) had the potential to produce better predictions than traditional machine learning models [27]. CNNs do not require precise tumor image segmentation, therefore, it is convenient to use in practice. A typical CNN framework for lung nodule classification is depicted in Figure 15.6.

The unsupervised clustering method and cox proportional hazard regression method are often used in radiomics study for prognostic analysis of lung cancer. In [1], researchers used an unsupervised clustering method to group NSCLC and head and neck cancer patients into three categories according to 440 radiomic features. The experimental results indicated that pathological staging and clinical staging of NSCLC were highly correlated with the three groups. Moreover, the radiomic signature could stratify the NSCLC patients into low-/high-risk groups with different overall survivals. Y. Q. Huang et al. found that the radiomic signature was significantly associated with the Disease-Free Survival (DFS) of NSCLC [28]. By combining the radiomic signature with other commonly used clinical factors, the individualized DFS could be estimated well in patients with early-stage NSCLC. Another study demonstrated that radiographic features showed good predictive value for both distant metastasis of lung cancer and survival of the patients, and the combination of radiomic features and clinical data could significantly improve the prediction performance [29]. T. P. Coroller et al. used radiomics method to predict pathological response of NSCLC patients after neoadjuvant chemoradiation [30]. Their study demonstrated that radiomic model could provide valuable information to estimate the treatment response in NSCLC. J. Song et al. found that radiomics could predict the Progression-Free Survival (PFS) of late-stage EGFR (Epidermal Growth Factor Receptor)-mutated NSCLC patients after tyrosine kinase inhibitors (TKI) therapy [31]. Delta-radiomics was also used to predict the treatment outcomes of lung cancer [24]. These above-mentioned studies suggested that radiomic model could be used as an indicator for lung cancer diagnosis, treatment outcome prediction, and prognosis.

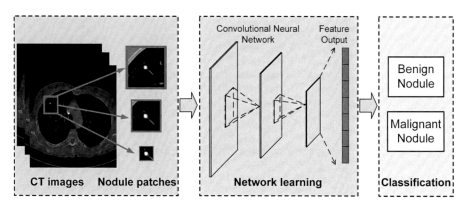

Figure 15.6 (See color insert.) Multi-scale CNN for lung nodule classification. A rectangle bounding box covering the whole lung nodule is manually selected as the input to the CNN. Afterwards, the CNN learned discriminative features from nodule images automatically. Finally, a classifier is used to classify the input nodule as malignant or benign [27]. (From Shen, W. et al., Multi-scale convolutional neural networks for lung nodule classification, in *IPMI 2015: Information Processing in Medical Imaging*, pp. 588–599, 2015.)

15.2 RADIOGENOMICS APPROACH FOR LUNG CANCER

Acquiring gene profiles for lung cancer is very important because the treatment decision-making and the prognostic prediction of cancer increasingly relies on gene profiles [32]. For instance, identifying EGFR gene mutation status can guide treatment decision-making for lung adenocarcinoma. If EGFR mutation is detected in patients with lung adenocarcinoma, targeted therapies can be applied, such as the use of erlotinib and gefitinib drugs [33–34].

Traditional methods for acquiring gene profiles often use biopsy and gene sequencing. However, due to the high heterogeneity of tumors, the biopsy results may include false negative diagnosis. For instance, when detecting EGFR mutation status by biopsy, EGFR-mutant tissues may not be biopsied, and therefore cause false negative diagnosis. In addition, biopsy testing is usually performed only once, but the gene mutation status may change through time. Tissue quality requirements and the relatively high costs also raise another challenge for biopsy testing in some situations.

Radiogenomics is a technology used for discovering the relationship between gene profiles and imaging phenotypes. Different gene mutations or gene pathway abnormalities may cause visual differences on tumor images such as tumor shape and texture variance. Current studies have shown progress in finding the relationships between image characteristics and gene pathways. Several studies also achieved gene mutation prediction through CT images. In the radiogenomics pipeline, machine learning models or statistical models are used to predict gene mutations or find the associations between image features and gene pathways.

15.2.1 METAGENE ANALYSIS

M. Zhou et al. used radiogenomics method to identify relationships between gene profiles and CT imaging phenotypes in NSCLC [35]. They studied a cohort of 113 patients with NSCLC who had both preoperative CT images and gene profiles. For each tumor, 87 semantic image features were extracted by a thoracic radiologist. Most semantic features were binary values reflecting the absence or presence of the corresponding visual characteristics. For example, internal air bronchograms indicated whether the air bronchograms existed inside tumor. Similarly, nodule calcification, attachment to pleura and primary patterns of nodule margins reflected the texture, shape, and the relationship between nodule and surrounding microenvironment. Afterwards, gene expression data was acquired from each tumor by RNA sequencing technology. The top 10 metagenes with the highest cluster homogeneities in five external gene expression data sets were selected [36]. Finally, they built a radiogenomics map to associate semantic image features with metagenes. The t statistic and Spearman correlation metric were used to assess significant associations between semantic features and gene expression clusters. To correct multiple testing, the false discovery rate was also used. The statistically significant associations between metagenes and image features were identified by p<0.05 and a false discovery rate value less than 0.01. In addition, the directional relationships between metagenes and semantic CT features were indicated by feature-to-metagene correlation coefficients. Due to the high-dimensionality of semantic features, the authors removed features with variance <10% and finally 35 semantic image features were reserved, including nodule location, nodule margins, nodule attenuation, nodule ground-glass composition, and the presence of emphysema for subsequent statistical analysis. In this study, 32 statistically significant correlations between semantic features and metagenes were found. For example, nodule attenuation and nodule margins were highly related with the metagene capturing the late cell cycle. When this metagene was active, the lesion tended to be solid (metagene 38), whereas when this metagene was inactive, the lesion tended to have poorly defined margins. Similarly, this study found that a normal lung background was positively correlated to a mesothelioma survival signature (metagene 60), which indicated that active genes in this pathway were associated with underlying lung parenchymal abnormalities. Specifically, when this pathway was under-expressed, the patients tended to have emphysema; when this pathway was overexpressed, the lung parenchymal morphologic structure tended to be normal. In addition, metagene 65 included genes associated with the epithelial-to-mesenchymal transition and had several significant associations with image features. For example, overexpression of metagene 65 might cause poorly defined margins, and under-expression of metagene 65 might cause smooth margins. In conclusion, the experimental results for radiogenomics analysis demonstrated that there were multiple associations between CT semantic features and molecular pathway activity in NSCLC.

15.2.2 GENE MUTATION PREDICTION

In the diagnosis of gene mutation in lung cancer, E.R. Velazquez et al. used quantitative CT image features to predict EGFR mutation [37]. In their study, they adjusted CT images to isometric voxels (3mm) via cubic interpolation method. Afterwards, radiomic features were extracted using 3D-Slicer software. These radiomic features included shape features, tumor intensity features, texture features, wavelet features, and Laplacian of Gaussian features. In order to ensure the robustness and repeatability of the features, a NSCLC test-retest dataset was used to remove the features with low (under 0.8) intra-class correlation coefficients. On the basis of this, the authors used the method of principal component analysis to select the features based on Pearson coefficient greater than 0.9. The remaining 26 characteristics were used to study the differences among EGFR / KRAS / others. In each case, six clinical features and 20 features selected by minimum Redundancy Maximum Relevance (mRMR) were added into the final random forest classification model. For the development and validation of radiomic and clinical based signatures, a training set of 353 cases and a validation set of 352 cases were acquired. After the steps of feature selection and univariate analysis, 16/14/10 radiomic features were found significantly different between EGFR+/EGFR-, EGFR+/KRAS+, and KRAS+/KRAS-, respectively. Combing the top 20 mRMR-ranked features, three radiomic signatures were constructed to address the three kinds of mutation problems. The experimental results demonstrated that the three signatures were all significant for mutation prediction in the validation set. Furthermore, a clinical model based on the clinical variables (age, gender, race, smoking status, and clinical stage) achieved better predictive performance, and yielded higher area under the receiver operating characteristic (ROC) curve (AUC) values than radiomic signatures. While for distinguishing KRAS-mutant from KRAS non-mutated tumors, the clinical model achieved the best results among all the predictors.

In a study of 180 patients with NSCLC, L.W. Zhang et al. [2] built a radiomic model for predicting EGFR mutation. To quantitate the phenotype characteristics of tumors, they extracted 485 radiomic features from the tumor ROI. These features included four categories: shape, intensity, texture, and wavelet. To remove the redundant features and retain the potential discriminative features, the authors used the least absolute shrinkage and selection operator to select the important features and utilized a logistic regression method to develop a radiomic signature. Then they used multivariable logistic regression to build a radiomic nomogram based on clinical factors and radiomic signature. The authors validated their models on a validation cohort of 40 patients, and found the radiomic nomogram yielded an AUC of 0.87 in predicting EGFR mutation, which was significantly higher than that obtained from other predictors.

15.2.3 DEEP LEARNING IN RADIOGENOMICS

In recent years, deep learning has emerged as a research hotspot in radiomics and radiogenomics. A recent study investigated the prediction of lung cancer EGFR mutation with a convolutional neural network (CNN) and demonstrated good classification performance [38]. The use of CNN in radiogenomics classification task could reduce time consumption in tumor segmentation and the influence caused by ROI delineation. The framework of CNN is given in Figure 15.7. Their study proposed a three-layer CNN to learn the mapping between tumor images and corresponding EGFR mutation status. Due to the limitation of data amount, they used the CNN only for feature learning. After a softmax layer, the last fully connected layer in this network was trained to act as a feature extractor. The 30 features extracted from CNN were then fed into an SVM classifier for final prediction. The deep learning model was validated on two independent datasets, revealing good predictive ability and outperforming conventional hand-crafted radiomic features. The quantitative results of this deep learning model are shown in Table 15.1. The ROC curves on the two datasets are illustrated in Figure 15.8.

Besides, S. Wang et al. [39] collected 844 patients with lung adenocarcinoma and built an end-to-end deep learning model to differentiate EGFR-mutant patients from EGFR-wild type patients. Considering the relatively small training size in this medical classification task, the authors leveraged the transfer learning method to prevent their deep convolutional neural network from over-fitting. Having 24 weighted layers, the deep network achieved satisfying predictive performance with AUCs of 0.85 and 0.81 in the training and validation cohorts respectively. Meanwhile, the authors also constructed three models for comparison based on the clinical factors, semantic features, and hand-crafted radiomic features, respectively. The experimental results revealed that deep learning model exhibited significantly better predictive ability than all above three comparative models in the validation cohorts. Furthermore, to identify the tumor regions with high probability of EGFR-mutant, they visualized the deep learning model to acquire the suspicious region that might help choosing biopsy position.

The concept of radiogenomics consistently builds bridges between quantitative information from medical images and gene expression. Given the rapidly development of deep mining methods of image features, the researches focusing on the connections among image features, gene expression, pathology and prognosis will become important research objectives in the future. A new era of personalized medicine has dawned with the combination of genomics and radiomics at the pathological and genetic level.

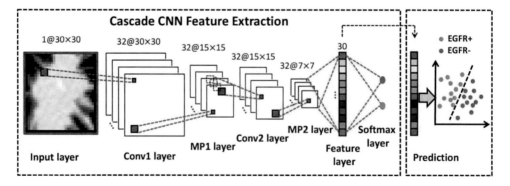

Figure 15.7 (See color insert.) Structure of the proposed deep learning model for EGFR-mutation prediction. (From Yu, D. et al., Convolutional neural networks for predicting molecular profiles of non-small cell lung cancer biomedical imaging, in *2017 IEEE 14th International Symposium on Biomedical Imaging (ISBI 2017)*, Melbourne, Australia, pp. 569–572, 2017.)

Table 15.1 Predictive performance on two datasets (mean±std). AUC (area under the ROC curve), ACC (accuracy), SEN (sensitivity), SPE (specificity)

	Cross validation dataset1				Testing on dataset2			
	AUC	ACC(%)	SEN(%)	SPE(%)	AUC	ACC(%)	SEN(%)	SPE(%)
STG30	0.726±0.006	68.55±0.80	68.49±1.22	68.61±1.06	0.494	58.55	25.93	72.76
STG60	0.734±0.007	70.27±0.79	69.07±1.16	71.33±1.07	0.655	63.66	55.56	67.19
STG100	0.721±0.007	69.22±0.87	68.50±1.36	69.84±1.14	0.636	55.85	76.65	46.80
STG150	0.722±0.007	68.71±0.85	67.79±1.38	69.50±1.24	0.609	50.51	77.78	38.64
Deep30	0.828±0.005	76.16±0.84	73.80±1.15	78.24±1.08	0.668	67.55	48.59	75.81

Source: Yu, D. et al., Convolutional neural networks for predicting molecular profiles of non-small cell lung cancer biomedical imaging, in *2017 IEEE 14th International Symposium on Biomedical Imaging (ISBI 2017)*, Melbourne, Australia, pp. 569–572, 2017.

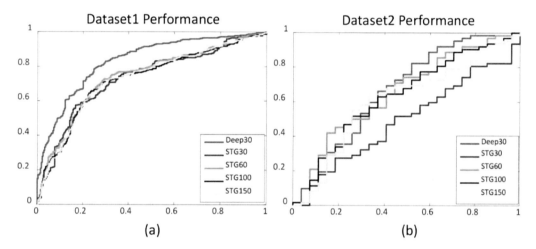

Figure 15.8 (See color insert.) ROC curves of EGFR mutation prediction based on feature sets with different number of features in Dataset1 (a) and Dataset2 (b). (From Yu, D. et al., Convolutional neural networks for predicting molecular profiles of non-small cell lung cancer biomedical imaging, in *2017 IEEE 14th International Symposium on Biomedical Imaging, (ISBI 2017)*, Melbourne, Australia, pp. 569–572, 2017.)

15.3 SUMMARY

We mainly introduced the previous studies of radiomics and radiogenomics in lung cancer in this chapter. In summary, radiomics has the potential to be the next frontier in lung cancer studies. The radiomics is an easy-to-use tool based on the routinely used imaging data, which could provide quantitative and deep-level imaging information of the entire tumor. The additional imaging information may promote the precision medicine of lung cancers and improve the clinical outcomes. For example, radiomics can preoperatively identify malignant tumors and predict treatment outcomes based on machine learning methods. The preoperative identification of patients with different prognosis may help avoid overtreatments or insufficient treatments. In addition, for the lung nodules that are difficult for biopsies, radiomics may provide an auxiliary non-invasive tool for clinicians. Therefore, radiomics is a good supplement to current diagnosis methods [40].

Despite of its promising future, radiomics still faces enormous challenges [41]. First of all, the most prominent challenge is the collection of huge amount of standard medical imaging data since the radiomic models are driven by the imaging data. Although several datasets, such as LIDC, NSLT, and LUNA, are already public available, there are still lack of open-access lung cancer datasets with detailed therapeutic information such as immunotherapy, targeted therapy, or radiotherapy. Moreover, the privacy protection and limitation of imaging data transfer make multi-center radiomics studies difficult. The distributed learning of radiomics, which needs only model parameters transfer rather than data transfer, may be a possible solution for future multi-center studies [23]. Besides, the radiomics is sensitive to imaging parameters and imaging systems, which makes the radiomic models often fail in multi-center and prospective applications. This is a great challenge for radiomics to achieve real clinical applications. Therefore, more and more researchers are focusing on developing robust features among different imaging systems or imaging parameters by using phantom study or deep learning feature extraction. In addition, the comprehensive evaluation of a lung cancer radiomics study is essential for its potential application. P. Lambin et al. proposed a standard quality score system (RQS, http://www.radiomics.world/) to aid the researchers in assessing the quality of radiomics studies [23]. The wide use of this scoring system will promote the radiomics study in lung cancer. In the future, it is a trend to combine radiomics with genomics, proteomics, and other -omics to provide valuable information for personalized medicine. With the emerging of deep learning algorithms, high quality imaging data, and multi-omics fusion technologies, radiomics will certainly make great progresses in lung cancer studies in the future [42].

REFERENCES

1. Aerts HJ, Velazquez ER, Leijenaar RT et al. (2014) Decoding tumour phenotype by noninvasive imaging using a quantitative radiomics approach. *Nat Commun* 5:4006.
2. Zhang LW, Chen BJ, Liu X et al. (2017) Quantitative biomarkers for prediction of epidermal growth factor receptor mutation in non-small cell lung cancer, *Transl Oncol* 11:94–101.
3. Song JD, Yang CY, Fan L et al. (2016) Lung lesion extraction using a toboggan based growing automatic segmentation approach. *IEEE Trans Med Imaging* 35(1):337–353.
4. Campos DM, Simões A, Ramos I, Campilho A (2014) Feature-based supervised lung nodule segmentation. In *The International Conference on Health Informatics*. Springer, Cham, Switzerland, pp. 23–26.
5. Diciotti S, Lombardo S, Falchini M, Picozzi G, Mascalchi M (2011) Automated segmentation refinement of small lung nodules in CT scans by local shape analysis. *IEEE Trans Biomed Eng* 58:3418–3428.
6. Dubey SR, Singh SK, Singh RK (2015) Local wavelet pattern: A new feature descriptor for image retrieval in medical CT databases. *IEEE Trans Image Process* 24:5892–5903.
7. Wu DJ, Lu L, Bi JB et al. (2010) Stratified learning of local anatomical context for lung nodules in CT images. *IEEE Conference on Computer Vision and Pattern Recognition (CVPR)*. doi:10.1109/Cvpr.2010.5540008:2791–2798.
8. Farag AA, Abd El Munim HE, Graham JH, Farag AA (2013) A novel approach for lung nodules segmentation in chest CT using level sets. *IEEE Trans Image Process* 22:5202–5213.
9. Sun SS, Guo Y (2013) Juxta-vascular nodule segmentation based on the flowing entropy and geodesic distance feature. *J Investig Med* 61:S8–S8.
10. Gu Y, Kumar V, Hall LO et al. (2013) Automated delineation of lung tumors from CT images using a single click ensemble segmentation approach. *Pattern Recognit* 46:692–702.
11. Bendtsen C, Kietzmann M, Korn R, Mozley PD, Schmidt G, Binnig G (2011) X-ray computed tomography: Semiautomated volumetric analysis of late-stage lung tumors as a basis for response assessments. *Int J Biomed Imaging* 2011:361589.
12. Athelogou M, Schmidt G, Schäpe A, Baatz M, Binnig G (2006) *Cognition Network Technology—A Novel Multimodal Image Analysis Technique for Automatic Identification and Quantification of Biological Image Contents*. Berlin, Germany: Springer.

13. Mansoor A, Bagci U, Xu ZY et al. (2015) A generic approach to pathological lung segmentation. *IEEE Trans Med Imaging* 33:2293–2310.

14. Hu SY, Hoffman EA, Reinhardt JM (2001) Automatic lung segmentation for accurate quantitation of volumetric X-ray CT images. *IEEE Trans Med Imaging* 20:490–498.

15. Armato SG, Sensakovic WF (2004) Automated lung segmentation for thoracic CT: Impact on computer- aided diagnosis. *Acad Radiol* 11:s1011–s1021.

16. Kakar M, Olsen DR (2009) Automatic segmentation and recognition of lungs and lesion from CT scans of thorax. *Comput Med Imaging Graph* 33:72–82.

17. Ding J, Li A, Hu Z, et al. (2017) Accurate pulmonary nodule detection in computed tomography images using deep convolutional neural networks. In *International Conference on Medical Image Computing and Computer-Assisted Intervention*. Springer, Cham, Switzerland, pp. 559–567.

18. Wang S, Zhou M, Liu Z et al. (2017) Central focused convolutional neural networks: Developing a datadriven model for lung nodule segmentation. *Med Image Anal* 40:172–183.

19. Messay T, Hardie RC, Rogers SK (2010) A new computationally efficient CAD system for pulmonary nodule detection in CT imagery. *Med Image Anal* 14:390–406.

20. Bavya K, Julian DP (2017) Feature extraction and classification of automatically segmented lung lesion using improved toboggan algorithm. *Int J Adv Eng Res Sci* 4:1–9.

21. Song J, Liu Z, Zhong W et al. (2016) Non-small cell lung cancer: Quantitative phenotypic analysis of CT images as a potential marker of prognosis. *Sci Rep* 6:38282.

22. Balagurunathan Y, Gu Y, Wang H et al. (2014) Reproducibility and prognosis of quantitative features extracted from CT images. *Transl Oncol* 7:72–87.

23. Lambin P, Leijenaar RTH, Deist TM et al. (2017) Radiomics: The bridge between medical imaging and personalized medicine. *Nat Rev Clin Oncol* doi:10.1038/nrclinonc.2017.141.

24. Fave X, Zhang L, Yang J et al. (2017) Delta-radiomics features for the prediction of patient outcomes in non-small cell lung cancer. *Sci Rep* 7:588.

25. Hawkins S, Wang H, Liu Y et al. (2016) Predicting malignant nodules from screening CT scans. *J Thorac Oncol* 11:2120–2128.

26. Kumar D, Shafiee MJ, Chung AG, Khalvati F, Haider MA, Wong A (2015) Discovery radiomics for computed tomography cancer detection. arXiv preprint arXiv:150900117.

27. Shen W, Zhou M, Yang F, Yang C, Tian J (2015) Multi-scale convolutional neural networks for lung nodule classification. In *IPMI 2015: Information Processing in Medical Imaging*, pp. 588–599.

28. Huang Y, Liu Z, He L et al. (2016) Radiomics signature: A potential biomarker for the prediction of disease-free survival in early-stage (I or II) non-small cell lung cancer. *Radiology* 281:947–957.

29. Coroller TP, Grossmann P, Hou Y et al. (2015) CT-based radiomic signature predicts distant metastasis in lung adenocarcinoma. *Radiother Oncol* 114:345–350.

30. Coroller TP, Agrawal V, Narayan V et al. (2016) Radiomic phenotype features predict pathological response in non-small cell lung cancer. *Radiother Oncol* 119:480–486.

31. Song J, Shi JY, Dong D, et al. (2018) A new approach to predict progression-free survival in stage IV EGFR-mutant NSCLC patients with EGFR-TKI therapy, *Clin Cancer Res*, 24: 3583–3592.

32. Garraway LA, Verweij J, Ballman KV (2013) Precision oncology: An overview. *J Clin Oncol* 31:1803–1805.

33. Jänne PA, Engelman JA, Johnson BE (2005) Epidermal growth factor receptor mutations in non-small cell lung cancer: Implications for treatment and tumor biology. *J Clin Oncol* 23:3227–3234.

34. Li T, Kung H-J, Mack PC, Gandara DR (2013) Genotyping and genomic profiling of non-small-cell lung cancer: Implications for current and future therapies. *J Clin Oncol* 31:1039–1049.

35. Zhou M, Leung A, Echegaray S et al. (2017) Non-small cell lung cancer radiogenomics map identifies relationships between molecular and imaging phenotypes with prognostic implications. *Radiology* 286:307–315.

36. Gevaert O, Xu J, Hoang CD et al. (2012) Non-small cell lung cancer: Identifying prognostic imaging biomarkers by leveraging public gene expression microarray data methods and preliminary results. *Radiology* 264:387–396.

37. Velazquez ER, Parmar C, Liu Y et al. (2017) Somatic mutations drive distinct imaging phenotypes in lung cancer. *Cancer Res* 77(14):3922–3930.

38. Yu D, Zhou M, Yang F et al. (2017) Convolutional neural networks for predicting molecular profiles of non-small cell lung cancer biomedical imaging. In *2017 IEEE 14th International Symposium on Biomedical Imaging (ISBI 2017)*, Melbourne, Australia, 2017, pp. 569–572.

39. Wang S, Shi JY, Ye ZX et al. (2019) Predicting EGFR Mutation Status in Lung Adenocarcinoma on Computed Tomography Image Using Deep Learning. *Eur Respir J* 53 (3): 1800986.

40. Gillies RJ, Kinahan PE, Hricak H (2015) Radiomics: Images are more than pictures, they are data. *Radiology* 278:563–577.

41. Shaikh FA, Kolowitz BJ, Awan O et al. (2017) Technical challenges in the clinical application of radiomics. *JCO Clin Cancer Inform* 1:1–8.

42. Limkin EJ, Sun R, Dercle L et al. (2017) Promises and challenges for the implementation of computational medical imaging (radiomics) in oncology. *Ann Oncol* 28:1191–1206.

The essence of R in head and neck cancer
Role of radiomics and radiogenomics from a radiation oncology perspective

HESHAM ELHALAWANI, ARVIND RAO, AND CLIFTON D. FULLER

16.1 INTRODUCTION

Radiation oncology possesses ingrained "big data" potential, with reference to its coalition of temporally oriented clinical characteristics, multi-imaging modalities, as well as various genomic and other tissue and liquid identifiers (El Naqa 2016). These "big data" elements have empowered the radiation oncology field to advocate biomarker discovery and validation over the past few years with the emergence of machine learning and artificial intelligence applications in precision medicine (Acharya et al. 2018). Radiomics is a rapidly growing field that extracts quantitative data from imaging scans to investigate spatial and temporal characteristics as well as phenotypic correlates of intrinsic tumor characteristics (Wong et al. 2016). To date, radiomics signatures have been proposed as multi-modality imaging biomarkers with predictive and prognostic capabilities in several types of cancer, including head and neck cancers (HNCs) (Huang et al. 2016, Liang et al. 2016, Vallieres et al. 2015).

Computed tomography (CT) quantifies tissue density per X-rays absorption, whereas ^{18}F-fluorodeoxyglucose positron emission tomography (^{18}F-FDG PET) measures cellular function in terms of glucose metabolism. Also, magnetic resonance imaging (MRI) is distinct in the way it depicts variations in proton density and relaxation characteristics. Different MRI sequences capture peculiar tumor heterogeneity whether it is attributed to differences in contrast enhancement (T1-weighted; T1W), water content (T2-weighted; T2W), or diffusivity (diffusion-weighted imaging; DWI) (Cook et al. 2018). HNCs are unique where all aforementioned imaging modalities, among others, are routinely acquired in clinical setting in abundance (Rumboldt et al. 2006). Moreover, head and neck cancer process exhibits frequent interplay between various biologic, molecular, and genomics attributes with well-established effects on disease prognosis and subsequent oncologic outcomes (Leemans et al. 2011). This provides an unprecedented opportunity to link computational imaging analysis to system biology via radiomics and radiogenomics applications.

In spite of some promising results in head and neck cancer studies, most of these studies were subject to the usual constraints owing to their retrospective nature, hindering reproducibility. These include, but are not limited to, small clinically heterogeneous patient cohorts, and many confounding variables, lack of external validation, among others. This further imposes critical barriers to progress in integration of biomarkers development and validation, both in research and clinical settings.

In this chapter, we present a review of milestones and recent updates of radiomics applications for head and neck cancer in the following domains:

1. Pattern recognition for lesions classification and segmentation
2. Risk stratification, in terms of development and validation of prognostic and/or predictive biomarkers
3. Radiotherapy-related normal tissue injury characterization and/or risk prediction
4. Imaging genomics, or "Radiogenomics."

16.2 RADIOMICS FOR PATTERN RECOGNITION IN TERMS OF CLASSIFICATION AND SEGMENTATION

Textural analysis has been vigorously investigated to design a pattern recognition (PR) system for discriminating between benign and malignant lesions or the so-called "imaging biopsy." Implementation of texture analysis to optimize automated (or semi-automated) segmentation and classification of pulmonary nodules is perceived by many to be one of the earliest efforts toward clinical utility of radiomics (Way et al. 2009). Interestingly, researchers in head and neck cancers spearheaded these endeavors as early as 1998. This was when Plant sought to differentiate between benign and malignant neck nodes using an ensemble of: edge, Fourier, and texture analyses, where the former two proved to be of more value (Plant 1998). Since then, preliminary evidence has compiled suggesting clinical utility of radiomics in head and neck cancers.

Texture-based classification of radiographic thyroid nodules is perhaps one of the most actively investigated research areas in head and neck. This approach might have been triggered by the relative success in pulmonary nodules characterization, both nodules classes having low malignancy event rate and high operator-dependence for diagnosis (Sayyouh et al. 2013, Singer 1996). Several groups have compiled data indicative of the potential capacity for radiomics/texture techniques applied to various imaging modalities to characterize thyroid nodules. A systematic review by Sollini et al. (2018) detailed myriad of studies exploring novel approaches for thyroid ultrasonography (US) pattern recognition that decipher texture and echogenicity features. We advise readers with interest in US texture analysis and MRI histogram analysis to read this systematic review that showcases this computer-assisted diagnosis (Way et al. 2009) system potential role in improving radiologists' accuracy (Ardakani et al. 2015, Gesheng et al. 2015, Kale and Punwatkar 2013, Schob et al. 2017, Yonghong et al. 2016).

From a functional imaging perspective, a study by Kim and Chang showed that heterogeneity factor [HF] (a surrogate for intra-tumoral metabolic heterogeneity) using ^{18}F-FDG PET fused with PET CT (PET-CT) can be a very specific diagnostic tool in case of inconclusive thyroid nodules per cytological testing (Kim and Chang 2015).

Exploring beyond US, Brown et al. examined thyroid nodules on 3.0 Tesla diffusion-weighted MRI (3T DW-MRI) acquired with spin echo echo planar imaging sequences. In contrast to mean apparent diffusion coefficient (ADC) values, a combination of 21 texture features could more accurately characterize thyroid nodules area under the curve (AUC: 0.97 versus 0.73). These results were further verified on an independent test set from another institution, using linear discriminant analysis (Jansen et al. 2016). In spite of the obvious need for larger scale prospective validation study (total sample size: 44), this study shows promise in clinical utility given the robustness of this classifier in presence of acquisition and image variability (Brown et al. 2016).

Similarly, one study showed 3.0 Tesla T1-weighted MRI-derived texture features yielded more significant discrimination between benign and malignant parotid gland lesions (within an accuracy of up to 84.5%

for linear discriminant analysis [LDA]) as compared to ADC values. Intrinsic pitfalls included various pathological entities per lesion class and wide range of patients' ages as potential confounders for textural variations. Albeit, mathematically computed texture features from contrast-enhanced (CE) (mainly "run-length matrix") and non-CE (solely from "co-occurrence matrix") T1-weighted images were shown to reflect underling histologic identity (Fruehwald-Pallamar et al. 2013). The same group conducted a more in-depth analysis on a larger more heterogeneous head and neck cohort from two institutes to determine the feasibility of streamlining cross-institutional MRI texture analysis studies. The results came not in favor of relying on texture analysis to differentiate benign and malignant lesions when varying MR protocols on different MR scanners are in use. Yet, this study draws attention to the importance of studying the effect of MR acquisition and imaging parameters on robustness of extracted features and subsequent standardization, similar to initiatives endorsed by the Radiological Society of North America (RSNA)-sponsored by Quantitative Imaging Biomarker Alliance (QIBA) (2017).

The list of anatomical and multi-parametric MR-derived radiomics classifiers of underlying tumor behavior and pathology keeps growing to include other applications. Texture-based machine-learning model successfully identified sinonasal squamous cell carcinomas (SCC) from benign inverted papilloma (IP) on a region of interest level outperforming the accuracy of neuroradiologists (89.1% versus 56.5%, $p = 0.0004$). The inputs to this model were mainly derived from: discrete orthonormal Stockwell transform (DOST) & Gabor filter banks (GFB) texture analyses, followed by support vector machine (SVM) technique to construct the predictive classifier. Though not tested in this study, the implementation of such classifiers into recognizing tumor sub volumes undergoing malignant transformation or necrosis could be of direct clinical utility (Ramkumar et al. 2017). Another instance of this application is the MR texture-based benign and malignant nasopharyngeal pattern recognition system with neural-fuzzy-based Adaboost classifier, proposed by Wu et al. (2016). They investigated a semi-automated tumor segmentation and gray-level co-occurrence matrix (GLCM) texture and geometric features extraction from 31 T2*-weighted MR sequences, with an ultimate classification accuracy of 92.8%.

Interestingly, GLCM texture-based analysis has been propagated to characterize suspicious-looking vocal fold lesions on laryngeal video stroboscopic images as an objective adjuvant to physician-based assessment. A study by Song et al. on 198 vocal fold lesions recognized clinically useful cut-offs for two GLCM features (entropy and variance) for diagnosing malignant-looking vocal fold lesions, i.e., leukoplakia and malignancy, with high sensitivity and specificity, 82.9% and 82.2% ,respectively (Song et al. 2013).

Furthermore, radiomics have been correlated to fundamental tumor biologic and microenvironment attributes beyond merely binary tumor behavior, i.e., benign versus malignant. Epstein Barr virus (EBV) infection has been recognized as a direct epigenetic driver in the pathogenesis of non-keratinizing nasopharyngeal cancer (NPC) (Raab-Traub 2002). A radiomics classifier extracted from baseline CE-CT imaging of pre-treated intact tumor was associated with EBV status, tested by serum immunoenzymatic assays, with receiver operating characteristic area under the curve (ROC AUC) of 0.73 [95% confidence interval (CI): 0.66–0.80]. The ensemble of features selected by the least absolute shrinkage and selection operator (LASSO) regression test were: sphericity, neighborhood gray-tone difference (NGTD)-based feature "texture strength," GLCM feature "correlation," and statistical feature "kurtosis" (Yang et al. 2018). This advent to linking imaging phenotypes to underlying tumor biology determinants have been also investigated in the oropharynx cancer (OPC) sphere.

In recent years, human papilloma virus (HPV) status has progressively been incorporated in clinical algorithms for the management of OPC and has led to distinct clinical staging systems for HPV positive (HPV+) and HPV negative (HPV–) disease entities in the latest version of the American Joint Committee on Cancer (AJCC)/Union for International Cancer Control (UICC) manual (Lydiatt et al. 2017). This has inspired many investigators to mine for characteristic quantitative imaging features of differential HPV status as a liaison to differential prognostic and predictive biomarkers rather than direct diagnostic value. A previous work by Cantrell et al. showed that qualitative image features of the primary OPC such as image attenuation and border definition can be broadly associated with HPV status (Cantrell et al. 2013). Subsequently, Aerts et al. interrogated an association between their 4-feature prognostic radiomics signature drawn out from baseline CT scans of 170 HNC patients and underlying HPV status. Even though the results were not significant, the consistency of prognostic performance of signature across HPV– subgroup encouraged others to investigate this imaging-biology interplay (Aerts et al. 2014).

Buch et al. extracted 42 first- and higher-order features from manually segmented primary OPCs belonging to 29 HPV+ and 11 HPV– patients on CE-CT images. Interestingly, only two histogram features (median and entropy) were found to be statistically significant between the two HPV phenotypes after correction for false discovery rate (FDR) (Buch et al. 2015). A follow-up study by the same group sought to quantitatively depict the CE-CT texture variations among 46 primary non-OPC based on HPV status applying the same previous feature set. Three textural features, two GLCM features (contrast and correlation), and one Law feature showed statistically significant difference between the two classes. Authors acknowledged for both studies that no machine learning techniques were applied to test the predictive capacity of these two feature sets or find individual features cut-off values for HPV status diagnosis, a limitation mostly due to small sample size (Fujita et al. 2016).

This unmet need has driven a group from MD Anderson Cancer Center in concert with the Medical Image Computing and Computer Assisted Intervention to organize the first ever head and neck radiomics challenges on Kaggle in Class (KiC) using a curated cohort of 288 OPC subjects (Elhalawani et al. 2017). The objective of one challenge was to assess the ability of participant-developed radiomic workflows to predict binary HPV status on a "Test" set, using a defined "Training" cohort as a "prior" dataset that includes all input (clinical data and manually segmented primary tumor) and outcome data, applying machine learning techniques (Elhalawani et al. 2018b). The winners were a team of academic biostatisticians with a radiomics-only model that achieved an AUC of 0.92 in the held-out test subset per the challenge rules. Their feature selection approach yielded the "shape" features of "mean breadth" and "spherical disproportion" from CE-CT scans as most predictive of HPV status, suggesting that HPV+ tumors tend to be smaller and more homogeneous (Yu et al. 2017).

Likewise, Bogowicz et al. concluded that HPV+ OPC display smaller width of CE-CT density distribution as compared to HPV– tumors in a retrospective study of 93 and 56 patients in the training and validation cohorts, respectively. This homogeneity was evidenced by a significantly lower coefficient of variation and L: low-pass filter; H: high-pass filter (LLL) standard deviation, as well as lower LLL small zone high gray-level emphasis. This might support the hypothesis of association between tumor homogeneity and better prognosis, given the more favorable survival outcomes of HPV+ subset of OPC. This radiomics classifier not only displayed satisfactory performance in HPV status prediction (AUC= 0.85, and 0.78 in training and validation sets, respectively), but also remained robust across varying CT scanners and acquisition parameters, namely, kilovoltage (kV) (Bogowicz et al. 2017). Yet the concern remains that all aforementioned studies were conducted on small populations without validation, or larger, but single institution datasets with detrimental effects on model development and validation (Larue et al. 2017).

Fortunately, the same group have further co-investigated the interrelation between OPC imaging and biology as a part of a larger scale retrospective multi-institutional study published later in 2018. They added three more independent cohorts accounting for a total of 778 OPC patients with baseline CE-CT and manually segmented primary tumors, with further 4:1 random splitting into training and validation sets, respectively. Besides good discrimination for HPV status of the model on validation sets (AUC ranging between 0.7 and 0.8), the model demonstrated robustness regardless of visible CT artifacts (mainly due to dental filling). Interestingly, Kaplan-Meier (KM) survival curves were plotted for HPV+ and HPV– patients classified per p16 test and their HPV radiomic classifier for all validation patients to further test prognostic potential of the model. This yielded a significantly consistent split between survival curves in either case with hazard ratio (HR) of 0.46 [95% CI (0.26–0.82)], and 0.5 [95% CI (0.31–0.97)] per the p16 and radiomics classifier, respectively (Leijenaar et al. 2018). Another advantage of this study is combining cohorts from North America and Europe that are known with inverse OPC association with HPV—being higher in the former—to mitigate class imbalance, an obstacle to many practical classification questions (Gillison et al. 2015, Fernndez-Delgado et al. 2014).

Vallieres et al. (2013) postulated that HPV-related differences in OPC biology can result in discernible metabolic changes that radiomics can capture from 18F-FDG-PET in a retrospective study, including 67

HNC patients. A primary tumor-derived 5-feature ensemble representing texture, SUV measures (% inactive volume), and shape best correlated to HPV status using multivariate models built via logistic regression (The Cancer Genome Atlas Research et al. 2013) and SVM, AUC = 0.64 and 0.72, respectively. This manifests the ability of a composite radiomics signature solely derived from baseline 18F-FDG-PET imaging to correlate to intrinsic tumor biology on a subcellular level.

As all of the aforementioned studies, this study did not take into consideration structural variations across various oropharynx subsites, e.g., base of tongue is very rich in lymphoid tissue, with subsequent intrinsic differences in CT density or FDG uptake as potential confounders (Trotta et al. 2011). Nevertheless, these studies were not presented as surrogates for routine tissue HPV testing, but rather as proof of concept of images enclosing "invisible data" that can be correlated to much more than what the eyes can see. This goes along the same lines with the innovative work by Crispin-Ortuzar et al. who developed a composite ^{18}F-FDG PET (first-order) and CE-CT (first- and higher-order) radiomics feature signature of 18F-Fluoromisonidazole (18F-FMISO)-assessed hypoxia in head and neck cancers, outperforming models based on 18F-FDG PET alone (AUC: 0.83 versus 0.76, respectively). Hypoxic tumors, defined in this study as exhibiting 18F-FMISO maximum tumor-to-blood uptake ratio >1.4, are known to have worse prognosis and treatment outcomes (Clavo et al. 2017). Again, this study proves the added benefit of combining various imaging modalities to best reflect the multi-faceted nature of physiological or pathological phenomena (Crispin-Ortuzar et al. 2018).

Texture analysis has also been implemented in yet another classification problem "tumor versus normal tissue" toward automation of segmentation process for head and neck cancers. Texture analysis has the capacity to provide synergistic benefit with human pattern recognition in identifying intrinsic tumor appearance, especially in small or infiltrating tumors (Siebers et al. 2010). A common area of research is automating the segmentation of suspected parotid lesions within otherwise normal gland. Wu et al. proposed a workflow that starts with localizing the suspected lesion tissues in low contrast tissue regions first via introduction of stationary wavelet transform (SWT). The wavelet coefficients in sub-bands were applied to derive the local texture features (energy, mean, and variance). Those along with gray level were then used for feature-based segmentation (FBS) and guiding modified active contour models (ACM). When comparing the performance of the proposed workflow with manual segmentations by clinical experts, the results were very promising with 94% true-positive (TP) rate on characterizing parotid lesions and over 93% dimension accuracy of delineation. This automated workflow for lesion characterization and delineation can provide valuable information for auto-segmentation and radiotherapy treatment planning (Wu and Lin 2013).

Another instance of this application was the study by Yu et al. (2009a) that involved texture analysis of co-registered 18F-FDG PET-CT in 20 HNC patients and 20 matched controls to recognize textural features capable of discriminating tumor from normal tissue in head and neck. Neighborhood gray-tone difference matrix (NGTDM) features including PET coarseness and contrast, along with CT coarseness showed satisfactory classification capacity, receiver operating characteristic AUC = 0.95 ± 0.007, when combined in a multivariate model, constructed via a decision tree-based K-nearest neighbor (DT-KNN) classifier. An ensemble of PET and CT texture features exhibited superior tumor versus normal tissue discrimination as compared to features solely derived from either PET or CT alone. Moreover, the classification accuracy based on NGTDM and spatial gray-level dependence method (SGLDM) features surpassed that of a human expert, hence advocating the integration of composite texture analysis from multi-imaging modalities in automated segmentation workflows in head and neck cancer. To that end, the same group published a follow-up hypothesis-testing paper on such an application, entitled "Co-registered multimodality pattern analysis segmentation system" for another 10 patients with HNC, again using baseline PET and CT images. Integrating 14 PET and 13 CT voxel-derived texture features, via decision tree-based K-nearest-neighbor classifier, co-registered multimodality pattern analysis segmentation system successfully auto-segmented all primary tumors and almost all positive lymph nodes (19/20). This surpassed other three simpler PET-based thresholding methods for tumor segmentation and yielded contours of accuracy comparable to manual expert segmentations (specificity 95% ± 2%, sensitivity 90% ± 12%) (Yu et al. 2009b).

16.3 RADIOMICS FOR RISK STRATIFICATION AS PROGNOSTIC AND PREDICTIVE BIOMARKERS

Radiomics has been intensively invested in development and validation of myriad of radiomics biomarkers of prognostic and predictive capacity in several types of cancer (Elhalawani et al. 2018a, Huang et al. 2016, Leijenaar et al. 2015, Liang et al. 2016, Vallieres et al. 2015). Biomarker can be defined as "A characteristic that is objectively measured and evaluated as an indicator of normal biologic processes, pathogenic processes, or pharmacological responses to a therapeutic intervention" (Hodgson et al. 2009). Prognostic marker is "a marker that predicts the prognosis of a patient (e.g., the likelihood of relapse, progression, and/or death) independent of future treatment effects," whereas predictive biomarkers are "measurements associated with response to or lack of response to a particular therapy." A factor can be both prognostic and predictive (Rosenthal et al. 2016). Multiple imaging modalities and techniques have been investigated for potential biomarker derivation.

Being the most widely used imaging modality for HNC diagnosis, radiotherapy (RT) planning, and follow up, investigators conducted multitude of CT quantitative imaging analysis early in the course of radiomics research (Wippold 2007). Surprisingly, one of the earliest radiomics-based predictive modeling of treatment outcomes was performed on PET imaging rather than CT scans. The 2009 El Naqa et al. study used pretreatment ^{18}F-FDG-PET images-derived intensity-volume histogram (IVH)-based V90 and shape extent to test a two-metric logistic regression model predicting treatment failure with subsequent AUC of 1.0. This represents an expected over-fitting given a very small sample of nine HNC patients (El Naqa et al. 2009).

The next leaps were taken in the CT imaging field when Zhang et al. explored—in 2013—the capacity of texture and histogram features to accurately predict the response of 72 patients with HNC to cisplatin, 5-fluorouracil, and docetaxel (TPF) induction chemotherapy regimen in terms of overall survival. Primary tumor entropy and skewness measurements with spatial filter of 1.0 were independent predictors of overall survival along with other clinical prognosticators. This was an initial endeavor to integrating quantitative tumor phenotypes into well-established clinical nomograms (Zhang et al. 2013).

However, the 2014 Aerts et al. study marked the actual kickoff of widespread radiomics applications in head and neck radiation oncology. The authors carried out an in-depth analysis of CT-derived 440 radiomics features in 1,019 patients with either lung or head and neck cancer. They selected the most significantly correlated feature from each of the four cardinal categories: intensity, shape, texture, and wavelet decomposition. This yielded a 4-feature radiomics signature (statistics energy, shape compactness, gray-level non-uniformity, and wavelet (L: low-pass filter; H: high-pass filter [HLH]) gray-level non-uniformity) of high prognostic value in terms of association with overall survival. Interestingly, the same radiomics signature that was trained on lung cancer dataset maintained its prognostic ability in HNC patient cohort, proposing universal prognostic value of intra-tumoral with potential linkage to intrinsic gene expression profiles (Aerts et al. 2014).

A follow-up study externally validated the same 4-feature radiomics signature on an independent cohort of 542 OPC patients. Besides showing satisfactory preservation of model discrimination (Harrell's c-index 0.628; $p = 2.72e^{-9}$), this signature maintained its prognostic capacity even in the presence of visible CT artifacts (Harrell's c-index 0.647; $p = 5.35e^{-6}$). This certainly increased confidence in the generalizability of this radiomics signature as well as its potential applicability in clinical settings where dental fillings-induced visible streaks are common (Leijenaar et al. 2015). Most radiomics studies have extracted radiomics features from primary tumor until Zhai et al. integrated CT radiomics features from both primary and nodal tumors aiming to boost the prognostic ability of well-established clinical variables. The combined clinical-radiomics model was trained on a 289 Chinese NPC patient cohort then externally validated on 298 Dutch HNC dataset. They reported favorable impact on the discriminatory ability of the clinical models when radiomics features were added. Albeit, the heterogeneity of the patient population, being from various demographic and biologic catchments, as well as not including radiomics features from all involved nodes (only the largest ones) was among the potential pitfalls (Zhai et al. 2017)

Afterward, investigators sought to include more clinically meaningful oncologic outcome endpoints other than overall survival. The Medical Image Computing and Computer Assisted Intervention Society (MICCAI)

2016 radiomics challenge identified "local tumor control" as a testable clinically oriented hypothesis with a binary endpoint that can be used for radiomics challenges using an existing dataset of 288 RT-treated OPC patients (Elhalawani et al. 2017, 2018b). A series of studies ensued where radiomics was more geared toward "predicting" the outcomes of RT in the definitive setting aiming at pre-emptively risk-stratifying the patients in terms of their subsequent local tumor control. A 3-feature radiomics signature (L: low-pass filter; H: high-pass filter [HHH] large size high gray-level emphasis, LLL sum entropy, and L: low-pass filter; H: high-pass filter [LLH] difference variance) was significantly associated with local control both in training ($n = 93$) and validation ($n = 56$) HNC cohorts, respectively (Bogowicz et al. 2017). Similarly, an ensemble of three histogram and four textural features composed a predictive CE-CT signature of local tumor recurrence following RT in a cohort of 62 patients with HNC.

Along the same lines, Ou et al. discovered a 24-feature CT radiomic signature of dual prognostic and predictive potential in a cohort of 121 HNC patients. After significantly predicting for overall survival (OS) (HR = 0.3, $p = 0.02$) and progression-free survival (PFS) (HR = 0.3, $p = 0.01$), the authors sought to explore the synergism between this signature and p16 status. Again, more significant improvement of prognostic and predictive performance of the combined model as compared to p16 solely was noticed for OS and PFS (AUC = 0.78 versus 0.64) and 0.78 versus 0.67, respectively. A larger scale single institute study by Elhalawani et al. (2018a) reported a 2-feature (intensity and neighborhood intensity difference features) CT radiomic signature that unfolded a consistent pattern of OPC patient stratification based on risk of post-RT local tumor recurrence. All aforementioned studies agreed that radiomics surrogates of baseline (untreated) tumor heterogeneity are associated with poorer response to subsequent RT. However, the unmet need is to validate different signatures across these institutes to help the radiomics field move past its infancy.

From metabolic imaging perspective, ^{18}F-FDG PET are—by far—the most investigated for potential prognostic and predictive radiomics biomarkers. Tumor FDG heterogeneity was assessed by histogram-based parameter skewness and most of second- or higher-order features were significantly correlated with OS and/ or PFS along with total lesion glycolysis (TLG) in a univariate analysis. Subsequent multivariable analysis confirmed the added prognostic value of second-order "uniformity" feature when combined with age and EBV DNA load for more reliable prediction of OS in a cohort of 101 NPC patients (Chan et al. 2017). On a predictive multi-imaging modality biomarker discovery level, Vallières et al. (2017a) explored the potential of integrating radiomics features from pre-treatment ^{18}F-FDG PET and co-registered CT into well-established clinical models for better reflection of underlying tumor behavior. In this multi-institutional study, 300 HNC patients from four institutes were stratified according to their risk of death, locoregional recurrences (LR), and distant metastases (DM) based on prediction models combining radiomics and clinical variables. These were constructed via random forests and class imbalance-adjustment strategies using two of the four cohorts, followed by independent validation of the prediction performance of the models using the other two cohorts (LR: AUC = 0.69; DM: AUC = 0.86). Subsequent Kaplan-Meier analysis demonstrated the potential of radiomics, either alone or combined with clinical variables, for assessing the risk of distinct oncologic outcomes.

Interestingly, Bogowicz et al. applied PET radiomics methodology toward prediction of local HNC control using the 3 months post-chemoradiotherapy ^{18}F-FDG PET images. Models encompassing histogram SUV range and GLCM difference entropy in the region of irradiated primary tumor significantly stratified patients into low- and high-risk for post-RT tumor recurrence, with a satisfactory concordance index (higher than 0.7) in both training and validation cohorts. However, this high correlation to local recurrence probability was not retained when two distinct software tools were implemented, highlighting the need for standardization before generalization.

Other PET tracers have been under investigation for potential inclusion in radiomics studies assessing tumor heterogeneity. A novel analysis by Lapa et al. scrutinized somatostatin receptor subtype II (SSTR) PET-derived biomarkers of response to peptide receptor radionuclide therapy (PRRT), a treatment modality for some thyroid cancer subtypes. Gray level non-uniformity demonstrated significant capacity to predict PFS (AUC = 0.93), whereas no textural feature correlated to OS. Although no direct conclusions can be drawn—maybe due to a small sample size of 12, this study should encourage researchers to mine novel imaging techniques for potential imaging correlates of known prognostic features, e.g., heterogeneity (Lapa et al. 2015).

In the meantime, the use of MRI in HNC is widely expanding, given its superior soft tissue discrimination and reduced dental amalgam interference in comparison with CT (Sumi et al. 2007). This has been reflected in growing number of MRI radiomics studies recently, especially NPC (Jethanandani et al. 2018). Liu et al. investigated the clinical potential of texture analysis using T1W, T2W, and DWI acquired using contrast-enhanced 3.0 Tesla MRI for discriminating "responders" and "non-responders" from a group of 53 NPC patients receiving definitive chemoradiotherapy. Texture features from all three sequences, especially T1W, demonstrated good discrimination between the two groups, both on internal and external validation, with corresponding AUCs: T1W: 0.95/0.94, T2W: 0.9/0.91, and DWI: 0.88/0.93, respectively (Liu et al. 2016).

A subsequent bigger study by Zhang et al. analyzed T2W and CE-T1W images for texture features candidates for combined clinical-imaging PFS nomogram. The discriminatory ability of a radiomics signature extracted from both sequences outperformed signatures derived from either sequence alone. Similarly, this combined T2W and CE-T1W radiomics signature synergistically ameliorated the predictive capacity of tumor-node-metastasis (TNM) staging system for post-treatment PFS (C-index: 0.76 versus 0.51) (Zhang et al. 2017). Similarly, a 15-feature radiomics signature derived both from T2W and CE-T1W successfully stratified 120 NPC patients receiving induction chemotherapy into "responders" and "non-responders." Again, the combined modality radiomics signature outperformed a 5-feature CE-T1W only signature, AUC: 0.82 versus 0.72 (Wang et al. 2018).

With the advancing role of multi-parametric MRI in HNC diagnosis, and potentially RT planning (Rumley et al. 2017), interest of radiomics researchers in this emerging imaging technology grew exponentially. A study by Jansen et al. of 19 HNC patients with available pre- and intra-treatment dynamic contrast-enhanced MRI (DCE-MRI) evaluated parametric maps derived from pharmacokinetic modeling, namely, Ktrans and ve. Furthermore, texture analysis (energy and homogeneity) on these parametric maps was performed. While delta-changes of mean and standard deviation for Ktrans and ve, energy feature from the ve map was shown to be significantly higher on intra-treatment scans compared to pre-treatment values (0.41 ± 0.22 versus 0.30 ± 0.11; $p < 0.04$). This study uncovered another application of radiomics, providing complementary quantitative information on tumor vascularity, and hence tumor response to treatment (Jansen et al. 2016).

16.4 RADIOTHERAPY-RELATED NORMAL TISSUE INJURY CHARACTERIZATION AND/OR RISK PREDICTION

In addition to quantifying intrinsic tumor characteristics, radiomics has shown promising potential in characterizing physiologic and structural alterations of normal tissues in the irradiated field. Numerous organs are at risk of radiation-induced injury given the complex anatomy of head and neck, with many vital structures crowded together in a relatively small area (Kulzer and Branstetter 2017). This is particularly beneficial for assessing and/or predicting radiation-induced injury to parotid salivary glands and subsequent detrimental effect on salivary function, namely, dry mouth or xerostomia (Deasy et al. 2010). Texture analysis has been sought to provide an additional dimension while assessing these structural alterations beyond the well-studied volumetric shrinkage (Pota et al. 2015, Sanguineti et al. 2015). Quantifying CT texture changes as surrogates for radiotherapy-related structural changes was first performed by Scalo et al. In their 2013 texture analysis study of 42 parotid glands from 21 radiotherapy-treated patients with NPC, a global decline in parotid tissue CT heterogeneity was reported at two different time points, second week (CT2), and last (CTlast) of RT course compared to baseline (CT1). Moreover, a combination of volume and fractal dimensionality was reported to early predict ultimate parotid shrinkage with 71.4% accuracy, using discriminant analysis, outperforming any other single feature (Scalco et al. 2013). In a bigger follow-up study by the same group, texture features dose-response was evaluated by extracting features from CT images performed at beginning, mid-RT, and end of RT course, i.e., incremental dose received by the glands. Volume and mean intensity variation were found to be most correlated with pre-treatment dosimetric attributes, hence quantifying relationship between dose plan and structural variation estimated after radiotherapy (Scalco et al. 2015). However, both studies have not delved into the implementation of texture analysis into characterizing imaging-derived pre-clinical signs of xerostomia.

van Dijk et al. explored the added value of incorporating CT radiomics features with known clinical and dosimetric factors for prediction of moderate-to-severe xerostomia at 12 months post-RT. Interestingly, including radiomic feature "short run emphasis"—a quantitative surrogate for parotid gland heterogeneity—to orthodox predictors, e.g., mean RT dose to contralateral parotid gland, pre-treatment xerostomia, and sticky saliva scores (correlates of baseline parotid function) was synergistic. This significantly boosted their predictive capacity in terms of AUC from 0.75 (0.69–0.81) to 0.77 (0.71–0.82), with and without adding the radiomic feature, respectively (van Dijk et al. 2017). High values of short run emphasis can be explained by fat saturation of irradiated parotid gland, a finding that has been correlated to impaired salivary function (Izumi et al. 2000).

This study has many strengths including a relatively big sample size ($n = 249$), well-balanced outcomes classes with 40% of the patients reporting late xerostomia according to the validated 35th version of the European Organization for Research and Treatment of Cancer Quality of Life Questionnaire Head and Neck Module (EORTC QLQ-H&N35) questionnaire, as well as including submandibular glands—the second biggest major salivary gland—in the analysis. Furthermore, the authors accounted for intrinsic limitations of multi-observer manual segmentations by assessing radiomics features stability against four other artificially created segmentations, setting a stability inter-class correlation coefficient cut-off of 0.7. They even excluded the maximum CT intensity feature of submandibular gland—though statistically significant on multivariable analysis—because of non-optimal stability when it came to inter-observer variability in delineations (van Dijk et al. 2017). One potential pitfall was the non-inclusion of dose parameters, like mean and V30, to the parotid and submandibular glands, which are recognized as key predictors of patient-rated xerostomia (Houweling et al. 2010). According to a study by Nardone et al. that reported univariate associations between CT texture features and acute and chronic xerostomia alike, adding radiomics features yields much higher value when mean dose and V30 are available. They reported a significant improvement of acute and chronic xerostomia predictive models AUCs as 0.61–0.86 and 0.87–0.98, respectively (Nardone et al. 2018).

Exploring beyond CT, van Dijk et al. again extracted intensity and textural features, this time from pre-treatment 18F-FDG PET images of 161 HNC patients and correlated (along with dose parameters) to prospectively collected patient-reported late xerostomia, i.e., at 12 months post-RT. Similar to their CT radiomics study, adding PET derived first- and second-order features significantly improved clinical-dose-only model predictive performance. 90th percentile SUV intensity and textural long-run high gray-level emphasis 3 (LRHG3E) from contralateral parotid glands boosted AUC (95% confidence interval) from 0.73 (0.65–0.81) to 0.77 (0.70–0.84), and 0.77 (0.69–0.84), respectively. This study suggested that high metabolic parotid gland activity is associated with lower risk of developing late xerostomia, possibly paving the way to integrating 18F-FDG-PET-derived functional and metabolic normal tissues attributes into normal tissue complication probability (NTCP) model development (van Dijk et al. 2018).

Amidst all previously discussed machine learning techniques, predictive modeling has been performed by providing inputs on previous associations with RT-induced structural or functional parotid gland injury to subsequently address the problem of timely predicting parotid shrinkage or xerostomia. Likelihood-fuzzy analysis (LFA), that involves representation of statistical information by fuzzy rule-based models, was applied. Various clinical attributes, dosimetric variables, and texture features derived from baseline and mid-RT CT scans and their relative variation (or: delta-change) served as inputs to models involved in this "learning approach." Various predictors including single time-point and longitudinally computed radiomics-based features were identified both for "parotid shrinkage" and "xerostomia at 12 months post-RT" with satisfactory classification accuracy (Pota et al. 2017). Of interest, similar results were reported with naïve Bayes classifier, both methods identifying the same best predictors for both clinical endpoints with more intrinsic interpretability provided by LFA (Pota et al. 2015).

This longitudinal granularity of radiomics analytics supports development of imaging biomarkers of kinetics of various tumors as well as normal tissues response to RT (Elhalawani et al. 2018c, 2018e). Elhalawani et al. (2018d) showed that a functional principal component analysis (FPCA) of CE-CT mandibular radiomic features extracted at multiple time points pre- and post-RT, can predict for development of osteoradionecrosis of the jaw bone in HNC patients. The FPCA-based radiomics signature captured the temporal trajectory of development of features, hence significantly outperformed prediction models based on pre-radiotherapy and delta radiomics features values. This FPCA screening tool—upon validation—can be used by radiologists and oncologists to characterize mandibular sub volumes at higher risk of manifesting post-RT bone necrosis.

Another instance of this application was a CE-CT radiomics signature that predicted chemoradiotherapy-induced sensorineural hearing loss (SNHL) in a cohort of 94 cochlea from 47 patients with head and neck and brain cancers. Out of 10 various machine learning techniques, decision stump and Hoeffding trees had the highest predictive performance, with 76.1% accuracy and 75.9% precision, respectively. Furthermore, a set of 10-radiomics features selected via LASSO penalized logistic modeling was highly associated with SNHL (AUC = 0.89). One drawback of this study is non-inclusion of cochlea-related dosimetric parameters (Abdollahi et al. 2018). This was accounted for in a subsequent MRI radiomics study by Thor et al. where 24 first- and higher-order T1-weighted post-contrast features were extracted from muscles of mastication to quantify RT-induced trismus, i.e., lock jaw. Mean dose to masseter and medial pterygoid muscles, and GLCM feature of the latter showed superior discriminative capacity with AUCs of 0.85, 0.77, and 0.78, respectively.

16.5 RADIOGENOMICS

Most studies exploring imaging-derived radiomics classifier of HPV status can be regarded as radiogenomics studies as they were considering p16 expression as surrogate for HPV status (Bogowicz et al. 2017, Leijenaar et al. 2018, Yu et al. 2017). Interestingly, some might not agree as the outcome of interest was HPV phenotype rather than the intrinsic signaling pathway directing the cancer initiation and growth. Perhaps the earliest undertaking in correlating non-radiomics imaging features to intrinsic tumor genomic expression profiles was the study by Pickering et al. (2012). Qualitative imaging criteria and simple bi-dimensional measures from untreated primary oral SCC, and largest metastatic lymph node (if present) were captured from pre-operative CE-CT scans. Level of lesion enhancement and mass effect were found to be correlated with vascular endothelial growth factor (VEGF) 1 and 2, and epidermal growth factor receptor (EGFR) receptors. These correlations—in spite of obvious limitations—proposed that inferring imaging-derived findings into categorical phenotypes may potentially correlate with specific genetic alterations and drug targets. It wasn't before 2015, when a dedicated imaging-genomics study—fulfilling the most recent definition of radiogenomics workflow—was conducted to investigate for an MRI classifier for the tumor suppressor p53 status on a small cohort of 16 OPC patients. Seven significant texture features extracted from post-gadolinium T1W, T2W, and ADC maps were integrated into a model that predicted p53 status with 81.3% accuracy and subsequent moderate level of agreement on cross-validation (Dang et al. 2015). Possible physiologic explanation for this finding is texture analysis can capture the differences in vascularity between p53-positive and p53-negative tumors, with higher microvascularity reported in the former (Guo et al. 2008). This promising texture-based tool not only accurately diagnosed a key prognostic factor in head and neck cancers, but also encouraged more investigators to delve more into radiogenomics (Poeta et al. 2007).

A similar, but smaller study by Meyer et al. searched for correlation between T1-pre contrast and T2-weighted MRI-derived thyroid carcinoma texture variables and Ki67 or p53 count. Several significant correlations between texture features and these histopathologic correlates of tumor proliferation and suppression were detected. Various sub-bands of wavelet transforms scored the highest correlation coefficients with Ki67, where WavEnLL_S2 feature computed from T1-weighted ($r = -0.8$) and WavEnHL_s-1 T2-weighted imaging feature ($r = -0.77$) correlated negatively with Ki67. Whereas co-occurrence matrix features were the top correlated variables to p53 count, namely, T1-weighted images-derived S(5;0)SumofSqs ($r = 0.65$) and T2-weighted images-derived S(1;-1)SumEntrp ($r = -0.72$) (Meyer et al. 2017).

Exploring beyond MRI, a retrospective study by Chen et al. investigated the potential interrelation between 31 textural correlates of heterogeneity of stage III–IV head and neck tumors on 18F-FDG PET and select gene-expression profiles acquired via IHC test. These were chosen to represent distinct markers of tumor biology and microenvironment, including: hypoxia (Glut1, CAIX, VEGF, and HIF-1α), radio-resistance (Bcl-2, CLAUDIN-4, YAP-1, and c-Met), and (Ki-67), progression (EGFR), as well as CDKN2A as a surrogate marker for HPV. A positive association was depicted between vascular endothelial growth factor (VEGF) overexpression and metabolic tumor heterogeneity quantified using the gray-level non-uniformity for zone (GLNUz) and run-length non-uniformity (RLNU). Similarly, Claudin-4 overexpression (a marker of radio-resistance) showed higher homogeneity, energy, busyness, long-zone emphasis (LZE), short-zone low gray-level emphasis (SZHGE), and long-zone low gray-level emphasis (LZLGE). Intensity of c-Met expression—another radioresistance marker—was associated with texture features energy and short-run emphasis (SRE) (Kataria et al. 2016) correlated with the intensity of c-Met expression. Additionally, SRE, high gray-level run emphasis (HGRE), short-run high gray-level emphasis (SRHGE), long-run high gray-level emphasis (LRHGE), and high gray-level zone emphasis (HGZE) along with SUVmax were correlated with the staining intensity of Ki-67, a tumor proliferation index. On the other hand, negative associations were found between other texture features expression levels of VEGF, Glut1, Claudin-4, Ki-67, and c-Met. Surprisingly, p16 or EGFR—two of the most commonly studied expression profiles in the field of radiogenomics—showed no statistically significant correlations with any of these textural indices. One interesting approach adopted in this study was to test the capacity of a combination of FDG-PET texture features, select genomic expression profiles to predict the outcome of radiotherapy in pharyngeal cancer patients. Intra-tumoral FDG uptake heterogeneity, determined using a matrix index of gray-level non-uniformity for zone (GLNUz; a measure of gray level size-zone matrix [GLSZM]) or run percentage (RP; a measure of gray-level run-length matrix [GLRLM]), along with overexpression of VEGF and HIF-1α can exhibit synergism with T-staging for predicting treatment outcomes (Chen et al. 2017). In spite of small study scale, the results clearly showed promise in multi-faceted decision making tools where—omics studies can represent an integral part in guiding personalized treatment algorithms.

With the advent of immunotherapy in head and neck cancers management, programmed death-ligand 1 (PD-L1) upregulation has been identified as a potential mechanism by which cancer cells evade human immune system, and hence a potential therapeutic target (Thompson et al. 2004). Radiomics has exhibited preliminary potential for linking 18F-FDG-PET textural features to PD-L1 expression, detected through immunohistochemistry (IHC) in a follow-up study by the same group on a cohort of 53 OPC patients. PD-L1 expression intensity was inversely associated with gray-level non-uniformity for run, run percentage, and short-zone low gray-level emphasis (SZLGE; $p = 0.04$, $\gamma = -0.28$). More specifically, some texture features could predict a PD-L1 expression intensity $\geq 5\%$ with a similar pattern noticed when PD-L1 expression was dichotomized at 1%. Interestingly, on setting a cut-off of 0.26% for GLCM (one of the negative predictors of PD-L1 expression), its accuracy for predicting expression levels <5% almost tripled. Naturally, interpretation of this study results would be restricted by the small cohort size given the much bigger number of tested features. However, this was the first study to test the ability of quantitative metabolic imaging phenotypes to predict tumor PD-L1 expression, as well as other protein biomarkers of tumor microenvironment elements including hypoxia, angiogenesis, radioresistance, and tumor proliferation (Chen et al. 2018). In addition, this study paves the way for developing textural features that can predict or assess the therapeutic effects of PD-1 or PD-L1 antibodies, i.e., immune checkpoint inhibitors.

16.6 CHALLENGES FACING GENERALIZABILITY AND REPRODUCIBILITY

In spite of promising results, most studies (including some of the studies included in this chapter) were subject to the usual constraints of retrospective studies, hindering reproducibility. These include, but are not limited to small clinically heterogeneous patient cohorts (with variable train-test distributions), variability in imaging acquisition and scanning protocols, variable image preprocessing methods, and several confounding variables, among others. Hence, we recommend these results to be cautiously interpreted in the context of pilot studies, suggestive of further hypothesis testing and validation. Ensuring equivalence of training-test distributions during predictive modeling is essential for generalizable inference models. Moreover, study of dimensionality reduction, i.e., feature selection or feature extraction, needs intensive investigation to optimize machine learning technique selection according to the problem of interest (Leger et al. 2017, Lu et al. 2016). The seminal work by Parmar et al. is a good example in head and neck radiomics studies where the authors reported that choice of classification method is the cardinal driver of variation in performance among different head and neck prognostic models (Parmar et al. 2015). Similarly, Leger et al. recognized a set of machine learning techniques that best serve toward constructing stable and clinically relevant predictive models for time-to-event survival endpoints, e.g., loco-regional control and overall survival

(Leger et al. 2017). To provide a balanced perspective on the nature of the derived prediction models, the transparent reporting of a multivariable prediction model for individual prognosis or diagnosis (TRIPOD): The TRIPOD statement, has been formulated to provide guidelines for "Reporting of studies developing, validating, or updating a prediction model, whether for diagnostic or prognostic purposes" (Moons et al. 2015). Such community efforts might be able to possibly standardize the development and interrogation of predictive models with diagnostic or prognostic intent.

From imaging perspective, head and neck radiomics are uniquely subject to the effects of image artifacts from intrinsic patient factors, such as metal dental implants and bone. The effects of resulting streak artifacts and beam-hardening artifacts on robustness of extracted radiomics features have been reported (Block et al. 2017, Leijenaar et al. 2015). Besides investigating for radiomics signatures that are not sensitive to the effect of artifacts, consensus is needed regarding universally accepted guidelines for managing these artifacts as a part of a bigger radiomics workflow (Aerts et al. 2014, Leijenaar et al. 2018).

Moreover, the multi-dimensionality of the radiomics studies data imitate real-life clinical situations where diverse clinical/imaging/tissue-derived data attributes are made available for clinicians to assimilate into decision-making workflows. Data repositories such as The Cancer Genome Atlas (The Cancer Genome Atlas Research et al. 2013), The Cancer Imaging Archive (Clark et al. 2013), and Oncospace (Nakatsugawa et al. 2017), in the radiation oncology domain, were created to facilitate sharing of larger-scale high-quality research data in a centralized, standardized fashion. Thanks to TCIA, there is now a centralized cancer imaging data repository for clinical researchers to upload, share, and easily search for de-identified, curated imaging datasets for various cancer sites, including head and neck (Grossberg et al. 2017a, 2017b, 2018, Vallières et al. 2017).

As big-data analysis becomes more cross-disciplinary, the need for collaboration between these distinct databases increases. For instance, Guo et al. used integrated data from The Cancer Imaging Archive and The Cancer Genome Atlas databases to build predictive models for clinical outcomes in invasive breast carcinoma (Guo et al. 2015). Similarly, Zhu et al. (2017) conducted a comprehensive study to discover imaging-genomics correlations and explore the potential for predicting oropharyngeal tumor genomic alternations using radiomics features. For such multi-disciplinary studies to become commonplace, different data repositories must facilitate the curation of matched imaging and clinical data pools. Promoting a culture of multi-disciplinary collaboration and communication within and between academic institutions as well as between data repositories will be critical to achieving these goals. Indeed, methodologies are currently in development that permit the sharing and online updating of predictive models, when direct sharing of the data underlying those models might be infeasible (Chang et al. 2018).

16.7 CONCLUSIONS AND FUTURE DIRECTIONS

In summary, radiomics and radiogenomics have already shown promising potential in revealing tumor and normal tissue information that is above and beyond visual analysis. Although caution should be taken while interpreting these results, radiomics provides rich and fertile environment for exploratory correlative biomarker analysis and establishing personalized multi-faceted decision-making nomograms. Integrating radiomics into clinical trials is essential for development of a translational pipeline for radiomics from methodology to clinical implementation.

REFERENCES

Abdollahi, H., Mostafaei, S., Cheraghi, S., Shiri, I., Rabi Mahdavi, S. & Kazemnejad, A. 2018. Cochlea CT radiomics predicts chemoradiotherapy induced sensorineural hearing loss in head and neck cancer patients: A machine learning and multi-variable modelling study. *Physica Medica*, 45, 192–197.

Acharya, U. R., Hagiwara, Y., Sudarshan, V. K., Chan, W. Y. & Ng, K. H. 2018. Towards precision medicine: From quantitative imaging to radiomics. *Journal of Zhejiang University: Science B*, 19, 6–24.

Aerts, H. J. W. L., Velazquez, E. R., Leijenaar, R. T. H., Parmar, C., Grossmann, P., Carvalho, S., Bussink et al. 2014. Decoding tumour phenotype by noninvasive imaging using a quantitative radiomics approach. *Nature Communications*, 5, 4006.

Ardakani, A. A., Gharbali, A. & Mohammadi, A. 2015. Application of texture analysis method for classification of benign and malignant thyroid nodules in ultrasound images. *Iranian Journal of Cancer Prevention*, 8, 116–124.

Block, A. M., Cozzi, F., Patel, R., Surucu, M., Hurst, N., Emami, B. & Roeske, J. C. 2017. Radiomics in head and neck radiation therapy: Impact of metal artifact reduction. *International Journal of Radiation Oncology Biology Physics*, 99, E640.

Bogowicz, M., Riesterer, O., Ikenberg, K., Stieb, S., Moch, H., Studer, G., Guckenberger, M. & Tanadini-Lang, S. 2017. Computed tomography radiomics predicts HPV status and local tumor control after definitive radiochemotherapy in head and neck squamous cell carcinoma. *International Journal of Radiation Oncology Biology Physics*, 99, 921–928.

Brown, A. M., Nagala, S., Mclean, M. A., Lu, Y., Scoffings, D., Apte, A., Gonen, M. et al. 2016. Multi-institutional validation of a novel textural analysis tool for preoperative stratification of suspected thyroid tumors on diffusion-weighted MRI. *Magnetic Resonance in Medicine*, 75, 1708–1716.

Buch, K., Fujita, A., Li, B., Kawashima, Y., Qureshi, M. M. & Sakai, O. 2015. Using texture analysis to determine human papillomavirus status of oropharyngeal squamous cell carcinomas on CT. *American Journal of Neuroradiology*, 36, 1343–1348.

Cantrell, S. C., Peck, B. W., Li, G., Wei, Q., Sturgis, E. M. & Ginsberg, L. E. 2013. Differences in imaging characteristics of HPV-positive and HPV-negative oropharyngeal cancers: A blinded matched-pair analysis. *American Journal of Neuroradiology*, 34, 2005–2009.

Chan, S. C., Chang, K. P., Fang, Y. D., Tsang, N. M., Ng, S. H., Hsu, C. L., Liao, C. T. & Yen, T. C. 2017. Tumor heterogeneity measured on F-18 fluorodeoxyglucose positron emission tomography/computed tomography combined with plasma Epstein-Barr virus load predicts prognosis in patients with primary nasopharyngeal carcinoma. *Laryngoscope*, 127, E22–E28.

Chang, K., Balachandar, N., Lam, C., Yi, D., Brown, J., Beers, A., Rosen, B., Rubin, D. L. & Kalpathy-Cramer, J. 2018. Distributed deep learning networks among institutions for medical imaging. *Journal of the American Medical Informatics Association*, 25, 945–954.

Chen, R. Y., Lin, Y. C., Shen, W. C., Hsieh, T. C., Yen, K. Y., Chen, S. W. & Kao, C. H. 2018. Associations of tumor PD-1 ligands, immunohistochemical studies, and textural features in 18F-FDG PET in squamous cell carcinoma of the head and neck. *Scientific Reports*, 8, 105.

Chen, S.-W., Shen, W.-C., Lin, Y.-C., Chen, R.-Y., Hsieh, T.-C., Yen, K.-Y. & Kao, C.-H. 2017. Correlation of pretreatment 18F-FDG PET tumor textural features with gene expression in pharyngeal cancer and implications for radiotherapy-based treatment outcomes. *European Journal of Nuclear Medicine and Molecular Imaging*, 44, 567–580.

Clark, K., Vendt, B., Smith, K., Freymann, J., Kirby, J., Koppel, P., Moore et al. 2013. The Cancer Imaging Archive (TCIA): Maintaining and operating a public information repository. *Journal of Digital Imaging*, 26, 1045–1057.

Clavo, B., Robaina, F., Fiuza, D., Ruiz, A., Lloret, M., Rey-Baltar, D., Llontop, P. et al. 2017. Predictive value of hypoxia in advanced head and neck cancer after treatment with hyperfractionated radio-chemotherapy and hypoxia modification. *Clinical and Translational Oncology*, 19, 419–424.

Cook, G. J. R., Azad, G., Owczarczyk, K., Siddique, M. & Goh, V. 2018. Challenges and promises of PET radiomics. *International Journal of Radiation Oncology Biology Physics*, 102, 1083–1089.

Crispin-Ortuzar, M., Apte, A., Grkovski, M., Oh, J. H., Lee, N. Y., Schoder, H., Humm, J. L. & Deasy, J. O. 2018. Predicting hypoxia status using a combination of contrast-enhanced computed tomography and [(18)F]-Fluorodeoxyglucose positron emission tomography radiomics features. *Radiotherapy and Oncology*, 127, 36–42.

Dang, M., Lysack, J. T., Wu, T., Matthews, T. W., Chandarana, S. P., Brockton, N. T., Bose, P. et al. 2015. MRI texture analysis predicts p53 status in head and neck squamous cell carcinoma. *American Journal of Neuroradiology*, 36, 166–170.

Deasy, J. O., Moiseenko, V., Marks, L., Chao, K. S. C., Nam, J. & Eilsbruch, A. 2010. Radiotherapy dose-volume effects on salivary gland function. *International Journal of Radiation Oncology Biology Physics*, 76, S58–S63.

El Naqa, I. 2016. Perspectives on making big data analytics work for oncology. *Methods*, 111, 32–44.

El Naqa, I., Grigsby, P., Apte, A., Kidd, E., Donnelly, E., Khullar, D., Chaudhari, S. et al. 2009. Exploring feature-based approaches in PET images for predicting cancer treatment outcomes. *Pattern Recognition*, 42, 1162–1171.

Elhalawani, H., Kanwar, A., Mohamed, A. S. R., White, A., Zafereo, J., Wong, A., Berends, J. et al. 2018a. Investigation of radiomic signatures for local recurrence using primary tumor texture analysis in oropharyngeal head and neck cancer patients. *Scientific Reports*, 8, 1524.

Elhalawani, H., Lin, T. A., Volpe, S., Mohamed, A. S. R., White, A. L., Zafereo, J., Wong, A. J. et al. 2018b. Machine learning applications in head and neck radiation oncology: Lessons from open-source radiomics challenges. *Frontiers in Oncology*, 8, 294.

Elhalawani, H., Mohamed, A. S. R., Kanwar, A., Dursteler, A., Rock, C. D., Eraj, S. E., Meheissen, M. et al. 2018c. EP-2121: Serial parotid gland radiomic-based model predicts post-radiation xerostomia in oropharyngeal cancer. *Radiotherapy and Oncology*, 127, S1167–S1168.

Elhalawani, H., Mohamed, A. S. R., White, A. L., Zafereo, J., Wong, A. J., Berends, J. E., Abohashem, S. et al. 2017. Matched computed tomography segmentation and demographic data for oropharyngeal cancer radiomics challenges. *Scientific Data*, 4, 170077.

Elhalawani, H., Volpe, S., Barua, S., Mohamed, A., Yang, P., Ng, S. P., Lai, S. et al. 2018d. Exploration of an early imaging biomarker of osteoradionecrosis in oropharyngeal cancer patients: Case-control study of the temporal changes of mandibular radiomics features. *International Journal of Radiation Oncology Biology Physics*, 100, 1363–1364.

Elhalawani, H. E., Mohamed, A. S. R., Volpe, S., Yang, P., Campbell, S., Granberry, R., Ger, R. et al. 2018e. PO-0991: Serial tumor radiomic features predict response of head and neck cancer treated with radiotherapy. *Radiotherapy and Oncology*, 127, S551.

Fernandez-Delgado, M., Cernadas, E., Barro, S. & Amorim, D. 2014. Do we need hundreds of classifiers to solve real world classification problems? *The Journal of Machine Learning Research*, 15, 3133–3181.

Fruehwald-Pallamar, J., Czerny, C., Holzer-Fruehwald, L., Nemec, S. F., Mueller-Mang, C., Weber, M. & Mayerhoefer, M. E. 2013. Texture-based and diffusion-weighted discrimination of parotid gland lesions on MR images at 3.0 Tesla. *NMR in Biomedicine*, 26, 1372–1379.

Fujita, A., Buch, K., Li, B., Kawashima, Y., Qureshi, M. M. & Sakai, O. 2016. Difference between HPV-positive and HPV-negative non-oropharyngeal head and neck cancer: Texture analysis features on CT. *Journal of Computer Assisted Tomography*, 40, 43–47.

Gesheng, S., Fuzhong, X. & Chengqi, Z. 2015. A model using texture features to differentiate the nature of thyroid nodules on sonography. *Journal of Ultrasound in Medicine*, 34, 1753–1760.

Gillison, M. L., Chaturvedi, A. K., Anderson, W. F. & Fakhry, C. 2015. Epidemiology of human papillomavirus–positive head and neck squamous cell carcinoma. *Journal of Clinical Oncology*, 33, 3235–3242.

Grossberg, A., Abdallah, M., Elhalawani, H., Bennett, W., Smith, K., Nolan, T., Chamchod, S., et al. 2017a. Data from head and neck cancer CT atlas. *The Cancer Imaging Archive*, doi:10.7937/K9/TCIA.2017. umz8dv6s.

Grossberg, A. J., Mohamed, A. S. R., Chamchod, S., Bennett, W., Smith, K., Nolan, T., Kantor, M., Browne, T., Rosenthal, D. I. & Fuller, C. D. 2017b. (P055) TCIA imaging database for head and neck squamous cell carcinoma patients treated with radiotherapy. *International Journal of Radiation Oncology Biology Physics*, 98, E29–E30.

Grossberg, A. J., Mohamed, A. S. R., Elhalawani, H., Bennett, W. C., Smith, K. E., Nolan, T. S., Williams, B. et al. 2018. Imaging and clinical data archive for head and neck squamous cell carcinoma patients treated with radiotherapy. *Scientific Data*, 5, 180173.

Guo, R., Li, Q., Meng, L., Zhang, Y. & Gu, C. 2008. P53 and vascular endothelial growth factor expressions are two important indices for prognosis in gastric carcinoma. *West Indian Medical Journal*, 57, 2–6.

Guo, W., Li, H., Zhu, Y., Lan, L., Yang, S., Drukker, K., Morris, E. et al. 2015. Prediction of clinical phenotypes in invasive breast carcinomas from the integration of radiomics and genomics data. *Journal of Medical Imaging*, 2, 041007.

Hodgson, D. R., Whittaker, R. D., Herath, A., Amakye, D. & Clack, G. 2009. Biomarkers in oncology drug development. *Molecular Oncology*, 3, 24–32.

Houweling, A. C., Philippens, M. E. P., Dijkema, T., Roesink, J. M., Terhaard, C. H. J., Schilstra, C., Ten Haken, R. K., Eisbruch, A. & Raaijmakers, C. P. J. 2010. A comparison of dose–response models for the parotid gland in a large group of head-and-neck cancer patients. *International Journal of Radiation Oncology Biology Physics*, 76, 1259–1265.

Huang, Y., Liu, Z., He, L., Chen, X., Pan, D., Ma, Z., Liang, C., Tian, J. & Liang, C. 2016. Radiomics signature: A potential biomarker for the prediction of disease-free survival in early-stage (I or II) non—small cell lung cancer. *Radiology*, 281, 947–957.

Izumi, M., Hida, A., Takagi, Y., Kawabe, Y., Eguchi, K. & Nakamura, T. 2000. MR imaging of the salivary glands in sicca syndrome. *American Journal of Roentgenology*, 175, 829–834.

Jansen, J. F. A., Lu, Y., Gupta, G., Lee, N. Y., Stambuk, H. E., Mazaheri, Y., Deasy, J. O. & Shukla-Dave, A. 2016. Texture analysis on parametric maps derived from dynamic contrast-enhanced magnetic resonance imaging in head and neck cancer. *World Journal of Radiology*, 8, 90–97.

Jethanandani, A., Lin, T. A., Volpe, S., Elhalawani, H., Mohamed, A. S. R., Yang, P. & Fuller, C. D. 2018. Exploring applications of radiomics in magnetic resonance imaging of head and neck cancer: A systematic review. *Frontiers in Oncology*, 8, 131.

Kale, S. D. & Punwatkar, K. M. 2013. Texture analysis of thyroid ultrasound images for diagnosis of benign and malignant nodule using scaled conjugate gradient backpropagation training neural network. *International Journal of Computational Engineering and Management*, 16, 33–38.

Kataria, T., Gupta, D., Goyal, S., Bisht, S. S., Basu, T., Abhishek, A., Narang, K., Banerjee, S., Nasreen, S., Sambasivam, S. & Dhyani, A. 2016. Clinical outcomes of adaptive radiotherapy in head and neck cancers. *The British Journal of Radiology*, 89, 20160085.

Kim, S.-J. & Chang, S. 2015. Predictive value of intratumoral heterogeneity of F-18 FDG uptake for characterization of thyroid nodules according to Bethesda categories of fine needle aspiration biopsy results. *Endocrine*, 50, 681–688.

Kulzer, M. H. & Branstetter, B. F. T. 2017. Chapter 1 neck anatomy, imaging-based level nodal classification and impact of primary tumor site on patterns of nodal metastasis. *Seminars in Ultrasound CT and MR*, 38, 454–465.

Lapa, C., Werner, R. A., Schmid, J. S., Papp, L., Zsoter, N., Biko, J., Reiners, C., Herrmann, K., Buck, A. K. & Bundschuh, R. A. 2015. Prognostic value of positron emission tomography-assessed tumor heterogeneity in patients with thyroid cancer undergoing treatment with radiopeptide therapy. *Nuclear Medicine and Biology*, 42, 349–354.

Larue, R. T., Defraene, G., De Ruysscher, D., Lambin, P. & Van Elmpt, W. 2017. Quantitative radiomics studies for tissue characterization: A review of technology and methodological procedures. *The British Journal of Radiology*, 90, 20160665.

Leemans, C. R., Braakhuis, B. J. & Brakenhoff, R. H. 2011. The molecular biology of head and neck cancer. *Nature Reviews Cancer*, 11, 9–22.

Leger, S., Zwanenburg, A., Pilz, K., Lohaus, F., Linge, A., Zöphel, K., Kotzerke, J. et al. 2017. A comparative study of machine learning methods for time-to-event survival data for radiomics risk modelling. *Scientific Reports*, 7, 13206.

Leijenaar, R. T., Bogowicz, M., Jochems, A., Hoebers, F. J., Wesseling, F. W., Huang, S. H., Chan, B. et al. 2018. Development and validation of a radiomic signature to predict HPV (p16) status from standard CT imaging: A multicenter study. *The British Journal of Radiology*, 91, 20170498.

Leijenaar, R. T., Carvalho, S., Hoebers, F. J., Aerts, H. J., Van Elmpt, W. J., Huang, S. H., Chan, B., Waldron, J. N., O'sullivan, B. & Lambin, P. 2015. External validation of a prognostic CT-based radiomic signature in oropharyngeal squamous cell carcinoma. *Acta Oncologica*, 54, 1423–1429.

Liang, C., Huang, Y., He, L., Chen, X., Ma, Z., Dong, D., Tian, J., Liang, C. & Liu, Z. 2016. The development and validation of a CT-based radiomics signature for the preoperative discrimination of stage I–II and stage III–IV colorectal cancer. *Oncotarget*, 7, 31401–31412.

Liu, J., Mao, Y., Li, Z., Zhang, D., Zhang, Z., Hao, S. & Li, B. 2016. Use of texture analysis based on contrast-enhanced MRI to predict treatment response to chemoradiotherapy in nasopharyngeal carcinoma. *Journal of Magnetic Resonance Imaging*, 44, 445–455.

Lu, L., Ehmke, R. C., Schwartz, L. H. & Zhao, B. 2016. Assessing agreement between radiomic features computed for multiple CT imaging settings. *PLoS One*, 11, e0166550.

Lydiatt, W. M., Patel, S. G., O'sullivan, B., Brandwein, M. S., Ridge, J. A., Migliacci, J. C., Loomis, A. M. & Shah, J. P. 2017. Head and neck cancers—Major changes in the American Joint Committee on cancer eighth edition cancer staging manual. *CA: A Cancer Journal for Clinicians*, 67, 122–137.

Meyer, H. J., Schob, S., Hohn, A. K. & Surov, A. 2017. MRI texture analysis reflects histopathology parameters in thyroid cancer—A first preliminary study. *Translational Oncology*, 10, 911–916.

Moons, K. G., Altman, D. G., Reitsma, J. B., Ioannidis, J. P., Macaskill, P., Steyerberg, E. W., Vickers, A. J., Ransohoff, D. F. & Collins, G. S. 2015. Transparent reporting of a multivariable prediction model for individual prognosis or diagnosis (TRIPOD): Explanation and elaboration. *Annals of Internal Medicine*, 162, W1–73.

Nakatsugawa, M., Cheng, Z., Hui, X., Choflet, A., Kiess, A. P., Bowers, M. R., Utsunomiya, K. et al. 2017. The value of continuous toxicity updates on the accuracy of prediction models within a learning health system. *International Journal of Radiation Oncology Biology Physics*, 99, E647–E648.

Nardone, V., Tini, P., Nioche, C., Mazzei, M. A., Carfagno, T., Battaglia, G., Pastina, P., Grassi, R., Sebaste, L. & Pirtoli, L. 2018. Texture analysis as a predictor of radiation-induced xerostomia in head and neck patients undergoing IMRT. *Radiologia Medica*, 123, 415–423.

Parmar, C., Grossmann, P., Rietveld, D., Rietbergen, M. M., Lambin, P. & Aerts, H. J. W. L. 2015. Radiomic machine-learning classifiers for prognostic biomarkers of head and neck cancer. *Frontiers in Oncology*, 5, 272.

Plant, R. L. 1998. Image analysis of benign and malignant neck masses. *Annals of Otology, Rhinology and Laryngology*, 107, 689–696.

Pickering, C. R., Shah, K., Ahmed, S., Rao, A., Fredrick, M. J., Zhang, J., Unruh, A. K., Wang, J., Ginsberg, L. E., Kumar, A. J., Myers, J. N. & Hamilton, J. D. 2013. CT imaging correlates of genomic expression for oral cavity squamous cell carcinoma. *American Journal of Neuroradiology*, 34, 1818–1822.

Poeta, M. L., Manola, J., Goldwasser, M. A., Forastiere, A., Benoit, N., Califano, J. A., Ridge, J. et al. 2007. TP53 mutations and survival in squamous-cell carcinoma of the head and neck. *New England Journal of Medicine*, 357, 2552–2561.

Pota, M., Scalco, E., Sanguineti, G., Cattaneo, G. M., Esposito, M. & Rizzo, G. 2015. Early classification of parotid glands shrinkage in radiotherapy patients: A comparative study. *Biosystems Engineering*, 138, 77–89.

Pota, M., Scalco, E., Sanguineti, G., Farneti, A., Cattaneo, G. M., Rizzo, G. & Esposito, M. 2017. Early prediction of radiotherapy-induced parotid shrinkage and toxicity based on CT radiomics and fuzzy classification. *Artificial Intelligence in Medicine*, 81, 41–53.

Raab-Traub, N. 2002. Epstein-Barr virus in the pathogenesis of NPC. *Seminars in Cancer Biology*, 12, 431–441.

Radiological Society Of North America, I. CTP-The RSNA Clinical Trial Processor. Radiological Society of North America, Inc. Available via http://mircwiki.rsna.org/index.php?title=CTP-The_RSNA_Clinical_Trial_Processor. Accessed December 2017.

Ramkumar, S., Ranjbar, S., Ning, S., Lal, D., Zwart, C. M., Wood, C. P., Weindling, S. M. et al. 2017. MRI-based texture analysis to differentiate sinonasal squamous cell carcinoma from inverted papilloma. *American Journal of Neuroradiology*, 38, 1019–1025.

Rosenthal, D. I., Harari, P. M., Giralt, J., Bell, D., Raben, D., Liu, J., Schulten, J., Ang, K. K. & Bonner, J. A. 2016. Association of human papillomavirus and p16 status with outcomes in the IMCL-9815 phase III registration trial for patients with locoregionally advanced oropharyngeal squamous cell carcinoma of the head and neck treated with radiotherapy with or without cetuximab. *Journal of Clinical Oncology*, 34, 1300–1308.

Rumboldt, Z., Gordon, L., Gordon, L., Bonsall, R. & Ackermann, S. 2006. Imaging in head and neck cancer. *Current Treatment Options in Oncology*, 7, 23–34.

Rumley, C. N., Lee, M. T., Holloway, L., Rai, R., Min, M., Forstner, D., Fowler, A. & Liney, G. 2017. Multiparametric magnetic resonance imaging in mucosal primary head and neck cancer: A prospective imaging biomarker study. *BMC Cancer*, 17, 475.

Sanguineti, G., Ricchetti, F., Wu, B., Mcnutt, T. & Fiorino, C. 2015. Parotid gland shrinkage during IMRT predicts the time to xerostomia resolution. *Radiation Oncology*, 10, 19.

Sayyouh, M., Vummidi, D. R. & Kazerooni, E. A. 2013. Evaluation and management of pulmonary nodules: State-of-the-art and future perspectives. *Expert Opinion on Medical Diagnostics*, 7, 629–644.

Scalco, E., Fiorino, C., Cattaneo, G. M., Sanguineti, G. & Rizzo, G. 2013. Texture analysis for the assessment of structural changes in parotid glands induced by radiotherapy. *Radiotherapy and Oncology*, 109, 384–387.

Scalco, E., Moriconi, S. & Rizzo, G. 2015. Texture analysis to assess structural modifications induced by radiotherapy. *Annual International Conference of the IEEE Engineering in Medicine and Biology Society*, 2015, 5219–5222.

Schob, S., Meyer, H., Dieckow, J., Pervinder, B., Pazaitis, N., Höhn, A., Garnov, N., Horvath-Rizea, D., Hoffmann, K.-T. & Surov, A. 2017. Histogram analysis of diffusion weighted imaging at 3T is useful for prediction of lymphatic metastatic spread, proliferative activity, and cellularity in thyroid cancer. *International Journal of Molecular Sciences*, 18, 821.

Siebers, S., Zenk, J., Bozzato, A., Klintworth, N., Iro, H. & Ermert, H. 2010. Computer aided diagnosis of parotid gland lesions using ultrasonic multi-feature tissue characterization. *Ultrasound in Medicine and Biology*, 36, 1525–1534.

Singer, P. A. 1996. Evaluation and management of the solitary thyroid nodule. *Otolaryngologic Clinics of North America*, 29, 577–591.

Sollini, M., Cozzi, L., Chiti, A. & Kirienko, M. 2018. Texture analysis and machine learning to characterize suspected thyroid nodules and differentiated thyroid cancer: Where do we stand? *European Journal of Radiology*, 99, 1–8.

Song, C. I., Ryu, C. H., Choi, S. H., Roh, J. L., Nam, S. Y. & Kim, S. Y. 2013. Quantitative evaluation of vocal-fold mucosal irregularities using GLCM-based texture analysis. *Laryngoscope*, 123, E45–E50.

Sumi, M., Kimura, Y., Sumi, T. & Nakamura, T. 2007. Diagnostic performance of MRI relative to CT for metastatic nodes of head and neck squamous cell carcinomas. *Journal of Magnetic Resonance Imaging*, 26, 1626–1633.

The Cancer Genome Atlas Research Network, Weinstein, J. N., Collisson, E. A., Mills, G. B., Shaw, K. R. M., Ozenberger, B. A., Ellrott, K., Shmulevich, I., Sander, C. & Stuart, J. M. 2013. The Cancer genome atlas pan-cancer analysis project. *Nature Genetics*, 45, 1113.

Thompson, R. H., Gillett, M. D., Cheville, J. C., Lohse, C. M., Dong, H., Webster, W. S., Krejci, K. G. et al. 2004. Costimulatory B7-H1 in renal cell carcinoma patients: Indicator of tumor aggressiveness and potential therapeutic target. *Proceedings of the National Academy of Sciences of the United States of America*, 101, 17174–17179.

Trotta, B. M., Pease, C. S., Rasamny, J. J., Raghavan, P. & Mukherjee, S. 2011. Oral cavity and oropharyngeal squamous cell cancer: Key imaging findings for staging and treatment planning. *Radiographics*, 31, 339–354.

Vallieres, M., Freeman, C. R., Skamene, S. R. & El Naqa, I. 2015. A radiomics model from joint FDG-PET and MRI texture features for the prediction of lung metastases in soft-tissue sarcomas of the extremities. *Physics in Medicine & Biology*, 60, 5471–5496.

Vallières, M., Kay-Rivest, E., Perrin, L. J., Liem, X., Furstoss, C., Aerts, H. J., Khaouam, N. et al. 2017a. Radiomics strategies for risk assessment of tumour failure in head-and-neck cancer. *Scientific Reports*, 7, 10117.

Vallières, M., Kayrivest, E., Perrin, L. J., Liem, X., Furstoss, C., Khaouam, N., Nguyen-Tan, P. F., Wang, C., S. & Sultanem, K. 2017b. Data from Head-Neck-PET-CT. The Cancer Imaging Archive.

Vallieres, M., Kumar, A., Sultanem, K. & El Naqa, I. 2013. FDG-PET image-derived features can determine HPV status in head-and-neck cancer. *International Journal of Radiation Oncology Biology Physics*, 87, S467.

van Dijk, L. V., Brouwer, C. L., Van Der Schaaf, A., Burgerhof, J. G. M., Beukinga, R. J., Langendijk, J. A., Sijtsema, N. M. & Steenbakkers, R. 2017. CT image biomarkers to improve patient-specific prediction of radiation-induced xerostomia and sticky saliva. *Radiotherapy and Oncology*, 122, 185–191.

van Dijk, L. V., Noordzij, W., Brouwer, C. L., Boellaard, R., Burgerhof, J. G. M., Langendijk, J. A., Sijtsema, N. M. & Steenbakkers, R. 2018. (18)F-FDG PET image biomarkers improve prediction of late radiation-induced xerostomia. *Radiotherapy and Oncology*, 126, 89–95.

Wang, G., He, L., Yuan, C., Huang, Y., Liu, Z. & Liang, C. 2018. Pretreatment MR imaging radiomics signatures for response prediction to induction chemotherapy in patients with nasopharyngeal carcinoma. *European Journal of Radiology*, 98, 100–106.

Way, T. W., Sahiner, B., Chan, H. P., Hadjiiski, L., Cascade, P. N., Chughtai, A., Bogot, N. & Kazerooni, E. 2009. Computer-aided diagnosis of pulmonary nodules on CT scans: Improvement of classification performance with nodule surface features. *Medical Physics*, 36, 3086–3098.

Wippold, F.J. 2007. Head and neck imaging: The role of CT and MRI. *Journal of Magnetic Resonance Imaging*, 25, 453–465.

Wong, A. J., Kanwar, A., Mohamed, A. S. & Fuller, C. D. 2016. Radiomics in head and neck cancer: From exploration to application. *Translational Cancer Research*, 5, 371–382.

Wu, M. C., Chin, W. C., Tsan, T. C. & Chin, C. L. 2016. The benign and malignant recognition system of nasopharynx in MRI image with neural-fuzzy based Adaboost classifier. *Institute of Electrical and Electronics Engineers Inc*, 47–51.

Wu, T. Y. & Lin, S. F. 2013. A method for extracting suspected parotid lesions in CT images using feature-based segmentation and active contours based on stationary wavelet transform. *Measurement Science Review*, 13, 237.

Yang, P., Mackin, D., Chen, C., Mohamed, A. S. R., Elhalawani, H., Shi, Y., Yao, J. et al. 2018. Discrimination of Epstein-Barr virus status in NPC using CT-derived radiomics features: Linking imaging phenotypes to tumor biology. *International Journal of Radiation Oncology Biology Physics*, 100, 1361.

Yonghong, H., Chu, P., Weiwei, C., Tao, L., Wenzhen, Z. & Jianpin, Q. 2016. Differentiation between malignant and benign thyroid nodules and stratification of papillary thyroid cancer with aggressive histological features: Whole-lesion diffusion-weighted imaging histogram analysis. *Journal of Magnetic Resonance Imaging*, 44, 1546–1555.

Yu, H., Caldwell, C., Mah, K. & Mozeg, D. 2009a. Coregistered FDG PET/CT-based textural characterization of head and neck cancer for radiation treatment planning. *IEEE Transactions on Medical Imaging*, 28, 374–383.

Yu, H., Caldwell, C., Mah, K., Poon, I., Balogh, J., Mackenzie, R., Khaouam, N. & Tirona, R. 2009b. Automated radiation targeting in head-and-neck cancer using region-based texture analysis of PET and CT images. *International Journal of Radiation Oncology Biology Physics*, 75, 618–625.

Yu, K., Zhang, Y., Yu, Y., Huang, C., Liu, R., Li, T., Yang, L., Morris, J. S., Baladandayuthapani, V. & Zhu, H. 2017. Radiomic analysis in prediction of human papilloma virus status. *Clinical and Translational Radiation Oncology*, 7, 49–54.

Zhai, T. T., van Dijk, L. V., Huang, B. T., Lin, Z. X., Ribeiro, C. O., Brouwer, C. L., Oosting, S. F. et al. 2017. Improving the prediction of overall survival for head and neck cancer patients using image biomarkers in combination with clinical parameters. *Radiotherapy and Oncology*, 124, 256–262.

Zhang, B., Tian, J., Dong, D., Gu, D., Dong, Y., Zhang, L., Lian, Z. et al. 2017. Radiomics features of multiparametric MRI as novel prognostic factors in advanced nasopharyngeal carcinoma. *Clinical Cancer Research*, 23, 4259–4269.

Zhang, H., Graham, C. M., Elci, O., Griswold, M. E., Zhang, X., Khan, M. A., Pitman, K. et al. 2013. Locally advanced squamous cell carcinoma of the head and neck: CT texture and histogram analysis allow independent prediction of overall survival in patients treated with induction chemotherapy. *Radiology*, 269, 801–809.

Zhu, Y., Mohamed, A. S., Lai, S. Y., Yang, S., Kanwar, A., Wei, L., Kamal, M et al. 2017. Imaging-genomics study of head-neck squamous cell carcinoma: Associations between radiomic phenotypes and genomic mechanisms via integration of TCGA and TCIA. *bioRxiv*, doi:10.1101/214312.

Gastrointestinal cancers

ZAIYI LIU

17.1 INTRODUCTION

Gastrointestinal (GI) cancer refers to a spectrum of cancers of the GI tract and accessory organs of diges-
tion, including cancers of the esophagus, stomach, small intestine, colon, rectum, pancreas, and gall-
bladder. Gastric cancer (GC) and colorectal cancer (CRC) are the two most common gastrointestinal
malignancies and the third and fourth leading causes of cancer-related fatality globally, respectively.

Esophagus cancer (EC) is the sixth most common gastrointestinal cancer in the world, and its incidence is increasing. Hepatocellular carcinoma (HCC) is the sixth most common malignant tumor and the third most frequent reason for cancer-related mortality (Forner et al. 2012). Meanwhile, the liver is the most common metastasis site for CRC (developed in approximately 50% of patients) (Al Bandar and Kim 2017). Pancreatic adenocarcinoma (PA) is the most common malignant tumor of the pancreas, with a low overall 5-year survival rate of 8% (Chiaravalli et al. 2017). Intraductal papillary mucinous neoplasm (IPMN) is the precursor of PA (Fong and Fernandez-Del Castillo 2016).

Radiomics has emerged as a powerful technique that enables quantification of the tumor phenotype through medical imaging and has deepened our understanding of tumor characteristics. Rapid progress in radiomics for GI cancers has been made, with a vast majority of radiomics studies that have revealed important information for diagnosis, prognosis, and prediction for EC, GC, and CRC, as well as liver and pancreatic malignancies.

The role of image-based phenotyping for precision medicine is currently being investigated. As the availability of robust and validated biomarkers is essential to move precision medicine forward, radiomics analyses, capable of offering a nearly limitless supply of imaging biomarkers in gastrointestinal cancers, epitomize the pursuit of precision medicine and promise to increase precision in diagnosis, assessment of prognosis, and prediction of therapy response (Gillies et al. 2016).

17.2 CLINICAL APPLICATION

Radiomics can be applied to a large number of conditions in gastrointestinal cancers, with emerging research that aims to aid in tumor characterization, staging, treatment planning, assessment of prognosis, and longitudinal response monitoring. In this section, a detailed review of the clinical application of radiomics features in gastrointestinal cancers will be summarized and categorized according to its major topics. The largest proportion of these studies have concentrated on the therapeutic response or clinical outcome predictions in patients with a specific treatment because heterogeneity of the diseases is a potential mechanism (Holzel et al. 2013, Junttila and de Sauvage 2013, McGranahan and Swanton 2017) that accounts for recurrence or treatment failure in a specific therapy and requires patient selection for better management. Histological grading and subtyping using radiomics features are also common because they are associated with tumor characterization and risk stratification. While ostensibly for the application of differential diagnoses, most studies were preliminary research aimed at testing a new classifier or algorithm.

17.2.1 DIFFERENTIAL DIAGNOSIS, HISTOLOGICAL GRADING, AND SUBTYPING

In the diagnostic setting, the probability that a particular tumor is present can be used to inform the referral of patients for further testing and to initiate treatment directly. Pathological data are the gold standard for differential diagnosis, histological grading, and subtyping. However, pathological features of a tumor can only be obtained by invasive techniques such as surgical resection or biopsy. Medical imaging is a valuable data source that holds the potential to provide non-invasive diagnostic surrogates for tumors. Through extraction of quantitative features from medical images, radiomics has shown great promise for enabling diagnosis, including the identification or detection of a specific feature of a disease, such as the malignancy of a tumor and a physiological or pathological feature with clinical significance (Torre et al. 2015).

17.2.1.1 GASTRIC CANCER

The feasibility of texture analysis for the classification of gastric malignancies (gastric adenocarcinoma, lymphoma, and gastrointestinal stromal tumors) on contrast-enhanced hydrodynamic-multidetector computed tomography (MDCT) images was demonstrated early in 2013 (Ba-Ssalamah et al. 2013). Although the results showed that textural information may aid in establishing correct diagnosis in a pair-wise approach (in cases where the differential diagnosis can be narrowed down to two histological subtypes), it failed in distinguishing between all three subtypes at the same time. It is worth mentioning that there was a difference in the reconstruction section thickness for the arterial (1 mm) and portal-venous phase (4 mm) used in this study. Although there is a lack of research on the effects of reconstruction thickness in radiomics studies for GI cancers, which is certainly worthy of further investigation, recent studies have shown that the heterogeneity of acquisition and reconstruction protocols influences the robustness of radiomics analyses (Larue et al. 2017, Lu et al. 2016). A further limitation of this study was the sample size, with only five patients with lymphoma (arterial-phase scan) enrolled.

A further study by Ma et al. developed a radiomics signature based on pre-treatment CT images to distinguish Borrmann type IV gastric cancer (GC) from primary gastric lymphoma (PGL). The radiomics signature showed a higher area under the curve (AUC) (0.886), sensitivity, and specificity than the subjective imaging findings. Moreover, combining the radiomics data with subjective CT findings further improved the discrimination performance (AUC: 0.903; specificity: 1; sensitivity: 0.7) (Ma et al. 2017). In addition, as the first study using 3D radiomics features for differential diagnosis in gastric malignancy, the results of this study support the previous finding that using 3D radiomics features improved the discrimination accuracy compared with using 2D features. By taking all of the available slices into consideration, it provides a more comprehensive view of the whole tumor.

17.2.1.2 COLORECTAL CANCER

The histogram features of primary colon cancer and hepatic metastases derived from pre-treatment PET/CT and contrast-enhanced CT (CECT) images were retrospectively analyzed in a study by Wagner et al. (2017). Among the 50 patients with histopathology-proven primary CRC, 18 patients had liver metastases, and the others were in the M0 stage. When comparing the imaging modalities separately for histogram features, the results showed that the PET/CT and CECT were complementary in terms of the information they provide about tumor heterogeneity. A significant difference in histogram features was observed ($p < 0.05$) between primary colon cancer and hepatic metastases. However, these features were not able to differentiate primary colon cancer lesions of the patients with hepatic metastases from those without ($p > 0.05$). Unpublished data by the author of this chapter showed that CT image-based imaging biomarker could be applied to detect synchronous colorectal liver metastases as an independent predictor. Among four imaging biomarkers, the portal venous phase CT image-based imaging biomarker demonstrates better performance compared to that of the unenhanced CT image or arterial phase CT image-based imaging biomarker while the combined imaging biomarker, showed the best prediction performance (Table 17.1).

As for the grading of CRC, unpublished data by the author of this chapter showed that a CT-based radiomics approach enabled the pre-operative discrimination of high- from low-grade colorectal adenocarcinomas. With 366 patients retrospectively enrolled and 10959 3D radiomics features extracted from pre-operative CT images, this unpublished study developed and validated a radiomics signature (consisting of 14 features) that successfully discriminated high- from low-grade CRC (AUC: 0.812 in training cohort, 0.735 in validation cohort). The constructed radiomics signature showed good discrimination (concordance index [C-index]: 0.719; 95% confidence interval: 0.715–0.723) and classification performance (sensitivity = 0.726;

Table 17.1 AUC, sensitivity, specificity, PPV, NPV, and the best cut-off of imaging biomarkers for the prediction of liver metastasis in colorectal cancer patients

CT images-based biomarkers	AUC (95% CI)	Sensitivity	Specificity	PPV	NPV	Cut-off
Unenhanced images	0.754 (0.742, 0.766)	0.663	0.813	0.663	0.188	0.056
Arterial phase images	0.799 (0.788, 0.810)	0.838	0.638	0.838	0.363	−0.254
Portal venous phase images	0.814 (0.803, 0.825)	0.763	0.775	0.763	0.225	−0.041
Combined	0.912 (0.905, 0.919)	0.750	0.925	0.750	0.075	0.254

Note: CI: confidence interval, AUC: area under the curve, PPV: positive predicted value, NPV: negative predicted value.

specificity = 0.628). The proposed radiomics approach of these unpublished studies implies that radiomics-based biomarkers may serve as non-invasive and quantitative surrogates, through histological grading or by revealing prognostically relevant histopathological features pre-operatively, to assist in the individualized treatment of CRC patients.

17.2.1.3 LIVER

Gatos et al. (2017) tested the performance of a probabilistic neural network (PNN) based computer-aided diagnostic (CAD) system in the differentiation of 71 focal liver lesions (including benign lesions, hepatocellular carcinomas, and metastases) on T2-weighted MRI and found that the overall accuracy was 90.1%. Virmani et al. tested the capability of CAD classifiers in the discrimination of focal liver lesion (FLL) on ultrasound (US) images in a series of studies. Using an support vector machine (SVM)-based method, the sensitivity of detecting HCCs from normal and cirrhotic liver on US was 86.6% (Virmani et al. 2013a). In another study, the overall accuracies of identifying the normal liver, typical and atypical cases of cysts, hemangiomas, metastases, and HCCs using back propagation neural network (BPNN)-, PNN-, and K-nearest neighbour (KNN)-based CAD systems were 87.7%, 86.1%, and 85%, respectively (Virmani et al. 2013b). In comparison, the accuracy of an neural network ensemble (NNE)-based CAD system in the identification of normal liver, typical and atypical cases of cysts, hemangiomas, metastases, and HCCs was 95% (Virmani et al. 2014). In the study of Dankerl et al. (2013), a retrieval-based CAD diagnosis system on CT images was evaluated. It turned out that the AUC for differentiating benign lesions from malignancies was 91.4, and the AUC for characterizing 2325 FLLs was 95.5. Raman et al. found that the performance accuracy of a random-forest model of CT texture analysis was 91.2%, 94.4%, and 98.6% for hepatic adenoma, focal nodular hyperplasia, and HCC, respectively (Raman et al. 2015). In a study by Zhou et al. (2017a), the performance of texture features extracted from arterial phase MRI, including the mean intensity and gray-level run-length non-uniformity (GLN), were assessed in differentiating the malignancy of 46 HCC lesions, which were dichotomized into low-grade and high-grade subgroups according to their biological aggressiveness. The best performance, with an AUC of 0.918, was observed in the mean intensity, while the AUCs for GLN in four directions were 0.846 for 0°, 0.836 for 45°, 0.827 for 90°, and 0.838 for 135°.

In addition, accurate identification of microvascular invasion (MVI) with a non-invasive technique before surgery may be helpful for selecting the appropriate therapeutic methods in patient management in HCC. Banerjee et al. (2015) tested the role of the imaging biomarker radiogenomic venous invasion (RVI), which was based on pre-operative contrast-enhanced computed tomography in predicting the presence of MVI and overall survival (OS) of patients. The diagnostic performance of RVI had an accuracy of 89%, sensitivity of 76%, and specificity of 94%. In addition, patients with negative RVI scores demonstrated better OS than those with positive scores (147 versus 34 months, $p < 0.001$). Another study performed by Bakr et al. (2017) investigated the performance of the image features obtained from triphasic enhanced CT as well as the delta features (absolute difference and the ratio of features between different phases) in the identification of MVI in 28 patients with HCCs, which also showed the potential of features obtained from arterial and venous phases in predicting MVI in HCC.

17.2.1.4 IPMN

Hanania et al. (2016) extracted 360 features of intensity, texture, and shape from 53 patients with suspected IPMNs on arterial-phase CT images, and 14 features of gray-level co-occurrence matrix (GLCM) were identified by logistic regression analysis, which could differentiate low-grade lesions from high-grade lesions, with an AUC of 0.82 in the most predictive individual marker and an AUC of 0.96 in a cross-validated panel composed of 10 markers. Similarly, in the study by Permuth et al. (2016), the radiomics features were combined with miRNA genomic classifier (MGC) data in discriminating the malignancy of 38 IPMNs. There were 14 radiomics features selected, with an AUC of 0.77 in differentiation, while the AUC was even higher (0.92) when the features were combined with MGC, which showed the potential of radiogenomics in non-invasive histological grading.

17.2.2 TUMOR STAGING

Tumor staging forms the basis for understanding the extent of disease at its initial presentation and provides by far the most important benchmark for predicting prognosis and determining therapeutic decisions. The TNM classification system (Edge et al. 2010) established by the American Joint Committee on Cancer (AJCC) and the Union for International Cancer Control (UICC) is widely used as the cancer staging system in clinical settings and has been extensively demonstrated to be highly valuable for estimation of the clinical outcomes (e.g., disease-free survival, DFS; OS; progression-free survival, PFS) of a variety of cancers. The TNM staging system is established based on the tumor (T), lymph node (N), and metastasis (M) status, in which T refers to the depth of tumor invasion and invasion of or adherence to adjacent organs or structures, N refers to the number of local lymph node metastases, and M refers to the absence or presence of distant metastases. The TNM classification system is compatible with the Dukes cancer staging system, adding greater precision to the identification and risk stratification of prognostic subgroups, and it applies to both clinical and pathologic staging. The currently used TNM staging system for gastrointestinal cancers could provide more prognostic details than other clinically used cancer staging systems, however, most gastrointestinal cancers are post-operatively staged after pathologic examination of a resected specimen. Pre-operative knowledge of TNM status could provide valuable information for determining the need for adjuvant therapy and the adequacy of surgical resection, thus aiding in pre-treatment clinical decision-making. Although the TNM classification is proven to be very powerful, it only provides inadequate prognostic and predictive information, and clinically, the outcomes are observed to vary substantially among cancer patients, even within the same clinical TNM stage.

In clinical settings, CT, MRI, and PET are widely used for pre-operative workup and are important parts of the staging process of GI cancers. However, a variety of studies have shown the inadequacy of accurate cancer staging with pre-operative CT and MRI. The advent of radiomics has the potential to enable the pre-operative prediction of cancer staging by the conversion of medical images (pre-treatment CT or MRI) into high-dimensional, mineable data through the high-throughput extraction of quantitative imaging features and radiomics data analysis.

17.2.2.1 ESOPHAGUS CANCER

Most studies have pre-operatively evaluated cancers, including for cancer staging, treatment evaluation, and outcome prediction with PET-CT. In terms of esophagus cancer staging, Dong et al. have investigated the relationship of texture features extracted from ^{18}F-fluorodeoxyglucose (^{18}F-FDG) PET images with a maximum standardized uptake value (SUV_{max}) and tumor TNM staging. Their initial findings have shown that there were significant correlations between T stage and the texture features of entropy ($p < 0.001$) and energy ($p = 0.002$) and between texture features of entropy and energy and the N stage ($p = 0.001–0.04$). They reported an area under the receiver operating characteristic (ROC) curve of 0.789 when they used a cut-off value of 4.699 for entropy in detecting tumors above stage II(b). Their results suggested a complementary role of ^{18}F-FDG PET image-based texture features for staging and prognosis evaluation in esophageal squamous cell carcinoma (Dong et al. 2013). Ma et al. also demonstrated the clinical potential of ^{18}F-FDG PET image-based texture analysis in esophageal cancer staging in which they found significant correlation between the SUV_{max}, SUV_{mean}, length (LEN), and eccentricity and TNM staging ($p < 0.01$). They also demonstrated that texture features extracted from ^{18}F-FDG PET images allowed for better stratification of TNM staging than those extracted from fluorine-18 fluorothymidine (^{18}F-FLT) PET in esophageal cancers (Ma et al. 2015). Their results showed the advantage of ^{18}F-FDG over ^{18}F-FLT PET.

17.2.2.2 GASTRIC CANCER

For the staging of gastric cancer, a recent study investigated the role of texture analysis based on whole volume apparent diffusion coefficient (ADC) maps in the pre-operative assessment of gastric cancer's aggressiveness (Liu et al. 2017a). This retrospective study enrolled 64 patients with gastric cancers, with seven whole-lesion entropy-related parameters extracted from ADC maps. The results showed that all these parameters, especially the first-order entropy, were significantly correlated and associated with T, N, and overall stages. Although bias may have been introduced in this study due to the relatively small sample size, with only a few patients at the M1 ($n = 5$) and IV stages ($n = 2$), this study provides initial evidence that ADC-based texture analysis may offer prognostic information for gastric cancer.

17.2.2.3 COLORECTAL CANCER

For the pre-operative overall clinical staging of CRC, a study by Liang et al. (2016) proposed a 16-feature-based radiomics signature. The radiomics signature could successfully categorize CRC into stages I–II and stages III–IV ($p < 0.0001$), which achieved an AUC of 0.708 (95% confidence interval: 0.698–0.718) in the validation dataset, with a sensitivity of 0.611, and specificity of 0.680. Focusing more specifically on the N stage (lymph node [LN] metastasis) of CRC, Huang et al. (2016) developed and validated a radiomics nomogram (see Figure 17.1, cited from the article). Although LN metastasis is difficult to evaluate through traditional interpretation of pre-operative CT, the proposed CT-based radiomics nomogram demonstrated good discrimination and good calibration through validation (see Figure 17.2, cited from the article), which improved the predictive efficacy by 23% compared with a traditional CT evaluation. The proposed approach built a radiomics signature with 24 selected features extracted from pre-operative CT images, followed by building the prediction model to incorporate the radiomics signature and clinical pathological predictor. Clinical utility in decision-making and application prospects in the management of CRC patients were also demonstrated using the decision curve analysis (see Figure 17.3, cited from the article). It is worth mentioning that during the radiomics extraction process, 2D analysis of the largest axial slice was considered in the above two studies. Given that there was a study reporting that 3D analysis appeared to be more representative of tumor heterogeneity than 2D analysis (Ng et al. 2013a), there is doubt as to whether the features extracted from the largest single cross-sectional slice may adequately describe the characteristics of CRC, but it is certainly worthy of further investigation.

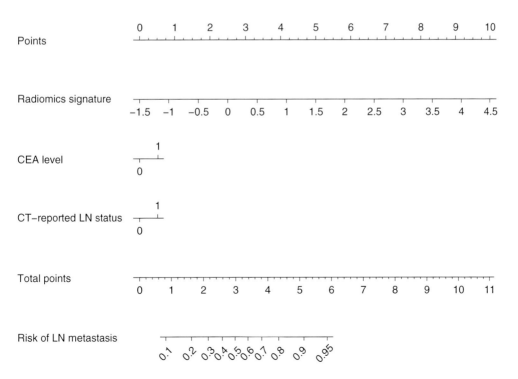

Figure 17.1 Developed radiomics nomogram. The radiomics nomogram was developed in the primary cohort, with the radiomics signature, carcinoembryonic antigen (CEA) level, and CT–reported LN status incorporated.

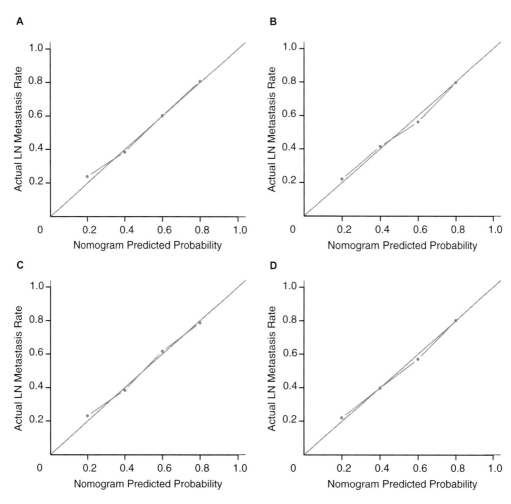

Figure 17.2 (See color insert.) Calibration curves of the radiomics nomogram and the model with the addition of histologic grade prediction in each cohort. (A) Calibration curve of the radiomics nomogram in the primary cohort. (B) Calibration curve of the radiomics nomogram in the validation cohort. (C) Calibration curve of the model with addition of histologic grade in the primary cohort. (D) Calibration curve of the model with addition of histologic grade in the validation cohort. Calibration curves depict the calibration of each model in terms of the agreement between the predicted risks of LN metastasis and observed outcomes of LN metastasis. The y-axis represents the actual LN metastasis rate. The x-axis represents the predicted LN metastasis risk. The diagonal dotted line represents a perfect prediction by an ideal model. The pink solid line represents the performance of the nomogram, of which a closer fit to the diagonal dotted line represents a better prediction. The calibration curve was drawn by plotting $\hat{P}$ on the x-axis and $P_C = [1 + exp -(y_0 + y_1 L)]^{-1}$ on the y-axis, where P_C is the actual probability, $L = logit\ (\hat{P}), \hat{P}$ is the predicted probability, y_0 is the corrected intercept, and y_1 is the slope estimates.

17.2.3 PREDICTING PROGNOSIS

In medicine, therapeutic decisions for tumors are made on the basis of an estimated probability that a specific outcome or event (for example, mortality, disease recurrence, or complication) will occur in the future in an individual, which is the context of the tumor prognosis. The probability estimates are commonly based on combining information from multiple observed or measured predictors, with multivariable prediction models being developed, validated, updated, and implemented to assist doctors and individuals in estimating probabilities and to potentially influence their decision-making. Predictors may range from demographic characteristics (age and sex), medical history-taking, and physical examination results to imaging findings, laboratory tests, pathologic examinations, and disease stages or characteristics, or results from genomics, proteomics, transcriptomics, pharmacogenomics,

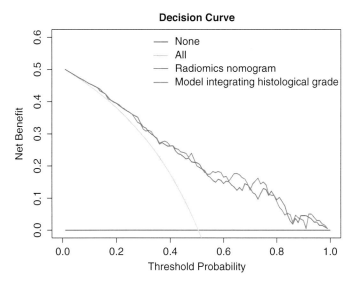

Figure 17.3 (See color insert.) Decision curve analysis for the radiomics nomogram and the model with addition of histological grade. The Y-axis measures the net benefit. The pink line: radiomics nomogram. The blue line: the model with addition of histological grade. The green line: assume all patients have LN metastases. Thin black line: assume no patients have LN metastases. The net benefit was calculated by subtracting the proportion of all patients who are false-positive from the proportion who are true-positive, weighting by the relative harm of forgoing treatment compared with the negative consequences of an unnecessary treatment. In here, the relative harm was calculated by $\left(\frac{p_t}{1-p_t}\right)$. "$p_t$" (threshold probability) is where the expected benefit of treatment is equal to the expected benefit of avoiding treatment, at which a patient will opt for treatment is informative of how a patient weighs the relative harms of false-positive and false-negative results $((a-c)/(b-d) = (1-p_t)/p_t)$. a–c is the harm from a false-negative result; b–d is the harm from a false-positive result. a, b, c, and d give, respectively, the value of true positive, false positive, false negative, and true negative. The decision curve showed that if patient's or doctor's threshold probability is higher than 10%, then using the radiomics nomogram in the current study to predict LN metastases adds more benefit than treat all patients, or treat none scheme. For example, if a patient's personal threshold probability is 60% (i.e., the patient would opt for treatment if his probability of cancer was greater than 60%), then the net benefit is 0.145 when using the radiomics nomogram to make the decision to undergo treatment, with added benefit than treat all scheme, or treat none scheme. Besides, the net benefit was comparable (with several overlaps) basing on the radiomics nomogram and the model integrating histological grade.

metabolomics, and other new biological measurement platforms that continuously emerge (Collins et al. 2015). Radiomics is such a new field that promises to increase precision in the assessment of tumor prognoses. The mining of quantitative image features based on intensity, shape, size or volume, and texture offers information on tumor phenotype and microenvironment, which, in conjunction with the other information, have been correlated with clinical outcomes and investigated for tumor prognosis over the past few years (Gillies et al. 2016).

17.2.3.1 ESOPHAGUS CANCER

Foley et al. (2017) developed and validated a prognostic model using clinical variables and texture features extracted from patients with ECs who underwent PET before treatment. TNM stage, age, treatment, log(TLG), log(histogram energy), and histogram kurtosis were determined by Cox regression analysis (p all < 0.05) and integrated into a prognostic score that stratified patients into quartiles. A Kaplan-Meier analysis showed significant differences in the OS of different quartiles of patients ($p < 0.001$). Ganeshan et al. (2012) performed a pilot study using CT textures to predict the prognosis of EC and determined coarse uniformity as the most significant predictor of OS [odds ratio, 4.45 (95% confidence interval 1.08, 18.37); $p = 0.039$]. Using a cut-off value of 0.8477 determined by ROC analysis, there was a significant difference between the two Kaplan–Meier (KM) curves ($p = 0.0006$). Another study performed by Yip et al. (2014) investigated the role of whole-tumor textures extracted from enhanced CT before and after chemoradiotherapy (CRT) in predictions of patient survival. Post-treatment medium entropy, coarse entropy, and medium uniformity were identified as significant predictors of survival (p all < 0.05), and a combination of these features and maximal wall thickness in the survival model outperformed the conventional morphologic feature-based model (AUC 0.664–0.802 versus 0.659).

17.2.3.2 GASTRIC CANCER

The association between CT-based texture analysis and OS in patients with gastric cancer has been investigated recently by Giganti et al. (2017a). The following parameters were identified to be significantly associated with a negative prognosis and were shown to be independently predictive of OS: energy [no filter]; entropy [no filter]; entropy [filter 1.5]; maximum Hounsfield unit value [filter 1.5]; skewness [filter 2]; root mean square [filter 1]; and mean absolute deviation [filter 2]. Aiming to identify an imaging biomarker that can provide additional prognostic information for patient selection, another recent study reported that CT texture features were related to the survival rate of human epidermal growth factor receptor 2 (HER2) positive gastric cancer after trastuzumab treatment (Yoon et al. 2016). Using a threshold of >265.8480 for contrast, >488.3150 for variance, and 0.1319×10^{-3} for correlation, all of the AUCs were higher than 0.7 for the identification of good survival group (patients were dichotomized into a good and poor survival group based on cut-off points of OS of 12 months). Although there were limitations contained within the study, including the small sample size (26 patients) and the lack of whole-tumor analysis, the finding that heterogeneous texture features on CT images were associated with better survival in patients with HER2-positive advanced gastric cancer who received trastuzumab-based treatment implies that texture analysis based on CT images may represent a promising, non-invasive prognostic tool for gastric cancer.

17.2.3.3 COLORECTAL CANCER

As the first study applying whole-tumor volumetric textural analysis directly to address primary colorectal cancer heterogeneity, the study by Ng et al. (2013b) aimed to determine whether the CT-based texture features of primary CRC are related to survival. Results showed that tumors demonstrating less heterogeneity at fine filter levels were associated with poorer survival, and tumor entropy less than or equal to 7.89, uniformity greater than 0.01, kurtosis less than or equal to 2.48, skewness greater than 20.38, and standard deviation of the pixel distribution histogram less than or equal to 61.83 were associated with a poorer 5-year OS. These findings suggest that texture features may augment current characterization of CRC and may be promising as a prognostic biomarker. Based on another modality, as the first study to assess the prognostic significance of texture features in addition to morphological MRI and histopathological parameters of rectal cancer undergoing CRT, a more recent study showed that textural features of rectal cancer can predict long-term survival in patients treated with long-course CRT (Jalil et al. 2017). The results of this study showed that several texture features and morphological MRI features independently predicted OS and DFS, while only texture features independently predicted relapse free survival (RFS) on pre- or post-treatment analyses. This study shows that MRI-based textural analysis of rectal cancer can act as a complementary prognostic predictor of survival in patients with locally advanced rectal cancer, which could contribute to disease risk stratification and allow for individualized therapy. The heterogeneity in primary tumor 18F-FDG and 18F-FLT distribution has also been investigated to predict the prognosis of patients with colorectal cancer who underwent surgery (Nakajo et al. 2017a). This study showed that high ^{18}F-FDG-IV (intensity variability) and high ^{18}F-FDG-SZV (size-zone variability) were significant predictors for 5-year PFS. Although prospective studies with a large population are needed to confirm the validity of the present findings, this study implies that the heterogeneity parameters extracted from PET/CT have a potential to assist in tumor prognosis for CRC.

17.2.3.4 LIVER CANCER

HCC: A study performed by Rosenkrantz et al. (2016) collected 20 HCC patients who underwent gadoxetic acid-enhanced MRI and found lesions demonstrating hypointensity on the hepatobiliary phase without enhancement on the arterial phase. Using the whole-lesion histogram analysis, they determined four features (standard deviation, coefficient variation, skewness, and entropy) with the highest AUCs (0.655–0.750 when images were acquired at a flip angle of 12 and 0.686–0.800 at a flip angle of 25) in predicting the progression of these lesions into HCCs. Chen et al. (2017) examined whether texture features extracted from the arterial phase and portal venous phase CT can predict the OS and DFS after hepatectomy in patients with single HCCs. In this cohort of 61 patients, they revealed that the variation of Gabor and wavelet features between the arterial phase and portal venous phase could be used as the best predictor of OS and DFS in patients with single HCCs after hepatectomy. Zhou et al. (2017b) collected 215 patients with HCCs and built a radiomics signature with an AUC of 0.817 based on features extracted from pre-operative arterial/portal venous phase CT images to predict early recurrence after resection.

Colorectal liver metastasis: Simpson et al. (2017) evaluated whether pre-operative CT texture analysis could predict OS and DFS in 198 patients who underwent colorectal metastasis (CRLM) resection. It turned out that the tumor signal (a predictor integrated by two texture features extracted from the tumor, namely, correlation and contrast) and the future liver remnant (FLR) signal (a predictor integrated by two texture features extracted from the FLR, namely, energy and entropy) were independent predictors of OS (hazard ratio 2.35 and 2.15, $p = 0.013$ and 0.029, respectively), while the FLR signal was an independent predictor of DFS (hazard ratio 2.21, $p = 0.01$). In three other studies (Ahn et al. 2016, Lubner et al. 2015, Rao et al. 2016), the capability of texture features in the prediction of clinical outcomes or treatment responses of chemotherapy in patients with CRLM was assessed. In Lubner et al. (2015) study, entropy at coarse filtration levels could negatively predict the OS of CRLM (hazard ratio for death 0.65, $p = 0.03$). Ahn et al. (2016) analyzed the histogram features, volumetric features, and morphologic features in 235 patients with CRLM who received 5-fluorouracil and leucovorin and oxaliplatin (FOLFOX) and 5-fluorouracil/levofolinate/irinotecan (FOLFIRI) chemotherapy and found that skewness extracted in 2D analysis and standard deviation in 3D were independent predictors of therapeutic response (odds ratios 6.739 and 3.163, $p = 0.025$ and 0.002, respectively). Similarly, Rao et al. (2016) collected 21 patients with CRLMs and extracted the mean intensity, entropy, and uniformity under different filters in an attempt to determine the best predictors of chemotherapy response. In their study, the relative changes in entropy and uniformity without filtration between pre- and post-treatment turned out to be the significant predictors (odds ratios 1.34 and 0.95, respectively, p values not provided). In the study of Beckers et al. (2017), 165 patients with colorectal cancer were retrospectively collected, and whole-liver CT texture analysis determined that uniformity with a filter of $\sigma = 0.5$ as a significant predictor of early liver metastasis, but no predictors of intermediate/late metastases were found.

17.2.3.5 PANCREATIC CANCER

Cassinotto et al. (2017) used tumor attenuation parameters and heterogeneity parameters extracted from the pre-operative CT (portal-venous phase) to predict tumor aggressiveness and DFS in 99 patients with PA. The study showed that tumor attenuation parameters were associated with the tumor differentiation grade and lymph node invasion, while the most hypoattenuating area within the tumor was a significant predictor of DFS (hazard ratio 0.98, p value not provided). Another study about pre-operative prediction of the OS of PA was performed by Eilaghi et al. (2017), which enrolled 30 patients and used CT-derived texture features such as uniformity, entropy, dissimilarity, correlation, and normalized inverse difference. Cox regression and Kaplan-Meier tests showed that tumor dissimilarity (B value-0.1292, $p = 0.045$) and normalized inverse difference (B value 71.81, $p = 0.046$) were significant predictors of OS. For patients with unresectable PA, Yue et al. (2017) identified several texture features for the prediction of OS after radiation therapy. The variation (marked as Δ) of 37 extracted GLCM features between pre- and post-radiation therapy PET images were calculated and underwent dimension reduction. Finally, Δhomogeneity, Δvariance, and Δcluster tendency were identified as predictors of OS (hazard ratios 0.033, 0.731, 1.087, $P = 0.065$, 0.078, and 0.081, respectively).

17.2.4 TREATMENT RESPONSE EVALUATION

An effective treatment will lead to a reduced probability of the event and thus improve tumor prognosis. The direct translation of the average treatment effect from therapeutic trials to individuals is often difficult because not all individuals respond to treatments similar to the average. In clinical practice, some individuals benefit more than average from treatment, whereas others do not or may even be harmed. In fact, it is widely recognized that an average benefit can mask both positive and negative effects in different patient subgroups, which leads to treatment of patients who do not benefit and may suffer harm. Prediction of the treatment effect for individuals may enable precision medicine in an evidence-based manner, which could help the doctors to make better-informed treatment decisions and perhaps motivate patients to adhere to treatment regimens. Advocates of the evidence-based approach have stressed the importance of taking into account a person's absolute risk of an outcome and the need to weigh potential benefits and harms when applying the evidence from therapeutic trials for individuals (Dorresteijn et al. 2011, Moynihan et al. 2014).

17.2.4.1 ESOPHAGUS CANCER

van Rossum et al. (2016) investigated the potential of subjective assessment, metabolic features, and texture/geometry features extracted from ^{18}F-FDG PET in predicting the therapeutic response of pre-operative CRT in 217 patients with esophageal adenocarcinomas. When combined with an original clinical prediction model of pathoCR (corrected C-index 0.67), their study demonstrated the highest corrected C-index in texture/geometry features (0.77), followed by conventional metabolic features (0.73), and subjective assessment (0.72). However, no clear incremental value of the above features was found in the decision curve analysis at a threshold ≥ 0.9. In the study of Tan et al. (2013), the features extracted from pre-CRT ^{18}F-FDG PET and post-CRT ^{18}F-FDG PET and the changes between them were used to predict the pathologic response to neoadjuvant CRT in 20 patients. Their study showed that spatial-temporal intensity features and texture features, including changes in SUV_{mean} decline, pre-CRT skewness, post-CRT inertia, post-CRT correlation, and post-CRT cluster prominence, could significantly predict the therapeutic response (AUCs all ≥ 0.76).

In a retrospective study of 52 patients with ECs who underwent ^{18}F-FDG PET before CRT, Nakajo et al. (2017b) found that metabolic tumor volume and total lesion glycolysis, as well as texture features, including intensity variability and size-zone variability, were significant predictors of tumor response rather than predictors of patient prognosis. Desbordes et al. (2017) used a random forest classifier to select the best predictors of tumor response to CRT and prognosis in patients with ECs who underwent ^{18}F-FDG PET before CRT. A subset of features composed of metabolic tumor volume and homogeneity was reported to be predictive of tumor response, with an AUC of 0.836 ± 0.105, while no significant prognostic features were detected in the Kaplan-Meier analysis. Paul et al. (2017) compared the performance of a genetic algorithm based on the random forest (GARF) classifier and other classifiers, including the sequential forward selection (SFS) method, the hierarchical forward selection (HFS) method, the recursive feature elimination (RFE) method, and the least absolute shrinkage and selection operator (LASSO) method, for classification of therapeutic responses and clinical outcomes using the features obtained from PET before and after treatment in 65 patients with ECs. Their study showed that the GARF classifier outperformed other feature selection methods in the therapeutic response and clinical outcome classification (AUC 823 ± 0.032 and 0.750 ± 0.108, respectively).

Based on the radiomics features extracted from 107 patients with ECs who underwent ^{18}F-FDG PET before neoadjuvant chemotherapy, Yip et al. (2016a) compared the performance of a 3S-convolutional neural network (3S-CNN) and other statistical classifiers (logistic regression, gradient boosting, random forest (RF) and support vector machines) and found that the features extracted by the 3S-CNN were the best predictors of therapeutic response among all the classifiers (with an average sensitivity of 80.7% and an average specificity of 81.6%).

Yip et al. (2016b) investigated the temporal changes of PET-based features during pre-operative CRT and revealed that entropy and run-length matrix (RLM) textures could significantly differentiate non-responders from partial and complete responders, with an AUC ranging from 0.71 to 0.79 (p all < 0.05). Significant differences were also noted in the Kaplan-Meier analysis when patients were dichotomized by the medians of entropy and RLM textures (log-rank p < 0.02), which indicated the correlation between temporal changes in entropy and RLM textures and the OS of the patients with ECs.

17.2.4.2 GASTRIC CANCER

The role of CT-based texture analysis in predicting the response rate of patients with gastric cancers to neoadjuvant therapy has been investigated in a recent study by Giganti et al. (2017b). Entropy, range, and root mean square were found to be able to independently identify non-responders. The successful differentiation of responders from non-responders suggests that texture analysis holds promise for predicting treatment response in gastric cancer, though further multi-center studies with larger sample sizes are necessary before its widespread application in common clinical practice is warranted.

17.2.4.3 COLORECTAL CANCER

Effort has been made to apply texture or radiomics analysis to evaluate the therapeutic responses to chemoradiotherapy using various pre- or post-treatment imaging data, including T2-weighted MRI (T2WI) (Liu et al. 2017b, De Cecco et al. 2015), dynamic contrast-enhanced MRI (Nie et al. 2016), and diffusion weighted image (DWI) (Liu et al. 2017b, Liu et al. 2017c). An early study by De Cecco et al. enrolled 15 consecutive patients with rectal cancer who underwent both pre-treatment and mid-treatment 3-T MRI to determine whether texture features on T2WI can predict tumoral response in patients treated with CRT (De Cecco et al. 2015). The results showed that there were significant differences in pre-treatment medium texture-scale kurtosis (SSF4) and mid-treatment kurtosis without filtration (SSF0) between the pathological complete response (pCR) subgroup and the partial response (PR) + non-response (NR) subgroup. For the change between pre-treatment and mid-treatment parameters, the changes in SSF0 and SSF5 were significantly greater in the PR + NR subgroup than in the pCR subgroup. ROC analysis to discriminate between pCR and PR + NR showed that pre-treatment kurtosis demonstrated a higher AUC compared with post-CRT kurtosis and the absolute change in kurtosis from pre-treatment. Although the number of patients in this study was limited (n: pCR = 6, PR = 4, NR = 5), these results demonstrate the possible use of MRI-based texture analysis (TA) to predict the response of rectal cancer to CRT, in particular, stratifying patients with pCR from those with PR or NR.

In the first study to integrate anatomical, perfusion, and diffusion MRI using both volume-averaged and voxel-based heterogeneity analysis to predict pCR, Nie et al. showed that a voxelized heterogeneity analysis could provide additional information compared with conventional volume-averaged analyses in assessing treatment outcomes (Nie et al. 2016). For pCR prediction, the results showed that using the voxelized heterogeneity analysis will yield improved AUC compared with using a volume average-based analysis. After combining all categories of image features together into artificial neural network (ANN) training, the AUC could be further improved to 0.84. For a good response prediction based on the tumor regression grading (TRG) evaluation, similar findings have shown that the voxelized heterogeneity analysis had better prediction ability. When combining all categories of image information together, the final AUC could be improved to 0.89. These results suggest the potential limitation of only using summary mean values to evaluate treatment responses and highlight the potential of voxelized heterogeneity analyses. In addition, the finding that combining different imaging modalities provided improved prognostic performance emphasized the value of image registration.

A more recent study by Liu et al. (2017c) demonstrated the TA-based treatment response prediction on pre-therapy ADC mapping. Eighteen of 133 radiomics parameters extracted from an ADC map showed a significant difference between the responsive group and non-responsive group, with two independent predictors reported (namely, energy variance and SdGa47). The AUC of the prediction model was reported to be 0.908. These preliminary findings indicated that texture features derived from pre-therapy ADC mapping could potentially be helpful to assist in treatment response prediction. A further study was designed to incorporate both the pre-treatment and post-treatment MRI data for the prediction, with 2252 radiomics features extracted from pre- and post-treatment imaging (Liu et al. 2017b). A radiomics model was proposed for the prediction of pCR that incorporated the radiomics signature and tumor length and showed good discrimination (AUC: 0.9756) and good calibration in the validation cohort. A decision curve analysis was presented to demonstrate the clinical utility of the model. The proposed radiomics nomogram showed a potential to serve as an effective tool for evaluating chemoradiotherapeutic outcomes in patients with locally advanced rectal cancer (LARC), though a larger dataset from multiple centers with a considerably larger sample of patients with pCR ought to be investigated to validate the robustness and reproducibility of this analysis.

17.2.5 Others

Simpson et al. (2015) performed a retrospective study and compared the GLCM features extracted from pre-operative CT between two groups of patients, namely, patients who developed liver insufficiency after a major hepatic resection and patients without post-operative liver insufficiency. They found that patients who developed liver insufficiency after major hepatic resection tended to have more varied, less symmetric/homogeneous features on CT, which may assist in pre-operative stratification.

The relationships between radiomics and other subjects, e.g., histopathology and genetics, were also investigated recently. Liu et al. explored the ability of CT-based texture analysis in predicting histopathological features in patients with GC (Liu et al. 2017d). The results show that CT texture analysis held a great advantage in predicting histopathological features of GC. Lovinfosse et al. evaluated SUVs, texture analysis, and volume-based parameters to investigate the relationship with the mutational status of rectal cancer (Lovinfosse et al. 2016). All the features of tumors show considerably higher glucose metabolism than wild-type tumors. However, for selecting an optimal cut-off value, the accuracy of the results of those available parameters has little clinical relevance. At present, there are few studies using radiomics approach combined with other disciplines on GI cancers, with further investigation warranted to provide new perspective into cancer care.

17.3 CONCLUSIONS AND FUTURE DIRECTIONS

Recent breakthroughs in radiomics with applications in gastrointestinal tumors have been encouraging in the fields of tumor diagnosis, staging, prognosis, and longitudinal treatment response monitoring. Radiomics, along with several advanced disciplines, including genetics, molecular technologies, and quantitative patho-logical analysis, has led to a more comprehensive study of tumor heterogeneity. The deepened understanding of tumor heterogeneity and its environment directly relates to the development of new targeted therapies. In addition, radiomics enables the visualization of the change of tumor biology and metabolism during therapy.

However, before radiomics can be implemented within standard-of-care imaging protocols in clinical practice, there are several issues that should be addressed.

First, in spite of a promising methodological approach, there are still many challenges for radiomics studies to overcome: (1) most of the published studies are retrospective studies with small sample sizes; (2) the image acquisition protocols are not standardized between different studies, and sometimes even within the same study, images from different protocols are pooled together for analysis; (3) the variation in methods of image segmentation may, in a way, make the results from different studies incomparable; (4) at present, ensuring the robustness of radiomics is challenging, given that the majority of radiomics studies in gastrointestinal cancers lack external validation, which is crucial for future clinical implication. While encouraging results have been achieved, further progress is still required to make the radiomics approach acceptable for clinical practice.

Second, in gastrointestinal tumors, there are few studies investigating the value of radiomics in combina-tion with other disciplines, and the association between radiomics features and pathophysiological changes, genomics, and even proteomics has not been fully investigated and validated. Therefore, the application of multidisciplinary combinations has great potential, and a series of well-designed prospective studies with full attention on these issues are required before its extensive application in clinical practice.

Third, the heart of radiomics is the extraction of high-dimensional quantitative features to describe tumor heterogeneity (Gillies et al. 2016). Hand-crafted features are mathematically defined and extracted quantitative descriptors. More and more hand-crafted features can be defined currently. However, more does not necessarily mean better. Too many unrelated features were subjected to the "curse of dimensional-ity" in following the model building process. Moreover, hand-crafted radiomics features were dependent on domain knowledge and experience. They only reflect a part of the information in medical images. Recently, the increase in deep learning as a new frontier in machine learning has advanced medical image analysis. Deep learning technologies such as convolutional neural networks (CNNs) have been applied to the field of computer vision, where they have produced results comparable to and in some cases superior to human experts. In contrast with the hand-crafted features, CNNs can learn features in raw images automatically without the hand-engineered process. This independence from prior knowledge and human effort in feature design is a major advantage. CNN models can be considered as feature extractors. Deep learning features can be extracted from the last fully connected layer. Compared with natural image sets, medical image sets have only a limited size to train the CNN model. Transfer learning and fine-tuning methods are needed to improve the model's performance under small dataset conditions. We can follow the routine approach where all CNN layers apart from the last are fine-tuned with a learning rate smaller than the default. The last fully connected layer was randomly initialized to be trained to suit for new object categories. One advantage of deep learning features is that the CNN feature extraction method does not need accurate tumor segmentation, which saves a lot of time compared with hand-crafted features that need a clearly delineated tumor boundary. There is an appeal in attempting to fully integrate these two distinct types of feature generation strategies that can potentially outperform either type of feature extraction strategy, which is certainly worthy of further investigation.

REFERENCES

Ahn SJ, Kim JH, Park SJ, Han JK. 2016. Prediction of the therapeutic response after FOLFOX and FOLFIRI treatment for patients with liver metastasis from colorectal cancer using computerized CT texture analysis. *Eur J Radiol* 85: 1867–1874.

Al Bandar MH, Kim NK. 2017. Current status and future perspectives on treatment of liver metastasis in colorectal cancer (Review). *Oncol Rep* 37: 2553–2564.

Bakr S, Echegaray S, Shah R et al. 2017. Noninvasive radiomics signature based on quantitative analysis of computed tomography images as a surrogate for microvascular invasion in hepatocellular carcinoma: A pilot study. *J Med Imaging* (Bellingham) 4: 041303.

Banerjee S, Wang DS, Kim HJ et al. 2015. A computed tomography radiogenomic biomarker predicts microvascular invasion and clinical outcomes in hepatocellular carcinoma. *Hepatology* 62: 792–800.

Ba-Ssalamah A, Muin D, Schernthaner R et al. 2013. Texture-based classification of different gastric tumors at contrast-enhanced CT. *Eur J Radiol* 82: e537–e543.

Beckers RCJ, Lambregts DMJ, Schnerr RS et al. 2017. Whole liver CT texture analysis to predict the development of colorectal liver metastases-A multicentre study. *Eur J Radiol* 92: 6471.

Cassinotto C, Chong J, Zogopoulos G et al. 2017. Resectable pancreatic adenocarcinoma: Role of CT quantitative imaging biomarkers for predicting pathology and patient outcomes. *Eur J Radiol* 90: 152–158.

Chen S, Zhu Y, Liu Z, Liang C. 2017. Texture analysis of baseline multiphasic hepatic computed tomography images for the prognosis of single hepatocellular carcinoma after hepatectomy: A retrospective pilot study. *Eur J Radiol* 90: 198–204.

Chiaravalli M, Reni M, O'Reilly EM. 2017. Pancreatic ductal adenocarcinoma: State-of-the-art 2017 and new therapeutic strategies. *Cancer Treat Rev* 60: 32–43.

Collins GS, Reitsma JB, Altman DG, Moons KG. 2015. Transparent reporting of a multivariable prediction model for individual prognosis or diagnosis (TRIPOD): The TRIPOD statement. *BMJ* 350: g7594.

Dankerl P, Cavallaro A, Tsymbal A et al. 2013. A retrieval-based computer-aided diagnosis system for the characterization of liver lesions in CT scans. *Acad Radiol* 20: 1526–1534.

De Cecco CN, Ganeshan B, Ciolina M et al. 2015. Texture analysis as imaging biomarker of tumoral response to neoadjuvant chemoradiotherapy in rectal cancer patients studied with 3-T magnetic resonance. *Invest Radiol* 50: 239–245.

Desbordes P, Ruan S, Modzelewski R et al. 2017. Predictive value of initial FDG-PET features for treatment response and survival in esophageal cancer patients treated with chemo-radiation therapy using a random forest classifier. *PLoS One* 12: e0173208.

Dong X, Xing L, Wu P et al. 2013. Three-dimensional positron emission tomography image texture analysis of esophageal squamous cell carcinoma: Relationship between tumor 18F-fluorodeoxyglucose uptake heterogeneity, maximum standardized uptake value, and tumor stage. *Nucl Med Commun* 34: 40–46.

Dorresteijn JA, Visseren FL, Ridker PM et al. 2011. Estimating treatment effects for individual patients based on the results of randomised clinical trials. *BMJ* 343: d5888.

Edge SB, Byrd DR, Compton CC, Fritz AG, Greene FL, Trotti A, (Eds.), *AJCC Cancer Staging Manual*. 7th ed. New York: Springer 2010.

Eilaghi A, Baig S, Zhang Y et al. 2017. CT texture features are associated with overall survival in pancreatic ductal adenocarcinoma—A quantitative analysis. *BMC Med Imaging* 17: 38.

Foley KG, Hills RK, Berthon B et al. 2017. Development and validation of a prognostic model incorporating texture analysis derived from standardised segmentation of PET in patients with oesophageal cancer. Eur Radiol. [Epub ahead of print]

Fong ZV, Fernandez-Del Castillo C. 2016. Intraductal papillary mucinous neoplasm of the pancreas. *Surg Clin North Am* 96: 1431–1445.

Forner A, Llovet JM, Bruix J. 2012. Hepatocellular carcinoma. *Lancet* 379: 1245–1255.

Ganeshan B, Skogen K, Pressney I, Coutroubis D, Miles K. 2012. Tumour heterogeneity in oesophageal cancer assessed by CT texture analysis: Preliminary evidence of an association with tumour metabolism, stage, and survival. *Clin Radiol* 67: 157–164.

Gatos I, Tsantis S, Karamesini M et al. 2017. Focal liver lesions segmentation and classification in nonenhanced T2-weighted MRI. Med Phys. 44: 3695–3705. [Epub ahead of print]

Giganti F, Antunes S, Salerno A et al. 2017a. Gastric cancer: Texture analysis from multidetector computed tomography as a potential preoperative prognostic biomarker. *Eur Radiol* 27: 1831–1839.

Giganti F, Marra P, Ambrosi A et al. 2017b. Pre-treatment MDCT-based texture analysis for therapy response prediction in gastric cancer: Comparison with tumour regression grade at final histology. *Eur J Radiol* 90: 129–137.

Gillies RJ, Kinahan PE, Hricak H. 2016. Radiomics: Images are more than pictures, they are data. *Radiology* 278: 563–577.

Hanania AN, Bantis LE, Feng Z et al. 2016. Quantitative imaging to evaluate malignant potential of IPMNs. *Oncotarget* 7: 85776–85784.

Holzel M, Bovier A, Tuting T. 2013. Plasticity of tumour and immune cells: A source of heterogeneity and a cause for therapy resistance? *Nat Rev Cancer* 13: 365–376.

Huang YQ, Liang CH, He L et al. 2016. Development and validation of a radiomics nomogram for preoperative prediction of lymph node metastasis in colorectal cancer. *J Clin Oncol* 34: 2157–2164.

Jalil O, Afaq A, Ganeshan B et al. 2017. Magnetic resonance based texture parameters as potential imaging biomarkers for predicting long-term survival in locally advanced rectal cancer treated by chemoradiotherapy. *Colorectal Dis* 19: 349–362.

Junttila MR, de Sauvage FJ. 2013. Influence of tumour micro-environment heterogeneity on therapeutic response. *Nature* 501: 346–354.

Larue R, van Timmeren JE, de Jong EEC et al. 2017. Influence of gray level discretization on radiomic feature stability for different CT scanners, tube currents and slice thicknesses: A comprehensive phantom study. *Acta Oncol* 1–10.

Liang C, Huang Y, He L et al. 2016. The development and validation of a CT-based radiomics signature for the preoperative discrimination of stage I-II and stage III-IV colorectal cancer. *Oncotarget* 7: 31401–31412.

Liu M, Lv H, Liu LH et al. 2017c. Locally advanced rectal cancer: Predicting non-responders to neoadjuvant chemoradiotherapy using apparent diffusion coefficient textures. *Int J Colorectal Dis* 32: 1009–1012.

Liu S, Liu S, Ji C et al. 2017d. Application of CT texture analysis in predicting histopathological characteristics of gastric cancers. Eur Radiol. [Epub ahead of print]

Liu S, Zheng H, Zhang Y et al. 2017a. Whole-volume apparent diffusion coefficient-based entropy parameters for assessment of gastric cancer aggressiveness. J Magn Reson Imaging. [Epub ahead of print]

Liu Z, Zhang XY, Shi YJ et al. 2017b. Radiomics analysis for evaluation of pathological complete response to neoadjuvant chemoradiotherapy in locally advanced rectal cancer. *Clin Cancer Res* 23: 7253–7262.

Lovinfosse P, Koopmansch B, Lambert F et al. 2016. (18)F-FDG PET/CT imaging in rectal cancer: Relationship with the RAS mutational status. *Br J Radiol* 89: 20160212.

Lu L, Ehmke RC, Schwartz LH, Zhao B. 2016. Assessing agreement between radiomic features computed for multiple CT imaging settings. *PLoS One* 11: e0166550.

Lubner MG, Stabo N, Lubner SJ et al. 2015. CT textural analysis of hepatic metastatic colorectal cancer: Pre-treatment tumor heterogeneity correlates with pathology and clinical outcomes. *Abdom Imaging* 40: 2331–2337.

Ma C, Li D, Yin Y, Cao J. 2015. Comparison of characteristics of 18F-fluorodeoxyglucose and 18F-fluorothymidine PET during staging of esophageal squamous cell carcinoma. *Nucl Med Commun* 36: 1181–1186.

Ma Z, Fang M, Huang Y et al. 2017. CT-based radiomics signature for differentiating borrmann type IV gastric cancer from primary gastric lymphoma. *Eur J Radiol* 91: 142–147.

McGranahan N, Swanton C. 2017. Clonal heterogeneity and tumor evolution: Past, present, and the future. *Cell* 168: 613–628.

Moynihan R, Henry D, Moons KG. 2014. Using evidence to combat overdiagnosis and overtreatment: Evaluating treatments, tests, and disease definitions in the time of too much. *PLoS Med* 11: e1001655.

Nakajo M, Jinguji M, Nakabeppu Y et al. 2017b. Texture analysis of 18F-FDG PET/CT to predict tumour response and prognosis of patients with esophageal cancer treated by chemoradiotherapy. *Eur J Nucl Med Mol Imaging* 44: 206–214.

Nakajo M, Kajiya Y, Tani A et al. 2017a. A pilot study for texture analysis of 18F-FDG and 18F-FLT-PET/CT to predict tumor recurrence of patients with colorectal cancer who received surgery. Eur J Nucl Med Mol Imaging. [Epub ahead of print]

Ng F, Ganeshan B, Kozarski R, Miles KA, Goh V. 2013b. Assessment of primary colorectal cancer heterogeneity by using whole-tumor texture analysis: Contrast-enhanced CT texture as a biomarker of 5-year survival. *Radiology* 266: 177–184.

Ng F, Kozarski R, Ganeshan B, Goh V. 2013a. Assessment of tumor heterogeneity by CT texture analysis: Can the largest cross-sectional area be used as an alternative to whole tumor analysis? *Eur J Radiol* 82: 342–348.

Nie K, Shi L, Chen Q et al. 2016. Rectal cancer: Assessment of neoadjuvant chemoradiation outcome based on radiomics of multiparametric MRI. *Clin Cancer Res* 22: 5256–5264.

Paul D, Su R, Romain M et al. 2017. Feature selection for outcome prediction in oesophageal cancer using genetic algorithm and random forest classifier. *Comput Med Imaging Graph* 60: 42–49.

Permuth JB, Choi J, Balarunathan Y et al. 2016. Combining radiomic features with a miRNA classifier may improve prediction of malignant pathology for pancreatic intraductal papillary mucinous neoplasms. *Oncotarget* 7: 85785–85797.

Raman SP, Schroeder JL, Huang P et al. 2015. Preliminary data using computed tomography texture analysis for the classification of hypervascular liver lesions: Generation of a predictive model on the basis of quantitative spatial frequency measurements—A work in progress. *J Comput Assist Tomogr* 39: 383–395.

Rao SX, Lambregts DM, Schnerr RS et al. 2016. CT texture analysis in colorectal liver metastases: A better way than size and volume measurements to assess response to chemotherapy? *United European Gastroenterol J* 4: 257–263.

Rosenkrantz AB, Pinnamaneni N, Kierans AS, Ream JM. 2016. Hypovascular hepatic nodules at gadoxetic acid-enhanced MRI: Whole-lesion hepatobiliary phase histogram metrics for prediction of progression to arterial-enhancing hepatocellular carcinoma. *Abdom Radiol* 41: 63–70.

Simpson AL, Adams LB, Allen PJ et al. 2015. Texture analysis of preoperative CT images for prediction of postoperative hepatic insufficiency: A preliminary study. *J Am Coll Surg* 220: 339–346.

Simpson AL, Doussot A, Creasy JM et al. 2017. Computed tomography image texture: A noninvasive prognostic marker of hepatic recurrence after hepatectomy for metastatic colorectal cancer. Ann Surg Oncol. 24: 2482–2490. [Epub ahead of print]

Tan S, Kligerman S, Chen W et al. 2013. Spatial-temporal [(1)(8)F]FDG-PET features for predicting pathologic response of esophageal cancer to neoadjuvant chemoradiation therapy. *Int J Radiat Oncol Biol Phys* 85: 1375–1382.

Torre LA, Bray F, Siegel RL et al. 2015. Global cancer statistics, 2012. *CA Cancer J Clin* 65: 87–108.

van Rossum PS, Fried DV, Zhang L et al. 2016. The incremental value of subjective and quantitative assessment of 18F-FDG PET for the prediction of pathologic complete response to preoperative chemoradiotherapy in esophageal cancer. *J Nucl Med* 57: 691–700.

Virmani J, Kumar V, Kalra N, Khandelwal N. 2013a. SVM-based characterization of liver ultrasound images using wavelet packet texture descriptors. *J Digit Imaging* 26: 530–543.

Virmani J, Kumar V, Kalra N, Khandelwal N. 2013b. A comparative study of computer-aided classification systems for focal hepatic lesions from B-mode ultrasound. *J Med Eng Technol* 37: 292–306.

Virmani J, Kumar V, Kalra N, Khandelwal N. 2014. Neural network ensemble based CAD system for focal liver lesions from B-mode ultrasound. *J Digit Imaging* 27: 520–537.

Wagner F, Hakami YA, Warnock G et al. 2017. Comparison of contrast-enhanced CT and [18F]FDG PET/CT analysis using kurtosis and skewness in patients with primary colorectal cancer. Mol Imaging Biol. 19: 795–803. [Epub ahead of print]

Yip C, Landau D, Kozarski R et al. 2014. Primary esophageal cancer: Heterogeneity as potential prognostic biomarker in patients treated with definitive chemotherapy and radiation therapy. *Radiology* 270: 141–148.

Yip SS, Coroller TP, Sanford NN et al. 2016a. Relationship between the temporal changes in positron-emission-tomography-imaging-based textural features and pathologic response and survival in esophageal cancer patients. *Front Oncol* 6: 72.

Yip SS, Coroller TP, Sanford NN et al. 2016b. Use of registration-based contour propagation in texture analysis for esophageal cancer pathologic response prediction. *Phys Med Biol* 61: 906–922.

Yoon SH, Kim YH, Lee YJ et al. 2016. Tumor heterogeneity in human epidermal growth factor receptor 2 (HER2)-positive advanced gastric cancer assessed by CT texture analysis: Association with survival after trastuzumab treatment. *PLoS One* 11: e0161278.

Yue Y, Osipov A, Fraass B et al. 2017. Identifying prognostic intratumor heterogeneity using pre- and post-radiotherapy 18F-FDG PET images for pancreatic cancer patients. *J Gastrointest Oncol* 8: 127–138.

Zhou W, Zhang L, Wang K et al. 2017a. Malignancy characterization of hepatocellular carcinomas based on texture analysis of contrast-enhanced MR images. *J Magn Reson Imaging* 45: 1476–1484.

Zhou Y, He L, Huang Y et al. 2017b. CT-based radiomics signature: A potential biomarker for preoperative prediction of early recurrence in hepatocellular carcinoma. *Abdom Radiol* 42: 1695–1704.

Radiomics in genitourinary cancers
Prostate cancer

SATISH E. VISWANATH AND ANANT MADABHUSHI

18.1 INTRODUCTION

In the field of genitourinary cancers, prostate cancer has been one of the earliest diseases to see wide use of high-resolution imaging *in vivo* (Fusco et al. 2017). For instance, T2-weighted (T2w) MRI has been shown to enable excellent visualization of internal prostate structures and tissue detail (Westphalen et al. 2016). Recent guidelines such as the Prostate Imaging Reporting and Data System (PI-RADS) v2 describe clear morphologic characteristics for visual identification of suspicious lesions on T2w MRI (Purysko et al. 2016; Weinreb et al. 2016). While multi-parametric MR imaging (i.e., addition of diffusion or perfusion scans) further bolsters the confidence of radiologists in making a diagnosis of prostate cancer, the T2w MRI scan remains the mainstay of clinical evaluation.

However, PI-RADS v2 has been associated with only moderate inter-observer reproducibility [average of 40%–50% agreement (Muller et al. 2015; Rosenkrantz et al. 2016)], especially with regards to the exact location of the lesion within the PI-RADS v2 map [agreement of only 20% (Greer et al. 2018)]. This has led to the advent and wide-spread use of radiomics in prostate cancer for different applications via MR imaging. Radiomics refers to the high-throughput, computerized extraction of quantitative descriptors from radiographic imaging and the subsequent analysis of these mineable features for decision support and classification (Gillies et al. 2016; Kumar et al. 2012; Lambin et al. 2012). These descriptors could include texture, shape, appearance, volume, as well as other quantifiable characteristics from imaging.

18.1.1 Radiomic approaches for prostate T2w MRI

The most widely used suite of radiomic features for prostate cancer characterization are texture features (Lemaître et al. 2015; Liu et al. 2016; Wang et al. 2014), which quantify local variations and dependencies of intensity values within small neighborhoods on the MR image. Texture operators are used in radiomic analysis include Haralick [co-occurrence (Haralick et al. 1973)], Gabor [steerable wavelets (Jain and Farrokhnia 1991)], Laws [filter combinations (Laws 1980)], among others. Applications of radiomic texture features have been examined in prostate cancer detection and diagnosis (Lemaître et al. 2015; Litjens et al. 2014; Liu et al. 2016; Viswanath et al. 2012; Vos et al. 2012; Wang et al. 2014), grading and clinical significance (Fehr et al. 2015; Litjens et al. 2016; Tiwari et al. 2013; Vignati et al. 2015; Wang et al. 2017), treatment planning (Shiradkar et al. 2016), and more recently, treatment response prediction and prognosis (Gnep et al. 2017; Litjens et al. 2014; Shiradkar et al. 2018; Viswanath et al. 2014). Most recently, radiomics tools have been shown to accurately identify both presence or absence of clinically significant prostate cancer (Algohary et al. 2018) and thus improve overall performance of expert observers using PI-RADS v2 specifically (Wang et al. 2017).

Critically, radiomic texture features are directly dependent on the nature of the underlying grayscale intensity values in the MR image, which are in turn a function of properties such as tissue relaxation and density. The routine T2-weighted (T2w) MR sequence highlights differences in the T2 relaxation time of tissues and requires long echo (TE) and repetition (TR) times. T2w MRI is excellent for visualizing pathology, but as the degree of "weighting" is not the same for all tissues, there is a lack of reproducible T2w MR signal intensity values associated with disease. This is in contrast to apparent diffusion coefficient maps (from diffusion-weighted MRI, based on the diffusion of water molecules in a tissue region) where a specific range of ADC values have been determined to be associated with cancer presence (Weinreb et al. 2016). Instead, T2w MR intensities have been found to vary within a patient, across patient images acquired on the same scanner, as well as across multiple scanners (Nyúl and Udupa 1999). In more quantitative imaging modalities such as positron emission tomography (PET) or CT (where intensity values correspond to tissue-specific radiodensity and attenuation in Hounsfield units), radiomic features have been found to be reproducible over a wide range of acquisition settings (van Velden et al. 2016; Zhao et al. 2016). By contrast, initial studies in organs such as the brain suggest that even the underlying range of grayscale MR intensities can have an impact on textural radiomic features (Molina et al. 2017).

18.1.2 Acquisition-related artifacts in prostate T2w MRI and impact on radiomics

The underlying variability in T2w MR signal intensities has also been termed *intensity drift*. This problem has been addressed through the development of *intensity standardization*, where T2w signal intensity distributions from different patient datasets are mapped to a common intensity range via a post-processing transformation (Nyul et al. 2000). As a result, signal intensities across different scanners or datasets have a more consistent, tissue-specific meaning. Intensity standardization has been shown to improve segmentation accuracy (Zhuge and Udupa 2009) as well as classifier performance (Shah et al. 2011).

We note here the discrimination between intensity drift from the other known artifact associated with MR imaging: presence of a bias field or *intensity inhomogeneity*, referring to the wide variation of grayscale intensities across the MR image (Hou 2006). This artifact is exacerbated in prostate imaging due to use of an endorectal coil, where regions proximal to the coil appear brighter than regions farther away (Liney et al. 1998). While a number of methods exist for correction of the bias field in prostate T2w MR images (Vovk et al. 2007), an additional operation is required to correct for the problem of intensity drift. In a comprehensive study of the interplay between bias correction and intensity standardization (Madabhushi and Udupa 2005), it was found that bias correction actually introduced intensity drift in MR images. Further, the most optimal image quality was achieved when standardization was performed after bias correction, indicating it should be one of the final post-processing steps performed prior to radiomic feature analysis.

While bias field correction has been employed in different prostate MR radiomics applications, intensity standardization has not been as widely explored in this regard (Lemaître et al. 2015). This is likely because most studies of radiomic features for prostate MRI have been limited to data from a single site or scanner (Lemaître et al. 2015; Liu et al. 2016; Wang et al. 2014). These studies have instead utilized statistical or tissue-based normalization, where MR intensity values of the prostate are scaled based on the z-score technique [dividing by the mean (Artan et al. 2013)] or with respect to reference tissue regions [muscle, bladder (Niaf et al. 2012)]. More recently, in a multi-institutional study of radiomic features across three different sites, intensity standardization was applied to identify robust set of prostate tumor-specific set of features which were shown to result in good cross-site detection accuracy (Ginsburg et al. 2017).

In fact, reproducibility of radiomic features from MRI has primarily been studied in a single-site setting even for other diseases (Gourtsoyianni et al. 2017; Molina et al. 2017). These studies have primarily focused on measures such as concordance correlation coefficient, coefficient of variation, or intra-class correlation, all of which quantify how reproducible a radiomic feature is between test and retest imaging data (i.e., repeated scans). However, these measures may not be as relevant when quantifying reproducibility of radiomic features in a multi-site setting. Recently, Leo et al. (2016) presented two new measures of feature reproducibility: one for intra-site evaluation and another for multi-site evaluation. The latter, termed preparation-induced instability, allows for quantitative comparison of feature reproducibility across multiple sites irrespective of the number of individual samples from each institution. This measure enables feature reproducibility testing using clinically acquired imaging data from multiple sites.

18.1.3 OVERVIEW OF BOOK CHAPTER

In this work, we present a case study on evaluating the effect of applying intensity standardization to the original T2w signal intensity values as well as associated radiomic features, in a multi-site setting. We will utilize a retrospectively curated cohort of clinical prostate T2w MRI datasets curated from across four different sites and three different manufacturers. All datasets will be processed in an identical manner for annotation, segmentation, and bias field correction, culminating in intensity standardization. 131 radiomic texture features from across six different feature families will then be extracted and evaluated in terms of cross-site reproducibility. This analysis will be performed for both the original MR intensities as well as derived radiomic features. Unlike previous studies that have more narrowly focused on the reproducibility of MRI radiomic features in the context of specific sources of variation such as voxel size in a controlled single-site cohort, we will examine radiomic feature reproducibility in a broader multi-site setting. As a result, our data cohort encompasses multiple sources of intensity non-standardness including differences in acquisition protocols, scanners, and resolutions. By comparing pre- and post-standardization T2w MRI data, we will attempt to quantify and evaluate how intensity standardization impacts the reproducibility of radiomic features within both tumor and non-tumor regions in the prostate.

18.2 EXPERIMENTAL DESIGN

This study utilized 147 T2w prostate MRI datasets from four different institutions. Each MRI dataset was acquired pre-operatively using an endorectal coil on a 3T MR scanner, from a patient confirmed with prostate cancer and who later underwent radical prostatectomy. T2w MRI data from every patient were acquired as a series of DICOM images, which were directly saved from the scanner (acquisition parameters summarized in Table 18.1). Datasets from each site were then annotated by a different expert radiologist for tumor

Table 18.1 Summary of three Tesla prostate T2w MRI data, curated from four different sites

Site	[x, y, z] voxel dimensions	Scanner manufacturer	TR/TE	Number of datasets
S1	[0.27, 0.27, 2.20]	GE	4216-8266/155-165	15
S2	[0.41, 0.41, 3.00]	Siemens	2840-7500/107-135	11
S3	[0.27, 0.27, 3.00]	Siemens	4754/115	56
S4	[0.36, 0.36, 2.97]	Philips	4000/120-122	65

extent in the peripheral zone (based on available pathology sections and reports) as well as for the outer boundary of the peripheral zone (PZ). Analysis was limited to the mid-gland of the prostate, which ensured maximal confidence of annotation as well as the most consistent appearance of the PZ region across patients.

18.2.1 POST-PROCESSING TO CORRECT FOR BIAS FIELD ARTIFACTS ON MRI

Due to use of an endorectal coil during MR image acquisition, a non-linear signal intensity variation was visually discernible across the prostate MR image (see top row of Figure 18.1). This is termed the *bias field* and can be modeled as the multiplicative effect at lower frequencies in the MR image. Correcting for the bias field has been shown to enable more accurate radiomic feature-based detection of prostate cancer via MRI (Viswanath et al. 2011; Vos et al. 2010). For sites S_1, S_3, and S_4, N4ITK was employed to correct the bias field inhomogeneity (see bottom row of Figure 18.1). Data from S_2 were found to have been bias corrected on the scanner and did not require further processing in this regard [compare Figure 18.1(b) to Figures 18.1(a), (c) and (d)].

18.2.2 CORRECTING FOR INTENSITY DRIFT VIA STANDARDIZATION

T2-weighted MR intensity values are known to suffer from intensity drift, resulting in MR image intensities lacking tissue-specific numeric meaning within the same MRI protocol, for the same body region,

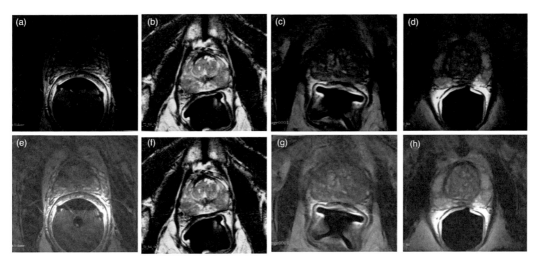

Figure 18.1 Representative prostate T2w MR images (a–d) before bias field correction and (e–h) after bias field correction. Note the markedly improved contrast and visibility of internal prostatic structures for images in the bottom row, as a result of correcting image inhomogeneity. Each column corresponds to a dataset from a different site. Note that (b,f) correspond to data from site S_2, which was corrected on the scanner and was not subjected to further processing.

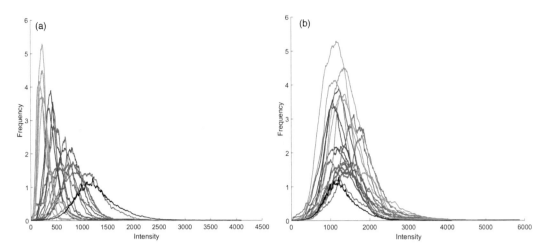

Figure 18.2 (See color insert.) T2w signal intensity distributions (a) before standardization, and (b) after standardization for five representative MRI datasets from each site. Different colors correspond to each of the four different sites, and black corresponds to the template to which all the datasets were standardized. Note the improved alignment between intensity distributions in (b), implying tissue specific-meaning across sites, scanners, and patients as a result of standardization.

or for images of the same patient obtained on the same scanner (Nyúl and Udupa 1999). Landmark-based histogram transformation (Nyul et al. 2000) was used to align T2w signal intensity distributions across all patient datasets. Five patients from S_1 were selected at random to generate a template distribution for the prostate as a whole [depicted in black in Figures 18.2(a), (b)]. Distributions for all patient volumes from all four sites were then non-linearly mapped to the template distribution, using deciles as landmarks on both target and template distributions. As a result, distributions for all patient datasets were brought into alignment [Figure 18.2(b)]. This ensured that the signal intensities were in prostate tissue-specific correspondence across sites.

18.2.3 IDENTIFICATION OF TUMOR AND NON-TUMOR REGION OF INTEREST

The largest tumor annotation per 2D section in each T2w MRI dataset was selected as the tumor region of interest (ROI). The tumor ROI was dilated by 7 pixels and removed from the PZ. The largest remaining contiguous region remaining per 2D section was selected as the non-tumor ROI. This was repeated for every mid-gland slice in each prostate T2w MRI dataset.

18.2.4 RADIOMIC FEATURE EXTRACTION

Previous studies have widely demonstrated that prostate appearance within the PZ can be modeled using image texture features (Lemaître et al. 2015; Liu et al. 2016; Viswanath et al. 2012; Wang et al. 2014). A total of 131 radiomic features from across six different families were extracted on a per voxel basis from each T2w MRI dataset. Features were extracted on a pixel-wise basis in 2D from each mid-gland section in each prostate MR dataset. Figure 18.3 depicts representative radiomic features for a single 2D section from one of the post-standardization T2w MRI datasets, while Table 18.2 summarizes their significance for prostate tumor characterization. Note that radiomic features were derived from the pre-standardization volume as well as the post-standardization volume, associated with each MRI dataset.

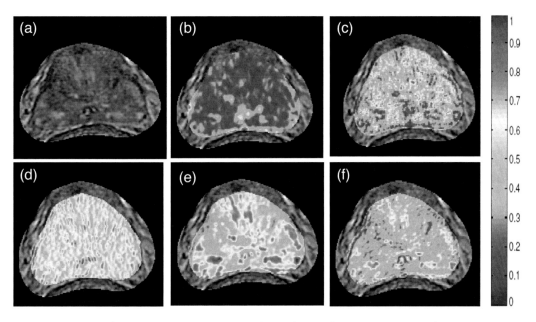

Figure 18.3 (See color insert.) Representative radiomic feature heatmaps corresponding to (a) T2w MR image: (b) 1st-order statistics, (c) 2nd-order statistics, (d) gradient, (e) Laws, and (f) Gabor.

Table 18.2 Summary of different classes of radiomic features, associated parameters, and significance for characterizing prostate tumor appearance

Feature type	Parameters	Window sizes	Significance for prostate tumor appearance
Non-steerable gradient	Sobel, Kirsch, gradient operators in X, Y directions	–	Accurately detect region boundaries, macro-scale edges
Steerable gradient (Gabor)	X-Y orientations	3, 5, 7	Quantify multi-scale, multi-oriented visual processing features used by radiologists when examining appearance of the carcinoma
First-order statistical	Mean, median, variance, range	3, 5, 7	May help localize regions of significant differences on T2w MR image
Second-order statistical (Haralick)	Co-occurrence features for contrast energy, contrast inverse moment, contrast average, contrast variance, contrast entropy, intensity average, intensity variance, intensity entropy, entropy, energy, correlation, info. measure of correlation 1	3, 5, 7	Useful in differentiating homogeneous low signal intensity regions from more hyper-intense appearance of normal prostate
Laws	Edge (E), ripple (R), spot (S), level (L), wave (W) operators	3, 5	Pattern-specific responses which can differentiate normal gland structure from cancer
Haar	–	3, 5	Differentiate the amorphous nature of the non-CaP regions within foci of low SI

- **Gradient:** Non-steerable gradient features were obtained using Sobel and Kirsch edge filters and first-order spatial derivative operations (Russ 2016). The Gabor filters, comprising the steerable class of gradient features, were defined by the convolution of a 2D Gaussian function with a sinusoid (Jain and Farrokhnia 1991; Ou et al. 2016). The wavelength of the sinusoid controls the spatial frequency (scale) and was defined as a function of the width of the Gaussian envelope. Gabor filter responses were directionally computed at different orientations, as dictated by co-ordinate transformations of the sinusoid.
- **First-order statistics:** The mean, median, standard deviation, and range were calculated for the signal intensity values of pixels, using a sliding window across the image. These were calculated at different window sizes.
- **Second-order statistics:** Using a sliding window, a spatial gray-level co-occurrence matrix was calculated at every pixel after quantizing the MR image to 128 gray levels. This represented the frequency with which two distinct signal intensity values were adjacent within the same 8-pixel sliding window neighborhood (Haralick et al. 1973). A total of 12 second-order statistical features were extracted based off this co-occurrence matrix. This was repeated for different window sizes.
- **Laws:** These responses were captured via convolution of the image with local masks that were obtained from vectors which capture local average, edge, spot, wave, and ripple patterns (Laws 1980).
- **Haar:** These transform decomposed signal intensity values in the discrete space while offering localization in the time and frequency domains (Busch 1997). These features comprised the coefficients obtained via wavelet decomposition at multiple scales.

18.2.5 Computing per-ROI radiomic features

For each of the 131 radiomic features and the T2w MR intensity associated with each dataset, a normalized value (Madabhushi and Udupa 2005) was calculated within each tumor and non-tumor ROI across all the 2D sections per dataset on which these ROIs had been identified. The normalized radiomic feature value $\mu(c)$ for a given ROI c that is associated with a radiomic feature $F(c)$, is given by:

$$\mu(c) = \frac{F_{50}}{F_0 - F_{99.8}},$$

where F_{50} corresponds to the 50th percentile of $F(c)$ (i.e., the mean), and F_0 and $F_{99.8}$ correspond to the 0th and 99.8th percentiles of $F(c)$, respectively.

18.2.6 Evaluating cross-site radiomic feature reproducibility

Leo et al. recently presented a new inter-site feature evaluation measure termed *Preparation-Induced Instability* (IS). IS was computed as the percentage of bootstrapped pairwise comparisons in which a radiomic feature was significantly different between each pair of sites. To consider a few scenarios: (a) IS = 0.01 corresponds to a feature that is unstable only 1% of the time (i.e., is reproducible in 99% of cross-site comparisons), (b) IS = 0.1 corresponds to a feature that is unstable 10% of the time (i.e., is reproducible in 90% of all cross-site comparisons), and (c) IS = 0.5 corresponds to a feature that is unstable 50% of the time (i.e., is reproducible in 50% of all cross-site comparisons), and so on. In general, features with an IS closer to 1 were considered to be more unstable, and hence less reproducible across sites. Further algorithmic details for the implementation of IS are provided in the original paper (Leo et al. 2016). This instability measure was applied here to quantify radiomic feature sensitivity to the standardization procedure. IS was quantified on a cross-site basis for each of 131 radiomic features as well as the T2w MR intensity value, and for tumor and non-tumor regions separately.

18.2.7 EXPERIMENT 1: EVALUATING PRE- AND POST-STANDARDIZATION T2W MR SIGNAL INTENSITY VALUES

To evaluate the hypothesis that intensity standardization will likely impact the cross-site reproducibility of the original T2w intensity values within the prostate PZ region, the effect of standardization on the original T2w MR intensity values was evaluated. Qualitatively, box-and-whisker plots were utilized to examine the trends in the normalized T2w MR intensities for each site individually, for tumor and non-tumor ROIs separately. This trend was also quantified by comparing the cross-site instability score of normalized T2w MR intensity values within tumor and non-tumor ROIs, before and after standardization.

18.2.8 EXPERIMENT 2: EVALUATING PRE- AND POST-STANDARDIZATION RADIOMIC FEATURES

Here, differences in the cross-site instability score in each of the 131 radiomic features were evaluated between pre- and post-standardization datasets. This was done in order to evaluate the hypothesis that standardization impacts the reproducibility of the radiomic features within tumor and non-tumor regions in the prostate PZ. Trends were visualized via pre- and post-standardization "instability score heatmaps", for tumor and non-tumor ROIs separately. This visualization involves two rows of cells where each feature was represented by a single cell, and the rows corresponded to pre-standardization and post-standardization data, respectively. Each cell which was then colored based on how high (denoted by red) or low (denoted by blue) the cross-site instability score was for the corresponding feature. Additionally, the number of radiomic features in each feature family that demonstrated: (a) a marked difference (increase or decrease) and (b) negligible or no change in their instability scores between pre- and post-standardization data were counted. This was done separately for tumor and non-tumor ROIs.

18.3 RESULTS AND DISCUSSION

18.3.1 EXPERIMENT 1: EFFECT OF STANDARDIZATION ON T2W MR SIGNAL INTENSITY VALUES

Figure 18.4 depicts box-and-whisker plots of normalized T2w signal intensity values within PZ non-tumor and PZ tumor regions, before and after standardization. Across all four sites, standardization appears to result in a numeric consistency in normalized T2w intensity values in both tumor and non-tumor regions, observed by comparing the median, upper, and lower bounds, as well as number of outliers. Interestingly, T2w intensity values within non-tumor regions appear to benefit more from standardization, based on the improved alignment of box-and whisker plots in Figure 18.4(b) versus Figure 18.4(a).

This is quantitatively reflected in the IS scores for normalized T2w signal intensity values (Table 18.3). Pre-standardization T2w intensity values were reproducible in ~75% of cross-site comparisons (IS = 0.248), while post-standardization T2w intensity values were reproducible in over 99% of cross-site comparisons (IS = 0.008). In other words, non-tumor regions exhibit markedly less instability in their signal intensity values as a result of standardization. By contrast, the T2w intensity values within tumor regions exhibit a marginal change in their IS scores: from being reproducible in 99.5% of cross-site comparisons prior to standardization to being reproducible in 98.5% of cross-site comparisons. However, the box-and-whisker plots in Figure 18.4(d) suggest that cross-site comparisons involving site S_2 likely caused the observed change in post-standardization IS scores, as tumor regions from site S_2 appear to have marginally different T2w intensity values and ranges in comparison to sites S_1, S_3, and S_4. The latter three sites can be seen to have more consistent medians as well as upper and lower bounds after standardization [Figure 18.4(d)], as compared to before standardization [Figure 18.4(c)]. Notably, this trend can be observed within non-tumor regions as well [Figure 18.4(b)], albeit in a less exaggerated fashion.

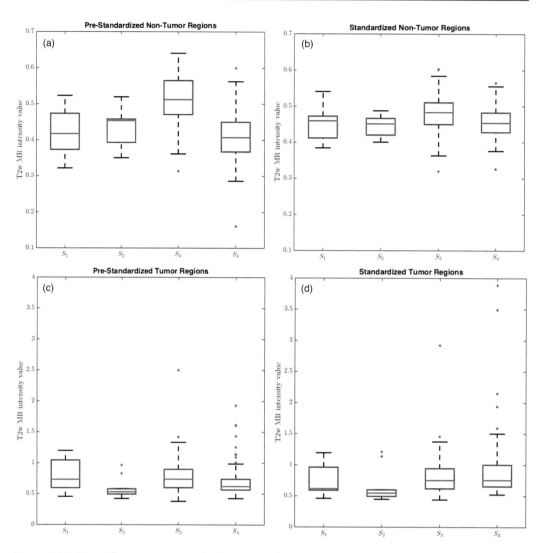

Figure 18.4 (See color insert.) Box-and-whisker plots for normalized T2w signal intensity values within (a,b) non-tumor regions, and (c,d) tumor regions. Each box-and-whisker plot corresponds to data from a single site. T2w MR intensities appear less consistent before standardization (left column) than after standardization (right column).

Table 18.3 IS Scores for normalized T2w signal intensity values before and after standardization, within tumor and non-tumor ROIs. Lower values indicate less instability, and hence better reproducibility

	Non-tumor	Tumor
Pre-standardized	0.248	0.008
Standardized	0.005	0.015

Previously, a relative intensity normalization method was shown to improve classification accuracy for PZ tumors in a cross-scanner setting (Artan et al. 2013), albeit on a limited cohort of 18 patients which were imaged on two different MRI scanners. This resonates with our findings in experiment 1, which suggests that correcting for intensity differences between sites and scanners is critical for reproducible identification of tumor extent on prostate MRI. A study of a specialized T2 map sequence for imaging the prostate on two different scanners (Hoang Dinh et al. 2015) did not demonstrate significant differences between scanners for T2 map values associated with both PZ tumor and non-tumor regions. As we did not have access to this specialized sequence in the current study, we opted to post-process retrospectively acquired clinical T2w MR images to correct for intensity drift via standardization. This has similarly enabled us to demonstrate that the standardized T2w intensity value may not significantly differ across site or scanners, for PZ tumor and non-tumor regions.

18.3.2 EXPERIMENT 2: EFFECT OF STANDARDIZATION ON RADIOMIC FEATURE VALUES

Figures 18.5 and 18.6 depict instability score heatmaps to visualize the cross-site reproducibility trends across all 131 radiomic features, for non-tumor and tumor regions, respectively. Blue cells indicate radiomic features with instability scores that are in the range 0–0.2, corresponding to reproducibility in 80%–100% of all cross-site comparisons. The bottom rows of both Figures 18.5 and 18.6 (corresponding to post-standardization data) can be seen to comprise a greater proportion of stable features (i.e., deep blue cells) than the top rows. These trends are summarized in Tables 18.4 and 18.5, which report changes in instability scores between pre- and post-standardization data within each radiomic feature family. Across all 131 features, 37% of non-tumor radiomic features and 18% of tumor features improved to being reproducible in 92% of cross-site comparisons after standardization (IS = 0.08). Thus, a significant proportion of radiomic features are more reproducible within both tumor and non-tumor regions when extracted from standardized data, as compared to pre-standardization data.

Interestingly, 54% of the radiomic features within non-tumor regions and 81% of features within tumor regions actually demonstrated a negligible difference (<5%) in instability scores between pre- and post-standardization data. Resonating with our findings from experiment 1, a smaller proportion of tumor radiomic

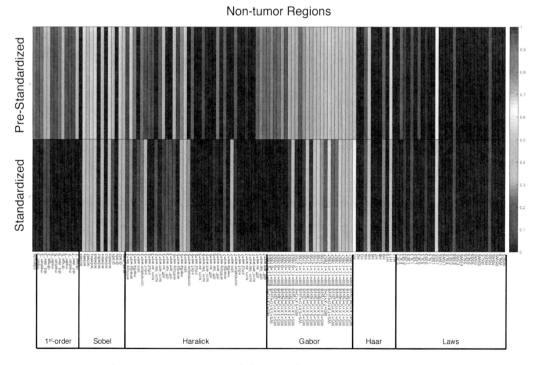

Figure 18.5 (See color insert.) Cross-site instability score heatmap for non-tumor regions, where blue indicates IS score = 0 (i.e., highly reproducible), while red indicates IS score = 1 (i.e., highly variable). A majority of features (e.g., Haralick, Gabor) appear less reproducible pre-standardization (top row, predominantly green), compared to after standardization (bottom row, same cells are now deep blue).

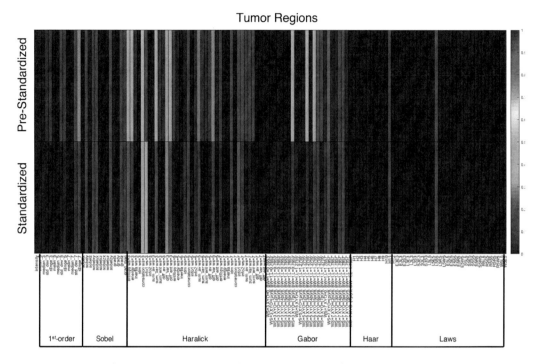

Figure 18.6 (See color insert.) Cross-site instability score heatmap for tumor regions, where blue indicates IS score = 0 (i.e., highly reproducible), while red IS score = 1 (i.e., highly variable). A majority of features are reproducible pre-standardization (top row comprises largely blue cells) and are even more reproducible after standardization (bottom row, deeper shade of blue for most cells).

Table 18.4 Trends in radiomic feature instability scores within non-tumor regions, indicating changes in standardized data in comparison to pre-standardized data

	Standardized non-tumor regions		
	Increased cross-site reproducibility	Negligible difference	Worse cross-site reproducibility
1st-order statistics (12)	12	–	–
Gradient (13)	4	9	–
2nd-order statistics (36)	15	17	7
Gabor (24)	15	6	3
Haar (12)	–	11	1
Laws (31)	3	28	–

Table 18.5 Trends in radiomic feature instability scores within tumor regions, indicating changes in standardized data in comparison to pre-standardized data

	Standardized tumor regions		
	Increased cross-site reproducibility	Negligible difference	Worse cross-site reproducibility
1st-order statistics (12)	4	8	–
Gradient (13)	–	13	–
2nd-order statistics (36)	15	23	1
Gabor (24)	5	19	–
Haar (12)	–	12	–
Laws (31)	–	31	–

features exhibited differences in instability scores between pre- and post-standardization data. After standardization, radiomic features in non-tumor regions improved their instability scores by 21% (pre-standardization IS = 0.3 versus post-standardization IS = 0.08) and by 14% in tumor regions (pre-standardization IS = 0.22 versus post-standardization IS = 0.08), on average. Thus, the magnitude of change in the instability scores between pre- and post-standardization data was also markedly larger in non-tumor regions as compared to tumor regions. Overall, it appears radiomic features of tumor regions may be less affected by the standardization procedure as compared to non-tumor regions.

Further examining instability score trends within radiomic feature families (Tables 18.4 and 18.5) reveals which of them are most impacted by standardization. For instance, kernel-based gradient and filter responses (Laws, Sobel, Kirsch) as well as wavelet decomposition (Haar) are not markedly affected by standardization. Almost all of these features exhibit negligible differences in their instability scores (IS change <5%) between pre- and post-standardization data, within both tumor and non-tumor regions. By contrast, 1st-order statistical operators are more reproducible on post-standardized data than on pre-standardized data, within both tumor and non-tumor regions (average IS increase of 9% and 23%, respectively). Given direct dependence of 1st-order statistics (mean, median, variance, range) on the actual T2w intensity values in the MR image, it is intuitive that these features are more likely to benefit from cross-site standardization and numeric consistency. By contrast, gradient or kernel operators are likely to be more resilient, as they are based on filter convolution responses.

Haralick radiomic features appear to be most differentially affected by standardization. For both tumor and non-tumor regions, ~42% of Haralick features exhibit better cross-site reproducibility after standardization (though different features are associated with each region). The only tumor radiomic feature with worse reproducibility within standardized tumor regions also exhibits poor reproducibility within standardized non-tumor regions (Haralick information measure 1, non-tumor IS = 0.49, tumor IS = 0.47). Similar to other Haralick features which demonstrate higher instability within standardized non-tumor regions, this feature is a function of the Haralick entropy, but is computed based on specific directional co-occurrences. While Haralick entropy itself demonstrates better reproducibility after standardization (IS increase of 10% compared to pre-standardization data), entropy variants (sum variance, difference entropy and variance, and inverse difference moment) demonstrate worse reproducibility within standardized non-tumor regions (average IS decrease of 18% compared to pre-standardized data). This may be due to the complex nature of these specific Haralick operators, which were also noted to be sensitive in a previous phantom study across multiple CT scanners (Shafiq-Ul-Hassan et al. 2017) as well as in a study of brain tumor MRIs (Molina et al. 2017).

Finally, Gabor radiomic features also largely benefit from standardization, with 62% of non-tumor features and 21% of tumor radiomic features demonstrating better cross-site reproducibility on standardized regions. The few Gabor features with poorer instability scores within non-tumor regions after standardization correspond to those extracted at larger window sizes (5×5, 7×7). By contrast, Haralick features which perform poorly after standardization are those extracted at smaller window sizes (3×3). This suggests that highly localized feature responses (Haralick) as well as very macro-scale wavelet kernel responses (Gabor) may be most sensitive to the effects of standardization. A previous study of global and local radiomic features on rectal MRIs also identified similar trade-offs between different scales of responses and their repeatability (Gourtsoyianni et al. 2017).

Previous studies have largely examined the reproducibility of radiomic features extracted from CT imaging, albeit primarily in a test/re-test setting (Leijenaar et al. 2013; Zhao et al. 2016). Similar to our findings, they reported that a majority of radiomic features were reproducible in a cross-site setting after post-processing for acquisition differences. While a majority of studies in prostate cancer have not examined multi-site data extensively, two previous studies (Ginsburg et al. 2017; Wang et al. 2017) did utilize cross-site intensity standardization to successfully develop cross-site radiomic classifiers that could accurately detect PZ tumors. While we did not explicitly study how classifiers perform as a function of standardization, previous work has suggested that standardization of MRI data results in more accurate classifier performance (Shah et al. 2011). Our results indicate that improved machine learning performance is likely due to more reproducible radiomic features being extracted from standardized data, than otherwise.

18.4 CONCLUDING REMARKS

In this work, we have evaluated the effect of applying intensity standardization to the prostate T2w MR signal intensity values as well as associated radiomic features in a multi-site setting. Our case study utilized retrospectively curated clinical imaging data from across four different sites which used three different scanner manufacturers and thus exhibited a variety of MR acquisition differences factors which contribute to intensity drift across sites. The impact of intensity standardization (to correct for this intensity drift) was uniquely evaluated in terms of *instability*, a novel measure that has been demonstrated to provide a quantitative measure of cross-site feature reproducibility. 131 different radiomic features were evaluated for cross-site reproducibility within expert-annotated tumor and non-tumor regions, both before and after standardization.

We demonstrated that intensity standardization largely improves the reproducibility and stability of both T2w MR signal intensities and radiomic features within both tumor and non-tumor regions. Notably, radiomic characterization of tumor regions appeared to be less affected by standardization than non-tumor regions. Among the most resilient radiomic features within both tumor and non-tumor regions were kernel and gradient-based operators such as Laws, Sobel, and Kirsch features. The popular Gabor and Haralick radiomic features largely benefited from standardization, except for very local complex co-occurrence statistics and very macro-scale wavelet responses (within non-tumor regions alone).

Our study did have its limitations. We utilized retrospectively curated T2-weighted MRI data and expert annotations of prostatic regions. There may thus be some level of bias in our cohort, though we did limit our study to the mid-gland and the PZ where experts are most confident in their annotations and T2w MR signal intensity values tend to be more representative of prostatic tissue (Purysko et al. 2016; Weinreb et al. 2016). As whole-mount pathology data were only available for a subset of our cohort, we did not attempt more rigorous radiology-pathology co-registration to map cancer regions onto MRI (Chappelow et al. 2011; Li et al. 2017). Instead, our experts utilized pathology images and reports when annotating the MR imaging data. We also did not examine how standardization affects other MR protocols such as diffusion-weighted imaging, as these were not available for all the sites. In a previous study, we did determine that standardization of other MRI sequences enabled tissue-specific treatment response to be identified (Viswanath et al. 2014). This suggests that our findings regarding the effects of standardization of prostate T2w MRI are likely to be representative of other MR imaging protocols as well. Using the clinically routine T2-weighted imaging protocol did enable our data cohort to be one of the larger cross-site and cross-scanner studies performed of radiomic features of prostate cancer (Lemaître et al. 2015; Liu et al. 2016; Viswanath et al. 2012; Wang et al. 2014). Additional post-processing operations such as noise correction and resampling voxel resolutions were also not examined by us in this work. However, an initial study we performed on the interplay between noise filtering, bias correction, and standardization did reveal that combining all three operations could impact the reproducibility and discriminability of radiomic features (Palumbo et al. 2011).

In conclusion, applying intensity standardization to prostate T2w MRI ensured that both T2w signal intensities as well as radiomic features had more consistent tissue-specific meaning in a cross-site setting, within both tumor and non-tumor regions. From a machine learning perspective, radiomic classifiers that utilize standardized T2w MR images are likely to be more confident in differentiating between tumor and non-tumor regions. It may thus be critical to first apply standardization prior to examining radiomic features of prostate MRI, especially in a multi-site setting.

ACKNOWLEDGMENTS

The authors would like to acknowledge the assistance of Michael Yim and Prathyush Chirra in implementing the experiments and analysis reported in this work.

Research reported in this publication was supported by the National Cancer Institute of the National Institutes of Health under award numbers 1U24CA199374-01, R01CA202752-01A1, R01CA208236-01A1, R01 CA216579-01A1, and R01 CA220581-01A1; National Center for Research Resources under award number 1 C06 RR12463-01; the Department of Defense (DOD) Prostate Cancer Idea Development Award; the DOD Lung Cancer Idea

Development Award; the DOD Peer Reviewed Cancer Research Program W81XWH-16-1-0329; the Cleveland Digestive Diseases Research Core Center; the NIH/NIDDK 1P30DK097948 DDRCC Pilot/Feasibility Award Program; the Ohio Third Frontier Technology Validation Fund; the Wallace H. Coulter Foundation Program in the Department of Biomedical Engineering; and the Clinical and Translational Science Award Program (CTSA) at Case Western Reserve University.

Opinions, interpretations, conclusions, and recommendations are those of the authors and are not necessarily endorsed by the Department of Defense and do not necessarily represent the official views of the National Institutes of Health.

REFERENCES

Algohary, A., Viswanath, S., Shiradkar, R., Ghose, S., Pahwa, S., Moses, D., Jambor, I. et al., 2018. Radiomic features on MRI enable risk categorization of prostate cancer patients on active surveillance: Preliminary findings: Radiomics categorizes PCa patients on AS. *J. Magn. Reson. Imaging* 48, 818–828. doi:10.1002/jmri.25983.

Artan, Y., Oto, A., Yetik, I.S., 2013. Cross-device automated prostate cancer localization with multiparametric MRI. *IEEE Trans. Image Process. Publ.* 22, 5385–5394.

Busch, C., 1997. Wavelet based texture segmentation of multi-modal tomographic images. *Comput. Graph.* 21, 347–358. doi:10.1016/s0097-8493(97)00012-5.

Chappelow, J., Bloch, B.N., Rofsky, N., Genega, E., Lenkinski, R., DeWolf, W., Madabhushi, A., 2011. Elastic registration of multimodal prostate MRI and histology via multiattribute combined mutual information. *Med. Phys.* 38, 2005–2018. doi:10.1118/1.3560879.

Fehr, D., Veeraraghavan, H., Wibmer, A., Gondo, T., Matsumoto, K., Vargas, H.A., Sala, E., Hricak, H., Deasy, J.O., 2015. Automatic classification of prostate cancer Gleason scores from multiparametric magnetic resonance images. *Proc. Natl. Acad. Sci. USA.* 112, E6265–6273. doi:10.1073/pnas.1505935112.

Fusco, R., Sansone, M., Granata, V., Setola, S.V., Petrillo, A., 2017. A systematic review on multiparametric MR imaging in prostate cancer detection. *Infect. Agent. Cancer* 12, 57. doi:10.1186/s13027-017-0168-z.

Gillies, R.J., Kinahan, P.E., Hricak, H., 2016. Radiomics: Images are more than pictures, they are data. *Radiology* 278, 563–577. doi:10.1148/radiol.2015151169.

Ginsburg, S.B., Algohary, A., Pahwa, S., Gulani, V., Ponsky, L., Aronen, H.J., Boström, P.J. et al., 2017. Radiomic features for prostate cancer detection on MRI differ between the transition and peripheral zones: Preliminary findings from a multi-institutional study. *J. Magn. Reson. Imaging* 46, 184–193. doi:10.1002/jmri.25562.

Gnep, K., Fargeas, A., Gutiérrez-Carvajal, R.E., Commandeur, F., Mathieu, R., Ospina, J.D., Rolland, Y. et al., 2017. Haralick textural features on T2-weighted MRI are associated with biochemical recurrence following radiotherapy for peripheral zone prostate cancer. *J. Magn. Reson. Imaging* 45, 103–117. doi:10.1002/jmri.25335.

Gourtsoyianni, S., Doumou, G., Prezzi, D., Taylor, B., Stirling, J.J., Taylor, N.J., Siddique, M., Cook, G.J.R., Glynne-Jones, R., Goh, V., 2017. Primary rectal cancer: Repeatability of global and local-regional MR imaging texture features. *Radiology* 284, 552–561. doi:10.1148/radiol.2017161375.

Greer, M.D., Shih, J.H., Barrett, T., Bednarova, S., Kabakus, I., Law, Y.M., Shebel, H. et al., 2018. All over the map: An interobserver agreement study of tumor location based on the PI-RADSv2 sector *map. J. Magn. Reson. Imaging* 48, 482–490. doi:10.1002/jmri.25948.

Haralick, R.M., Shanmugam, K., Dinstein, I., 1973. Textural features for image classification. *IEEE Trans. Syst. Man Cybern.* 3, 610–621. doi:10.1109/TSMC.1973.4309314.

Hoang Dinh, A., Souchon, R., Melodelima, C., Bratan, F., Mège-Lechevallier, F., Colombel, M., Rouvière, O., 2015. Characterization of prostate cancer using T2 mapping at 3T: A multi-scanner study. *Diagn. Interv. Imaging* 96, 365–372. doi:10.1016/j.diii.2014.11.016.

Hou, Z., 2006. A review on MR image intensity inhomogeneity correction. *Int. J. Biomed. Imaging* 2006, 49515. doi:10.1155/IJBI/2006/49515.

Jain, A.K., Farrokhnia, F., 1991. Unsupervised texture segmentation using gabor filters. *Pattern Recognit.* 24, 1167–1186. doi:10.1016/0031-3203(91)90143-S.

Kumar, V., Gu, Y., Basu, S., Berglund, A., Eschrich, S.A., Schabath, M.B., Forster, K. et al., 2012. Radiomics: The process and the challenges. *Magn. Reson. Imaging* 30, 1234–1248. doi:10.1016/j.mri.2012.06.010.

Lambin, P., Rios-Velazquez, E., Leijenaar, R., Carvalho, S., van Stiphout, R.G.P.M., Granton, P., Zegers, C.M.L. et al., 2012. Radiomics: Extracting more information from medical images using advanced feature analysis. *Eur. J. Cancer Oxf. Engl.* 48, 441–446. doi:10.1016/j.ejca.2011.11.036.

Laws, K.I., 1980. Textured Image Segmentation. USC Image Processing Institute Report 940. https://apps.dtic.mil/docs/citations/ADA083283

Leijenaar, R.T.H., Carvalho, S., Velazquez, E.R., van Elmpt, W.J.C., Parmar, C., Hoekstra, O.S., Hoekstra, C.J. et al., 2013. Stability of FDG-PET radiomics features: An integrated analysis of test-retest and inter-observer variability. *Acta Oncol.* 52, 1391–1397. doi:10.3109/0284186X.2013.812798.

Lemaître, G., Martí, R., Freixenet, J., Vilanova, J.C., Walker, P.M., Meriaudeau, F., 2015. Computer-aided detection and diagnosis for prostate cancer based on mono and multi-parametric MRI: A review. *Comput. Biol. Med.* 60, 8–31. doi:10.1016/j.compbiomed.2015.02.009.

Leo, P., Lee, G., Shih, N.N.C., Elliott, R., Feldman, M.D., Madabhushi, A., 2016. Evaluating stability of histo-morphometric features across scanner and staining variations: Prostate cancer diagnosis from whole slide images. *J. Med. Imaging Bellingham Wash* 3, 047502. doi:10.1117/1.JMI.3.4.047502.

Li, L., Pahwa, S., Penzias, G., Rusu, M., Gollamudi, J., Viswanath, S., Madabhushi, A., 2017. Co-registration of ex vivo surgical histopathology and in vivo T2 weighted MRI of the prostate via multi-scale spectral embedding representation. *Sci. Rep.* 7, 8717. doi:10.1038/s41598-017-08969-w.

Liney, G.P., Turnbull, L.W., Knowles, A.J., 1998. A simple method for the correction of endorectal surface coil inhomogeneity in prostate imaging. *J. Magn. Reson. Imaging* 8, 994–997.

Litjens, G., Debats, O., Barentsz, J., Karssemeijer, N., Huisman, H., 2014. Computer-aided detection of prostate cancer in MRI. *IEEE Trans. Med. Imaging* 33, 1083–1092. doi:10.1109/TMI.2014.2303821.

Litjens, G.J.S., Elliott, R., Shih, N.N., Feldman, M.D., Kobus, T., Hulsbergen-van de Kaa, C., Barentsz, J.O., Huisman, H.J., Madabhushi, A., 2016. Computer-extracted features can distinguish noncancerous confounding disease from prostatic adenocarcinoma at multiparametric MR imaging. *Radiology* 278, 135–145. doi:10.1148/radiol.2015142856.

Litjens, G.J.S., Huisman, H.J., Elliott, R.M., Shih, N.N., Feldman, M.D., Viswanath, S., Futterer, J.J., Bomers, J.G.R., Madabhushi, A., 2014. Quantitative identification of magnetic resonance imaging features of prostate cancer response following laser ablation and radical prostatectomy. *J. Med. Imaging Bellingham Wash* 1, 035001. doi:10.1117/1.JMI.1.3.035001.

Liu, L., Tian, Z., Zhang, Z., Fei, B., 2016. Computer-aided detection of prostate cancer with MRI: Technology and applications. *Acad. Radiol.* 23, 1024–1046. doi:10.1016/j.acra.2016.03.010.

Madabhushi, A., Udupa, J.K., 2005. Interplay between intensity standardization and inhomogeneity correction in MR image processing. *IEEE Trans. Med. Imaging* 24, 561–576. doi:10.1109/TMI.2004.843256.

Molina, D., Pérez-Beteta, J., Martínez-González, A., Martino, J., Velasquez, C., Arana, E., Pérez-García, V.M., 2017. Lack of robustness of textural measures obtained from 3D brain tumor MRIs impose a need for standardization. *PLoS One* 12, e0178843. doi:10.1371/journal.pone.0178843.

Muller, B.G., Shih, J.H., Sankineni, S., Marko, J., Rais-Bahrami, S., George, A.K., de la Rosette, J.J.M.C.H. et al., 2015. Prostate cancer: Interobserver agreement and accuracy with the revised prostate imaging reporting and data system at multiparametric MR imaging. *Radiology* 277, 741–750. doi:10.1148/radiol.2015142818.

Niaf, E., Rouvière, O., Mège-Lechevallier, F., Bratan, F., Lartizien, C., 2012. Computer-aided diagnosis of prostate cancer in the peripheral zone using multiparametric MRI. *Phys. Med. Biol.* 57, 3833–3851. doi:10.1088/0031-9155/57/12/3833.

Nyúl, L.G., Udupa, J.K., 1999. On standardizing the MR image intensity scale. *Magn. Reson. Med.* 42, 1072–1081.

Nyul, L.G., Udupa, J.K., Xuan Zhang, 2000. New variants of a method of MRI scale standardization. *IEEE Trans. Med. Imaging* 19, 143–150. doi:10.1109/42.836373.

Ou, X., Pan, W., Zhang, X., Xiao, P., 2016. Skin image retrieval using gabor wavelet texture feature. *Int. J. Cosmet. Sci.* 38, 607–614. doi:10.1111/ics.12332.

Palumbo, D., Yee, B., O'Dea, P., Leedy, S., Viswanath, S., Madabhushi, A., 2011. Interplay between bias field correction, intensity standardization, and noise filtering for T2-weighted MRI. *Conf. Proc. Annu. Int. Conf. IEEE Eng. Med. Biol. Soc. IEEE Eng. Med. Biol. Soc. Annu. Conf.* 2011, 5080–5083. doi:10.1109/IEMBS.2011.6091258.

Purysko, A.S., Rosenkrantz, A.B., Barentsz, J.O., Weinreb, J.C., Macura, K.J., 2016. PI-RADS Version 2: A pictorial update. *Radiogr. Rev. Publ. Radiol. Soc. N. Am. Inc* 36, 1354–1372. doi:10.1148/rg.2016150234.

Rosenkrantz, A.B., Ginocchio, L.A., Cornfeld, D., Froemming, A.T., Gupta, R.T., Turkbey, B., Westphalen, A.C., Babb, J.S., Margolis, D.J., 2016. Interobserver reproducibility of the PI-RADS version 2 lexicon: A multicenter study of six experienced prostate radiologists. *Radiology* 280, 793–804. doi:10.1148/radiol.2016152542.

Russ, J.C., 2016. *The Image Processing Handbook*, Seventh Edition, 7th ed. Boca Raton, FL: CRC Press.

Shafiq-Ul-Hassan, M., Zhang, G.G., Latifi, K., Ullah, G., Hunt, D.C., Balagurunathan, Y., Abdalah, M.A. et al., 2017. Intrinsic dependencies of CT radiomic features on voxel size and number of gray levels. *Med. Phys.* 44, 1050–1062. doi:10.1002/mp.12123.

Shah, M., Xiao, Y., Subbanna, N., Francis, S., Arnold, D.L., Collins, D.L., Arbel, T., 2011. Evaluating intensity normalization on MRIs of human brain with multiple sclerosis. *Med. Image Anal.* 15, 267–282. doi:10.1016/j.media.2010.12.003.

Shiradkar, R., Ghose, S., Jambor, I., Taimen, P., Ettala, O., Purysko, A.S., Madabhushi, A., 2018. Radiomic features from pretreatment biparametric MRI predict prostate cancer biochemical recurrence: Preliminary findings: Prostate cancer recurrence prediction. *J. Magn. Reson. Imaging* 48, 1626–1636. doi:10.1002/jmri.26178.

Shiradkar, R., Podder, T.K., Algohary, A., Viswanath, S., Ellis, R.J., Madabhushi, A., 2016. Radiomics based targeted radiotherapy planning (Rad-TRaP): A computational framework for prostate cancer treatment planning with MRI. *Radiat. Oncol. Lond. Engl.* 11, 148. doi:10.1186/s13014-016-0718-3.

Tiwari, P., Kurhanewicz, J., Madabhushi, A., 2013. Multi-kernel graph embedding for detection, Gleason grading of prostate cancer via MRI/MRS. *Med. Image Anal.* 17, 219–235. doi:10.1016/j.media.2012.10.004.

van Velden, F.H.P., Kramer, G.M., Frings, V., Nissen, I.A., Mulder, E.R., de Langen, A.J., Hoekstra, O.S., Smit, E.F., Boellaard, R., 2016. Repeatability of radiomic features in non-small-cell lung cancer [(18)F]FDG-PET/CT studies: Impact of reconstruction and delineation. *Mol. Imaging Biol. Off. Publ. Acad. Mol. Imaging* 18, 788–795. doi:10.1007/s11307-016-0940-2.

Vignati, A., Mazzetti, S., Giannini, V., Russo, F., Bollito, E., Porpiglia, F., Stasi, M., Regge, D., 2015. Texture features on T2-weighted magnetic resonance imaging: New potential biomarkers for prostate cancer aggressiveness. *Phys. Med. Biol.* 60, 2685–2701. doi:10.1088/0031-9155/60/7/2685.

Viswanath, S., Toth, R., Rusu, M., Sperling, D., Lepor, H., Futterer, J., Madabhushi, A., 2014. Identifying quantitative multi-parametric MRI features for treatment related changes after laser interstitial thermal therapy of prostate cancer. *Neurocomputing* 144, 13–23. doi:10.1016/j.neucom.2014.03.065.

Viswanath, S.E., Bloch, N.B., Chappelow, J.C., Toth, R., Rofsky, N.M., Genega, E.M., Lenkinski, R.E., Madabhushi, A., 2012. Central gland and peripheral zone prostate tumors have significantly different quantitative imaging signatures on 3 Tesla endorectal, in vivo T2-weighted MR imagery. *J. Magn. Reson. Imaging* 36, 213–224. doi:10.1002/jmri.23618.

Viswanath, S.E., Palumbo, D., Chappelow, J, Patel, P., Bloch, B.N., Rofsky, N.M., Lenkinski, R.E., Genega, E.M., Madabhushi, A., 2011. Empirical evaluation of bias field correction algorithms for computer-aided detection of prostate cancer on T2w MRI, In: *SPIE Medical Imaging, Computer-Aided Diagnosis*. Bellingham, WA: SPIE, p. 79630V. doi.10.1117/12.878813.

Vos, P.C., Barentsz, J.O., Karssemeijer, N., Huisman, H.J., 2012. Automatic computer-aided detection of prostate cancer based on multiparametric magnetic resonance image analysis. *Phys. Med. Biol.* 57, 1527. doi:10.1088/0031-9155/57/6/1527.

Vos, P.C., Hambrock, T., Barenstz, J.O., Huisman, H.J., 2010. Computer-assisted analysis of peripheral zone prostate lesions using T2-weighted and dynamic contrast enhanced T1-weighted MRI. *Phys. Med. Biol.* 55, 1719–1734. doi:10.1088/0031-9155/55/6/012.

Vovk, U., Pernus, F., Likar, B., 2007. A review of methods for correction of intensity inhomogeneity in MRI. *IEEE Trans. Med. Imaging* 26, 405–421. doi:10.1109/TMI.2006.891486.

Wang, H., Viswanath, S., Madabhushi, A., 2017. Discriminative scale learning (DiScrn): Applications to prostate cancer detection from MRI and needle biopsies. *Sci. Rep.* 7, 12375. doi:10.1038/s41598-017-12569-z.

Wang, J., Wu, C.-J., Bao, M.-L., Zhang, J., Wang, X.-N., Zhang, Y.-D., 2017. Machine learning-based analysis of MR radiomics can help to improve the diagnostic performance of PI-RADS v2 in clinically relevant prostate cancer. *Eur. Radiol.* doi:10.1007/s00330-017-4800-5.

Wang, S., Burtt, K., Turkbey, B., Choyke, P., Summers, R.M., 2014. Computer aided-diagnosis of prostate cancer on multiparametric MRI: A technical review of current research. *BioMed Res. Int.* 2014, 789561. doi:10.1155/2014/789561.

Weinreb, J.C., Barentsz, J.O., Choyke, P.L., Cornud, F., Haider, M.A., Macura, K.J., Margolis, D. et al., 2016. PI-RADS prostate imaging—Reporting and data system: 2015, Version 2. *Eur. Urol.* 69, 16–40. doi:10.1016/j.eururo.2015.08.052.

Westphalen, A.C., Noworolski, S.M., Harisinghani, M., Jhaveri, K.S., Raman, S.S., Rosenkrantz, A.B., Wang, Z.J., Zagoria, R.J., Kurhanewicz, J., 2016. High-resolution 3-T endorectal prostate MRI: A multireader study of radiologist preference and perceived interpretive quality of 2D and 3D T2-weighted fast spin-echo MR images. *Am. J. Roentgenol.* 206, 86–91. doi:10.2214/AJR.14.14065.

Zhao, B., Tan, Y., Tsai, W.-Y., Qi, J., Xie, C., Lu, L., Schwartz, L.H., 2016. Reproducibility of radiomics for deciphering tumor phenotype with imaging. *Sci. Rep.* 6, 23428. doi:10.1038/srep23428.

Zhuge, Y., Udupa, J.K., 2009. Intensity standardization simplifies brain MR image segmentation. *Comput. Vis. Image Underst.* 113, 1095–1103. doi:10.1016/j.cviu.2009.06.003.

Radiomics analysis for gynecologic cancers

HARINI VEERARAGHAVAN

19.1 INTRODUCTION

Radiomics analysis is the process of extracting high-dimensional quantitative characterize of imaging heterogeneity within cancers. Many solid cancers exhibit widespread genomic tumor heterogeneity within and between cancers [1], and increased heterogeneity has been implicated to be associated with worse outcomes including in ovarian cancers [2]. Unlike biopsy-based quantification of heterogeneity that computes a static measurement of the tumor state at one time, radiomics analysis is non-invasive and therefore can be used to quantify and trace the evolution of image-based tumor heterogeneity over time. Furthermore, biopsy-based quantification of tumor heterogeneity only examines a small portion of the tumor while radiomics biomarkers can non-invasively quantify the tumor heterogeneity over the whole tumor extent. Whereas morphologic heterogeneity within cancers is subjectively described in radiological reports [3,4], radiomic analysis can produce potentially limitless number of quantitative imaging biomarkers that are distinct from subjective assessment and add to clinical decision-making [5]. Radiomics analysis has shown feasibility in characterization of disease aggressiveness and outcomes [6–8], as well as in predicting outcomes to treatment in a variety of cancers [9].

Majority of the works in gynecologic cancers have studied the feasibility of using subjectively assessed and semi-quantitative imaging parameters to predict outcomes, tumor grade, and to correlate such features with underlying genetic expression. Subjective radiologic assessments from computed tomography (CT) imaging have been shown to be associated with aggressive mesenchymal transcriptomic profile (Classification of Ovarian Cancer [CLOVAR]) of high-grade serous ovarian cancer (HGSOC) [10,11], and with BRCA mutation status in HGSOC [12] and survival [11,12]. Tumor volume combined with apparent diffusion coefficient (ADC) computed from diffusion-weighted magnetic resonance (MR) images have been shown to be correlated with outcomes in endometrial [13] cancers. Halle [14] identified a dynamic contrast enhanced (DCE)-MR parameter A(Brix), where low A(Brix) was significantly associated with upregulation of genes related to hypoxia. Semi-quantitative DCE-MRI parameters have been shown to predict chemoradiotherapy [15], radiation response from pre-treatment imaging for cervical cancers [16,17]. ADC computed from diffusion-weighted (DW-MRI) quantifies the diffusion in the tissues. ADC has been shown to be associated with clinical prognostic factors [18] and predictive of outcomes to therapy in cervical [19,20] and ovarian cancers [21].

Recently, hybrid imaging combining contrast-enhanced CT and positron emission tomography (PET)/CT imaging [22] as well as PET/MR imaging [23,24] have demonstrated feasibility in predicting outcomes from pre-treatment images for cervical and ovarian cancers. The study in [25] combined MRI and PET images and showed that tumor heterogeneity computed from semi-quantitative and first-order histogram-based features from DWI-MRI, DCE-MRI, and fludeoxyglucose (18F) (FDG) PET and CT images differed between modalities and showed longitudinal changes during treatment with radiation therapy in cervical cancer patients.

In this chapter, we will present some of the recent works that have employed radiomics analysis for gynecologic cancers specific to MRI, CT, and PET/CT images, discuss some of the challenges involved in performing robust radiomics analysis, and suggest potential solutions.

19.2 RADIOMICS IN GYNECOLOGIC CANCERS

Radiomics analysis using CT, MRI, and PET images in gynecologic cancers is relatively recent compared to other cancers. Table 19.1 outlines the various works that employed radiomics of ovarian, endometrial, and cervical cancers. Predominant number of methods used 18F-FDG PET images and texture measures extracted from those images for assessing outcomes including in [26–29]. Three methods to our knowledge applied radiomics analysis from MR images for predicting outcomes in gynecologic cancers including the studies in [30–32]. CT-based radiomics analysis has been used in [33,34] to predict outcomes to therapy in ovarian cancers.

Outcomes: Investigated outcomes were typically treatment response [28,29,31,33,34], tumor aggressiveness [26,35], histology [27], and survival [28,34]. The study in [30] investigated the feasibility of developing radiomic biomarkers from multi-parametric MRI for detecting lymphovascular invasion (LVSI), deep myometrial invasion (DMI), and non-invasively classifying tumor grade, all of which are associated with worse prognosis in endometrial cancers [13]. One study in [32] assessed the textural differences within the tumors from the normal tissue to investigate the utility of the entropy texture in extracting intra-tumoral heterogeneity from MRI. The study in [34] also explored the association of CT texture features with gene expression.

Segmentation technique: Manual volumetric delineation of tumors has been used by a majority of methods [30–32,34,35], particularly those using CT and MRI for performing radiomics. Manual delineation of single slice ROI was used in [30]. Automatic and semi-automatic tumor segmentation using fuzzy clustering or intensity thresholding has been used by a few works including in [26–28].

Feature extraction: Features have been extracted either by standardizing (through center scaling) the intensity values within the tumor volume [26] or min-max normalization of signal intensities within the tumor to fixed number of discrete levels [27,28]. Z-score normalization of features [31] was used to enable appropriate generalization using the support vector machine (SVM) algorithm. In [35], all images were preprocessed with bias field correction and histogram standardization using [36] so as to use images extracted from multiple institutions and different scanners for extracting textures.

Feature selection and outcomes modeling: In [26], no specific feature selection was performed. Instead, features were grouped based on Pearson correlation, and the features within the individual groups that resulted in the highest area under the curve (AUC) using receiver operating curve analysis were used to train individual SVM classifiers. The work in [31] studied the utility of DCE MRI-based

Table 19.1 Summary of radiomics analysis in gynecologic cancers

Outcome	Modality	Cancer	Features	# Fs	#Pt	Result	Conclusion	References
Treatment outcome, Gene expression	CT	Ovary	Inter-site tumor heterogeneity measures	12	38	Low- vs. high- risk patient groups were classified with an accuracy of 0.71 (true positive rate [TPR] 0.6, true negative rate [TNR] 0.86) using inter-site heterogeneity metrics. FO and GLCM-based textures produced an accuracy of 0.56 (TPR 0.4, TNR 0.79). 8 out of 13 patients with amplification to *CCNE1* occurred in clusters with high inter-site heterogeneity.	Extracting heterogeneity between the different sites of disease is feasible and measures of inter-site heterogeneity show better accuracy than the conventional single site measures in predicting outcomes.	[34]
Tumor aggressiveness: LMS vs. ALM	T2w MRI	Uterine leiomyosarcoma	First order (FO), Gray level correlation matrix (GLCM), (GAB), Gabor Gabor image GLCM	21	24	Feature clustering combining intensity and Gabor texture measures separated LMS from benign ALM with accuracy of 0.75 (sensitivity: 0.70, specificity: 0.79).	Texture measures are feasible for distinguishing malignant leiomyosarcoma from benign and atypical leiomyomas. Qualitative assessment was more accurate and there was large inter-reader agreement.	[35]
Treatment outcome	MRI	Cervix	FO features, GLCM computed from DCE-MR parametric maps	127	81	FO features did not predict outcomes; GLCM features significantly predicted outcomes with 70% accuracy similar to clinical factors (69% accuracy).	Texture measures computed from DCE-MR parametric maps are similarly accurate as the clinical variables in predicting outcomes.	[31]
Tumor stage: early (ES) vs. late (LS)	PET	Cervix	SUV(max, mean, peak), MTV, GLCM, GLSZM, GLRL, NGLDM	58	42	Non-linear SVM classifier trained using Run percentage (RP) achieved an AUC of 0.88.	Textures computed from GLRL could differentiate early from late stage cervical cancers.	[26]
Histology	PET	Cervix	SUV histogram, MTV, TLG, NGLCM, NGTDM	18	83	Squamous cell carcinoma (SCC) had significantly higher correlation (0.7 ± 0.07) compared to non-squamous cell carcinoma (NSCC) (0.64 ± 0.07; $p = 0.003$).	Texture measures can potentially distinguish tumor histology.	[27]

(Continued)

Table 19.1 (Continued) Summary of radiomics analysis in gynecologic cancers

Outcome	Modality	Cancer	Features	# Fs	#Pt	Result	Conclusion	References
Tumor grade, DMI, LVSI	MRI	Endometrial	FO features from DCE-MR, LoG features, FO from LoG	137	180	RF model produced AUC of 0.84 for DMI; AUC of 0.80 for LVSI; 0.83 for tumor grade.	MR texture-based models resulted in similar accuracy in detecting DMI, LVSI, and tumor grade as radiologist assessment.	[30]
Treatment Local control (LC) Overall survival (OS)	PET	Cervix	SUV, TLG, MTV, GLCM, GLRL, GLZLM	11	108	SUV^{peak} was the best predictor of LC (AUC 0.75). Combined neutrophilia and texture score of 2 was an independent predictor of LC (HR = 7.5, $p < 0.001$) and OS (HR = 5.8, $p = 0.001$).	Combining PET radiomics measures with neutrophil grade is feasible to predict treatment outcome.	[28]
Feature reproducibility	MRI	Cervix	ADC intensity-histogram entropy, GLCM entropy (0°, 45°,90°, 135°), Entropy mean, SD, kurtosis	8	51	All features showed excellent inter-observer agreement (intra-class correlation coefficient [ICC] > 0.9). All entropy features except range and SD were significantly higher in cancer vs. normal tissues ($p < 0.0001$).	Entropy-based texture measures are reproducible between different observers and show significant differences from normal structures.	[32]
Recurrence	PET	Cervix	GLCM, GLZLM, GLRLM	11	118	Eight features were statistically significant features of recurrence in one dataset G1 ($p < 0.05$). Multivariate signature trained in G2 validated on G1 (AUC = 0.76, $p < 0.001$) and were more accurate than SUV^{max} ($p = .02$). Four features were significantly different between G1 and G2.	Radiomic features can predict local recurrence. However, features show significant difference between scanners.	[29]
Treatment response	CT	Ovary	Pre-Post treatment features, FO, Wavelet, shape	159	91	Features combined from pre- and post-chemotherapy treatment CTs predicted early recurrence with AUC of 0.80 compared with 0.74 using response evaluation criteria in solid tumors (RECIST) measures.	Combining radiomic features from pre and post-treatment CT images improves outcome prediction compared to conventional RECIST measures.	[33]

FO: first order; MTV: metabolic tumor volume; TLG: total lesion glycolysis; GLCM: gray-level co-occurrence matrix; GLSZM: gray-level size-zone matrix; GLRL: gray-level run length; NGLCM: normalized gray-level difference matrix; NGTDM: neighborhood gray-tone difference; DMI: deep myometrial invasion.

functional parameters in comparison and in combination with the second-order texture measures using SVM following feature selection that extracted features that best explained the total variance in the data. Highly correlated features were removed in [30], and the remaining features were then used in the outcomes modeling. Random forest classifiers using pre-selected features were used in [30] for distinguishing between multiple outcomes. In [29], univariate analysis was used to eliminate uninformative features, following which a multi-variate statistical analysis was used with stepwise model selection using Akaike information criterion to select the best model for distinguishing patients by recurrence. Additionally, the same work also applied the models on two different datasets obtained through different scanners as validation and found that features derived from one scanner did not validate the data obtained through a different scanner.

Novel features: The study in [28] developed a novel radiomics score that combined radiomics features with underlying tumor biology, namely, the neutrophil count to predict local control and overall survival in advanced cervical cancers. A single radiomics feature that best differentiated the patients by outcomes through univariate analysis was selected and combined with neutrophilia (or neurotrophil count) to produce a neutrophil standard uptake value (SUV) grade score (NSG = 0,1,2) and used to predict outcomes. Of all the texture and semi-quantitative measures, SUVpeak was the most relevant measure and was also independent of neutrophilia. In [33], features from pre and post-treatment CT scans were combined and the delta features were used to predict outcomes to chemotherapy. In our study in [34], we developed novel measures of inter-site tumor heterogeneity to correlate with outcomes including survival and gene expression. Exploration of measures beyond the conventionally used measures can improve the outcome prediction as described in the following example.

19.2.1 IMPROVING OUTCOMES PREDICTION THROUGH INCORPORATION OF INTER-SITE TUMOR HETEROGENEITY: EXAMPLE FROM HIGH-GRADE SEROUS OVARIAN CANCER

A different texture analysis approach to outcomes prediction was undertaken in [37], where a novel approach was developed to extract the textural heterogeneity between metastatic sites with the goal of quantifying the tumor heterogeneity in a patient as a whole. The same work was motivated by the fact that most HGSOCs are metastatic at the time of presentation and that most cancers, including ovarian cancers exhibit intense within and between tumor heterogeneity [1,2,38], and it is feasible to image and model the heterogeneity between the various tumor sites [39].

The inter-site heterogeneity metrics as described in [34] are computed by measuring the textural dissimilarity between the individual metastatic sites and consist of four features, namely: (i) site entropy, which measures the entropy in the distribution of dissimilarity values, (ii) site variance that measures that variance in the dissimilarities, and (iii) variants of cluster shade and (iv) cluster prominence texture measures applied to the dissimilarity map and called site cluster shade and site cluster.

Such inter-site heterogeneity measures when combined with single site Haralick texture measures can better distinguish high-risk from low-risk patients. High-risk patients were defined as those with overall survival <60 months and with either amplification to the 19q12 involving cyclin E1 gene *CCNE1* or incomplete surgical resection. Figure 19.1 shows the relative importance of features extracted using an elastic net regression classifier trained with leave-one-out cross-validation (LOOCV) (Figure 19.1a), the comparison in receiver operating curves (ROC) curves between classifiers trained with all features versus just the Haralick texture measures computed from the largest tumor site, and difference between the site entropy (SE) for the high- and low-risk patients (Figure 19.1c) and the difference in the most relevant texture measure, namely, entropy between those two groups of patients (Figure 19.1d). As seen, incorporating the disease heterogeneity across all the tumor sites helps to obtain a better differentiation of the patients. All Haralick textures were computed after rescaling the images to the range 0–255 and extracting textures from 128 bins using an offset distance of 1 pixel and averaging across all directions. A software wrapper in C++ using the Insight ToolKit software was used to generate the Haralick textures while the inter-site texture heterogeneity measures were computed using in-house software implemented in Matlab.

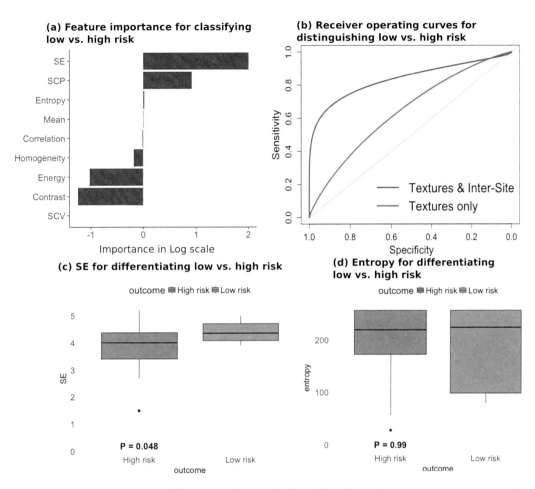

Figure 19.1 (See color insert.) Differentiating between high-risk from low-risk HGSOC patients using single-site and inter-site texture measures. The relative importance of features when using both single-site and inter-site measures in a model trained using leave-one-out cross-validation using elastic net regression identified the inter-site measures as most relevant (a), comparison of receiver operating characteristic curves show better performance of the combined model (single-site textures and inter-site textures) compared to single-site texture (b), and comparison in differentiating high-risk from low-risk patients using inter-site entropy (SE) (c), and single-site entropy (d).

The study in [37] explored the association of image-based radiomics measures with gene expression, in particular *CCNE1* amplification, and found that clusters associated with higher inter-site heterogeneity measures were associated with *CCNE1* amplification at a higher proportion compared to clusters with lower measures of inter-site tumor heterogeneity. *CCNE1* amplification has been implicated with development of platinum resistance in women treated with platinum-based chemotherapy in ovarian cancers [40].

While the radiomics measures themselves showed no significant correlation to the CLOVAR subtype cancers in [37], the study using qualitative measures [11] found that the presence of mesenteric infiltration and diffuse peritoneal involvement was significantly associated with the mesenchymal subtype. This suggests that combining radiomics analysis of the tumors with qualitatively descriptive morphological traits could add value to obtaining overall outcomes prediction.

Summary: Almost all works except in [37] only studied primary tumors for assessing relation of tumor heterogeneity with outcomes. Reasons for studying only the primary tumors include small metastatic lesions [26] or difficultly in generating volumetric segmentations. Although single slice region of interests (ROIs) are easier to generate [30], such methods can lose the textural differences within the entire tumor. Similarly, overly restricting the number of features as in [26] may under-utilize the full capacity of a radiomic analysis. Comparison of algorithm's performance with radiologist assessment [30,33,35] or clinical parameters [31] is essential to benchmark the performance of radiomics methods and to understand their potential in relation to existing clinical measures. Although radiologist assessment can be subject to inter-rater variability, combining clinically used quantitative assessments or histological measures as in [28] can potentially aid in improving radiologist assessment and improving the accuracy of diagnosis.

19.3 ROBUST RADIOMICS ANALYSIS

Radiomics analysis involves multiple steps starting from data acquisition, segmentation, feature extraction, and outcomes modeling. In this chapter, we will focus on two important issues in the radiomic analysis for gynecologic cancers, in particular, segmentation and robust feature extraction. Both of these steps are critical for achieving robust outcomes modeling [5,9].

19.3.1 SEGMENTATION

Segmentation is the critical step in radiomics analysis. Repeatable and reasonably accurate segmentation are essential to obtain repeatable radiomic analysis. Most radiomic analyses employ manual delineation owing to difficulty of obtaining reasonably accurate automatic or semi-automatic segmentations. However, manual segmentations are prone to high inter-rater variability compared to semi-automatic methods. For example, [41] has shown that radiomics measures extracted from semi-automatic grow-cut [42] were more repeatable compared to features extracted from manually delineated contours. Furthermore, fixed threshold-based methods used in PET segmentation particularly for the extraction of metabolic tumor volume (MTV) have been shown to be less accurate compared to iterative adaptive threshold selection for cervical cancers [43].

As the focus of majority of radiomics methods are in finding associations between radiomics-based markers and outcomes, developing appropriate segmentation algorithms themselves for each application is difficult. We suggest some open-source segmentation methods available in 3D-Slicer software, which may be suitable for some of the radiomics applications. Previously, grow-cut has been applied to brain glioma from MRI [42], lung tumor [41] segmentations from CT images, and has found to be relatively simple to use and useful for radiomics analysis. Figure 19.2 shows segmentation of two different uterine leiomyosarcomas using the grow-cut segmentation together with the expert delineated ground truth. As seen, the algorithm achieved reasonably accurate segmentations for both tumors with accuracies over 0.8 Dice overlap coefficient.

Besides the grow-cut method, 3D Slicer has multiple algorithms including marker-controlled watershed, fast grow-cut [44], robust statistics segmenter [45], and basic methods including Otsu thresholding and simple region growing, all of which can use the same inputs consisting of foreground (tumor) and background labels provided by the user. Additional methods are available as Slicer extensions. The Editor module in 3D Slicer, shown in Figure 19.3 provides interface to multiple segmentation methods. Finally, 3D Slicer also provides segmentation-editing tools with functionalities for performing morphological open and close operations to remove excess segmentations and fill in holes in the segmentation.

All the aforementioned algorithms in 3D Slicer are applicable to MR and CT images. 3D Slicer also has an algorithm to specifically segment tumors from PET images called PET Tumor Segmentation Effect available in the Editor module and downloadable as an extension that can segment tumors and lymph nodes from PET images using a semi-automated method [46].

As any segmentation method, all open source methods mentioned above have been developed for different cancers and typically evaluated using small datasets. Therefore, it is necessary to first evaluate how these algorithms perform to the specific application before using them directly for radiomics analysis.

19.3.2 ROBUST FEATURE EXTRACTION

Radiomics analysis involves extracting quantitative measures of tumor biology to characterize the phenotypic differences between tumor subtypes and outcomes. Despite promising results for a variety of cancers, large number of methods are lacking in standardized evaluation of their performance, reproducibility

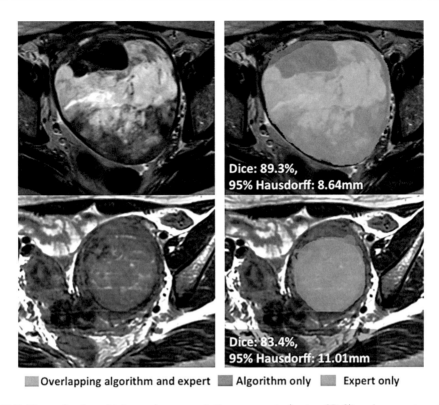

■ Overlapping algorithm and expert ■ Algorithm only ■ Expert only

Figure 19.2 (See color insert.) Example segmentations generated using 3D Slicer (grow-cut method) for uterine leiomyosarcomas from T2w MRI. The overlapping portion between the algorithm and the expert delineated manual segmentation (blue), the algorithm only (in brown), and the expert only portion (in green) are also shown along with the segmentation concordance computed as Dice overlap score and 95% Hausdorff metric between the algorithm and manual delineation.

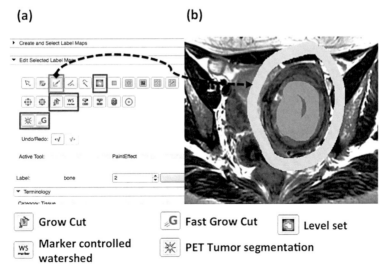

Figure 19.3 (See color insert.) Example segmentation methods available through the Editor effect in 3D Slicer. (a) Editor panel in 3D Slicer and (b) user inputs produced using paint effect.

across multiple datasets, and ultimately limited in clinical utility [47]. This in turn diminishes confidence in implementing radiomics measures as biomarkers in the clinic. Larger studies involving data across multiple institutions require repeatable and reproducible measures. Image features with higher reproducibility will be able to distinguish small differences between cancers better than features with poor reproducibility [48].

19.3.2.1 INFLUENCE OF IMAGE ACQUISITION AND SEGMENTATION ON TEXTURE ROBUSTNESS

One source of variability is the segmentation process itself. Employing automatic or semi-automatic methods can reduce the variability resulting from fully manual segmentations. However, semi-automatic segmentation methods are impacted by user differences as they try to adjust the segmentation according to the user preference. Therefore, for robust texture measurement, one approach might be to combine multiple segmentation methods and use a consensus measure or perturb the same algorithm using noise to generate multiple hypotheses segmentations. However, depending on tumors themselves, most importantly the size, small changes to segmentations may not impact the radiomics measures that are computed as an average over the whole tumor, while robust segmentation is essential for small tumors where small changes can alter the radiomics measures dramatically.

A second source of variation in the radiomics measures results from differences in the images themselves due to differences in acquisition protocols across different institutions and other factors including variable slice thickness, position, and use of different body coils in the case of MR image acquisitions, contrast administration, and reconstruction algorithms. Recent studies have shown that radiomics analysis performed with variable slice thicknesses have a large impact on radiomic features [48], and that resampling CT images to a common voxel size or normalizing features by voxel size has been shown to improve feature robustness [49]. Feature standardization is particularly important when using images produced from different scan manufacturers and with widely different reconstruction methods. The study in [50] showed that feature robustness was most affected by the reconstruction methods for cervical cancers imaged using 18F-FDG PET scans. The study in [51] reported a similar finding where the radiomic features were most influenced by differences in slice thickness and CT scanners compared to variation of the bin width used for discretization of gray levels to compute the higher order texture measures. The study in [50] also investigated the effect of different segmentation methods including fully automatic unsupervised clustering, supervised machine-learning-based segmentation, and semi-automatic interactive methods on the robustness of the radiomics measures and found that gray-level size-zone measures (GLSZM) were most impacted by differences in segmentation. Furthermore, 55% of the frequently used gray-level co-occurrence matrix (GLCM) features were not reproducible with different segmentations.

19.3.2.2 INFLUENCE OF IMAGE PREPROCESSING ON TEXTURE ROBUSTNESS

Besides resampling of images to produce a common voxel size, image preprocessing to reduce noise can also impact the robustness of the texture measures. In particular, the choice of preprocessing technique can impact the robustness of extracted features. The study in [52] analyzed 165 texture measures including the typically used intensity-based histogram features (IHF), Haralick textures, gray-level run-length (GLRL), Law's textures, 2D-wavelet transform-based features, local binary patterns (LPB), and other features including orthonormal Stockwell transform, and 2D Gabor edge features using planning CT (pCT) and cone beam CT (cbCT) images. As shown in [52], whereas low-pass filtering using Gaussian filters and noise did not impact feature robustness, features robustness was impacted when using edge-based feature images computed using Laplacian-of-Gaussian (LoG) filters. Furthermore, Law's and GLRL features were the least robust in both the imaging modalities while the local binary patterns features were the most insensitive to image variations.

19.3.2.3 IMPACT OF IMAGE PREPROCESSING ON CLASSIFICATION ACCURACY: EXAMPLE FROM CLASSIFICATION OF ATYPICAL LEIOMYOMA AND LEIOMYOSARCOMA

Although MRI offers better soft-tissue contrast compared to CT images, radiomics analysis using MR images is challenging particularly because MR images are less standardized compared to the CT images and show variations from a range of factors including scan manufacturers, magnet strengths, field of view, and imaging matrix size. Therefore, using robust features is extremely important when performing radiomics analysis using MRI. We demonstrate the impact of image preprocessing in improving the classification performance using T2w MRI for distinguishing between patients with benign atypical leiomyoma from malignant leiomyosarcoma as used in [35].

We analyzed T2w MR images from 24 patients with uterine leiomyosarcoma (LMS) ($N = 12$) and uterine leiomyoma (ALM) ($N = 12$) scanned using GE ($N = 16$) or Siemens ($N = 8$) scanners and acquired from multiple institutions. The images were affected by motion resulting in mild to moderate motion artifacts in some or most slices containing the tumor. We studied the feasibility of distinguishing the tumors using texture measures extracted with: (a) no preprocessing, (b) preprocessing using bias field correction with histogram standardization, (c) low-pass Gaussian filtering with a bandwidth ($\lambda = 1.414$), and (d) mean filtering using a radius of ($1 \times 1 \times 1$). Both Gaussian and mean filtering were applied to images preprocessed using bias field correction followed by histogram standardization.

We chose low-pass filters such as the Gaussian and mean filters to reduce the impact of sharp edges from motion. Edge-preserving smoothing filters including bilateral filtering [53] and anisotropic diffusion [54] eliminate soft textures as all low-pass filters, but can result in "gradient reversal" that leads to spurious edges near strong edges due to lack of pixels with similar intensity distribution adjacent to edge pixels. Therefore, care needs to be taken in the choice of smoothing filters depending on the particular problem.

Histogram standardization is a process for normalizing the histogram of image signal intensities in one image using the signal intensity distribution in a reference image. Insight ToolKit (ITK) has a histogram standardization method based on the method in [36] that was originally developed for standardizing MR signal intensities.

Figure 19.4 shows T2w MRI from two patients with LMS and scanned using different scan manufacturers and the effect of preprocessing using histogram standardization in (A) and the effect of image smoothing using Gaussian and mean filters in (B). As shown, histogram standardization is performed using a target image (shown in green solid line in Figure 19.4a). Both Gaussian and mean filtering reduce high frequency variations from both images (Figure 19.4b). Whereas both Gaussian and mean filtering reduce the effect of motion artifacts, mean filtering preserves some of the edge and textural information within the tumors in comparison with excessive smoothing of those edges when using Gaussian filtering (Figure 19.4b).

Texture features consisting of intensity image histogram features ($n = 4$), Haralick textures (energy, entropy, contrast, homogeneity, correlation), Gabor edge features at two different orientations ($\Theta = 0°,90°$), and Haralick texture measures from the Gabor edge images were computed for all patients subjected to various image preprocessing. Table 19.2 shows the impact of the various preprocessing methods on the features computed from images acquired from different scanners. As shown, preprocessing with mean filtering following the bias field correction and histogram standardization results in lowest overall coefficient of variation (CV = 770) compared with no preprocessing (CV = 1680), BC/HS (CV = 959), and BC/HS/GF (CV = 794) for features extracted from patients with benign ALM. Furthermore, with the exception of skewness, none of the texture measures computed using BC/HS/Mean filtering approach showed any significant differences between the scanners. In comparison, all of the intensity-based histogram measures except the mean signal intensity and both Gabor edge-based entropy textures showed significant differences between scanners for textures extracted from unprocessed images.

We demonstrate the utility of image preprocessing on an example radiomics task for differentiating between ALM from LMS cancers. First, principal component analysis was used to reduce the dimensionality

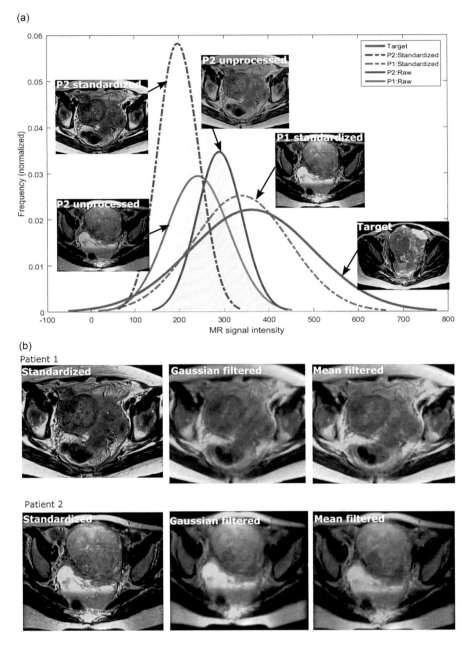

Figure 19.4 (See color insert.) Effect of image preprocessing including (a) histogram standardization and (b) smoothing.

of features computed from each preprocessing step. Next, a linear support vector machine classifier was trained with leave one-out-cross validation using the five main principal components. Table 19.3 shows the result of classification using each of the preprocessing steps. We chose a linear SVM to limit the number of hyper-parameters used for classifying a small dataset. As seen, addition of preprocessing improves the classification accuracy compared with using the unprocessed image features.

Table 19.2 Differences in texture features extracted from ALM patients due to preprocessing. Coefficient of variation in features extracted using various preprocessing methods for patients scanned from GE scanners (a), and results of Wilcoxon tests applied to texture measures extracted from various preprocessing methods between patients scanned with GE and Siemens scanners (b). No adjustment for multiple comparisons was performed to assess the impact of scanner differences on the features

Feature	(a) Coefficient of variation for features from ALM patients scanned with GE scanners				(b) GE vs. Siemens p values (unadjusted)			
	NP	BC/HS	BC/HS/ GF	BC/HS/ Mean	NP	BC/HS	BC/ HS/GF	BC/HS/ Mean
Energy	36.7	37.0	20.7	31.0	0.71	0.53	0.27	0.45
Contrast	100.5	142.8	145.5	173.1	0.40	0.40	**0.04**	0.38
Homogeneity	69.7	77.1	58.4	65.9	0.27	0.27	**0.04**	0.61
Entropy	26.3	36.0	40.8	30.5	1.0	0.71	0.53	0.45
Correlation	6.3	7.0	2.9	5.2	0.40	0.40	0.27	0.32
Gab (0°) Mean	40.9	21.8	23.7	24.7	0.40	0.89	0.89	0.57
Gab (0°) Energy	15.4	50.1	26.1	31.3	0.40	0.71	0.53	0.93
Gab (0°) Contrast	11.7	11.0	8.6	10.5	0.71	0.40	0.53	0.38
Gab (0°) Homogeneity	9.5	13.0	10.8	14.3	0.89	0.89	0.27	0.38
Gab (0°) Entropy	8.0	4.6	6.0	4.5	**0.04**	0.53	1.0	0.70
Gab (0°) Correlation	6.3	13.3	10.3	11.9	0.53	0.71	0.40	0.88
Gab (90°) Mean	44.7	22.9	18.1	17.4	0.71	0.53	0.53	0.83
Gab (90°) Energy	13.1	46.0	40.8	43.0	0.18	1.0	1.0	0.61
Gab (90°) Contrast	8.5	13.1	8.6	10.3	0.53	0.71	0.53	0.98
Gab (90°) Homogeneity	11.0	15.5	10.8	13.7	0.71	0.31	0.27	0.14
Gab (90°) Entropy	8.1	3.7	11.9	6.6	**0.04**	**0.09**	0.71	0.53
Gab (90°) Correlation	5.7	15.0	13.5	14.1	0.40	0.18	0.53	0.45
MR Mean	106.2	53.4	54.0	54.9	0.27	0.53	0.71	0.93
MR SD	92.5	61.6	64.1	56.4	**0.09**	0.71	0.71	0.98
MR Kurtosis	20.2	36.2	29.1	39.5	**0.04**	**0.04**	**0.04**	0.17
MR Skewness	1038.7	278.0	188.6	121.0	**0.09**	**0.04**	**0.09**	**0.03**

Table 19.3 Classification performances for distinguishing between ALM vs. LMS using linear support vector machine classifiers trained with leave one-out cross validation using PCA features computed with no preprocessing, bias field correction and histogram standardization (BC/HS), and bias field correction, histogram standardization and gaussian filtering (BC/HS/GF), and with bias field correction, histogram standardization and mean filtering (BC/HS/MF)

Method	TPR	TNR	FPR	FNR	AUC 95% CI
No preprocessing	1.0	0.58	0.42	0.0	0.79 (0.65–0.94)
BC/HS	0.83	0.83	0.17	0.17	0.83 (0.68–0.99)
BC/HS/GF	0.83	0.75	0.25	0.17	0.79 (0.62–0.96)
BC/HS/MF	0.92	0.75	0.25	0.08	0.83 (0.68–0.98)

19.3.3 SUMMARY

In summary, it is essential to use robust features to ensure repeatable performance of radiomics analysis. Number of factors starting from differences in image acquisition, stability of image segmentation methods, and image preprocessing methods impact the accuracy and repeatability of radiomics analysis. Although it is relatively easy to extract several hundred features from images, computing stable features and performing feature selection prior to outcomes modeling is crucial. Furthermore, semi-automatic and fully automatic segmentation methods are preferable to manual segmentations as such methods reduce inter-rater variability and can potentially speed up volumetric delineation when analyzing larger datasets.

Particular care needs to taken to prevent over-fitting when using small datasets. Z-score normalization of extracted features prior to analysis especially when using support vector machine classifiers [31] is essential to ensure meaningful generalization using SVM classifiers. Although radial basis function (RBF) or Gaussian kernels are known to produce better accuracy compared to linear SVM classifiers, as RBF kernels need one additional parameter for fitting, it's very easy to produce overfitting when using small datasets with large number of features. The choice of parameters for generating textures including bin sizes such as for gray-level co-occurrence matrices can lead to different results [31]. Selecting the best bin size for any single application would require using a calibration or discovery dataset, which is not always available. In general, smaller number of bins results in over-averaging of the textural differences within a structure while large number of bins may capture noise in texture calculations.

Finally, where possible, performing internal and external validation of the learned models will help determine the clinical potential of the radiomics methods particularly when using MR images particularly when using images obtained from different scanners as in [28].

19.4 FUTURE WORK AND CONCLUSIONS

Most radiomics analysis is restricted to extracting features from pre-treatment images. Measuring temporal changes in image heterogeneity can be leveraged to obtain better prediction of outcomes as shown in [33]. Radiomic analysis using combination of multiple imaging modalities including MR and PET can help to combine features that characterize the anatomic and functional variations in those images. Almost all works in radiomics focus on extracting image-based features for predicting treatment outcomes or histology. However, very little progress has been made in understanding the relation between the imaging heterogeneity and the underlying tumor biology. Works on examining the relation between imaging heterogeneity and gene expression are limited mostly due to data size limitations. There is a need for reliable markers of early treatment response for novel treatments including immunotherapy. Works such as in [55,56] are a step toward combining image-based measures with underlying tumor biology and ultimately correlating with treatment outcomes. There is a significant variability in the workflow and the features used in radiomics analysis. As pointed out in [50], the formulae used for computing features with the same names differ, which makes it difficult to interpret and compare results from different studies. More importantly, robust and repeatable features should be used for the analysis. Robust approaches for feature extraction including using semi-automatic segmentation, performing reproducibility analysis, and using cross-validation and internal and external validation where possible [47] is essential to translate radiomics from research into clinical practice. Comparison of radiomics measures using clinically used measures as benchmarks is important to help assess the real utility of radiomics analysis for different applications. Finally, combining clinically used measures including qualitative measures with radiomics measures may be potentially useful in improving outcomes prediction.

REFERENCES

1. Gerlinger M., Rowan A.J., Horswell S., Larkin J., Endesfelder D., Gvonroos E. et. al, Intratumor heterogeneity and branched evolution revealed by multiregion sequencing. *N Engl J Med* , 2012. **366**(10): p. 883–892.

2. Cooke S.L., Ng C.K.Y., Melnyk N., Garcia M.J., Hardcastle T., Temple J., Langdon S., Huntsman D., Brenton J.D., Genomic analysis of genetic heterogeneity and evolution in high-grade serous ovarian cancer. *Oncogene*, 2010. **29**(35): p. 4905–4913.

3. Gatenby R.A., Grove O., Gillies R.J., Quantitative imaging in cancer evolution and ecology. *Radiology*, 2013. **269**(1): p. 8–15.

4. Sala E. Mema E., Himoto Y., Veeraraghavan H., Brenton J.D., Snyder A., Weigelt B., Vargas H.A., Unravelling tumour heterogeneity using next-generation imaging: Radiomics, radiogenomics, and habitat imaging. *Clin Radiol*, 2017. **72**(1): p. 3–10.

5. Gillies R.J., Kinahan P., Hricak H., Radiomics: Images are more than pictures, they are data. *Radiology*, 2016. **278**(2): p. 563–577.

6. Aerts H.J.W.L., Velazquez E.R., Leijenaar R.T.H., Parmar C., Grossman P., Carvalho S., Bussink J. et al., Decoding tumor phenotype by noninvasive imaging using a quantitative radiomics approach. *Nat Commun*, 2014. **5**: p. 4006 EP.

7. Fehr D., Veeraraghavan H., Wibmer A., Gondo T., Matsumoto K., Vargas H.A., Sala E., Hricak H., Deasy J.O., Automatic classification of prostate cancer Gleason scores from multiparametric magnetic resonance images. *Proc Nat Academy Sciences*, 2015. **112**(46): p. E6265–E6273.

8. Grove O., Berglund A.E., Schabath M.B., Aerts H.J.W.L., Dekker A., Wang H., Velazquez E.R. et al., Quantitative computed tomographic descriptors associate tumor shape complexity and intratumor heterogeneity with prognosis in lung adenocarcinoma. *PLoS One*, 2015. **10**(3): p. e0118261.

9. Lambin P., Rios-Velazquez E., Leijenaar R., Carvalho S., Van Stiphout R.G. Granton P., Zegers C.M. et al., Radiomics: Extracting more information from medical images using advanced feature analysis. *Eur J Cancer*, 2012. **48**(4): p. 441–446.

10. Vargas H.A., Huang E.P., Lakhman Y., Ippolito J.E., Bhosale P., Mellnick V., Shinagare A.B. et al., Radiogenomics of high-grade serous ovarian cancer: Multireader multi-institutional study from the cancer genome atlas ovarian cancer imaging research group. *Radiology*, 2017. **285**(2): p. 482–492.

11. Vargas H.A., Miccò M., Hong S.I., Goldman D.A., Dao F., Weigelt B., Soslow R.A., Hricak H., Levine D.A., Sala E., Association between morphologic CT imaging traits and prognostically relevant gene signatures in women with high-grade serous ovarian cancer: A hypothesis-generating study. *Radiology*, 2015. **274**(3): p. 742–751.

12. Nougaret S., Lakhman Y., Gönen M., Goldman D.A., Miccò M., D'anastasi M., Johnson S.A. et al., High-grade serous ovarian cancer: Associations between BRCA mutation status, CT imaging pheno-types, and clinical outcomes. *Radiology*, 2017. **285**(2): p. 472–481.

13. Nougaret S., Reinhold C., Alsharif S.S., Addley H., Arceneau J., Molinari N., Guiu B., Sala E., Endometrial cancer: Combined MR volumetry and diffusion-weighted imaging for assessment of myo-metrial and lymphovascular invasion and tumor grade. *Radiology*, 2015. **276**(3): p. 797–808.

14. Halle C., Andersen E., Lando M., Aarnes E.K., Hasvold G., Holden Syljuåsen R.G., Sundfør K., Kristensen G.B., Holm R., Malinen E., Lyng H., Hypoxia-induced gene expression in chemoradioresistant cervical cancer revealed by dynamic contrast-enhanced MRI. *Cancer Res*, 2012. **72**(20): p. 5285–5295.

15. Andersen E.K., Hole K.H., Lund K.V., Sundfør K., Kristensen G.B., Lyng H., Malinen E., Pharmacokinetic parameters derived from dynamic contrast-enhanced MRI of cervical cancers predict chemoradiother-apy outcome. *Radiother Oncol*, 2013. **107**(1): p. 117–122.

16. Lund K.V., Simonsen T.G., Hompland T., Kristensen G.B., Rofstad E.K., Short-term pretreatment DCE-MRI in prediction of outcome in locally advanced cervical cancer. *Radiother Oncol*, 2015. **115**(3): p. 379–385.

17. Zahra M.A., Tan L.T., Priest A.N., Graves M.J., Arends M., Crawford R.A., Brenton J.D., Sala E., Semiquantitative and quantitative dynamic contrast-enhanced magnetic resonance imaging measure-ments predict radiation response in cervix cancer. *Int J Radiat Oncol Biol Phys*, 2009. **74**(3): p. 766–773.

18. Dappa E., Eiger T., Hasenburg A., Düber C., Battista M.J., Hötker A.M., The value of advanced MRI techniques in the assessment of cervical cancer: A systematic review. *Insights Imaging*, 2017 **8**(5): p. 471–481.

19. Yang W., Qiang J.W., Tian H.P., Chen B., Wang A.J., Zhao J.G., Multi-parametric MRI in cervical cancer: Early prediction of response to concurrent chemoradiotherapy in combination with clinical prognostic factors. *Eur Radiol*, 20182 **8**(1): p. 437–445.

20. Harry V.N., Semple S.I., Gilbert F.J., Parkin D.E., Diffusion-weighted magnetic resonance imaging in the early detection of response in chemoradiation in cervical cancer. *Gynecol Oncol*, 2008. **111**(2): p. 213–220.

21. Sala E., Kataoka M.Y., Priest A.N., Gill A.B., Mclean M.A., Joubert I., Graves M.J. et al., Advanced ovarian cancer: Multiparametric MR imaging demonstrates response- and metastasis-specific effects. *Radiology*, 2012. **263**(1): p. 149–159.

22. Sala E., Kataoka M.Y., Pandit-Taskar N., Ishill N., Mironov S., Moskowitz C.S., Mironov O., Collins M.A., Chi D.S., Larson S., Hricak H., Recurrent ovarian cancer: Use of contrast-enhanced CT and PET/CT to accurately localize tumor recurrence and to predict patients' survival. *Radiology*, 2010. **257**(1): p. 125–134.

23. Sarabhai T., Schaarschmidt B.M., Wetter A., Kirchner J., Aktas B., Forsting M., Ruhlmann V., Herrmann K., Umutlu L., Grueneisen J., Comparison of 18F-FDG PET/MRI and MRI for pre-therapeutic tumor staging of patients with primary cancer of the uterine cervix. *Eur J Nucl Med Mol Imaging*, 2018 **45**(1): p. 67–76.

24. Miccò M., Vargas H.A., Burger I.A., Kollmeier M.A., Goldman D.A., Park K.J., Abu-Rustum N.R., Hricak H., Sala E., Combined pre-treatment MRI And 18F-FDG PET/CT parameters as prognostic biomarkers in patients with cervical cancer. *Eur J Radiol*, 2014. **83**(7): p. 1169–1176.

25. Bowen S.R. Yuh W.T.H., Hippe D.S., Wu W., Patridge S.C., Elias S., Jia G. et al., Tumor radiomic heterogeneity: Multiparametric functional imaging to characterize variability and predict response following cervical cancer radiation therapy. *J Magn Reson Imaging*, 2018. **47**(5): p. 1388–1396.

26. Mu W., Chen Z., Liang Y., Shen W., Yang F., Dai R., Wu N., Tian J., Staging of cervical cancer based on tumor heterogeneity characterized by texture features on 18F-FDG PET images. *Plys Med Biol*, 2015. **60**(13): p. 5123–5139.

27. Tsujikawa T., Rahman T., Yamamoto M., Yamada S., Tsuyoshi H., Kiyono Y., Kimura H., Yoshida Y., Okazawa H., 18F-FDG PET radiomics approaches: Comparing and clustering features in cervical cancer. *Ann Nucl Med*, 2017. **31**(9): p. 678–685.

28. Schernberg A., Reuze S., Orlhac F., Buvat I., Dercle L., Sun R., Limkin E. et al., A score combining baseline neutrophilia and primary tumor SUVpeak measured from FDG PET is associated with outcome in locally advanced cervical cancer. *Eur J Med Mol Imaging*, 2018. **45**(2): p. 187–195.

29. Reuze S., Orlhac F., Chargari C., Nioche C., Limkin E., Riet F., Escande A. et al., Prediction of cervical cancer recurrence using textural features extracted from 18F-FDG PET images acquired with different scanners. *Oncotarget*, 2017. **8**(26): p. 43169–43179.

30. Ueno Y., Forghani B., Forghani R., Dohan A., Zeng X.Z., Chamming's F., Arseneau J. et al., Endometrial carcinoma: MR imaging-based texture model for preoperative risk stratification-A preliminary analysis. *Radiology*, 2017. **284**(3): p. 748–757.

31. Torheim T., Malinen E., Kvaal K., Lyng H., Indahl U.G., Andersen E.K., Futsaether C.M., Classification of dynamic contrast MR images of cervical cancers using texture analysis and support vector machines. *IEEE Trans on Med Imaging*, 2014. **33**(8): p. 1648–1656.

32. Guan Y., Li W., Jiang Z., Chen Y., Liu S., He J., Zhou Z., Ge Y., Whole-lesion apparent diffusion coefficient-based entropy-related parameters for characterizing cervical cancers: Initial findings. *Acad Radiol*, 2016. **23**(12): p. 1559–1567.

33. Danala G., Thai T., Camille C.G., Moxley K.M., Moore K., Mannel R.S., Liu H., Zheng B., Qiu Y., Applying quantitative CT image feature analysis to predict response of ovarian cancer patients to chemotherapy. *Acad Radiol*, 2017. **24**(10): p. 1233–1239.

34. Vargas H.A, Veeraraghavan H., Micco M., Nougaret S., Lakhman Y., Meier A.A., Sosa R. et al., A novel representation of inter-site tumor heterogeneity from pre-treatment computed tomography textures classifies ovarian cancers by clinical outcome. *Eur Radiology*, 2017. 27(9): p. 3991–4001.

35. Lakhman Y., Veeraraghavan H., Chaim J., Feier D., Goldman D.A., Moskowitz C.S., Nougaret S. et al., Differentiation of uterine leiomyosarcoma from atypical leiomyoma: Diagnostic accuracy of qualitative MR imaging features and feasibility of texture analysis. *Eur Radiol*, 2017. **27**(7): p. 2903–2915.

36. Laszlo G.N., Udupa J.K., and Zhang X., New variants of a method of MRI scale standardization. *IEEE Trans on Med Imaging*, 2000. **19**(2): p. 143–150.

37. Vargas H.A., Veeraraghavan H., Micco M., Nougaret S., Lakhman Y., Meier A.A., Sosa R. et al., A novel representation of inter-site tumour heterogeneity from pre-treatment computed tomography texture classifies ovarian cancers by clinical outcome. *Eur Radiol*, 2017. **27**(9): p. 3991–4001.

38. Schwarz R.F., Ng C.K.Y., Cooke S.L., Newman S., Temple J., Piskorz A.M. et. al, Spatial and temporal heterogeneity in high grade serous ovarian cancer: A phylogenetic analysis. *PLoS Med*, 2015. **12**(2): p. e1001789.

39. O'Connor J.P.B., Rose C.J., Waterton J.C., Carano R.A.D., Parker G.J.M., Jackson A. Imaging intratumor heterogeneity: Role in therapy response, resistance, and clinical outcome. *Clin Cancer Res*, 2015. **21**(2): p. 249–257.

40. Patch A.M., Christie E.L., Etemadmoghadam D., Garsed D.W., George J., Fereday S., Nones K. et al., Whole-genome characterization of chemoresistant ovarian cancer. *Nature*, 2015. **521**(7553): p. 489–494.

41. Parmar C., Rios-Velazquez E., Leijenaar R., Jermoumi M., Carvalho S., Mak R.H., Mitra S. et al., Robust radiomics feature quantification using semiautomatic volumetric segmentation. *PLoS One*, 2014. **9**(7): p. e102107.

42. Egger J., Kapur T., Fedorov A., Pieper S., Miller J.V., Veeraraghavan H., Friesleben B., Golby A.J., Nimsky C., Kikinis R., GBM volumetry using 3D Slicer medical image computing platform. *Sci Rep*, 2013. **3**: p. 1364.

43. Xu W., Yu S., Ma Y., Liu C., Xin J., Effect of different segmentation algorithms on metabolic tumor volume measured on 18F-FDG PET/CT of cervical primary squamous cell carcinoma. *Nucl Med Commun*, 2017. **38**(3): p. 259–265.

44. Zhu L., Kolesov I., Gao Y., Kikinis R., Tannenbaum A. An effective interactive medical image segmentation method using fast growcut. In *Intl Conf Med Image Comput Assist Interv Workshop on Interactive Methods*. 2014. Boston, MA: MICCAI 2014.

45. Gao Y., Kikinis R., Bouix S., Shenton M., Tannenbaum A., A 3D interactive multi-object segmentation tool using local robust statistics driven active contours. *Med Image Anal*, 2012. **16**(6): p. 1216–1227.

46. Beichel R.R., Van Tol M., Ulrich E.J., Bauer C., Chang T., Plichta K.A., Smith B.J., Sunderland J.J., Graham M.M., Sonka M., Buatti J.M., Semiautomated segmentation of head and neck cancers in 18F-FDG PET Scans: A just-enough-interaction approach. *Med Phys*, 2016. **43**(6): p. 2948–2964.

47. Lambin P., Leijenaar R.T.H., Diest T.M., Peerlings J., De Jong E.E.C., Van Timmeren J., Sanduleanu S. et al., Radiomics: The bridge between medical imaging and personalized medicine. *Nat Rev Clin Oncol*, 2017. 14(12): p. 749.

48. Zhao B., Tan Y., Tsai W.-Y., Qi J., Xie C., Lu L., Schwartz L.H., Reproducibility of radiomics for deciphering tumor phenotype with imaging. *Sci Rep*, 2016. **6**: p. 23428.

49. Shafiq-Ul-Hassan M., Zhang G.G., Latifi K., Ullah G., Hunt D.C., Balagurunathan Y., Abdalah M.A. et al., Intrinsic dependencies of CT radiomic features on voxel size and number of gray levels. *Med Phys*, 2017. **44**(3): p. 1050–1062.

50. Altazi B.A., Zhang G.G., Fernandez D.C., Montejo M.E., Hunt D., Wener J., Biagioli M.C., Moros E.G., Reproducibility of F18-FDG PET radiomic features for different cervical tumor segmentation methods, gray-level discretization, and reconstruction algorithms. *J Appl Clin Med Phys*, 2017. 18(6):p. 32–48.

51. Larue R.T.H.M., Van Timmeren J.E., De Jong E.E.C., Feliciani G., Leijennar R.T.H., Schereurs W.M.J., Sosel M.N. et al., Influence of gray level discretization on radiomic feature stability for different CT scanners, tube currents and slice thicknesses: A comprehensive phantom study. *Acta Oncol*, 2017: p. 1–10.

52. Bagher-Ebadian H., Siddiqui F., Liu C., Movsas B., Chetty I.J., On the impact of smoothing and noise on robustness of CT and CBCT radiomics features for patients with head and neck cancers. *Med Phys*, 2017. **44**(5): p. 1755–1770.

53. Tomasi C., Manduchi R.. Bilateral filtering for gray and color images. In *Sixth International Conference on Computer Vision*. 1998. Bombay, India: IEEE.

54. Perona P., Malik J., Scale space and edge detection using anisotropic diffusion. IEEE *Trans Pattern Anal Machine Intel*, 1990. **12**(7): p. 629–639.

55. Jiménez-Sánchez A., Memon D., Pourpe S., Veeraraghavan H., Li Y., Vargas H.A., Gill M.B. et al., Heterogeneous tumor-immune microenvironments among differentially growing metastases in an ovarian cancer patient. *Cell*, 2017. **170**(5): p. 927–938.

56. Reuben A., Spencer C.N., Prieto P.A., Gopalakrishnan V., Reddy S.M., Miller J.P., Mao X. et al., Genomic and immune heterogeneity are associated with differential responses to therapy in melanoma. *NPJ Genom Med*, 2017. **2**(1): p. 10.

Applications of imaging genomics beyond oncology

XIAOHUI YAO, JINGWEN YAN, AND LI SHEN

20.1 INTRODUCTION

Imaging genomics is an emerging research field that arises with the recent advances in high-throughput omics data and multi-modal imaging data. This chapter focuses on imaging genomics applications beyond oncology, specifically in brain diseases. The major task of brain imaging genomics is to perform integrative analysis of genomic data and structural, functional, and molecular brain imaging data. Pathological changes in the brain often precede the earliest clinical symptoms [1]. Therefore, brain imaging data can provide essential information regarding disease risk, status, and progression, as the earliest signs of disease may occur in the brain and can be measured before noticeable symptoms. Bridging imaging and genomic factors and exploring their associations have the potential to provide important new insights into the phenotypic characteristics and genetic mechanisms of normal and/or disordered neurobiological structures and functions, which in turn will impact the development of new diagnostic, therapeutic, and preventative approaches.

In this chapter, we first briefly discuss the fundamental principles and technical bases involved in brain imaging genomics, and then present their applications and significant findings in the context of several common neurological and psychiatric disorders.

20.2 FUNDAMENTAL PRINCIPLES AND TECHNICAL BASES

Brain imaging genomics investigates the influence of genetic variations on brain imaging phenotypes. Compared with using categorical disease status as the phenotype, genetic association analyses of quantitative phenotypes have shown distinct advantages in statistical power and mechanistic understanding, especially the usage of brain imaging quantitative traits (iQTs) due to their prominent performance on brain disease differentiation and prediction.

Bioinformatics strategies for brain imaging genomics, which is a relatively young field, have been rapidly evolving. In this section, we will briefly review a few commonly used strategies in brain imaging genomics. Shown in Figure 20.1 is a schematic summary of these strategies.

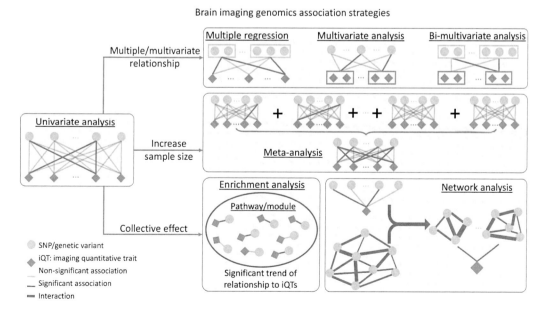

Figure 20.1 (See color insert.) Schematic summary of analytic strategies for association discovery in brain imaging genomics.

Univariate analysis: Early efforts start from examining pairwise univariate associations between single genetic marker and single neuroimaging phenotype. Typically, univariate regression model was used to evaluate the additive effect of each single variant on each single iQT. Genome-wide association study (GWAS), as a well-known implementation of univariate analysis, has been performed in a number of brain disease studies to identify genetic markers such as single nucleotide polymorphisms (SNPs) that are susceptible to brain iQTs [2]. In addition to the genome-wide study, candidate genetic association analysis of brain iQTs has also been investigated to increase statistical power and improve biological interpretation, by including prior knowledge of gene function into study design [3,4].

Pairwise univariate analysis could identify significant SNP-iQT associations in a straightforward way, which, however, ignores the underlying correlations among multiple SNPs and multiple iQTs, considering both genetic variants and brain regions are not independent functional units. In addition, the statistical power of pairwise univariate analysis has suffered from the increasing number of comparisons between high-dimensional imaging and genetic data.

Multivariate analysis: To identify more flexible associations involving multiple genetic markers and multiple imaging phenotypes, recent studies employed several multivariate statistical models, including multiple regression and multivariate multiple regression [5], principle component regression (PCA) [6], canonical correlation analysis (CCA), reduced rank regression (RRR), and so on [7]. Sometimes these models were coupled with powerful machine learning approaches and valuable prior knowledge to discover relevant imaging and genomic features [8–10]. Also, given the high-dimensionality of both imaging and genetic data, sparse models were studied. For example, sparse regression models were designed to select a small number of genetic risk variants [11–13], and sparse canonical correlation analysis models were presented to identify imaging genomic patterns with a small number of SNPs and iQTs [14,15]. In addition, prior knowledge, for example, gene-gene interactions, was incorporated to promote the identification of biologically meaningful imaging genomic associations [16].

Meta-analysis: To increase statistical power, meta-analysis studies can be performed to quantitatively synthesize imaging genomic findings from multiple independent analyses [17–19]. Meta-analysis provides an effective approach to scale up the result to a larger population via combined analysis of individual results instead of actual data across multiple cohorts. Meta-analysis typically increases the yield of signal detection because more samples are included. In meta-analysis, the study diversity (e.g., variability of participants, quality of data, and potential for underlying biases) should be carefully considered, otherwise, the increased statistical power and the meta-analysis results could be compromised.

Pathway enrichment analysis: To identify biologically meaningful findings with increased statistical power, pathway enrichment analysis can be performed to test for collective effects of multiple variants within the same biological pathway [20]. Recently, imaging genetic enrichment analysis (IGEA) [21] was further proposed to mine set-level associations in both imaging and genomic domains. IGEA uses prior knowledge, including meaningful gene sets (GSs) and brain circuits (BCs), to jointly explore the complex relationships between interlinked genetic variants and correlated iQTs. Pathway enrichment analysis provides additional power for extracting biological insights on neurogenetic associations at a systems biology level.

Network enrichment analysis: Genes often do not perform functions separately, instead, they interact with others to jointly affect complex diseases or traits. Given a functional interaction network, functionally interacted genes can form a network module. Module identification strategies have been studied to identify disease-relevant gene sets. A few recent imaging genomic studies have analyzed tissue-specific networks to identify network modules enriched by GWAS findings of iQTs, where tissue-specificity of the iQTs was taken into consideration [22,23].

20.3 CLINICAL APPLICATIONS TO NEUROLOGICAL DISORDERS

Neurological disorders, characterized by progressive nervous cell degeneration, loss, and death, affect up to one billion people worldwide (http://www.who.int/). The causes of neurological disorders can be quite diverse, and the symptoms also vary greatly. It is critical to improve the understanding of molecular mechanisms of neurological disorders due to their high prevalence, social and family burdens, and because they currently cannot be prevented, cured, or even slowed in many cases. A number of disease biomarkers have been identified and measured for indicating and predicting the diseases progression, among which brain imaging and genetic risk profiling have provided promising evidences for differentiating disease staging from normal controls.

In this section, we discuss the research progress of neuroimaging genomics on two most common neurological disorders: Alzheimer's disease and frontotemporal dementia.

20.3.1 ALZHEIMER'S DISEASE

Alzheimer's disease (AD) is a complex degenerative disease of the brain characterized by neurodegeneration, memory impairment, and cognitive problems. AD is the most common form of dementias that accounts for 60%–80% of dementia cases. It is the fifth-leading cause of death among those age 65 and older [24]. It is critical to improve the understanding of molecular mechanism of AD because of its high and increasing prevalence and burdens as well as no available cure. Below, we will briefly summarize genetic findings of case-control studies, and then review imaging genomic studies in AD research and present the representative findings grouped by various imaging modalities.

20.3.1.1 CASE-CONTROL STUDIES

Genetic factors have been investigated in AD research due to their important role in the development and progression of the disease, especially in late-onset AD (LOAD) with a heritability estimate of 58%–79% [25]. Genetic studies have discovered a number of risk and protective loci, from the very early genome-wide linkage studies (GWLS) and linkage disequilibrium (LD) mapping [26] to the recent genome-wide association study (GWAS) [2].

GWAS has been performed in case-control studies to evaluate the associations of genetic variations such as SNPs with AD using categorical diagnosis as the phenotype. These studies have identified a number of susceptible loci, including complement C3b/C4b receptor 1 (CR1), clusterin (CLU), and phosphatidylinositol binding clathrin assembly protein (PICALM) [27], epistatic interaction between transferrin (TF) and hemochromatosis gene (HFE) [28], apolipoprotein E (APOE), and methylenetetrahydrofolate dehydrogenase (NADP + dependent) 1 like (MTHFD1L) [29], CR1 [30], membrane-spanning 4-domain family, subfamily A (MS4A) gene cluster [31], ATP binding cassette subfamily A member 7 (ABCA7), membrane spanning 4-domains A6A (MS4A6A)/membrane spanning 4-domains A4E (MS4A4E), Ephrin receptor A1 (EPHA1), CD33 molecule (CD33) and CD2 associated protein (CD2AP) [32], bridging integrator 1 (BIN1) [33], MS4A4/MS4A6E, CD2AP, CD33, and EPHA [34].

Suggestive novel associations have also been discovered in protein phosphatase 1 regulatory subunit 3B (PPP1R3B) [35] and alpha-ketoglutarate dependent dioxygenase (FTO) [36]. Recently, a large scale meta-analysis of GWAS presented a list of "top" genes susceptible to LOAD, from which 11 genes were reported as new AD susceptibility loci in addition to eight genes previously identified in independent studies [18].

Also, efforts have been made in AD genetic research to investigate the role of various types of genetic variants including rare variants and copy number variants (CNVs) to help improve the understanding of genetic architecture of AD [37,38]. However, besides the well-known APOE genotype explaining an estimated 20% genetic variance of LOAD [18], other genetic findings have suggested only small size of disease effect and have been rarely replicated [39].

20.3.1.2 MULTI-MODAL NEUROIMAGING AS QUANTITATIVE PHENOTYPES

In the study of complex neurological disorders, evidences have shown that the pathological changes begin to develop years or even decades prior to the earliest clinical symptoms emerge [40]. This extended pre-symptomatic stage provides essential information regarding to disease progression, as the earliest signs of diseases may occur in brain and can be measured before noticeable symptoms developed. In AD research, a number of biomarkers have been identified and measured for indicating the disease progression, including brain imaging, protein (beta-amyloid and tau) levels in cerebrospinal fluid (CSF), proteins in blood, genetic risk profiling (e.g., amyloid beta precursor protein [APP], APOE allele ε4). Among various biomarkers, brain imaging—including structural and functional imaging—has provided promising evidences for differentiating disease from normal aging [41].

In the following, we discuss the imaging genomic analyses in AD studies, as well as relevant findings. Among these analyses, various types of imaging biomarkers have been used as endophenotypes in genetic association studies.

Structural neuroimaging: Structural neuroimaging is the most widely studied phenotype category in AD. GWAS of structural magnetic resonance imaging (sMRI) has successfully identified several novel disease-relevant genetic associations with imaging measures at multiple levels including voxel, region of interest (ROI), as well as the whole brain [2,41]. Stein et al. [42] proposed voxelwise GWAS (vGWAS), a massive GWAS to explore the relation between 448,293 SNPs with each of 31,622 brain voxels. They suggested several promising genes for further investigation including CUB and Sushi multiple domains 2 (*CSMD2*) and calcium dependent secretion activator 2 (*CADPS2*).

GWAS of sMRI measures in AD-relevant ROIs also suggested a number of susceptible loci. For example, the hippocampus has been known as being involved at the onset of neuropathological pathways of AD, and GWAS of hippocampal imaging measure has been performed in a number of studies [8–11]. These studies have identified about a dozen genes relevant to volume changes of the hippocampus. These genes include *APOE*, translocase of outer mitochondrial membrane 40 (*TOMM40*), ephrin A5 (*EFNA5*), cullin associated and neddylation dissociated 1 (*CAND1*), membrane associated guanylate kinase, WW and PDZ domain containing 2 (*MAGI2*), arylsulfatase B (*ARSB*), and *PRUNE2* [43]; dipeptidyl peptidase 4 (*DPP4*), astrotactin 2 (*ASTN2*), methionine sulfoxide reductase B3 (*MSRB3*), Wnt inhibitory (*WIF1*), and harakiri, BCL2 interacting protein (*HRK*) [44]; *APOE*, F5, selectin P (*SELP*), LHFPL tetraspan subfamily member (*LHFP*), GC-rich sequence DNA-binding factor 2 (*GCFC2*), PICALM, synaptoporin (*SYNPR*), and tetratricopeptide repeat domain 27 (*TTC27*) [45]; high mobility group AT-hook 2 (*HMGA2*) [46]; and *PICALM* [39]. In addition, Stein et al. [47] performed a GWAS on bilateral temporal lobe volume and identified two genetic associations including rs10845840 from glutamate ionotropic receptor NMDA type subunit 2B (*GRIN2B*) and rs2456930.

GWAS analyses also identified genes associated with other AD-relevant ROIs. These findings include association of WD repeat domain 41 (*WDR41*) with caudate volume [19], association of flavin containing monooxygenase (*FMO*) with lentiform nucleus volume [48], and associations of eight SNPs (including rs945270, rs62097986, rs6087771, rs683250, rs1318862, rs77956314, rs61921502, and rs17689882) with seven subcortical regions [49]. Bakken et al. [50] performed a GWAS on cortical surface and identified common genetic variants in glycerophosphocholine phosphodiesterase 1 (*GPCPD1*) associated with visual cortical surface area. Shen et al. [51] performed a brain-wide ROI level GWAS for investigating genome-wide associations with gray matter (GM) density, volume, and cortical thickness in an AD cohort and confirmed the associations of several known AD genes (e.g., *APOE*, *TOMM40*) with multiple brain regions.

In addition to the genome-wide analysis, targeted or candidate genetic association analysis of brain structural imaging quantitative traits has also been widely investigated to increase statistical power and improve biological interpretation. For example, the associations of *APOE* with multiple structural MRI phenotypes have been examined, since *APOE* is the best-known risk genetic factor for AD [52–59]. Biffi et al. [60] investigated the association of AD candidate SNPs with AD related sMRI measures and confirmed the influence of four genes (*APOE*, *CLU*, *CR1*, and *PICALM*) on multiple regional measures including hippocampal volume, amygdala volume, and several others.

There are also other targeted genetic analyses that have suggested the role of candidate genes on brain structural changes or brain atrophy rates including methylenetetrahydrofolate reductase (*MTHER*) for promoting brain deficits in cognitively impaired elderly [61], *MS4A6A* associated with the atrophy rates of AD-relevant brain regions [62], and the effects of unc-5 netrin receptor C (*UNC5C*) [63] and Major Histocompatibility Complex (*HLA*) [64] variants on brain structures.

Functional neuroimaging: In addition to structural imaging, a number of genetic studies have been performed using neuroimaging measures from functional MRI (fMRI) and positron emission tomography (PET) including 18F-Fludeoxyglucose (FDG-PET), 18F-Florbetapir (AV45-PET), and 11C-Pittsburgh compound B (PiB-PET) imaging as phenotypes to help discover disease risk variants.

Ramanan et al. [65] performed a GWAS using AV45-PET imaging as phenotype in an AD study and identified associations of *APOE* and butyrylcholinesterase (*BCHE*) with deposition of amyloid-β (Aβ) in the cerebral cortex. In addition, Ramanan et al. [66] performed another GWAS on AV45-PET imaging to evaluate the genetic effect on longitudinal amyloid accumulation in an AD study, and discovered interleukin 1 receptor accessory protein (*IL1RAP*).

Given the role of *APOE* ε4 as a risk allele for LOAD, it has also been widely evaluated in functional imaging genomics studies. Murphy et al. [67] identified that the effect of *APOE* ε4 on Aβ plaque density measured from AV45-PET imaging in brain cortical region was twice than its effect on the AD diagnosis. Damoiseaux et al. [68] investigated the influence of *APOE* genotype on brain connectivity in an fMRI study of healthy older subjects. They presented that carriers of *APOE* ε4 allele showed significantly decreased connectivity in the default mode network compared to carriers of *APOE* ε3 homozygotes. Another study of cognitively normal participants from Alzheimer's Disease Neuroimaging Initiative (ADNI) [69] also confirmed that *APOE* ε4 was associated with increased deposition of β-amyloid in an AV45-PET study and with decreased glucose metabolism in an FDG-PET study. Jack et al. [56] examined the shape of five well-established AD biomarkers as a function of Mini-Mental State Examination (MMSE), including CSF t-tau and Aβ42, amyloid deposition from PiB-PET, glucose metabolic rate from FDG-PET, and hippocampal volume from sMRI. This study concluded that the complexity of these biomarker trajectories were affected by age and *APOE* ε4 status. Kiddle et al. [70] evaluated the effect of *APOE* ε4 on plasma *APOE* level and found the associations of *APOE* ε4 status with both the APOE protein level in plasma and brain amyloid burden. In addition, *APOE* ε4 genotype had presented a higher conversion probability from mild cognitive impairment (MCI) to AD and had been suggested for participant selection for clinical trials [71,72].

In addition to *APOE*, several other AD risk genes have also been investigated in a few studies. For example, Xu et al. [73] evaluated the association of Val66Met polymorphism of the brain derived neurotrophic factor (*BDNF*) gene with regional glucose metabolism measured from FDG-PET and identified significant differences in several brain regions in carriers of Met than in non-carriers. In a genetic candidate study of PiB-PET, Thambisetty et al. [74] identified decreased deposition of amyloid in carriers of protective allele of the rs3818361 SNP of the *CR1* gene compared to non-carriers. They also observed the association of *CR1-APOE* interaction with brain amyloid burden. A different study of PiB-PET focused on another AD risk gene, *CD33*, and identified that it was associated with the deposition of amyloid [75]. These imaging genomic findings can help provide more insights on the disease pathogenesis.

High-level imaging genomic association analysis: In addition to the univariate strategy that examines the association between a single genetic variant and a single imaging QT, a substantial amount of effort has been made to explore the high-level imaging genomic associations and to test the collective effect of multiple genes on one or more imaging QTs. Using whole genome sequencing data, Nho et al. [76] investigated the collective effect of presenilin 1 (*PSEN1*) variants on brain atrophy measured by sMRI using a gene-based analysis strategy named sequence kernel association test (SKAT) [77]. The authors identified a significant association of rare, but not common variants of *PSEN1* with bilateral entorhinal cortical thickness.

Pathway enrichment analysis has been applied in the AD-related GWAS and successfully confirmed a few GWAS findings with a few AD-relevant pathways [20]. In addition, several pathway- and network-based GWAS strategies have been recently presented to improve the statistical power of the high-level imaging genomic associations by involving multiple genes and/or multiple ROIs in modern machine learning models [21,22]. By jointly considering the modularity on both imaging and genetic domains, two-dimensional imaging genomic enrichment analysis [21] extracted complex relationships between correlated brain regions and interlinked genetic variants. It may provide more insights into neurogenomic associations from a system biological level.

Recently, network-based analysis has been applied in imaging genomic association analysis and promoted the identification of disease-relevant genes by integrating genetic pathway or interaction network with GWAS analysis. For example, Yao et al. [21,23] performed network-based GWAS by integrating brain tissue-specific network and corresponding tissue-specific GWAS findings, and identified several genetic modules relevant to AD.

20.3.2 FRONTOTEMPORAL DEMENTIA

Frontotemporal dementia (FTD) is the second most common form of dementia characterized by progressive neuronal loss in the brain frontal (the areas behind forehead) or temporal lobes (the regions behind ears). FTD accounts for approximately 20% of young-onset dementia cases that occur more commonly between the ages of 55 and 65 [78]. The symptoms of FTD include significant problems in social and personal behavior, executive dysfunction, emotion, and language. There is also no current cure for FTD.

There are three types of FTD categorized by distinct syndromes, including behavioral variant frontotemporal dementia (bvFTD), primary progressive aphasia (PPA), and motor neuron dementia (FTD-MND). The FTD-MND has overlaps with other three neurological disorders including amyotrophic lateral sclerosis (ALS), corticobasal syndrome (CBS), and progressive supranuclear palsy (PSP). The term frontotemporal lobar degeneration (FTLD) has been used as the neuropathological concept of the FTD and currently been categorized as three subtypes according to specific proteinaceous inclusions including tau (FTLD-tau), TAR DNA-binding protein 43 (FTLD-TDP), and fused in sarcoma (FUS) inclusions (FTLD-FUS).

20.3.2.1 GENETICS OF FRONTOTEMPORAL DEMENTIA

The clinical and pathological heterogeneity of FTD presents challenges for pre-diagnosis and prediction of FTD with its progression. FTD is highly heritable, with approximately 50% cases are familial from the genetic analysis. Furthermore, it has been shown that 10%–20% of FTD cases are caused by common mutations from three genes: microtubule associated protein tau (*MAPT*) [79,80] that is commonly associated with bvFTD [81], progranulin (*GRN*) [82,83] that is also typically associated with bvFTD, and chromosome 9 open reading frame 72 (*C9orf72*) [84,85] that is associated with pure FTD or FTD-MND as well as pure MND [81,86–88].

One GWAS has identified a novel gene transmembrane protein 106B (*TMEM106B*) associated with FTD-TDP, as well as increased risk of FTLD-TDP in individuals with *GRN* mutations [89]. Several other rare FTD risk genes, including charged multivesicular body protein 2B (*CHMP2B*), valosin containing protein (*VCP*), sequestosome 1 (*SQSTM1*), transactive response DNA-binding protein (*TARDP*), *FUS*, Tank-binding kinase 1 gene (*TBK1*), and coiled-coil-helix-coiled-coil-helix domain containing 10 (*CHCHD10*), have also been reported in small family studies [81, 86–88, 90–92]. Diekstra et al. [93] performed a meta-analysis of GWAS and reported two shared genetic associations of FTD and ALS including *C9orf72* and a novel gene unc-13 homolog A (*UNC13A*). Although the above common and rare genetic findings have contributed to the risk of different types of FTLD, some portion of disease heritability is still unexplained. In addition, the molecular mechanism of disease is not fully understood due to limited statistical power of case-control study and complex pathology of FTD.

20.3.2.2 MULTI-MODAL NEUROIMAGING AS QUANTITATIVE PHENOTYPES

Given the potential of neuroimaging biomarkers for capturing presymptomatic abnormality, efforts have been made for discovering and understanding genetic associations with multiple brain imaging modalities in FTD research. A number of imaging genomic studies have shown that the common FTD risk genes (i.e., *MAPT*, *GRN*, and *C9orf72*) related to distinct brain imaging patterns.

Structural neuroimaging: In several sMRI-based imaging genomic studies of FTD, variants of *MAPT* were identified to be associated with symmetrical atrophy of anteromedial temporal and orbitofrontal lobes [94,95], *GRN* variants were associated with asymmetrical atrophy in regions of frontal, temporal, and inferior parietal lobes [94,95], and *C9orf72* expansions were associated with relatively symmetrical atrophy in thalamus and superior cerebellum [96–98]. In addition, a longitudinal sMRI study of FTD showed that atrophy rate in *GRN* mutation group was faster than *C9orf72* expansion group [99]. A large cross-sectional sMRI imaging genomic study demonstrated that brain atrophy of three genetic FTD subgroups all occurred at least 10 years prior to the expected onset of clinical symptoms, where *MAPT* carriers started from hippocampus and amygdala regions, then temporal lobe, and thenceforth the insula; *GRN* carriers started from insula, followed by temporal and parietal lobes, and finally the striatum; and *C9orf72* carriers began from subcortical areas, insula and occipital cortex, thereafter frontal and temporal lobes, and lastly the cerebellum [86,100].

Diffusion tensor imaging: Two diffusion tensor imaging (DTI)-based genetic studies of bvFTD have reported that *MAPT* carriers presented changes in left uncinate fasciculus, whereas *C9orf72* subgroup was associated with corpus callosum, cingulum bundle, and right paracallosal cingulum [101,102].

Functional neuroimaging: FDG-PET-based imaging genomic studies of FTD have shown variable association patterns between common FTD risk genes with brain glucose metabolism: *MAPT* mutations were associated with reduced use of glucose in medial temporal lobe, frontal, and parietal regions; *GRN* mutations were associated with hypometabolism in frontal and temporal regions, and *C9orf72* expansions were associated with reduced metabolism in limbic system, basal ganglia, and thalamus [86].

Arterial spin labeling: Dopper et al. [103] performed a longitudinal arterial spin labeling (ASL) imaging-based genetic study of FTD to examine the associations between cerebral blood flow (CBF) and risk genes and identified the associations of *MAPT* and *GRN* mutations with decreased CBF in frontal, temporal, parietal, and subcortical regions. Recently, genetic analysis of brain functional connectivity measured by rest-state MRI (RS-fMRI) has been evaluated, which identified the association of *C9orf72* expansions with reduced connectivity in salience and sensorimotor networks in bvFTD [17].

Rare risk genes: In addition to the three FTD common risk genes, several studies have been performed to explore the influence of rare FTD genes on brain imaging for improving the understanding of disease pathology. For example, Rohrer et al. [104,105] identified the association of *CHMP2B* mutations with increased whole brain atrophy rate compared to healthy controls (HC). However, most rare risk genes could not be replicated in either case-control studies or imaging genomic studies.

20.4 CLINICAL APPLICATIONS TO PSYCHIATRIC DISORDERS

In this section, we present an overview of neuroimaging genomics research in the context of three major psychiatric disorders, including schizophrenia, bipolar disorder, and depression.

20.4.1 SCHIZOPHRENIA

Schizophrenia is a chronic and severe psychiatric illness characterized by distortion of thoughts and perception, and it usually starts between ages 16 and 30. Common symptoms include hallucination, delusion, failure to concentrate, and confused thoughts. Schizophrenia patients usually also manifest other mental problems such as anxiety and depression [106].

20.4.1.1 GENETICS OF SCHIZOPHRENIA

Genetic factors were found to play a major role in the etiology of schizophrenia. A meta-analysis using pooled data from 12 twin studies estimated the heritability of schizophrenia to be approximately 80% [107]. To date, around 30 schizophrenia-associated loci have been identified through GWAS to play a role in conferring the risk of schizophrenia, such as catechol-O-methyltransferase (*COMT*), Disrupted In Schizophrenia 1 (*DISC1*), regulator of G protein signaling 4 (*RGS4*), neuregulin 1 (*NRG1*), dystrobrevin binding protein 1 (*DTNBP1*), D-amino acid oxidase activator (*DAOA*), phosphodiesterase 4B (*PDE4B*), Dopamine- and cAMP-regulated phosphoprotein, Mr 32 kDa (*DARPP-32*) protein phosphatase 1 regulatory subunit 3B and glutamate metabotropic receptor 3 (*GRM3*) [108]. There are also growing evidences from exome sequencing studies indicating that some risk genes and pathways are affected by both common and rare variants [109], which implies large effects of rare variants on individual risk. This can be best exemplified by 11 large, rare recurrent CNVs and loss-of-function variants in set domain containing 1A, histone lysine methyltransferase (*SETD1A*) [109,110]. Evidences from other exome sequencing studies imply more other rare variants conferring substantial individual risk [111,112]. Despite the remarkable progress in the search for risk genes associated with schizophrenia, translation of genetic associations into targetable mechanisms related to disease pathogenesis remains poorly understood.

20.4.1.2 NEUROIMAGING OF SCHIZOPHRENIA

In addition to genetic data, brain imaging measurements are constantly studied in schizophrenia as intermediate phenotype. Significant differences in both brain structure and function have been observed in people with schizophrenia when compared to healthy individuals. For example, schizophrenia patients have abnormal subcortical brain volumes, e.g., smaller hippocampus, amygdala, thalamus, nucleus accumbens, and intracranial volumes, along with larger pallidum and lateral ventricle volumes. Structural and functional connectivity in the brain is also found to be affected with schizophrenia [113]. It is hypothesized in some theories that schizophrenia is a result of abnormal brain connectivity, which may ultimately lead to failure of some integrative processes in the brain. The perturbation of tract organization in schizophrenia brain is also reflected in some research findings that reduced fractional anisotropy and increased radial diffusivity are usually accompanied with the development of schizophrenia (e.g., [113–115]).

20.4.1.3 MULTI-MODAL NEUROIMAGING AS QUANTITATIVE PHENOTYPES

Considering that both genetic and brain imaging measures are highly associated with schizophrenia, substantial efforts have been later dedicated to imaging genomics of schizophrenia, which aim to explore the genetic effect on the brain structure and function measured by multiple imaging modalities. Since the brain imaging measures intermediate phenotypes that are closer to the molecular mechanisms than disease manifestation and its continuity enables better characterization of the disease progression, it has been increasingly acknowledged that associating genetics with imaging phenotypes is more statistically powerful and could potentially lead to more insights of the underlying neurobiology mechanisms.

Structural neuroimaging: Structural anatomic changes are among those imaging phenotypes that have been widely studied for exploration of the underlying genetic effect. Mata et al. examined the possible interactive effects of two risk genes, *NRG1* and *DISC1*, on brain volumes of schizophrenia patients [4]. They found that rs2793092 SNP in *DISC1* was significantly associated with increased lateral ventricle (LV) volume. However, after accounting for rs6994992 SNP in the *NRG1* gene, the *DISC1* SNP only predicted LV enlargement among those patients carrying the T allele in the *NRG1* SNP. Those patients with the combination of risk alleles in both genes show much greater LV volume than those with none of the allelic combinations. *NRG1* was also found to be related with temporal region volume [116]. Another risk allele, Val158 in *COMT* is found to be associated with smaller hippocampal gray matter volume [116].

However, in a recent large-scale meta-analysis study, based on results from common variant studies of schizophrenia and volumes of several brain structures (mainly subcortical) [117], they failed to find evidence of genetic overlap between schizophrenia risk and subcortical volume measures either at the level of common variant genetic architecture or for single genetic markers.

Functional neuroimaging: Genetic factors were also identified to play a major role in mediating the functional imaging phenotypes. Among those risk genes, *COMT* is one of the most frequently studied genes in functional imaging genomics analysis. Healthy subjects who carry the Val allele in *COMT* gene have been found to have greater activation during working memory tasks than Met variant carriers in the dorsolateral prefrontal cortex (DLPFC) and other frontal areas that are previously known to be associated with schizophrenia (e.g., [118,119]). For schizophrenia patients, Val/Val carriers show greater prefrontal activation than Met carriers [120] and less normalization of hyperactivation during neuroleptic treatment than Met/Met carriers [121]. In addition, Val/Val carriers were observed to have a reduced degree of deactivation in both normal subjects and schizophrenia patients in the n-back working memory task [122].

SNPs rs821616 and rs6675281 in *DISC1* are another two genotypes whose effect on functional imaging phenotypes have been extensively studied. Two studies, [123,124], investigated prefrontal function of different genotype groups using verbal fluency tasks. Compared to Cys704 carriers, healthy individuals with Ser704 homozygotes demonstrated bilateral prefrontal cortex activation with greater engagement in the left hemisphere and greater activation in the left middle and left superior frontal gyri. However, when pooling healthy and illness subjects together, no significant effect of Ser704Cys on cortical activation was observed. SNPs from *BDNF*, *DAT*, and human serotonin transporter (*SLC6A4*) were also identified to have an effect on functional connectivity.

In addition to these targeted analysis, some other studies employed genome-wide analysis of functional imaging QTs and yielded a few new targets. For example, using the mean blood oxygen level-dependent signal in the dorsolateral prefrontal cortex during a memory task as a QT, Potkins et al. [125] found six genes that harbor SNPs significant for the interaction between the imaging QT and the diagnosis (roundabout guidance receptor 1 [*ROBO1*]-*ROBO2*, Traf2 and Nck interacting kinase [*TNIK*], cortexin 3 [*CTXN3*]-*SLC12A*, POU class 3 homeobox 2 [*POU3F2*], Tnf Receptor Associated Factor [*TRAF*], and glypican 1 [*GPC1*]. Rietschel et al. performed GWAS and identified autophagy and beclin 1 regulator 1 (*AMBRA1*), diacylglycerol kinase zeta (*DGKZ*), cholinergic receptor muscarinic 4 (*CHRM4*), and midkine (*MDK*) as schizophrenia risk genes [126]. In their subsequent imaging genomics analysis, healthy carriers of the risk allele exhibited altered activation in the cingulate cortex during a cognitive control task.

Zinc finger protein 804A (*ZNF804*) is another risk gene which has been extensively studied to examine its effect on the functional brain imaging phenotypes, particularly rs1344706 [127]. It has been reported to be linked with reduced functional connectivity within inter-hemispheric prefrontal areas, reduced functional connectivity within ipsilateral and contralateral DLPFC [128], increased amygdala coupling with HF [129], reduced DLPFC coupling with PFC, and increased DLPFC coupling with HF [130]. Potkin et al. examined 27 SNPs in arginine and serine rich coiled-coil 1 (*RSRC1*) and 19 SNPs in Rho GTPase activating protein 18 (*ARHGAP18*) and found that they significantly affected the activation in the DLPF [131]. In addition, there is also a substantial interest in resting state fMRI, which also appears to be perturbed with the development of schizophrenia [132–134]. In [135], they found that two SNPs from doublecortin domain containing 2 (*DCDC2*) were related to the Broca-Medial Parietal network in both healthy controls and schizophrenia patients, but another two SNPs in *KIAA0319* were only associated with the left Broca-superior/inferior parietal network in healthy controls.

20.4.2 Bipolar disorder

20.4.2.1 GENETICS OF BIPOLAR DISEASE

Bipolar disorder, also known as manic depression, is a brain disorder characterized by dramatic and unpredictable shift in mood, energy, and activity levels. It is now affecting approximately 5.7 million American adults aged 18 years and older, which accounts for roughly 2.6% of the United States population. Unlike schizophrenia, bipolar disorder patients manifest different symptoms from person to person, which typically further change during the progression of disease. Twins and family studies have produced overwhelming evidences that genetic factor is an important contributor affecting the susceptibility of bipolar disease, where the overall heritability is estimated to reach as high as 93% [136,137].

As with many other complex diseases, a substantial progress in the discovery of genetic factors contributing to the pathology of bipolar disease has been achieved by genome-wide association studies. Testing 1.8 million variants in 4,387 cases and 6,209 controls, Ferreira et al. identified the region of ankyrin 3 (*ANK3*) and calcium voltage-gated channel subunit alpha1 C (*CACNA1C*) in strong association with disease risk [138]. The Psychiatric Genome-Wide Association Study Consortium Bipolar Disorder Working Group (PGC-BD) later reported results with an even larger sample size [139] where genotype data were assembled from a combined analysis of 16,731 samples. They successfully confirmed the genome-wide significance of *CACNA1C* and identified several other genetic regions, such as teneurin transmembrane protein (TENM4 [*ODZ4*]), *ANK3*, and spectrin repeat containing nuclear envelope protein 1 (*SYNE1*), to be related with bipolar. However, in the subsequent replication study with 46,912 samples, only 18 SNPs from *CACNA1C* and *ODZ4* showed significant signals with the same effect direction. Results from a recent GWAS based on Japanese population support nuclear factor I X (*NFIX*), mitotic arrest deficient 1 like 1 (*MAD1L1),* tetratricopeptide repeat and ankyrin repeat containing 1 (*TRANK1*), and *ODZ4* as susceptible genes associated with bipolar disease risk [140].

Accumulating evidences from both targeted and genome-wide association studies have consistently showed that bipolar disease share both susceptibility genetic loci, such as *COMT* [141], *CACNA1C* [142], polybromo 1 (*PBRM1*) [143], and polygenic background [144] with schizophrenia. Strengthened signals with genome-wide significance were identified when combining results from schizophrenia and bipolar disorder for *ZNF804A*, inter-alpha-trypsin inhibitor heavy chain 3 (*ITIH3*)-*ITIH4, ANK3, CACNA1C*, and mitogen-activated protein kinase 3 (*MAPK3*) [145].

20.4.2.2 MULTI-MODAL NEUROIMAGING AS QUANTITATIVE PHENOTYPES

The genetic findings from GWA studies have led to many imaging genomics studies in the past few years focusing on examining the effect of candidate genetic variants on brain imaging phenotypes.

Structural neuroimaging: As the most replicated risk gene of bipolar, *CACNA1C* has been extensively studied. Kempton et al. found the risk-associated SNP rs1006737 in *CACNA1C* to be significantly associated with increased cerebral gray matter volume in healthy subjects [146]. In [147], carriers of the *CACNA1C* rs1006737 risk allele exhibited increased gray matter density in the right amygdala and right hypothalamus in both healthy subjects and bipolar patients. They also identified the interaction effect between genotype and diagnosis on the left putamen volume. Frazier et al. focusing only on bipolar patients reported that A allele carriers had larger bilateral caudate, insula, globus pallidus, frontal pole, and nucleus accumbens volumes [112]. However, the association between rs1006737 and gray matter volume in healthy subjects could not be confirmed. The same group identified a new connection between rs1006737 and brain stem volume alterations [148].

Later, Tesli et al. examined 298 psychosis cases patients (121 bipolar cases, 116 schizophrenia cases, and 61 other psychosis cases) and 219 controls for connection between nine risk SNPs in *CACNA1C*, *ANK3*, *ODZ4*, and *SYNE1* and eight bipolar-altered structural brain measures, and reported no effect of these SNPs on frontal, parietal, temporal, or total cortical thickness, neither for SNP genotypes or for polygenic scores [149]. In a recent study of 117 bipolar patients [150], carriers of the *CACNA1C* gene polymorphism rs1006737 minor allele A showed greater left medial orbitofrontal cortex thickness compared to non-carriers. In the same study, they observed age-related cortical thinning of the left caudal anterior cingulate cortex in risk allele carriers, but not non-carriers. Taken together, the connection between sMRI and bipolar risk genes are less explored and existing findings are inconsistent across studies.

Functional neuroimaging: Many imaging genomics studies in bipolar depression focus on the genetic effect on brain function, and rs1006737 in *CACNA1C* has been a major focus in these studies. Bigos et al. found that rs1006737 is in significant association with both increased hippocampal activity during emotional processing and increased prefrontal activity during executive cognition [151]. Erk et al. also identified hippocampal activity to be altered in healthy risk allele carriers [3]. They reported that healthy subjects carrying the risk allele had a significant reduction of bilateral hippocampal activation during episodic memory recall, decreased functional connectivity between left and right hippocampus, and abnormal activation in the anterior cingulate cortex, a region previously associated with bipolar disorder. The same group later found that this risk allele also resulted in decreased hippocampal and perigenual anterior cingulate activation in relatives of bipolar patients [152]. They further replicated this discovery finding using another large dataset and additionally observed diminished activation of the dorsolateral prefrontal cortex [153].

Tesli et al. performed an fMRI study to explore the effect of bipolar candidate SNP rs1006737 on brain activity and showed significantly increased left amygdala activity in carriers of risk allele A compared to non-carriers in both the total studied subjects (123 controls, 66 bipolar cases, and 61 schizophrenia cases) and the subset of bipolar subjects (66 bipolar cases) [154]. In another study of 91 healthy subjects, they investigated the association of rs1006737 genotype in *CACNA1C* with prefrontal activation and fronto-hippocampal connectivity [155]. Subjects with two copies of rs1006737 minor allele A showed decreased activation in right-hemispheric DLPFC compared to others. Moreover, the fronto-hippocampal connectivity was found to be positively associated with the number of rs1006737-A alleles.

20.4.3 DEPRESSION

Major depressive disorder (MDD) is a common psychiatric disorder characterized by at least two weeks of depressive mood, loss of interest, and low energy. It is identified as a leading cause of disability, which affects 322 million people or 4.4% of the world's population (http://www.who.int/) and is more than twice as prevalent in women than in men [156]. Depression is also known to be highly recurrent, with 75% of depressed individuals having more than a single episode of depression within 2 years of recovery [157].

20.4.3.1 GENETICS OF MAJOR DEPRESSIVE DISORDER

Based on the results of twin studies, MDD is estimated to have a partly genetic etiology with moderate heritability ranging from 31% to 42% [158]. Given this high heritability, substantial effort has been dedicated to identifying the risk genetic factors underlying MDD. A variable repeat sequence in the 5' promoter region (*5-HTTLPR*) of the *SLC6A4* has been a major focus of many genetic studies in MDD, and it is hypothesized to play a key role in the pathophysiology. Individuals carrying short allele (s-allele) are more likely to develop MDD than those carrying homozygous long allele [159], and tend to exhibit increased anxiety related temperamental traits, which is associated with increased risk for MDD [160]. Other widely studied risk genes include *BDNF* [161] and tryptophan hydroxylase 2 (*TPH2*) [162].

In addition, more susceptibility genes are identified through genome-wide association analysis. In a GWA study of 2,431 cases and 3,673 screened controls, they examined more than 1 million SNPs for their association with the disease status, but none of them achieved the genome-wide significance. It is estimated that there may be fewer variants with intermediate or large effect in MDD compared with schizophrenia. Therefore, more samples may be required to detect risk variants in MDD that explain the same proportion of total variance [163].

To solve this problem, some studies tried meta-analysis to dissect the genetic basis of depression, where the detection power of risk alleles is expected to be improved by combining several underpowered small studies into one large study. Narrowing down the search space is another alternative strategy so that the final results will suffer less from the multiple comparison correction. One example study is [164], in which they performed a meta-analysis of the previously published findings, including 20 polymorphisms in 18 genes. They replicated the previous findings of *SLC6A4* and identified other significant genetic regions from *APOE*, G protein subunit beta 3 (*GNB3*), *MTHFR*, and *SLC6A3* to be associated with MDD.

20.4.3.2 MULTI-MODAL NEUROIMAGING AS QUANTITATIVE PHENOTYPES

Imaging genomic studies focusing on *5-HTTLPR* have consistently found that s-allele carriers tend to have abnormal amygdala volume. Significantly smaller perigenual anterior cingulate cortex and amygdala were observed in s-allele carriers than in l-allele homozygotes [165]. The same study also reported that individuals with s-allele had a reduced strength of negative functional connectivity between amygdala and subgenual anterior cingulate cortex during a perceptual processing task. In addition, the genetic effect of 5-HTTLPR amygdala function in MDD has also been widely examined. It is consistently reported that the homozygous l-allele carriers exhibited decreased amygdala activation compared to those who carry at least one copy of the s allele, in both resting state [166] and response to certain stimuli [167,168].

Compared to amygdala, there is very limited work investigating the effect of *5-HTTLPR* on prefrontal cortical function in MDD patients and existing findings of their associations are mostly ambiguous. Friedel et al. observed a strong association of the s-alleles with the increased activation in ventromedial prefrontal cortex during the processing of negatively valenced images [169]. This association was only identified in non-depressed individuals, but not in individuals with MDD. However, another study failed to confirm this association, but observed an opposite pattern where MDD patients carrying s-allele showed high resting-state activity in the ventromedial prefrontal cortex [166]. It is postulated that this inconsistency may due to the medication effect since the participants in the later study have been treated with medications while those in the first one are not [170].

20.5 DISCUSSION

In this section, we summarize the imaging genomic discoveries in various brain diseases, show the differences of imaging genomic studies between neurology/psychiatry and oncology, and discuss future directions of imaging genomics from both data and methodology aspects.

20.5.1 SUMMARY OF BRAIN IMAGING GENOMICS

The technological advances in acquiring high-throughput genomic data and multi-modal imaging data have successfully promoted biomarker discovery and mechanistic understanding in the study of brain disorders. In this chapter, we have systematically reviewed the applications and discoveries of imaging genomics in the study of several neurological and psychiatrical disorders.

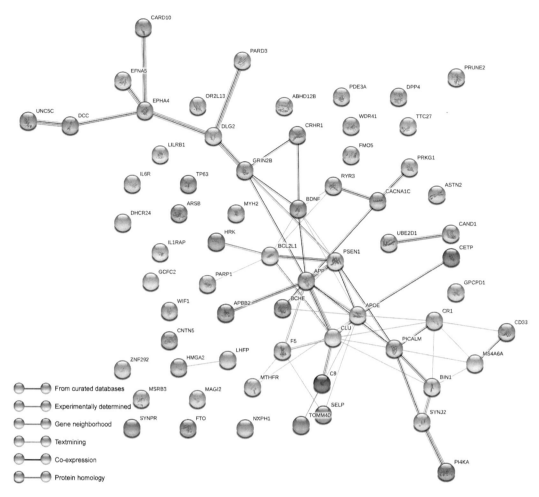

Figure 20.2 (See color insert.) Interaction network among imaging genomic findings in the study of Alzheimer's disease. Functional interactions are shown using the STRING database. Nodes represent genes, and edges denote the interactions. Content inside node shows known or predicted 3D structure of corresponding protein, and color of the edges indicates the source of interaction evidence as shown in legend.

Alzheimer's disease: Here, we examine the relationship among genetic findings from AD study with different imaging quantitative traits, given that there exist a large number of discoveries. Figure 20.2 shows the interactions among the AD-relevant imaging genomic findings based on STRING database (https://string-db.org/). It appears that several genes have dense connections with others while some have no interaction evidence with any other genes. Densely connected genes include most of top 10 AD genes [41] (*APOE*, *BIN1*, *CLU*, *ABCA7*, *CR1*, *PICALM*, *MS4A6A*, *CD33*, *MS4A4E*, *CD2AP*) as well as a few others, demonstrating the effectiveness of using imaging quantitative traits as phenotypes for identifying disease-relevant genes.

To better understand the functional implications of these imaging genomic findings, we further test their functional association using Kyoto Encyclopedia of Genes and Genomes (KEGG) pathway database (http://www.genome.jp/kegg/) and biological process (BP) terms from Gene Ontology (http://www.geneontology.org/). Table 20.1 (a and b) shows the functional annotation results from KEGG pathway and GO-BP terms, respectively, which include significantly enriched (false discovery rate [FDR] corrected) KEGG pathways and top 20 enriched GO-BP terms. Alzheimer's disease, as expected, is enriched in the KEGG pathway analysis. This demonstrates the disease-relevance of imaging genomic findings from AD studies and indicates brain imaging can effectively serve as promising biomarker and intermediate phenotypes to study AD.

Table 20.1 Functional annotation of imaging genetic findings in AD studies

(a) KEGG pathway enrichment analysis result

KEGG pathway	# of hits	FDR p-value
Alzheimer's disease	6	3.88E-03
Circadian entrainment	4	2.77E-02
Axon guidance	4	4.55E-02
Amyotrophic lateral sclerosis (ALS)	3	4.55E-02

(b) Top 20 enriched gene ontology biological process terms

GO Biological process	# of hits	FDR p-value
Regulation of neuron death	10	3.51E-05
Regulation of cellular component organization	21	1.12E-03
Positive regulation of beta-amyloid formation	3	1.12E-03
Positive regulation of cell death	11	3.42E-03
Cell surface receptor signaling pathway	19	3.47E-03
Learning	6	3.47E-03
Regulation of cell death	16	3.47E-03
Cyclic GMP (cGMP) mediated signaling	3	3.47E-03
Fear response	4	3.47E-03
Single-organism behavior	9	3.47E-03
Positive regulation of cellular process	29	3.47E-03
Regulation of synapse structure or activity	7	3.47E-03
Negative regulation of amyloid precursor protein catabolic process	3	3.47E-03
Positive regulation of response to stimulus	18	4.11E-03
Neuron apoptotic process	4	4.11E-03
Negative regulation of cellular component organization	10	4.25E-03
Regulation of response to stimulus	24	5.22E-03
Smooth endoplasmic reticulum calcium ion homeostasis	2	6.23E-03
Positive regulation of neurofibrillary tangle assembly	2	6.23E-03
Response to stress	24	6.39E-03

Note: (a) Significantly enriched KEGG pathways with FDR corrected $p < 0.05$. (b) Top 20 enriched gene ontology biological process terms.

Frontotemporal dementia: Current implementations of imaging genomic analysis in FTD typically focus on evaluating the differential imaging patterns associated with three common risk genes (*MAPT*, *GRN*, and *C9orf72*). We have reviewed the genetic studies of FTD and summarized the findings in Table 20.2. We can find that the statistical power of the imaging genomic strategy has not been sufficiently reflected in FTD research. Several other genes have been reported in FTD studies, but have rarely been replicated or validated in imaging genomic analyses. Given the advantage of neuroimaging QTs as continuous phenotypes, imaging genomic association analysis would be expected to discover more FTD-relevant genes as well as help mechanistic understanding. Moreover, with the development of high-level imaging genomic association mining strategies, the examination of collective effects of multiple genes would also be expected to gain promising pathological understanding for FTD.

Table 20.2 Major case-control and imaging genetic findings in the study of frontotemporal dementia

Gene	Chr	Phenotype	Modality	Paper
MAPT	17	bvFTD	—	[79–81]
		Anteromedial temporal lobe	sMRI	[95]
		Whole-brain volume, whole-brain boundary shift integral (BSI) atrophy rate, left and right hemisphere volume	sMRI	[94]
		Brain changes start from hippocampus and amygdala, then temporal lobe, and then the insula	sMRI	[86,100]
		Left uncinate fasciculus	DTI	[101,102]
		Medial temporal lobe, frontal, and parietal regions	FDG-PET	[86]
		Frontal, temporal, parietal, and subcortical regions	ALS-MRI	[103]
GRN	17	bvFTD	—	[82,83]
		Posterior temporal and parietal lobes	sMRI	[95]
		Whole-brain volume, whole-brain BSI atrophy rate, left and right hemisphere volume	sMRI	[94]
		Brain changes start from insula, then temporal and parietal lobes, and then the striatum	sMRI	[96–98]
		Frontal and temporal regions	FDG-PET	[86]
		Frontal, temporal, parietal, and subcortical regions	ALS-MRI	[103]
C9orf72	9	FTD-MND	—	[81,84–88]
		FTD		[93]
		Symmetrical atrophy in thalamus and superior cerebellum	sMRI	[96–98]
		Brain changes start subcortical areas, insula and occipital cortex, then frontal and temporal lobes, and then the cerebellum	sMRI	[86,100]
		Corpus callosum, cingulum bundle, right paracallosal cingulum	DTI	[101,102]
		Limbic system, basal ganglia, and thalamus	FDG-PET	[86]
		Salience and sensorimotor networks	—	[17]
CHMP2B	3	FTD	—	[87,88,91]
		Whole brain atrophy rate	sMRI	[104,105]
CHCHD10	22	FTD	—	[87]
FUS	16	FTD	—	[87,88,90,92]
SQSTM1	5	FTD	—	[87]
TARDP	1	FTD	—	[87,88]
TBK1	12	FTD	—	[87]
TMEM106B	7	FTD-TDP	—	[89]
		FTD	—	[87,88]
UNC13A	19	FTD	—	[93]
VCP	9	FTD	—	[87,88]

Schizophrenia: Table 20.3 summarizes the major findings of schizophrenia from both diagnostic and imaging genomic analyses. A number of candidate SNPs suggested from GWA case-control studies are examined for their associations with different brain regions. Several regions, including PFC, lateral ventricle, hippocampus, as well as connectivity among them, show association with schizophrenia candidate genes. These imaging genomic findings warrant further investigation to reveal the underlying molecular mechanisms of the functional and/or structural alterations of these candidate regions and their roles during the development and progression of schizophrenia.

Table 20.3 Major case-control and imaging genetic findings in the study of schizophrenia

SNP	A1/A2	Gene	Chr	Phenotype	Modality	Method	P-value	Cohort	Paper
rs11819869	C/T	AMBRA1	11	Schizophrenia (SCZ) vs HC	—	GWAS	3.89E-09	3,738 SCZ, 7,802 HC	[126]
rs7112229	C/T			SCZ vs HC	—	GWAS	7.38E-09	3,738 SCZ, 7,802 HC	
rs7130141	C/T			SCZ vs HC	—	GWAS	6.96E-09	3,738 SCZ, 7,802 HC	
rs12574668	C/A			SCZ vs HC	—	GWAS	1.02E-08	3,738 SCZ, 7,802 HC	
rs11815869	C/T			SCZ vs HC	—	analysis of variance (ANOVA)	2.90E-03	4,734 SCZ, 18,472 HC	
rs11819869	C/T			Medial PFC	fMRI	ANOVA	<0.05(corrected [corr])	121 HC	
Multiple SNPs	—	ARHGAP18	6	Left DLPFC	fMRI	GWAS	<1.00E-07	28 SCZ	[131]
rs4309482	A/G	CCDC68/TCF4	18	SCZ vs HC	—	GWAS	9.68E-07	3,738 SCZ, 7,802 HC	[126]
rs4680	A/G	COMT	22	HF, VLPFC, HF-VLPFC	fMRI	ANOVA	<0.05 (corr)	27 HC	[118, 120]
rs468C	A/G	COMT		Left IFG	fMRI	ANOVA	<0.005(corr)	80 HC	
rs468C	A/G	COMT		Left precentral gyrus (PreCG), MFG	fMRI	ANOVA	0.002(corr)	62 HC	
rs4680-variable number of tandem repeat (VNTR)	—	COMT-SLC6A3	—	PreCG, MFG	fMRI	ANOVA	0.02(corr)	62 HC	[120]
rs4680-VNTR	—		—	ACC	fMRI	ANOVA	0.01(corr)		
rs245178	G/A	CTXN3/SLC12A2	5	rDLPFC	fMRI	GWAS	1.22E-06	64 SCZ, 74 HC	[125]
rs245201	G/A	SLC12A2		rDLPFC	fMRI	GWAS	9.31E-08		
rs821516	T/A	DISC1	1	PFC	sMRI	ANOVA	<0.05(corr)	53 HC	[123]
rs2793092	C/T	DISC1		Left LV	fMRI	ANOVA	5.00E-03	94 SCZ	[4]
rs2793092	C/T			Left LV	fMRI	ANOVA	1.00E-03	94 SCZ	
rs2793092	C/T			Total LV	fMRI	ANOVA	2.00E-02	94 SCZ	
rs2793092	C/T			Total LV	fMRI	ANOVA	4.00E-03	94 SCZ	
rs2793092	C/T			Right LV	fMRI	ANOVA	3.00E-02	94 SCZ	

(Continued)

Table 20.3 (Continued) Major case-control and imaging genetic findings in the study of schizophrenia

SNP	A1/A2	Gene	Chr	Phenotype	Modality	Method	P-value	Cohort	Paper
rs1801028	C/G	DRD2	11	SCZ vs HC	—	Meta	8.30E-09	34,241 SCZ, 45,604 HC	[108]
rs1574192	T/C	GPC1	2	rDLPFC	fMRI	GWAS	3.92E-06	64 SCZ, 74 HC	[125]
rs2228595	T/C	GRM3	7	SCZ vs HC	—	Meta	1.00E-10	34,241 SCZ, 45,604 HC	[108]
rs367398	A/G	NOTCH4	6	SCZ vs HC	—	Meta	1.10E-18	34,241 SCZ, 45,604 HC	[108]
rs6994992-rs2793092	T/C-C/T	NRG1-DISC1	-	Right LV	sMRI	ANOVA	4.00E-02	57 SCZs	[4]
rs6994992-rs2793092	T/C-C/T			Left LV	sMRI	ANOVA	4.00E-03		
rs6994992-rs2793092	T/C-C/T			Total LV	sMRI	ANOVA	7.00E-03		
rs9491640	A/G	POU3F2	6	rDLPFC	fMRI	GWAS	9.23E-06	64 SCZ, 74 HC	[125]
rs7610746	T/C	ROBO2/	3	rDLPFC	fMRI	GWAS	7.56E-06	64 SCZ, 74 HC	[125]
rs9836484	A/G	ROBO1		rDLPFC	fMRI	GWAS	4.23E-06		
Multiple SNPs	—	RSRC1	3	lDLPFC	fMRI	GWAS	<1.00E-07	28 SCZ	[131]
Multiple SNPs	—			SCZ vs HC	—	ANOVA	<1.00E-04	82 SCZ and 91 HC	
LOF mutations	—	SETD1A	16	SCZ vs HC	—	Meta	3.30E-09	4,264 SCZ, 9,343 HC, 1,077 Tri	[110]
VNTR	9R/10R	SLC6A3	5	Left MFG	fMRI	ANOVA	0.04(corr)	62 HC	[120]
VNTR	9R/10R			Right MFG	fMRI	ANOVA	0.01(corr)		
VNTR	9R/10R			ACC	fMRI	ANOVA	0.05(corr)		
rs1800629	A/G	TNF	6	SCZ vs HC	—	Meta	1.70E-18	34,241 SCZ, 45,604 HC	[108]
rs2088885	A/C	TNIK	3	rDLPFC	fMRI	GWAS	6.24E-06	64 SCZ, 74 HC	[125]
rs7627954	A/C			rDLPFC	fMRI	GWAS	6.24E-06		
rs10133111	A/G	TRAF3	14	rDLPFC	fMRI	GWAS	4.77E-06		
rs1344706	C/T	ZNF804A	2	lDLPFC	fMRI	ANOVA	0.001(corr)	111 HC	[128]
rs1344706	C/T			rDLPFC	fMRI	ANOVA	0.003(corr)		
rs1344706	C/T			rDLPFC-lHIP	fMRI	ANOVA	0.007(corr)		
rs1344706	C/T			rDLPFC-rHIP	fMRI	ANOVA	0.012(corr)		

Table 20.4 Functional annotation imaging genetic findings in the study of schizophrenia

GO Biological process term	# of hits	FDR p-value
Central nervous system development	8	3.08E-06
Autonomic nervous system development	6	1.43E-05
Ganglion development	6	1.43E-05
Nervous system development	6	1.43E-05
Stomatogastric nervous system development	6	1.43E-05
Corticospinal neuron axon guidance	5	1.43E-05
VEGF-activated neuropilin signaling pathway involved in axon guidance	5	1.43E-05
Commissural neuron axon guidance	5	1.43E-05
Retinal ganglion cell axon guidance	5	1.43E-05
Anterior/posterior axon guidance	5	1.43E-05
Axon choice point recognition	5	1.43E-05
Corticospinal neuron axon guidance through spinal cord	5	1.43E-05
Corticospinal neuron axon guidance through the basilar pons	5	1.43E-05
Corticospinal neuron axon guidance through the cerebral peduncle	5	1.43E-05
Corticospinal neuron axon guidance through the cerebral cortex	5	1.43E-05
Roundabout signaling pathway involved in axon guidance	5	1.43E-05
Corticospinal neuron axon guidance through the internal capsule	5	1.43E-05
Dopaminergic neuron axon guidance	5	1.43E-05
Corticospinal neuron axon guidance through the medullary pyramid	5	1.43E-05
Dorsal/ventral axon guidance	5	1.43E-05

Note: Top 20 enriched gene ontology biological process terms are shown.

In addition, we explore the functional association of the genetic findings of schizophrenia. Shown in Table 20.4 are the top 20 enriched GO-BP terms. However, there are no significantly enriched KEGG pathways. We observe that a large number of enriched GO-BP terms are related to nervous system development (e.g., central nervous system development, autonomic nervous system development). Most of the enriched terms have direct or indirect relationships with psychiatric diseases or phenotypes.

We have also examined the associations for the genetic findings of schizophrenia, including risk genes from both case-control and imaging genomic analyses. Figure 20.3 illustrates the functional interactions from STRING database. Interactions appear to be among most of schizophrenia findings, and it warrants further investigation to discover their combined or interactive effects on development and progression of schizophrenia.

Bipolar disorder: Several genome-wide analyses have been performed to explore the genetic architecture of bipolar disorder (BPD) and identified several disease risk genes like *ANK3*, *CACNA1C*, *ODZ4*, and several others, which, however, use only disease statuses as phenotypes. Imaging genomics have also been applied in bipolar disorder, exploring the associations between BPD candidate SNPs/genes and brain regions. Shown in Table 20.5 are the genetic findings of bipolar disorder, where *CACNA1C* is the one that has been mostly explored and confirmed in imaging genomic studies. Several SNPs of *CACNA1C* show significant associations with different brain regions, including prefrontal cortex, hippocampus,

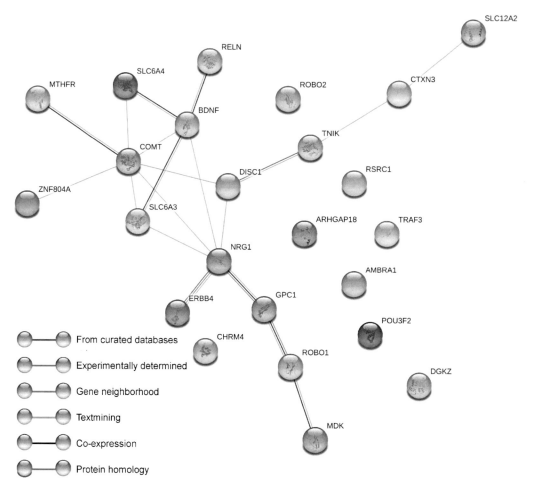

Figure 20.3 (See color insert.) Interaction network among imaging genomic findings in the study of Schizophrenia. Functional interactions are shown using the STRING database. Nodes represent genes, and edges denote the interactions. Content inside node shows known or predicted 3D structure of corresponding protein, and color of the edges indicates the source of interaction evidence as shown in legend.

caudate, amygdala, and some others. Besides, there are also a few genes reported or confirmed from the imaging genomics studies relevant to bipolar disorder, schizophrenia, as well as depression. It may promote the understanding of the underlying common mechanisms among the complex psychiatric disorders.

Depression: Table 20.6 summarizes imaging genomic findings in the study of major depressive disorder. Only a few of the genetic markers have been reported from depression association analysis, of which *SLC6A4* has been mostly explored in several imaging genomic studies. *SLC6A4* has been reported to be associated with both structural and functional changes of amygdala. More samples may be required to detect additional risk variants with smaller effect sizes for depression.

Table 2C.5 Major case-control and imaging genetic findings in the study of bipolar disorder

SNP	A1/A2	Gene	Chr	Phenotype	Modality	Method	P-value	Cohort	Paper
rs1099-336	T/C	ANK3	10	BPD vs HC	—	GWAS	9.10E-09	4,387 BPD, 6,209 HC	[138]
rs10994397	T/C			BPD vs HC	—	GWAS	5.50E-10	7,481 BPD, 9,250 HC	[139]
rs10994397	T/C			cross 5-PD	—	Meta	7.08E-09	6,990 BPD, 27,888 HC	[142]
rs10994359	C/T			BPD vs HC	—	GWAS	2.45E-08	16,374 BPD/SCZ, 14,044 HC	[145]
rs1006737	A/G	CACNA1C	12	BPD vs HC	—	GWAS	7.00E-08	4,387 BPD, 6,209 HC	[138]
rs4765914	T/C			cross 5-PD	—	Meta	1.52E-08	6,990 BPD, 27,888 HC	[142]
rs4765705	C/G			BPD vs HC	—	GWAS	7.01E-09	16,374 BPD/SCZ, 14,044 HC	[145]
rs1006737	A/G			Mean GM volume	sMRI	ANOVA	<0.03	77 HC	[146]
rs1006737	A/G			Right amygdala and right hypothalamus GM density	sMRI	ANOVA	<0.03	41 BPD, 40 HC	[147]
rs1006737	A/G			Left putamen GM density	sMRI	ANOVA	0.035	41 BPD, 40 HC	
rs1006737	A/G			Caudate volume	sMRI	ANOVA	<1.00E-03	91 BPD, 45 MDD, 123 HC	[112]
rs2051992	A/G			Brain stem volume	sMRI	ANOVA	5.71E-05	585 HC	[148]
rs2239050	A/G			Brain stem volume	sMRI	ANOVA	5.91E-05	585 HC	
rs7959938	A/G			Brain stem volume	sMRI	ANOVA	3.62E-05	585 HC	
rs1006737	A/G			Left amygdala	fMRI	ANOVA	0.026	66 BPD, 61 SCZ, 123 HC	[149]
rs1006737	A/G			Left amygdala	fMRI	ANOVA	0.041	66 BPD	
rs1006737	A/G			Left medial OFC	sMRI	ANOVA	0.003	117 BPD	[150]
rs1005737	A/G			PFC	fMRI	ANOVA	0.01(corr)	316 HC	[151]
rs1005737	A/G			Left/right HIP, subgenual anterior cingulate cortex (ACG), dorsal ACG, ventral striatum, left/right MTG, STG, SFG	fMRI	ANOVA	<0.05(corr)	110 HC	[3]
rs1006737	A/G			Left/right HIP, perigenual ACG	fMRI	ANOVA	<0.05(corr)	188 PD relatives, 110 HC	[152]

(Continued)

Table 20.5 (Continued) Major case-control and imaging genetic findings in the study of schizophrenia

SNP	A1/A2	Gene	Chr	Phenotype	Modality	Method	P-value	Cohort	Paper
rs1006737	A/G			Left HIP	fMRI	ANOVA	0.022(corr)	179 HC	[153]
rs1006737	A/G			Perigenual ACG	fMRI	ANOVA	0.011(corr)	179 HC	
rs1006737	A/G			lDLPFC	fMRI	ANOVA	0.007(corr)	179 HC	
rs1006737	A/G			Left HIP	fMRI	ANOVA	0.018(corr)	289 HC	
rs1006737	A/G			Right HIP	fMRI	ANOVA	0.006(corr)	289 HC	
rs1006737	A/G			Perigenual ACG	fMRI	ANOVA	0.016(corr)	289 HC	
rs1006737	A/G			lDLPFC	fMRI	ANOVA	0.004(corr)	289 HC	
rs1006737	A/G			rDLPFC	fMRI	ANOVA	0.037(corr)	94 HC	[155]
rs1006737	A/G			rDLPFC	fMRI	ANOVA	0.004(corr)	94 HC	
rs1006737	A/G			PFC-HIP	fMRI	ANOVA	<0.05(corr)	94 HC	
rs165599	G/A	COMT	22	BPD vs HC	—	Chi-square	9.80E-03	217 BPD, ~4,050 HC	[141]
rs28456	G/A	FADS2	11	BPD vs HC	—	Meta	6.40E-09	3,964 BPD, 61,887 HC	[140]
rs174576	A/C	FADS2	11	BPD vs HC	—	Meta	1.34E-10	10,445 SCZ, 71,137 HC	[145]
rs2239547	T/C	ITIH3-ITIH4	3	BPD vs HC	—	GWAS	7.83E-09	16,374 BPD/SZ, 14,044 HC	[140]
rs4332037	T/C	MAD1L1	7	BPD vs HC	—	Meta	1.91E-09	10,445 SCZ, 71,137 HC	[139]
rs7296288	C/A	-	12	BPD vs HC	—	GWAS	9.40E-09	7,481 SCZ, 9,250 HC	[140]
rs4926298	G/A	NFIX	19	BPD vs HC	—	Meta	5.78E-10	10,445 SCZ, 71,137 HC	[139]
rs12576775	G/A	ODZ4	11	BPD vs HC	—	GWAS	2.70E-08	7,481 SCZ, 9,250 HC	[140]
rs12576775	G/A	ODZ4	11	BPD vs HC	—	Meta	3.32E-09	10,445 SCZ, 71,137 HC	[142]
rs12576775	G/A	ODZ4	11	BPD vs HC	—	Meta	4.40E-08	6,990 BPD, 27,888 HC	[139]
rs9371601	T/G	SYNE1	6	BPD vs HC	—	GWAS	4.30E-09	7,481 SCZ, 9,250 HC	[142]
rs9371601	T/G	SYNE1	6	BPD vs HC	—	Meta	4.27E-09	6,990 BPD, 27,888 HC	[139]
rs9834970	C/T	TRANK1	3	BPD vs HC	—	Meta	2.08E-09	10,445 SCZ, 71,137 HC	[140]

Table 20.6 Major case-control and imaging genetic findings in the study of major depressive disorder

SNP	A1/A2	Gene	Chr	Clinical presentation/Phenotype	Modality	Method	P-value	Cohort	Paper
rs7412/rs429358	ε2/ε3	APOE	19	MDD vs HC	—	Meta	<1.0E-03	827 MDD, 1,616 HC	[164]
rs5443	T/C	GNB3	12	MDD vs HC	—	Meta	<0.01	375 MDD, 492 HC	
rs1801133	T/C	MTHFR	1	MDD vs HC	—	Meta	<0.01	875 MDD, 3,859 HC	
40 bp VNTR	—	SLC6A3	5	MDD vs HC	—	Meta	<0.01	151 MDD, 272 HC	
44-BP insertion/deletion	—	SLC6A4	17	MDD vs HC	—	Meta	<0.01	3,752 MDD, 5,707 HC	
5-HTTLPR	s/l			Perigenual ACC volume	sMRI	ANOVA	<0.05(corr)	114 HC	[165]
5-HTTLPR	s/l			Right amygdala volume	sMRI	ANOVA	<0.05(corr)	114 HC	
5-HTTLPR	s/l			Amygdala-perigenual ACC	fMRI	ANOVA	<0.05(corr)	94 HC	
5-HTTLPR	s/l			Medial PFC	rCBF	ANOVA	<0.05(corr)	89 MDD	[166]
5-HTTLPR	s/l			Left amygdala	rCBF	ANOVA	<0.05(corr)	89 MDD	
5-HTTLPR	s/l			Right amygdala	rCBF	ANOVA	<0.05(corr)	89 MDD	
5-HTTLPR	s/l			Medial PFC	rCBF	ANOVA	<0.05(corr)	89 MDD	
5-HTTLPR	s/l			Right amygdala	rCBF	ANOVA	<0.05(corr)	89 MDD	

20.5.2 STRENGTH AND LIMITATION OF BRAIN IMAGING GENOMICS

In genetic association analyses, imaging quantitative traits can help increase statistical power via serving as continuous phenotypes. In addition, using iQTs as intermediate phenotypes may be more sensitive for detecting the functional links related to the risk genes than diagnostic status, due to continuous phenotypes may capture the full subsyndromal changes in pathologic markers of brain diseases.

On the other hand, given the large number of available imaging phenotypes, the selection of imaging phenotypes is critical for improving statistical performance as well as help interpret genetic findings. The statistical power may suffer from the multiple correction when a large number of phenotypes are explored in one analysis. Such that prior knowledge or expert guidance for phenotype selection would help improve the detection of susceptible genes.

Multi-omics data analyses have been integrated in many cancer-related studies, including genomics, transcriptomics, proteomics, metabolomics, epigenomics, and so on. Compared with oncology studies, the availability of multi-omics data from brain tissues is far more limited in brain imaging genomic studies. It appears highly infeasible to collect brain region-specific omics data from the live subjects. Thus many strategies implemented in cancer imaging genomics cannot be directly applied to brain imaging genomics.

20.5.3 FUTURE DIRECTIONS

To bridge the gap between cancer imaging genomics and brain imaging genomics, future efforts can be made toward two directions: multi-omics data collection and novel approach development. On one hand, various brain banks have been created to collect brain tissues from neurological or psychiatric patients as well as normal controls. Once the size of a brain bank reaches certain scale, it will be possible to generate valuable multi-omics data using brain tissues for subsequent brain imaging genomics studies.

On the other hand, in case the direct collection of brain region-specific data is not available, advanced data science approaches could potentially be developed to incorporate relevant knowledge via analyzing multi-omics data from existing databases collected for other purposes, tissue ontology information and other related resources. For example, genome-wide functional interaction network has been constructed for a number of brain tissues by Greene et al. [171] that can reflect the functional interactions of genes in a tissue-specific context to some extent.

The existing approaches and findings provide a basis for continued research in the area of brain imaging genomics, which has a host of fundamental problems yet to be solved, especially for large scale and heterogeneous data covering both imaging and genomics domains. The increasing availability of large-scale biological data has the potential to further promote the development of data-driven research, increase our knowledge at a systems biology level, and benefit the mechanistic understanding of underlying biological processes.

REFERENCES

1. Morris, J.C., Early-stage and preclinical Alzheimer disease. *Alzheimer Dis Assoc Disord*, 2005. **19**(3): pp. 163–165.
2. Saykin, A.J. et al., Genetic studies of quantitative MCI and AD phenotypes in ADNI: Progress, opportunities, and plans. *Alzheimer's Dement*, 2015. **11**(7): pp. 792–814.
3. Erk, S. et al., Brain function in carriers of a genome-wide supported bipolar disorder variant. *Arch Gen Psychiatry*, 2010. **67**(8): pp. 803–811.
4. Mata, I. et al., Additive effect of NRG1 and DISC1 genes on lateral ventricle enlargement in first episode schizophrenia. *Neuroimage*, 2010. **53**(3): pp. 1016–1022.
5. Mitteroecker, P., J.M. Cheverud, and M. Pavlicev, Multivariate analysis of genotype-phenotype association. *Genetics*, 2016. **202**(4): pp. 1345–1363.
6. Hibar, D.P. et al., Voxelwise gene-wide association study (vGeneWAS): Multivariate gene-based association testing in 731 elderly subjects. *Neuroimage*, 2011. **56**(4): pp. 1875–1891.

7. Liu, J. and V.D. Calhoun, A review of multivariate analyses in imaging genetics. *Front Neuroinform*, 2014. **8**: p. 29.

8. Silver, M. et al., Identification of gene pathways implicated in Alzheimer's disease using longitudinal imaging phenotypes with sparse regression. *Neuroimage*, 2012. **63**(3): pp. 1681–1694.

9. Silver, M., G. Montana, and Alzheimer's Disease Neuroimaging Initiative, Fast identification of biological pathways associated with a quantitative trait using group lasso with overlaps. *Stat Appl Genet Mol Biol*, 2012. **11**(1): p. Article 7.

10. Wang, H. et al., Identifying quantitative trait loci via group-sparse multitask regression and feature selection: An imaging genetics study of the ADNI cohort. *Bioinformatics*, 2012. **28**(2): pp. 229–237.

11. Chi, E.C. et al., Imaging genetics via sparse canonical correlation analysis. *Proc IEEE Int Symp Biomed Imaging*, 2013. **2013**: pp. 740–743.

12. Vounou, M. et al., Sparse reduced-rank regression detects genetic associations with voxel-wise longitudinal phenotypes in Alzheimer's disease. *Neuroimage*, 2012. **60**(1): pp. 700–716.

13. Wan, J. et al., Hippocampal surface mapping of genetic risk factors in AD via sparse learning models. *Med Image Comput Comput Assist Interv*, 2011. **14**(Pt 2): pp. 376–383.

14. Du, L. et al., A novel structure-aware sparse learning algorithm for brain imaging genetics. *Med Image Comput Comput Assist Interv*, 2014. **17**(Pt 3): pp. 329–336.

15. Yan, J. et al., Transcriptome-guided amyloid imaging genetic analysis via a novel structured sparse learning algorithm. *Bioinformatics*, 2014. **30**(17): pp. i564–i571.

16. Hibar, D.P. et al., Genome-wide interaction analysis reveals replicated epistatic effects on brain structure. *Neurobiol Aging*, 2015. **36**(Suppl 1): pp. S151–S158.

17. Borroni, B. et al., Granulin mutation drives brain damage and reorganization from preclinical to symptomatic FTLD. *Neurobiol Aging*, 2012. **33**(10): pp. 2506–2520.

18. Lambert, J.C. et al., Meta-analysis of 74,046 individuals identifies 11 new susceptibility loci for Alzheimer's disease. *Nat Genet*, 2013. **45**(12): pp. 1452–1458.

19. Stein, J.L. et al., Discovery and replication of dopamine-related gene effects on caudate volume in young and elderly populations (*N* = 1198) using genome-wide search. *Mol Psychiatry*, 2011. **16**(9): pp. 927–937, 881.

20. Ramanan, V.K. et al., Genome-wide pathway analysis of memory impairment in the Alzheimer's disease neuroimaging initiative (ADNI) cohort implicates gene candidates, canonical pathways, and networks. *Brain Imaging Behav*, 2012. **6**(4): pp. 634–648.

21. Yao, X. et al., Two-dimensional enrichment analysis for mining high-level imaging genetic associations. *Brain Inform*, 2017. 4(1): p. 27.

22. Yao, X. et al., Tissue-specific network-based genome wide study of amygdala imaging phenotypes to identify functional interaction modules. *Bioinformatics*, 2017. **33**(20): pp. 3250–3257.

23. Yao, X. et al., Network-based genome wide study of hippocampal imaging phenotype in Alzheimer's disease to identify functional interaction modules. *Proc IEEE Int Conf Acoust Speech Signal Process*, 2017. **2017**: pp. 6170–6174.

24. Alzheimer's Assoc, Alzheimer's association report 2017 Alzheimer's disease facts and figures. *Alzheimer's Dement*, 2017. **13**: pp. 325–373.

25. Gatz, M. et al., Role of genes and environments for explaining Alzheimer disease. *Arch Gen Psychiatry*, 2006. **63**(2): pp. 168–174.

26. Bateman, R.J. et al., Autosomal-dominant Alzheimer's disease: A review and proposal for the prevention of Alzheimer's disease. *Alzheimer's Res Ther*, 2011. **3**(1): pp. 1.

27. Jun, G. et al., Meta-analysis confirms CR1, CLU, and PICALM as Alzheimer disease risk loci and reveals interactions with APOE genotypes. *Arch Neurol*, 2010. **67**(12): pp. 1473–1484.

28. Kauwe, J.S. et al., Suggestive synergy between genetic variants in TF and HFE as risk factors for Alzheimer's disease. *Am J Med Genet B Neuropsychiatr Genet*, 2010. **153B**(4): pp. 955–959.

29. Naj, A.C. et al., Dementia revealed: Novel chromosome 6 locus for late-onset Alzheimer disease provides genetic evidence for folate-pathway abnormalities. *PLoS Genet*, 2010. **6**(9): p. e1001130.

30. Antunez, C. et al., Genetic association of complement receptor 1 polymorphism rs3818361 in Alzheimer's disease. *Alzheimer's Dement*, 2011. **7**(4): pp. e124–e129.

31. Antunez, C. et al., The membrane-spanning 4-domains, subfamily A (MS4A) gene cluster contains a common variant associated with Alzheimer's disease. *Genome Med*, 2011. **3**(5): p. 33.

32. Hollingworth, P. et al., Common variants at ABCA7, MS4A6A/MS4A4E, EPHA1, CD33 and CD2AP are associated with Alzheimer's disease. *Nat Genet*, 2011. **43**(5): pp. 429–435.

33. Hu, X. et al., Meta-analysis for genome-wide association study identifies multiple variants at the BIN1 locus associated with late-onset Alzheimer's disease. *PLoS One*, 2011. **6**(2): p. e16616.

34. Naj, A.C. et al., Common variants at MS4A4/MS4A6E, CD2AP, CD33 and EPHA1 are associated with late-onset Alzheimer's disease. *Nat Genet*, 2011. **43**(5): pp. 436–441.

35. Kamboh, M.I. et al., Genome-wide association study of Alzheimer's disease. *Transl Psychiatry*, 2012. **2**: p. e117.

36. Reitz, C. et al., Genetic variants in the fat and obesity associated (FTO) gene and risk of Alzheimer's disease. *PLoS One*, 2012. **7**(12): p. e50354.

37. Jonsson, T. et al., A mutation in APP protects against Alzheimer's disease and age-related cognitive decline. *Nature*, 2012. **488**(7409): pp. 96–99.

38. Swaminathan, S. et al., Analysis of copy number variation in Alzheimer's disease in a cohort of clinically characterized and neuropathologically verified individuals. *PLoS One*, 2012. **7**(12): p. e50640.

39. Saykin, A.J. et al., Alzheimer's disease neuroimaging initiative biomarkers as quantitative phenotypes: Genetics core aims, progress, and plans. *Alzheimer's Dement*, 2010. **6**(3): pp. 265–273.

40. Morris, J.C., Early-stage and preclinical Alzheimer disease. *Alzheimer Dis Assoc Disord*, 2005. **19**(3): pp. 163–165.

41. Shen, L. et al., Genetic analysis of quantitative phenotypes in AD and MCI: Imaging, cognition and biomarkers. *Brain Imaging Behav*, 2014. **8**(2): pp. 183–207.

42. Stein, J.L. et al., Voxelwise genome-wide association study (vGWAS). *Neuroimage*, 2010. **53**(3): pp. 1160–1174.

43. Potkin, S.G. et al., Hippocampal atrophy as a quantitative trait in a genome-wide association study identifying novel susceptibility genes for Alzheimer's disease. *PLoS One*, 2009. **4**(8): p. e6501.

44. Bis, J.C. et al., Common variants at 12q14 and 12q24 are associated with hippocampal volume. *Nat Genet*, 2012. **44**(5): pp. 545–551.

45. Melville, S.A. et al., Multiple loci influencing hippocampal degeneration identified by genome scan. *Ann Neurol*, 2012. **72**(1): pp. 65–75.

46. Stein, J.L. et al., Identification of common variants associated with human hippocampal and intracranial volumes. *Nat Genet*, 2012. **44**(5): pp. 552–561.

47. Stein, J.L. et al., Genome-wide analysis reveals novel genes influencing temporal lobe structure with relevance to neurodegeneration in Alzheimer's disease. *NeuroImage*, 2010. **51**(2): pp. 542–554.

48. Hibar, D.P. et al., Genome-wide association identifies genetic variants associated with lentiform nucleus volume in *N* = 1345 young and elderly subjects. *Brain Imaging Behav*, 2013. **7**(2): pp. 102–115.

49. Hibar, D.P. et al., Common genetic variants influence human subcortical brain structures. *Nature*, 2015. **520**(7546): pp. 224–229.

50. Bakken, T.E. et al., Association of common genetic variants in GPCPD1 with scaling of visual cortical surface area in humans. *Proc Natl Acad Sci USA*, 2012. **109**(10): pp. 3985–3990.

51. Shen, L. et al., Whole genome association study of brain-wide imaging phenotypes for identifying quantitative trait loci in MCI and AD: A study of the ADNI cohort. *Neuroimage*, 2010. **53**(3): pp. 1051–1063.

52. Andrawis, J.P. et al., Effects of ApoE4 and maternal history of dementia on hippocampal atrophy. *Neurobiol Aging*, 2012. **33**(5): pp. 856–866.

53. Chang, Y.L. et al., APOE interacts with age to modify rate of decline in cognitive and brain changes in Alzheimer's disease. *Alzheimer's Dement*, 2014. **10**(3): pp. 336–348.

54. Desikan, R.S. et al., Apolipoprotein E epsilon4 does not modulate amyloid-beta-associated neurodegeneration in preclinical Alzheimer disease. *Am J Neuroradiol*, 2013. **34**(3): pp. 505–510.

55. Hostage, C.A. et al., Mapping the effect of the apolipoprotein E genotype on 4-year atrophy rates in an Alzheimer disease-related brain network. *Radiology*, 2014. **271**(1): pp. 211–219.

56. Jack, C.R., Jr. et al., Shapes of the trajectories of 5 major biomarkers of Alzheimer disease. *Arch Neurol*, 2012. **69**(7): pp. 856–867.

57. Risacher, S.L. et al., Longitudinal MRI atrophy biomarkers: Relationship to conversion in the ADNI cohort. *Neurobiol Aging*, 2010. **31**(8): pp. 1401–1418.

58. Vemuri, P. et al., Effect of apolipoprotein E on biomarkers of amyloid load and neuronal pathology in Alzheimer disease. *Ann Neurol*, 2010. **67**(3): pp. 308–316.

59. Wolk, D.A., B.C. Dickerson, and Alzheimer's Disease Neuroimaging Initiative, Apolipoprotein E (APOE) genotype has dissociable effects on memory and attentional-executive network function in Alzheimer's disease. *Proc Natl Acad Sci USA*, 2010. **107**(22): pp. 10256–10261.

60. Biffi, A. et al., Genetic variation and neuroimaging measures in Alzheimer disease. *Arch Neurol*, 2010. **67**(6): pp. 677–685.

61. Rajagopalan, P. et al., Common folate gene variant, MTHFR C677T, is associated with brain structure in two independent cohorts of people with mild cognitive impairment. *Neuroimage Clin*, 2012. **1**(1): pp. 179–187.

62. Ma, J. et al., MS4A6A genotypes are associated with the atrophy rates of Alzheimer's disease related brain structures. *Oncotarget*, 2016. **7**(37): p. 58779.

63. Sun, J.H. et al., The impact of UNC5C genetic variations on neuroimaging in Alzheimer's disease. *Mol Neurobiol*, 2016. **53**(10): pp. 6759–6767.

64. Wang, Z.X. et al., Genetic association of HLA gene variants with MRI brain structure in Alzheimer's disease. *Mol Neurobiol*, 2017. **54**(5): pp. 3195–3204.

65. Ramanan, V.K. et al., APOE and BCHE as modulators of cerebral amyloid deposition: A florbetapir PET genome-wide association study. *Mol Psychiatry*, 2014. **19**(3): pp. 351–357.

66. Ramanan, V.K. et al., GWAS of longitudinal amyloid accumulation on 18F-florbetapir PET in Alzheimer's disease implicates microglial activation gene IL1RAP. *Brain*, 2015. **138**(Pt 10): pp. 3076–3088.

67. Murphy, K.R. et al., Mapping the effects of ApoE4, age and cognitive status on 18F-florbetapir PET measured regional cortical patterns of beta-amyloid density and growth. *Neuroimage*, 2013. **78**: pp. 474–480.

68. Damoiseaux, J.S. et al., Gender modulates the APOE epsilon4 effect in healthy older adults: Convergent evidence from functional brain connectivity and spinal fluid tau levels. *J Neurosci*, 2012. **32**(24): pp. 8254–8262.

69. Jagust, W.J., S.M. Landau, and Alzheimer's Disease Neuroimaging Initiative, Apolipoprotein E, not fibrillar beta-amyloid, reduces cerebral glucose metabolism in normal aging. *J Neurosci*, 2012. **32**(50): pp. 18227–18233.

70. Kiddle, S.J. et al., Plasma based markers of [11C] PiB-PET brain amyloid burden. *PLoS One*, 2012. **7**(9): p. e44260.

71. Singh, N. et al., Genetic, structural and functional imaging biomarkers for early detection of conversion from MCI to AD. *Med Image Comput Comput Assist Interv*, 2012. **15**(Pt 1): pp. 132–140.

72. Yu, P. et al., Enriching amnestic mild cognitive impairment populations for clinical trials: Optimal combination of biomarkers to predict conversion to dementia. *J Alzheimer's Dis*, 2012. **32**(2): pp. 373–385.

73. Xu, C. et al., Effects of BDNF Val66Met polymorphism on brain metabolism in Alzheimer's disease. *Neuroreport*, 2010. **21**(12): pp. 802–807.

74. Thambisetty, M. et al., Effect of complement CR1 on brain amyloid burden during aging and its modification by APOE genotype. *Biol Psychiatry*, 2013. **73**(5): pp. 422–428.

75. Bradshaw, E.M. et al., CD33 Alzheimer's disease locus: Altered monocyte function and amyloid biology. *Nat Neurosci*, 2013. **16**(7): pp. 848–850.

76. Nho, K. et al., Integration of bioinformatics and imaging informatics for identifying rare PSEN1 variants in Alzheimer's disease. *BMC Med Genomics*, 2016. **9**(Suppl 1): p. 30

77. Wu, M.C. et al., Rare-variant association testing for sequencing data with the sequence kernel association test. *Am J Hum Genet*, 2011. **89**(1): pp. 82–93.

78. Snowden, J.S., D. Neary, and D.M. Mann, Frontotemporal dementia. *Br J Psychiatry*, 2002. **180**: pp. 140–143.

79. Hutton, M. et al., Association of missense and 5'-splice-site mutations in tau with the inherited dementia FTDP-17. *Nature*, 1998. **393**(6686): pp. 702–705.

80. Spillantini, M.G. et al., Mutation in the tau gene in familial multiple system tauopathy with presenile dementia. *Proc Natl Acad Sci USA*, 1998. **95**(13): pp. 7737–7741.

81. Lashley, T. et al., Review: An update on clinical, genetic and pathological aspects of frontotemporal lobar degenerations. *Neuropathol Appl Neurobiol*, 2015. **41**(7): pp. 858–881.

82. Baker, M. et al., Mutations in progranulin cause tau-negative frontotemporal dementia linked to chromosome 17. *Nature*, 2006. **442**(7105): pp. 916–919.

83. Cruts, M. et al., Null mutations in progranulin cause ubiquitin-positive frontotemporal dementia linked to chromosome 17q21. *Nature*, 2006. **442**(7105): pp. 920–924.

84. DeJesus-Hernandez, M. et al., Expanded GGGGCC hexanucleotide repeat in noncoding region of C9ORF72 causes chromosome 9p-linked FTD and ALS. *Neuron*, 2011. **72**(2): pp. 245–256.

85. Renton, A.E. et al., A hexanucleotide repeat expansion in C9ORF72 is the cause of chromosome 9p21-linked ALS-FTD. *Neuron*, 2011. **72**(2): pp. 257–268.

86. Meeter, L.H. et al., Imaging and fluid biomarkers in frontotemporal dementia. *Nat Rev Neurol*, 2017. **13**(7): pp. 406–419.

87. Pottier, C. et al., Genetics of FTLD: Overview and what else we can expect from genetic studies. *J Neurochem*, 2016. **138**(Suppl 1): pp. 32–53.

88. Seelaar, H. et al., Clinical, genetic and pathological heterogeneity of frontotemporal dementia: A review. *J Neurol Neurosurg Psychiatry*, 2011. **82**(5): pp. 476–486.

89. Van Deerlin, V.M. et al., Common variants at 7p21 are associated with frontotemporal lobar degeneration with TDP-43 inclusions. *Nat Genet*, 2010. **42**(3): pp. 234–239.

90. Kwiatkowski, T.J., Jr. et al., Mutations in the FUS/TLS gene on chromosome 16 cause familial amyotrophic lateral sclerosis. *Science*, 2009. **323**(5918): pp. 1205–1208.

91. Skibinski, G. et al., Mutations in the endosomal ESCRTIII-complex subunit CHMP2B in frontotemporal dementia. *Nat Genet*, 2005. **37**(8): pp. 806–808.

92. Vance, C. et al., Mutations in FUS, an RNA processing protein, cause familial amyotrophic lateral sclerosis type 6. *Science*, 2009. **323**(5918): pp. 1208–1211.

93. Diekstra, F.P. et al., C9orf72 and UNC13A are shared risk loci for amyotrophic lateral sclerosis and frontotemporal dementia: A genome-wide meta-analysis. *Ann Neurol*, 2014. **76**(1): pp. 120–133.

94. Rohrer, J.D. et al., Distinct profiles of brain atrophy in frontotemporal lobar degeneration caused by progranulin and tau mutations. *Neuroimage*, 2010. **53**(3): pp. 1070–1076.

95. Whitwell, J.L. et al., Voxel-based morphometry patterns of atrophy in FTLD with mutations in MAPT or PGRN. *Neurology*, 2009. **72**(9): pp. 813–820.

96. Lee, S.E. et al., Altered network connectivity in frontotemporal dementia with C9orf72 hexanucleotide repeat expansion. *Brain*, 2014. **137**(Pt 11): pp. 3047–3060.

97. Sha, S.J. et al., Frontotemporal dementia due to C9ORF72 mutations: Clinical and imaging features. *Neurology*, 2012. **79**(10): pp. 1002–1011.

98. Whitwell, J.L. et al., Neuroimaging signatures of frontotemporal dementia genetics: C9ORF72, tau, progranulin and sporadics. *Brain*, 2012. **135**(Pt 3): pp. 794–806.

99. Whitwell, J.L. et al., Brain atrophy over time in genetic and sporadic frontotemporal dementia: A study of 198 serial magnetic resonance images. *Eur J Neurol*, 2015. **22**(5): pp. 745–752.

100. Rohrer, J.D. et al., Presymptomatic cognitive and neuroanatomical changes in genetic frontotemporal dementia in the genetic frontotemporal dementia initiative (GENFI) study: A cross-sectional analysis. *Lancet Neurol*, 2015. **14**(3): pp. 253–262.

101. Mahoney, C.J. et al., Profiles of white matter tract pathology in frontotemporal dementia. *Hum Brain Mapp*, 2014. **35**(8): pp. 4163–4179.

102. Mahoney, C.J. et al., Longitudinal diffusion tensor imaging in frontotemporal dementia. *Ann Neurol*, 2015. **77**(1): pp. 33–46.

103. Dopper, E.G. et al., Cerebral blood flow in presymptomatic MAPT and GRN mutation carriers: A longitudinal arterial spin labeling study. *Neuroimage Clin*, 2016. **12**: pp. 460–465.

104. Rohrer, J.D. et al., Presymptomatic generalized brain atrophy in frontotemporal dementia caused by CHMP2B mutation. *Dement Geriatr Cogn Disord*, 2009. **27**(2): pp. 182–186.

105. Rohrer, J.D. et al., The heritability and genetics of frontotemporal lobar degeneration. *Neurology*, 2009. **73**(18): pp. 1451–1456.

106. Emsley, R.A. et al., Depressive and anxiety symptoms in patients with schizophrenia and schizophreniform disorder. *J Clin Psychiatry*, 1999. **60**(11): pp. 747–751.

107. Cardno, A.G. and Gottesman, II, Twin studies of schizophrenia: from bow-and-arrow concordances to star wars Mx and functional genomics. *Am J Med Genet*, 2000. **97**(1): pp. 12–17.

108. Farrell, M.S. et al., Evaluating historical candidate genes for schizophrenia. *Mol Psychiatry*, 2015. **20**(5): pp. 555–562.

109. Takata, A. et al., Loss-of-function variants in schizophrenia risk and SETD1A as a candidate susceptibility gene. *Neuron*, 2014. **82**(4): pp. 773–780.

110. Singh, T. et al., Rare loss-of-function variants in SETD1A are associated with schizophrenia and developmental disorders. *Nat Neurosci*, 2016. **19**(4): pp. 571–577.

111. International Schizophrenia Consortium, Rare chromosomal deletions and duplications increase risk of schizophrenia. *Nature*, 2008. **455**(7210): pp. 237–241.

112. Frazier, T.W. et al., Candidate gene associations with mood disorder, cognitive vulnerability, and fronto-limbic volumes. *Brain and Behavior*, 2014. **4**(3): pp. 418–430.

113. van den Heuvel, M.P. and A. Fornito, Brain networks in schizophrenia. *Neuropsychol Rev*, 2014. **24**(1): pp. 32–48.

114. Ellison-Wright, I. and E. Bullmore, Meta-analysis of diffusion tensor imaging studies in schizophrenia. *Schizophr Res*, 2009. **108**(1–3): pp. 3–10.

115. Skudlarski, P. et al., Diffusion tensor imaging white matter endophenotypes in patients with schizophrenia or psychotic bipolar disorder and their relatives. *Am J Psychiatry*, 2013. **170**(8): pp. 886–898.

116. Hashimoto, R. et al., Imaging genetics and psychiatric disorders. *Curr Mol Med*, 2015. **15**(2): pp. 168–175.

117. Franke, B. et al., Genetic influences on schizophrenia and subcortical brain volumes: Large-scale proof of concept. *Nat Neurosci*, 2016. **19**(3): pp. 420–431.

118. Bertolino, A. et al., Prefrontal-hippocampal coupling during memory processing is modulated by COMT Val158met genotype. *Biological Psychiatry* 2006. **60**(11): pp. 1250–1258.

119. Krug, A. et al., The effect of the COMT val(158)met polymorphism on neural correlates of semantic verbal fluency. *Eur Arch Psychiatry Clin Neurosci*, 2009. **259**(8): pp. 459–465.

120. Bertolino, A. et al., Additive effects of genetic variation in dopamine regulating genes on working memory cortical activity in human brain. *J Neurosci*, 2006. **26**(15): pp. 3918–3922.

121. Bertolino, A. et al., Interaction of COMT Val(108/158) met genotype and olanzapine treatment on prefrontal cortical function in patients with schizophrenia. *Am J Psychiatry* 2004. **161**(10): pp. 1798–1805.

122. Pomarol-Clotet, E. et al., COMT Val158Met polymorphism in relation to activation and de-activation in the prefrontal cortex: A study in patients with schizophrenia and healthy subjects. *Neuroimage*, 2010. **53**(3): pp. 899–907.

123. Prata, D.P. et al., DISC1 affects prefrontal cortical function in healthy individuals. *Bipolar Disord*, 2008. **10**: pp. 83–83.

124. Prata, D.P. et al., No association of Disrupted-in-Schizophrenia-1 variation with prefrontal function in patients with schizophrenia and bipolar disorder. *Genes Brain and Behavior*, 2011. **10**(3): pp. 276–285.

125. Potkin, S.G. et al., A genome-wide association study of schizophrenia using brain activation as a quantitative phenotype. *Schizophr Bull*, 2009. **35**(1): pp. 96–108.

126. Rietschel, M. et al., Association between genetic variation in a region on chromosome 11 and schizophrenia in large samples from Europe. *Mol Psychiatry*, 2012. **17**(9): pp. 906–917.

127. Tost, H., E. Bilek, and A. Meyer-Lindenberg, Brain connectivity in psychiatric imaging genetics. *Neuroimage*, 2012. **62**(4): pp. 2250–2260.

128. Esslinger, C. et al., Cognitive state and connectivity effects of the genome-wide significant psychosis variant in ZNF804A. *Neuroimage*, 2011. **54**(3): pp. 2514–2523.

129. Esslinger, C. et al., Neural mechanisms of a genome-wide supported psychosis variant. *Science*, 2009. **324**(5927): pp. 605–605.

130. Rasetti, R. et al., Altered cortical network dynamics a potential intermediate phenotype for schizophrenia and association with ZNF804A. *Arch Gen Psychiatry*, 2011. **68**(12): pp. 1207–1217.

131. Potkin, S.G. et al., Gene discovery through imaging genetics: Identification of two novel genes associated with schizophrenia. *Schizophrenia Bull*, 2009. **35**: pp. 194–194.

132. Argyelan, M. et al., Resting-State fMRI connectivity impairment in schizophrenia and bipolar disorder. *Schizophrenia Bull*, 2014. **40**(1): pp. 100–110.

133. White, R.S. and S.J. Siegel, Cellular and circuit models of increased resting-state network gamma activity in schizophrenia. *Neuroscience*, 2016. **321**: pp. 66–76.

134. Chahine, G. et al., Disruptions in the left frontoparietal network underlie resting state endophenotypic markers in schizophrenia. *Hum Brain Mapp*, 2017. **38**(4): pp. 1741–1750.

135. Jamadar, S. et al., Genetic influences of resting state fMRI activity in language-related brain regions in healthy controls and schizophrenia patients: A pilot study. *Brain Imaging and Behav*, 2013. **7**(1): pp. 15–27.

136. Craddock, N., and Sklar P., Genetics of bipolar disorder. *Lancet*, 2013. 381(9878): pp. 1654–1662.

137. McGuffin, P. et al., The heritability of bipolar affective disorder and the genetic relationship to unipolar depression. *Arch Gen Psychiatry*, 2003. **60**(5): pp. 497–502.

138. Ferreira, M.A.R. et al., Collaborative genome-wide association analysis supports a role for ANK3 and CACNA1C in bipolar disorder. *Nat Genet*, 2008. **40**(9): pp. 1056–1058.

139. Sklar, P. et al., Large-scale genome-wide association analysis of bipolar disorder identifies a new susceptibility locus near ODZ4. *Nat Genet*, 2011. **43**(10): pp. 977–983 U162.

140. Ikeda, M. et al., A genome-wide association study identifies two novel susceptibility loci and trans population polygenicity associated with bipolar disorder. *Mol Psychiatry*, 2018. **23**(3): p. 639.

141. Shifman, S. et al., COMT: A common susceptibility gene in bipolar disorder and schizophrenia. *Am J Med Genet B Neuropsychiatr Genet*, 2004. **128b**(1): pp. 61–64.

142. Cross-Disorder Group of the Psychiatric Genomics, Consortium, Identification of risk loci with shared effects on five major psychiatric disorders: A genome-wide analysis. *Lancet*, 2013. **381**(9875): pp. 1371–1379.

143. Williams, H.J. et al., Most genome-wide significant susceptibility loci for schizophrenia and bipolar disorder reported to date cross-traditional diagnostic boundaries. *Hum Mol Genet*, 2011. **20**(2): pp. 387–391.

144. Purcell, S.M. et al., Common polygenic variation contributes to risk of schizophrenia and bipolar disorder. *Nature*, 2009. **460**(7256): pp. 748–752.

145. Ripke, S. et al., Genome-wide association study identifies five new schizophrenia loci. *Nat Genet*, 2011. **43**(10): pp. 969–976.

146. Kempton, M.J. et al., Effects of the CACNA1C risk allele for bipolar disorder on cerebral gray matter volume in healthy individuals. *Am J Psychiatry*, 2009. **166**(12): pp. 1413–1414.

147. Perrier, E. et al., Initial evidence for the role of CACNA1C on subcortical brain morphology in patients with bipolar disorder. *Eur Psychiatry*, 2011. **26**(3): pp. 135–137.

148. Franke, B. et al., Genetic variation in CACNA1C, a gene associated with bipolar disorder, influences brainstem rather than gray matter volume in healthy individuals. *Biol Psychiatry*, 2010. **68**(6): pp. 586–588.

149. Tesli, M. et al., CACNA1C risk variant and amygdala activity in bipolar disorder, schizophrenia and healthy controls. *PLoS One*, 2013. **8**(2): p. E56970.

150. Soeiro-de-Souza, M.G. et al., The CACNA1C risk allele rs1006737 is associated with age-related prefrontal cortical thinning in bipolar I disorder. *Transl Psychiatry*, 2017. 7: p. e1086.

151. Bigos, K.L. et al., Genetic variation in CACNA1C affects brain circuitries related to mental illness. *Arch Gen Psychiatry*, 2010. **67**(9): pp. 939–945.

152. Erk, S. et al., Hippocampal and frontolimbic function as intermediate phenotype for psychosis: Evidence from healthy relatives and a common risk variant in CACNA1C. *Biol Psychiatry*, 2014. **76**(6): pp. 466–475.

153. Erk, S. et al., Replication of brain function effects of a genome-wide supported psychiatric risk variant in the CACNA1C gene and new multi-locus effects. *Neuroimage*, 2014. **94**: pp. 147–154.

154. Tesli, M. et al., No evidence for association between bipolar disorder risk gene variants and brain structural phenotypes. *J Affect Disord*, 2013. **151**(1): pp. 291–297.

155. Paulus, F.M. et al., Association of rs1006737 in CACNA1C with alterations in prefrontal activation and fronto-hippocampal connectivity. *Hum Brain Mapp*, 2014. **35**(4): pp. 1190–200.

156. Albert, P.R., Why is depression more prevalent in women? *J Psychiatry Neurosci*, 2015. **40**(4): pp. 219–221.

157. Boland, R.J. et al., Course and outcome of depression. *Handbook of Depression*, 2002. **2**: pp. 23–43.

158. Sullivan, P.F., M.C. Neale, and K.S. Kendler, Genetic epidemiology of major depression: Review and meta-analysis. *Am J Psychiatry*, 2000. **157**(10): pp. 1552–1562.

159. Caspi, A. et al., Influence of life stress on depression: Moderation by a polymorphism in the 5-HTT gene. *Science*, 2003. **301**(5631): pp. 386–389.

160. Lesch, K.P. et al., Association of anxiety-related traits with a polymorphism in the serotonin transporter gene regulatory region. *Science*, 1996. **274**(5292): pp. 1527–1531.

161. Castren, E. and T. Rantamaki, The role of BDNF and its receptors in depression and antidepressant drug action: Reactivation of developmental plasticity. *Dev Neurobiol*, 2010. **70**(5): pp. 289–297.

162. Zill, P. et al., Single nucleotide polymorphism and haplotype analysis of a novel tryptophan hydroxylase isoform (TPH2) gene in suicide victims. *Biol Psychiatry*, 2004. **56**(8): pp. 581–586.

163. Wray, N.R. et al., Genome-wide association study of major depressive disorder: New results, meta-analysis, and lessons learned. *Mol Psychiatry*, 2012. **17**(1): pp. 36–48.

164. Lopez-Leon, S. et al., Meta-analyses of genetic studies on major depressive disorder. *Mol Psychiatry*, 2008. **13**(8): pp. 772–785.

165. Pezawas, L. et al., 5-HTTLPR polymorphism impacts human cingulate-amygdala interactions: A genetic susceptibility mechanism for depression. *Nat Neurosci*, 2005. **8**(6): pp. 828–834.

166. Brockmann, H. et al., Influence of 5-HTTLPR polymorphism on resting state perfusion in patients with major depression. *J Psychiatr Res*, 2011. **45**(4): pp. 442–451.

167. Dannlowski, U. et al., Serotonergic genes modulate amygdala activity in major depression. *Genes Brain Behav*, 2007. **6**(7): pp. 672–676.

168. Costafreda, S.G. et al., Modulation of amygdala response and connectivity in depression by serotonin transporter polymorphism and diagnosis. *J Affect Disord*, 2013. **150**(1): pp. 96–103.

169. Friedel, E. et al., 5-HTT genotype effect on prefrontal-amygdala coupling differs between major depression and controls. *Psychopharmacology* (Berl), 2009. **205**(2): pp. 261–271.

170. Bigos, K.L., A.R. Hariri, and D.R. Weinberger, *Neuroimaging Genetics: Principles and Practices*. 2016. New York: Oxford University Press.

171. Greene, C.S. et al., Understanding multicellular function and disease with human tissue-specific networks. *Nat Genet*, 2015. **47**(6): pp. 569–576.

PART 4

FUTURE OUTLOOK

Quantitative imaging to guide mechanism-based modeling of cancer

DAVID A. HORMUTH II, MATTHEW T. MCKENNA, AND THOMAS E. YANKEELOV

21.1 INTRODUCTION

Mathematical models have been extensively utilized in the study of cancer. In general, mathematical modeling approaches abstract the key features of a physical system and describe those features in a series of mathematical equations. In this way, the system can be simulated *in silico* to further understand system behavior, generate hypotheses, and guide experimental design (Altrock et al. 2015). Cancer models can be built from first-order biological and physical principles, such as evolution (Gatenby et al. 2009) and diffusion (Swanson et al. 2000). Alternatively, statistical models without a specific biophysical basis can be constructed to describe observed data. For example, the linear quadratic model, which describes cell survival following radiation therapy, has been established as the most parsimonious description of cell survival data (Joiner 2009). At the patient level, mathematical models can leverage panels of patient-specific risk factors to predict survival [e.g., prediction of survival in renal cell carcinoma (Zisman et al. 2002)]. Mathematical models are tuned with any available data and simulated to discover system properties and predict response to various perturbations. However, the majority of mathematical models describing cancer growth and treatment response are not structured to leverage clinically available clinical data to make patient-specific predictions (Yankeelov et al. 2015). Indeed, these models have often been limited to *in silico* exploration as the model values are difficult (if not impossible) to measure with the requisite temporal and spatial resolution *in vivo*. In cases where those data are available (e.g., tumor size change, genetic risk factors, etc.), these models often provide a prediction on patient survival, notably, they do not provide specific treatment recommendations. Accordingly, the utility of mathematical models in guiding clinical decision making is currently limited.

Medical imaging provides a rich source of quantitative information on tumor biology. Imaging plays a critical role in cancer screening, diagnosis, and management. Image-based screening tests have been developed to detect pre-malignant lesions in populations with elevated risks of developing certain cancers. For example, mammography (Warner 2011) and low-dose computed tomography (CT) (Bertout et al. 2008) have been applied to detect developing breast and lung cancers, respectively. Following identification of a suspicious lesion, additional medical imaging modalities, including magnetic resonance imaging (MRI), positron emission tomography (PET), ultrasound, and CT, are leveraged in diagnosis and treatment planning. Specifically, these imaging data provide data used to stage the cancer, a description of the disease which incorporates tumor location, size, and metastatic sites (Takayasu et al. 1995; Law et al. 2003; Villers et al. 2009; Morrow et al. 2011; Omuro & DeAngelis 2013). The rich anatomical information provided by medical imaging data is critical in planning tumor biopsy (Pappa et al. 1996; Lewin et al. 2000; Berg 2004) and surgical approaches that seek to removing all cancerous tissue while sparing healthy tissue. Surgery (Yap et al. 2004; Barone et al. 2014), ablation (Goldberg 2002; Köhler et al. 2009; Lencioni et al. 2010), and radiotherapy (Jaffray et al. 2007; Zaidi et al. 2009; Jaffray 2012) all leverage imaging data. Further, medical imaging is used to assess therapeutic response by providing measurements of tumor size (Therasse et al. 2000; Eisenhauer et al. 2009; Wen et al. 2010; Chinot et al. 2013), metabolism (as quantified by standardized uptake value of glucose) (Kubota 2001), cellularity (Moffat et al. 2005; Koh & Collins 2007), and perfusion (Yankeelov & Gore 2009; Li & Padhani 2012) throughout a treatment course. Medical imaging can additionally be used in surveillance to identify tumor recurrence or metastases. Overall, imaging data from MRI, CT, and PET provide quantitative three-dimensional measurements of properties relevant to tumor growth and treatment response.

The incorporation of imaging data into mathematical models presents an opportunity to improve cancer care. The frequent, quantitative measurements of a tumor provided by medical imaging are well-suited for integration into mathematical models. Mathematical models can utilize the quantitative temporally and spatially sampled data describing tumor size, metabolism, and perfusion to predict tumor growth and optimize treatment response. Mathematical modeling tools could potentially assist in predicting response earlier in the course of therapy than existing response criteria. Alternatively, models that can accurately predict response could be used to optimize therapy (e.g., radiation therapy dose planning, chemotherapy agents, and schedule), predict recurrence, or minimize side effects for a given patient.

In this chapter, we will discuss how quantitative imaging measurements can be used to initialize and calibrate biophysical models of tumor growth and response. We will first provide a brief background on the imaging methods used in these mathematical models. We then review several approaches at incorporating data from MRI, CT, and PET into mathematical models. Notably, several of the imaging measurements discussed in this chapter are routinely collected clinically, facilitating the potential translation of patient-specific modeling approaches into clinical care. Finally, we will identify future directions for incorporating medical imaging data into mathematical models of cancer.

21.2 MEASUREMENTS MEDICAL IMAGING CAN PROVIDE

In this section, we will briefly discuss imaging measurements that have been used in biophysical models of tumor growth to provide estimates of tumor boundaries, tumor cellularity, and metabolism. Representative images for these characteristics are shown in Figure 21.1 for a rat injected with a C6 glioma.

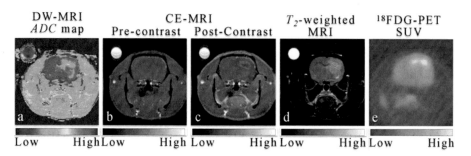

Figure 21.1 (See color insert.) Example images are shown for a rat injected with a C6 glioma. Panel a shows an example *ADC* map. Panels b and c show a CE-MRI image before and after the injection of a gadolinium-based contrast agent. Hyper-intense signal is observed at the periphery of the tumor volume. Panel d shows a T_2-weighted MRI. Panel e shows [18]FDG-PET SUV indicating high uptake in the tumor region.

21.2.1 TISSUE SEGMENTATION AND STRUCTURE

Contrast-enhanced MRI (CE-MRI) is typically used to segment tumor tissue from healthy tissue. In CE-MRI, T_1-weighted images are collected before and after the injection of (typically) a gadolinium-based contrast agent (Yankeelov & Gore 2009). As tumors typically have leaky blood vessels (Folkman et al. 1971), the contrast agent can escape the vasculature and accumulate in the extravascular space resulting in shortened T_1 values within this region. The decreased T_1 in tumor tissue results in an increase in signal relative to the pre-contrast agent image. CE-MRI is considered the gold standard for imaging of brain (Omuro & DeAngelis 2013) and other tumors. In brain tumors, other imaging methods such as T_2-weighted images or fluid-attenuated inversion recovery (FLAIR) images are often used to identify non-enhancing tumor regions (i.e., regions that are not enhanced in CE-MRI) and edema. T_2-weighted images often show hyper-intense signals in regions with increased water (i.e., cerebral spinal fluid, edematous, or inflamed regions), while FLAIR images are used to remove signal from cerebral spinal fluid. Using a segmentation algorithm such as the one described in Menze et al. (2015), it is possible to segment the tissue into normal tissue, edema region (hyper-intense T_2-weighted and hyper-intense FLAIR), non-contrast enhancing tumor (hyper-intense T_2-weighted and hypo-intense CE-MRI), contrast enhancing tumor (hyper-intense CE-MRI), and necrotic tumor (hypo-intense CE-MRI within hyper-intense CE-MRI region). Figure 21.1b and c show CE-MRI images before and after the injection of a contrast agent. After the injection of the contrast agent, the tumor shows a hyper-intense signal around the periphery of the tumor. Figure 21.1d also shows an example T_2-weighted image demonstrating hyper-intense signal in the tumor region.

Diffusion tensor imaging (DTI) is a variant of diffusion weighted-MRI (DW-MRI) that can be used to assess the magnitude and direction of water diffusion in tissue. In DTI, pulsed gradients are applied along at least six non-collinear directions to estimate the diffusion coefficient of water in those directions and build the diffusion tensor at each voxel. Structures such as white matter or muscles which are well organized exhibit anisotropic diffusion, while tumor regions typically exhibit more isotropic diffusion. DTI measurements are commonly used to characterize structural connectivity through tractography which follows the path of the dominant diffusion direction in highly anisotropic tissues (i.e., white matter tracts). The fractional anisotropy index which ranges from 0 (fully isotropic diffusion) to 1 (fully anisotropic diffusion) is often used to assess the degree of diffusion anisotropy within a region. The diffusion tensor is useful in biophysical models as the tensor can be used to define the direction of tumor cell movement or assist in defining the movement of tumor cells along white matter tracts. A detailed review of DTI can be found in (Sundgren et al. 2004).

21.2.2 CELLULARITY

DW-MRI is an imaging technique that has been used to assess cellularity (the number of cells within a region) *in vivo*. Briefly, DW-MRI is sensitive to the diffusion of water molecules within a sample. The diffusion of water in free solution is described by random, Brownian, motion which is due to thermal energy (Einstein 1905).

In tissue, however, water molecules are no longer freely diffusing, but are restricted by cells, macromolecules, and extracellular structures which serve to reduce the diffusion coefficient of water molecules. In DW-MRI, this reduced diffusion coefficient of water is termed the apparent diffusion coefficient (ADC). Typically, a pulsed gradient diffusion experiment is used to calculate the ADC in each voxel. Several studies have showed correlation between ADC and cellularity (Sugahara et al. 1999; Anderson et al. 2000; Guo et al. 2002; Humphries et al. 2007; Jiang et al. 2016) and correlation between ADC and extracellular space (Barnes et al. 2015). Other DW-MRI-based approaches such as VERDICT (Panagiotaki et al. 2014) or IMPULSED (Jiang et al. 2016) have been developed to provide estimates of cellularity and cell size. A more technical description of DW-MRI is found in Chapter 5, while a review of its application to cancer can be found in (Koh & Collins 2007). Figure 21.1a displays a representative ADC map which shows a higher ADC at the center of the tumor indicative of a region of necrosis.

Contrast-enhanced computed tomography (CE-CT) is another method that has been proposed to provide an estimate of cellularity or intracellular volume fraction. In CE-CT, a set of images are acquired before and after the injection of an iodine-based contrast agent. The contrast agent results in greater absorption of X-ray radiation resulting in a linear increase in signal proportional to the contrast agent concentration. During the first-pass of the contrast agent, the contrast agent remains within the blood vessels, however, over time (2–10 minutes) the contrast agent extravasates and starts to accumulate in the extracellular space. Images collected while the contrast agent is accumulating in the extracellular space can be used to calculate the extracellular volume fraction on a voxel by voxel basis. CE-CT measurements of extracellular volume fraction have shown strong correlation ($r = 0.71$, $p = 0.0007$) to histology measurements of extracellular collagen in heart muscle (Bandula et al. 2013). Wong et al. (2017) developed a novel approach to assign intracellular volume fraction (c) from CE-CT data. Briefly, CT data are collected before and after the injection of a contrast agent, and c is then calculated using Eq. (21.1):

$$c = \frac{HU_{tumor,post} - HU_{tumor,pre}}{E\left(HU_{blood,post} - HU_{blood,pre}\right)}\left(1 - Hct\right), \qquad (21.1)$$

where $HU_{tumor,\,pre}$ and $HU_{tumor,\,post}$ are the Hounsfield units of tumor tissue before and after the injection of a contrast agent, respectively, $HU_{blood,\,pre}$ and $HU_{blood,\,post}$ are the Hounsfield units of the aorta blood pool before and after the injection of a contrast agent, respectively, Hct is the hematocrit, and $E(\bullet)$ is the mean value. A more detailed description of CE-CT can be found in O'Connor et al. (2011). A more detailed description of CT can be found in Chapter 3.

21.2.3 TUMOR METABOLISM

PET can be used to provide estimates of, for example, glucose uptake and tumor hypoxia. A more detailed description of PET can be found in Chapter 4, while a review on usage of [18]FDG in oncology can be found in Kubota (2001) and Castell & Cook (2008). Briefly, ^{18}F-fludeoxyglucose ([18]FDG) is a common PET tracer that can provide estimates of glucose uptake. Cells uptake [18]FDG in a similar fashion to glucose, however, after phosphorylation [18]FDG is trapped within the cell resulting in an accumulation of the tracer intracellularly. In cancer, there is generally an overexpression of glucose transporters which results in increased concentrations of [18]FDG in tumor tissues relative to healthy tissues (Castell & Cook 2008). The observed signal is then proportional to the concentration of [18]FDG in a given voxel. PET data are often quantified by the standardized uptake value (SUV) which is the ratio of the [18]FDG concentration in tumor tissue to the total concentration in the body. Figure 21.1e shows a representative SUV map from [18]FDG-PET. A higher SUV is observed in the tumor region (yellow) indicating an increased accumulation of [18]FDG relative to contralateral brain tissue (green).

Another PET tracer that has been used to populate mathematical models is ^{18}F-fluoromisonidazole (^{18}F-MISO) which can be used to assess the level of hypoxia in tumors. A more detailed description of PET can be found in Chapter 4, while a review on usage of ^{18}F-MISO in oncology can be found in Eschmann et al. (2007); Padhani et al. (2007); and Rajendran & Krohn (2015). Briefly, once ^{18}F-MISO is internalized by cells, it is reduced to produce a radical anion (Krohn et al. 2008). In the presence of oxygen, oxygen accepts the electron from the radical anion allowing ^{18}F-MISO to leave the cell. In the absence of oxygen, the radical anion of ^{18}F-MISO eventually binds to other intracellular macromolecules trapping it in the cell. In this way, the concentration of ^{18}F-MISO (and thus there resulting PET signal) is inversely proportional to the oxygen concentration. ^{18}F-MISO is commonly analyzed by calculating either the SUV or the tumor to blood ratio (Muzi et al. 2015). Hypoxia levels can also be assessed *in vivo* using ^{61}Cu-copper(II)-diacetyl-di(N^4-methylthisemicarbazone) (^{61}Cu-ATSM) (Fujibayashi et al. 1997). ^{61}Cu-ATSM has a high membrane permeability and low redox potential allowing it to access intracellular space. Under normoxic conditions, ^{61}Cu-ATSM remains stable. In hypoxic conditions, the copper is reduced and separated from ATSM chelate. The copper remains trapped intracelluarly resulting in an accumulation of signal in hypoxic regions. A review on usage of ^{61}Cu-ATSM in oncology can be found in Lapi et al. (2015). ^{18}F-Flurodeoxythymidine PET (^{18}FLT-PET) can also be used to provide estimates of cellular proliferation. A more detailed description of PET can be found in Chapter 4, while a review on usage of ^{18}FLT-PET in oncology can be found in Soloviev et al. (2012) and Woolf et al. (2014). Briefly, FLT is transported into intracellular spaces and subsequently phosphorylated by thymidine kinase-1. The phosphorylated FLT is impermeable to the cell membrane and is metabolically trapped within the cell. During the DNA synthesis phase of the cell cycle, there is a large increase in the levels of thymidine kinase-1. The rapid proliferation rates of tumor cells result in an increased accumulation of FLT relative to healthy cells. ^{18}FLT-PET is commonly analyzed by calculating the SUV or with a four compartment kinetic model (Soloviev et al. 2012).

21.3 CURRENT APPROACHES

In this section, we will review the modeling literature which integrates the imaging measurements just described into a mathematical framework. We divide this section into models that either: (1) model tumor growth in the absence of therapy or implicitly include the effect of treatment into model parameters or (2) those that explicitly include the effects of treatment into model parameters.

21.3.1 MODELING UNTREATED TUMOR GROWTH OR IMPLICIT PARAMETERIZATION OF TREATMENT RESPONSE

Tumor growth is commonly modeled using a reaction-diffusion model which describes the proliferation (reaction) and movement (diffusion) of tumor cells (or other cell types). A standard reaction-diffusion model utilizing a logistic model of proliferation is shown in Eq. (21.2):

$$\frac{\partial N(x,y,z,t)}{\partial t} = \nabla \cdot \left(D\nabla N(x,y,z,t) \right) + k \cdot N(x,y,z,t) \left(1 - \frac{N(x,y,z,t)}{\theta} \right), \tag{21.2}$$

where $N(x, y, z, t)$ is the number of tumor cells at a given 3D position (x, y, z) and time t, D is their diffusion coefficient of tumor cells (describing their random movement), k is their proliferation rate, and θ is the carrying capacity [i.e., the maximum number of cells that can be fit within a volume of interest (e.g., a voxel)] of the tissue under investigation. Several groups have investigated methods to parameterize this model from imaging data. One approach by Swanson et al. (2000); Swanson and Harpold (2008); Baldock et al. (2013); and Neal et al. (2013) uses serial CE-MRI and T_2-weighted MRI to provide estimates of both the "visible" (i.e., enhancing tumor in CE-MRI) and "invisible" (i.e., hyper-intense tumor in T_2-weighted images) tumor margins in glioblastoma. $N(x, y, z, t)$ is assigned as 80% of θ in the enhancing CE-MRI region and assigned using a Gaussian-based curve within the hyper-intense region in T_2-weighted images. D and k are then calculated from the tumor volume changes in two pre-treatment time points. Additionally, based on observations that brain tumors advance in white matter much faster than gray matter (Scherer 1938), two separate values of D were calculated, where $D_{white\ matter} = 5 \times D_{gray\ matter}$. These estimated parameters are then used to generate an untreated virtual control (UVC) for each patient. Calibrated with pre-treatment measurements, the UVC provides a prediction of untreated tumor growth which can be compared to post-treatment tumor growth to assess treatment efficacy on an individual basis. This UVC can be used to calculate a "days-gained" metric which has shown to indicate improved progression free survival (days gained greater than 100 days) and improved overall survival (days gained greater than 117 days) in a study of 33 patients (Neal et al. 2013). Additionally, their approach showed a relationship between the D/k ratio and tumor grade (Harpold et al. 2007), correlation between the D/k ratio and patient survival (Wang et al. 2009), and predicting response to resection (Swanson et al. 2008), and radiation therapy (Rockne et al. 2010).

Several groups (Jbabdi et al. 2005; Clatz et al. 2005; Hogea et al. 2008; Bondiau et al. 2008; Konukoglu et al. 2010; Mosayebi et al. 2012; Painter & Hillen 2013; Swan et al. 2018) have investigated incorporating DTI information to allow for anisotropic tumor cell diffusion. The motivation for incorporating DTI information is to guide tumor cells along white matter fiber tracts. By incorporating anisotropic diffusion, the goal is to recapitulate some of the irregular tumor shapes observed *in vivo* that arise from the location of the tumor in the brain. In these approaches, the diffusion tensor is supplied either from a brain atlas (Jbabdi et al. 2005; Clatz et al. 2005; Hogea et al. 2008; Bondiau et al. 2008) or acquired on an individual basis (Konukoglu et al. 2010; Mosayebi et al. 2012; Painter & Hillen 2013; Swan et al. 2018). Techniques that use atlas data require registration of the patient data to the atlas to extract the diffusion tensors for each voxel, while DTI information collected on an individual basis can be used to provide subject specific diffusion tensors. In previous efforts (Swanson et al. 2000), tumor cell diffusion in white matter was assigned to be equal to 5x the rate of diffusion in gray matter. However, diffusion tensor information allows for tumor cells to grow along white matter fiber tracts. In the initial work by Jbabdi et al. (2005), simulated tumors were "seeded" and "grown" within a healthy brain atlas using a finite difference simulation of the reaction-diffusion model. When anisotropic diffusion was used, the tumors visually resembled the irregular shape and growth patterns typically observed *in vivo*, thereby indicating the importance of incorporating structural information of the brain into the modeling description. More recently, the work of Swan et al. (2018) applied their model incorporating anisotropic tumor cell diffusion to ten patients. This work used a patient-specific anisotropy parameter, κ, which is interpreted as the relative sensitivity of cancer cells to the underlying brain structure (i.e., white matter fiber tracts). A low κ was observed in patients with relatively spherical tumor growth indicating a low sensitivity to tissue structure and isotropic diffusion of tumor cells. Conversely, a larger κ indicated greater sensitivity to tissue structure and anisotropic diffusion resulting in irregular tumor shapes. Swan et al. (2018) compared results between models with both isotropic and anisotropic diffusion and observed that anisotropic diffusion model had a higher Jaccard index (a measure of the degree of overlap of the predicted and observed tumor shapes) for nine of the ten tumors and could account for the varying levels of anisotropy (κ) on a patient-specific basis. One proposed application for models incorporating anisotropic diffusion is in defining treatment volumes for radiation therapy. Currently, radiation treatment volumes are assigned from contrast-enhancing regions in T_1-weighted MRI with an additional 2 cm margin to account for the non-enhancing tumor region. Accurate models of tumor invasion could be used to identify regions with non-enhancing tumor cells and regions without tumor cells. A model guided approach could potentially target non-enhancing cells that fall outside of the 2 cm margin as well as potentially reduce exposure of healthy tissue to excess radiation.

The approach developed by Yankeelov et al. (2010); Atuegwu et al. (2011); Atuegwu et al. (2012); Weis et al. (2013); Weis et al. (2015); Hormuth et al. (2015); Weis et al. (2017); and Hormuth et al. (2017) uses DW-MRI measurements of tumor cellularity to calibrate tumor models. These measurements of tumor cellularity are then used to calibrate model parameters, which are then used to predict future tumor growth. An example of this modeling and image-processing framework applied to a murine model of glioma is shown in Figure 21.2. In this approach, the DW-MRI is used to estimate ADC at the voxel level (step 1), and then CE-MRI is used to segment the tumor from the surrounding healthy-appearing tissue (step 2). Tumor cell number (steps 3–4) is then calculated within the tumor region using Eq. (21.3):

$$N(x,y,z,t) = \theta\left(\frac{ADC_w - ADC(x,y,z,t)}{ADC_w - ADC_{min}}\right), \tag{21.3}$$

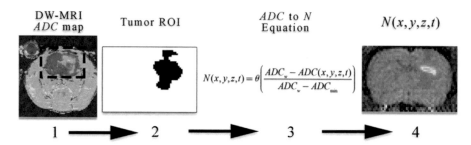

Figure 21.2 (See color insert.) An example flow chart for a rat inject with a C6 glioma is shown demonstrating the conversion of *ADC* to tumor cell number, $N(x,y,z,t)$. First, the *ADC* map is cropped to show only the brain. Second, the tumor region of interest (ROI) is determined from the CE-MRI data. Third, $N(x,y,z,t)$ is calculated from *ADC*. Fourth, the calculated $N(x,y,z,t)$ is displayed over a grayscale anatomical image.

where θ represents the maximum tumor cell carrying capacity for an imaging voxel, ADC_w is the *ADC* of free water at 37°C [i.e., $2.5 \times 10^{-3}\,mm^2/s$; (Whisenant et al. 2014)], $ADC(x,y,z,t)$ is the *ADC* value at a given 3D position (x, y, z) and time t, and ADC_{min} is the minimum *ADC* value observed within the tumor. ADC_{min} is assumed to correspond to the voxel with the largest number of cells. θ can be calculated based on the imaging voxel dimensions and assumptions on cell geometry and packing density. One caveat of this approach is the assumption that any change in *ADC* is directly related to changes in tumor cellularity (Barnes et al. 2015), when in practice, these changes may also result from changes in cell size, cell permeability, and cell tortuosity (Padhani et al. 2009). Using the 3D distribution of tumor cells at two or more time points, a least squares optimization algorithm [such as Levenberg-Marquardt (Levenberg 1944; Marquardt 1963)] can be used to calibrate model parameters. An example of this approach applied to a murine model of glioma is shown in Figure 21.3. Briefly, in the work of Hormuth et al. (2015), rats with C6 gliomas were imaged seven times with DW-MRI over 2 weeks. Three imaging time points (day 10, 12, and 14 after injection with C6 cells) were then used to calibrate model parameters (steps 1–3) using a Levenberg-Marquardt-weighted least

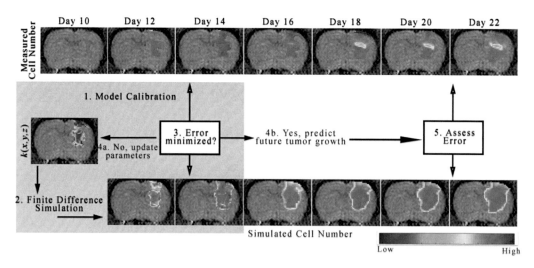

Figure 21.3 (See color insert.) An example optimization and model prediction framework is shown for a representative rat injected with a C6 glioma. The top row shows the measured cell number from day 10 to 22, while the bottom row shows the simulated cell number from day 12 to 22. First, the model is calibrated using data from days 10–14. An initial guess of the model parameters (k, D) is assigned. Second, a finite difference simulation of the reaction-diffusion model is performed. Third, the error is evaluated between the simulated and measured tumor cell number. Fourth, if the error is not minimized the model parameters are updated (4a), otherwise the current model parameters are used to predict future tumor growth (4b). Fifth, the prediction error is assessed between the simulated and measured tumor cell number.

squares algorithm. Initial guesses for k (assigned voxel-wise within the tumor) and D (assigned region-wise for white and gray matter) were used in a finite difference (FD) simulation of tumor growth at days 12 and 14 of Eq. (21.2) (step 2). Error was then assessed between the simulated and measured distribution of tumor cells at days 12 and 14 (step 3). The optimization algorithm continued (step 4a) until error is minimized. The calibrated model parameters were then used in a forward evaluation of the model system to predict future tumor growth (step 4b) at the remaining imaging time points. Prediction error was then assessed between the predicted and measured tumor cell distribution (step 5). This experimental and modeling framework allows for more experimental control in the development and validation of predictive biophysical models. In this work by Hormuth et al. (2015), the authors observed that the standard reaction-diffusion model poorly predicted the tumor volume and distribution of tumor cells and further model development was required.

One proposed solution (Hormuth et al. 2017) to reduce the high errors in predicting the temporal evolution of tumor volume was to incorporate a mechanically coupled diffusion coefficient (Garg & Miga 2008; Weis et al. 2013; Weis et al. 2015; Weis et al. 2017). The motivation to incorporate a mechanically coupled diffusion equation stemmed from experiments of tumor spheroids that demonstrated inhibited growth as the stiffness of the embedded matrix increased (Helmlinger et al. 1997). To implement this phenomenon, it was assumed that tumor cell diffusion, $D(x, y, z, t)$ in the absence of mechanical stress (D_0) would be damped exponentially as the von Mises stress increased:

$$D(x,y,z,t) = D_0 \cdot e^{-\lambda_1 \cdot \sigma_{vm}(x,y,z,t)}, \tag{21.4}$$

where λ_1 is an empirically derived stress-tumor cell diffusion coupling constant, and $\sigma_{vm}(x,y,z,t)$ is the von Mises stress (Garg & Miga 2008; Weis et al. 2013; Weis et al. 2015; Weis et al. 2017). (The von Mises stress is a term that reflects the total experienced stress for a given section of tissue. The von Mises stress is used to reflect the interaction between the growing tumor and its environment.) The von Mises stress was calculated by solving the mechanical equilibrium equation for a linear elastic material with an expansion force related to $N(x, y, z, t)$ shown in Eq. (21.5):

$$\nabla \bullet \sigma - \lambda_f \cdot \nabla N = 0, \tag{21.5}$$

where σ is the stress tensor, and λ_f is a tumor cell-force coupling constant. Through Hooke's law, the stress tensor could be written as a function of tissue strain and the mechanical properties of tissues within the simulation domain [i.e., white and gray matter in brain (Hormuth et al. 2017) or adipose and fibroglandular tissue (Weis et al. 2017)]. At the preclinical level in a murine model of glioma (Hormuth et al. 2017), a statistically significant decrease in tumor volume predictions was observed for models incorporating mechanical coupling (less than 8% error in tumor volume) compared to those without a mechanically coupled diffusion coefficient (greater than 16% error in tumor volume). At the clinical level in glioblastoma, several other groups have also investigated some form of mechanical interaction (Clatz et al. 2005; Hogea et al. 2008; Gooya et al. 2011). This model was also evaluated at the clinical level in breast cancer, where one pre-treatment image and one post-treatment image were used to calibrate the model parameters (Weis et al. 2013; Weis et al. 2015) which were then used to predict the final imaging time point. The effects of neoadjuvant therapy were implicitly included in the optimized model parameters. In Weis et al.'s studies (Weis et al. 2013; Weis et al. 2015), incorporating a mechanically coupled diffusion coefficient more accurately predicted complete pathological response with a receiver operating characteristic area under the curve (AUC) of 0.87 compared to the uncoupled model (AUC = 0.75), simple image analysis (AUC = 0.73), and the response evaluation criteria in solid tumors (RECIST) criteria (AUC = 0.71). Chen et al. (2013) introduced an alternative form of Eq. (21.6) where λ_f was replaced with $f(N(x, y, z, t))$ defined as follows:

$$f(N(x,y,z,t)) = \alpha \cdot \exp\left(-\beta \frac{\theta}{N(x,y,z,t)}\right), \tag{21.6}$$

where α and β are positive constants. $f(N(x,y,z,t))$ allows the amount of force exerted by the tumor cell mass to decrease spatially and temporally as cell density changes. That is, as $N(x,y,z,t)$ approaches θ proliferation is assumed to slow down resulting in invasive pressure. They evaluated this model in five patients with kidney tumors imaged at seven time points. The first six time points were used for model calibration, while the last time point was used to validate model predictions. The authors compared their mechanically coupled model to a non-coupled reaction-diffusion model and observed a decrease in volume error for all patients. A novel two-step coupling scheme was discussed in (Wong et al. 2017) where a biomechanical model for deformation was used to model tissue deformation due to both the tumor (elastic growth) and other organs (image derived motion) in pancreatic cancer. First, a reaction-diffusion model was used to simulate tumor growth. Secondly, the deformation field calculated from the biomechanical model was used to deform the simulated tumor growth to incorporate the effects of interaction of other tissues. The deformed simulated tumor growth was then compared directly to the measured tumor growth. Wong et al. (2017) observed that models that included both the elastic growth and image derived motion components in their biomechanical model resulted in reduced root mean square error compared to the reaction-diffusion model in both synthetic and clinical data.

Another shortcoming of the standard reaction diffusion model observed in Hormuth et al. (2015) was the poor prediction of intra-tumor heterogeneity in cell density. One proposed solution (Hormuth et al. 2017) to more accurately predict the heterogeneity of cell density within in tumors was the incorporation of a voxel-specific carrying capacity. In previous models, θ was calculated based on the physical limit of the number of cells that can fit with in a given region. In Hormuth et al. (2017), the carrying capacity was now assumed to be a function of both physical limitations (i.e., maximum number of cells within a voxel) and biological limitations (e.g., limited nutrient availability, poorly vascularized). Incorporating a voxel-specific carrying capacity resulted in a statistically significant decrease ($p < 0.05$) in the percent error between the measured and model predicted $N(x,y,z,t)$. An alternative approach for calibrating tumor cell proliferation rate is assigning it from [18]FDG-PET data (Liu et al. 2014; Wong et al. 2015; Wong et al. 2017). In [18]FDG-PET, the SUV can be related to proliferation rate through a model for metabolic energy allocation (West et al. 2001) shown in Eq. (21.7):

$$B = B_c\theta c + E_c\theta\left(kc(1-c)\right), \tag{21.7}$$

where B is the energy flow, B_c is the metabolic rate of a single cell, c is the normalized cell ratio (i.e., N/θ), and E_c is the energy needed to birth a daughter cell. SUV is included by assuming $\alpha SUV = B/E_c\theta$ and assigning $\beta = B_c/E_c$, where a and β are scalar parameters. Eq. (21.7) simplifies to Eq. (21.8):

$$k = \frac{\alpha SUV - \beta c}{c(1-c)}. \tag{21.8}$$

A benefit of this approach is that k can be assigned voxel-wise within the tissue accounting for regional variations in cellular proliferation. This approached was assessed using clinical data from pancreatic cancer patients. Briefly, one baseline and one follow-up image were used to calibrate their model system, leaving the second follow-up image to validate their model predictions. Tumor growth and response was modeled using a reaction-diffusion model [similar to Eq. (21.2)] with an additional advection term that coupled tissue movement (velocity) to cell movement. The incorporation of an [18]FDG-PET assigned k was compared to a calibrated k in six patients (Liu et al. 2014), where an improvement in the overall prediction c and error (root mean square difference) was observed for the [18]FDG-PET assignment of k in four of the six patients.

21.3.2 Explicit parameterization of treatment effects

The effects of radiation therapy have also been explicitly included in adaptations of the standard reaction diffusion model [Eq. (21.2)]. The effects of radiation therapy are typically described using the linear quadratic model (Douglas & Fowler 1976; Joiner 2009):

$$S(\alpha, \beta, Dose) = e^{-\alpha Dose - \beta Dose^2}, \tag{21.9}$$

where S is the surviving fraction, α and β are radiosensitivity parameters, and $Dose$ is the delivered radiation therapy dose. α and β have units of Gy^{-1} and Gy^{-2}, respectively, while $Dose$ is reported in units of Gy. (A Gy or gray is the SI unit of ionizing radiation defined as the absorption of one joule of radiation energy per one kilogram of matter.) The linear quadratic model was included in a reaction-diffusion model (Rockne et al. 2009; Rockne et al. 2010) to describe the rate of cell death during radiation therapy:

$$\frac{\partial N(x,t)}{\partial t} = \nabla \cdot \left(D \nabla N(x,t) \right) + k \cdot N(x,t) \left(1 - \frac{N(x,t)}{\theta} \right) - R(x,t,Dose) N(x,t) \left(1 - \frac{N(x,t)}{\theta} \right), \tag{21.10}$$

where $R(x,t,Dose(x,t))$ is the cell death term at a position x, time t, and the spatio-temporally varying dose, $Dose(x, t)$. $R(x,t,Dose(x,t))$ is defined as:

$$R(x,t,Dose(x,t)) = \begin{cases} 0 & \text{for } t \notin \text{ therapy} \\ 1 - S(\alpha, \beta, Dose(x,t)) & \text{for } t \in \text{ therapy} \end{cases}. \tag{21.11}$$

Nine patients with glioblastoma imaged a total of three times (two pre-treatment images and one post-treatment image) were used to evaluate this model. The pre-treatment images were used to calibrate patient specific diffusion and proliferation parameters as discussed in Harpold et al. (2007), while the patient's own dose schedule and treatment plan were used to assign $Dose(x, t)$. The post-treatment time point was then used to assess the effectiveness of therapy by estimating a patient-specific α for a fixed β. This work saw a correlation between the pre-treatment proliferation rate and the optimal α ($r = 0.89$, $p = 0.0007$). A leave-one-out cross-validation approach later showed that treatment response could be predicted from the pre-treatment proliferation rate. Corwin et al. (2013) expanded upon this approach by comparing a simulated intensity modulated radiation therapy (IMRT) treatment plan to the standard-of-care in glioblastoma patients. (IMRT is a treatment method that allows for spatially varying dose with very steep gradients between dose regions. A benefit of IMRT is that it may be able to provide higher doses to less sensitive tumors while sparing healthy tissues to that level of radiation.) In this effort, a multi-objective optimization algorithm was used to minimize the exposure of normal brain tissue to radiation, while minimizing the number of viable tumor cells remaining after radiation. A days-gained metric was calculated for both the received (standard of care) treatment and the simulated (optimized) treatment. (The days-gained metric is defined as the number of days between the post-radiation time point and the time-point on the model-simulated untreated growth trajectory where the tumor sizes are equal.) The simulated optimized treatment plan predicted an increase in days-gained from 21%–105% in 9 of 11 patients compared to the simulated standard of care therapy. Furthermore, they observed that the optimized therapy would decrease the normal tissue dose by 67%–93%.

The radiosensitivity parameter α could also be assigned spatially using ^{18}FMISO-PET images. Rockne et al. (2015) introduced a hypoxia-modulated radiation resistance factor based on the oxygen enhancing ratio (OER) to their reaction-diffusion model. Hypoxia is a common feature in aggressive tumors and has been known to reduce the efficacy of radiation therapy (Vaupel & Mayer 2007). OER is defined as the ratio of doses given to hypoxic and normoxic cells that produce an equivalent response (Chapman & Nahum 2015). ^{18}FMISO-PET tumor to blood ratios were used to assign an OER value between 0 and 3. To incorporate this phenomenon, α was assigned to vary spatially as a function of the OER using Eq. (21.12):

$$\alpha(x) = \begin{cases} \alpha/\text{OER} & \text{for } x \in \text{hypoxia region} \\ \alpha & \text{for } x \notin \text{hypoxia region} \end{cases}. \tag{21.12}$$

The predicted cell distribution was compared between models with and without the hypoxia-modulated radiation response. Results in one patient dataset demonstrated that incorporating the radiation resistance term more accurately characterized response in high-cellularity regions and reduced the percent error in tumor volume from 14.6% to 1.1%.

Titz et al. used two PET tracers to evaluate oxygenation status (^{61}Cu-ATSM) and proliferative potential (^{18}FLT) to model response in head and neck squamous cell carcinoma (Titz et al. 2008). ^{61}Cu-ATSM measurements of oxygenation status are used to alter both tumor cell proliferation and tumor cell response to radiation. Eq. (21.13) describes the probability of cell division as a function of pO_2, the oxygen partial pressure:

$$P(pO_2) = C \cdot \exp\left(-\exp\left(-B \cdot (pO_2 - M)\right)\right), \tag{21.13}$$

where C, B, and M are all assigned constants. In this model, cells undergo division when a sampled random number is less than or equal to the probability $P(pO_2)$. The probability of survival to radiation [Eq. (9)] is also a function of as described in Eq. (21.14):

$$P(Dose, pO_2) = \exp\left(-\alpha Dose \cdot OMF(pO_2) + \beta \left(Dose \cdot OMF(pO_2)\right)^2\right), \tag{21.14}$$

where OMF is an oxygen-dependent modification factor. OMR is determined by calculating the OER for a given pO_2 and normalizing that value by the maximum OER. While pO_2 is not directly measured by ^{61}Cu-ATSM uptake, a sigmoid relationship between ^{61}Cu-ATSM SUV and pO_2 is assumed. The initial proliferative cell population was estimated for each voxel by assuming that the measured ^{18}FLT SUV was directly proportional to the number of proliferative cells. Model parameters were assigned to match *in vitro* data of cell redistribution following radiation therapy (Johnson et al. 1997). The model was then applied to pre-treatment PET/CT images in a head and neck cancer patient. Radiation therapy was simulated to match the standard of care (2 Gy per weekday for 3 weeks). Incorporating ^{18}FLT and ^{61}Cu-ATSM into their model of radiation response demonstrated inhomogeneous response to fractionated therapy. The authors concluded that a uniform dose which fails to account for the varied response within a tumor may result in reduced treatment efficacy.

21.4 FUTURE DIRECTIONS

Opportunities exist to improve the dosing of anticancer therapeutics and the assessment of treatment response. In current clinical practice, treatment plans are often standardized with dosing adjusted to account for varying patient size or toxicities. Treatment response is subsequently assessed using the RECIST, (Eisenhauer et al. 2009) or the response assessment in neuro-oncology (RANO, Wen et al. 2010). These criteria rely on volumetric or tumor dimension changes, which occur temporally downstream from a tumor's physiological, cellular, or molecular responses to the therapy. The models discussed in this chapter have shown promising results in accurately predicting tumor size, shape, cellularity, and response to therapy. These models have potential to improve the treatment and assessment of cancer patients. In this section, we identify areas of future development needed to translate modeling efforts into clinical practice. Specifically, we will discuss: (1) incorporation of treatment response dynamics into models of radiation therapy, (2) application of models to systemic chemotherapies, and (3) requirements for model validation and selection.

Mathematical modeling will aid in dose optimization for a given patient. As discussed above, there have been promising efforts to model response to radiation therapy which can be calibrated on an individual basis (Rockne et al. 2010; Corwin et al. 2013). Optimizing radiation therapy plans is well suited for integration with personalized biophysical models. First, images are frequently acquired throughout the course of planning, positioning, and adapting therapy. These pre-treatment and during treatment images could be used to calibrate models, which in turn could be used to predict response or adapt therapy midcourse (Jaffray 2012). Second, radiation dose plans can be given precisely using methods such as gamma knife, IMRT, or stereotactic radiosurgery (REFS). The prescribed dose plans can provide a 3D map of dose distribution which act as an input to models of the treatment distribution. Models that consider intra-tumor heterogeneity (or resistance to therapy) could potentially identify regions that are less responsive and suggest alternative dose plans (Scott et al. 2017). However, more work is needed to capture the dynamics of both healthy and tumor tissue response to radiation therapy in order to provide optimal treatment. Current models of radiation response assume an instantaneous response (i.e., death) following exposure to radiation. However, the temporal response to radiotherapy is often bimodal with a group of cells that respond immediately to radiation exposure and another group responding later following several cell divisions (Eriksson & Stigbrand 2010). Capturing the dynamics of cell response may be important in determining the optimal dose and schedule for therapy. The importance of cellular dynamics in radiation therapy was demonstrated in Leder et al. (2014) where model designed dosing scheduled showed increased survival in mice with glioblastoma. Additionally, tissue oxygenation and distribution of tumor cells within the cell cycle are important factors in determining treatment response (Wind et al. 2012), and the dynamics of reoxygenation and redistribution following radiation therapy will likely impact optimal treatment schedules.

While there is a rich literature of models of response to systemic chemotherapies, there is a paucity of models that can be calibrated non-invasively on a patient-specific basis. In contrast to radiation therapy, in which the 3D dose distribution can be precisely estimated, the spatiotemporal distribution of chemotherapy agents in tumors is less well known. Indeed, patient-specific pharmacokinetic properties and tumor-specific perfusion both affect the distribution of drug within tumors (Trédan et al. 2007; Barbolosi et al. 2015). As reviewed above, several imaging measurements, such as DCE-MRI or DSC-MRI, can provide perfusion and blood volume maps that could be leveraged to estimate the delivery of pharmaceutical agents. Further work, however, is needed to validate these imaging techniques to estimate intra-tumoral drug distribution and transform existing chemotherapeutic models to leverage these data. These data could be incorporated with the ^{18}FDG-PET- and DW-MRI-based models described above to better quantify treatment response. Accurate and validated models of chemotherapy response could be used to improve the dosing and treatment schedule to reduce resistance and improve overall survival. As patients often receive both chemotherapy and radiation therapy, these modeling efforts may eventually be combined to provide a complete picture of patient response. A review of the combined chemo- and radiotherapy modeling efforts can be found in Grassberger & Paganetti (2016).

Prior to clinical implementation, developed models need to be validated to the relevant *in vivo* measurements or patient outcomes. This validation step provides information about the accuracy and precision of tumor growth predictions as well as identifying limitations or shortcomings of different modeling approaches. This is critical as it will allow for the most appropriate application of modeling approaches in patient care. An additional challenge as these model systems grow in complexity is the development of a model selection framework to select the most appropriate model for a given quantity of interest (e.g., tumor volume, cellularity, side effects). A model agnostic approach such as the one discussed in Lima et al. (2016) and Lima et al. (2017) could be used to select the most parsimonious model that accurately captures the tumor quantity of interest for a given patient. Just as tumors may demonstrate significant inter- and intra-tumoral heterogeneity, different mathematical models may be necessary to describe that heterogeneous behavior. Development along these lines should help bring the clinical translation of these modeling approaches to fruition.

21.5 SUMMARY AND CONCLUSIONS

We have discussed the emerging field at the interface of medical imaging and biophysical modeling of tumors wherein quantitative imaging data are used to calibrate and initialize predictive models of tumor growth. Medical imaging provides quantitative measurements on properties such as cellularity, vascularity, metabolism, relative oxygenation, and tumor shape. Coupling medical imaging with biophysical models is a natural approach which has shown promising results in brain, breast, pancreatic, and kidney cancers. The various levels of complexity demonstrated by the models discussed above indicate the importance for model selection to select the most appropriate model for the desired quantity of interest. Further development of models that can be initialized or calibrated from quantitative imaging measurements have the potential to dramatically transform the current clinical care of cancer on an individual level.

ACKNOWLEDGMENTS

This work was supported through funding from the National Cancer Institute R01CA138599, U01CA174706, U01 CA142565, and F30 CA203220; National Institute for General Medical Sciences T32 GM007347; and from the Cancer Prevention Research Institute of Texas RR160005. T.E.Y. is a CPRIT Scholar of Cancer Research.

REFERENCES

Altrock, P.M., Liu, L.L. & Michor, F., 2015. The mathematics of cancer: Integrating quantitative models. *Nature Reviews Cancer*, 15(12), 730.

Anderson, A.W. et al., 2000. Effects of cell volume fraction changes on apparent diffusion in human cells. *Magnetic Resonance Imaging*, 18(6), 689–695.

Atuegwu, N.C. et al., 2011. Integration of diffusion-weighted MRI data and a simple mathematical model to predict breast tumor cellularity during neoadjuvant chemotherapy. *Magnetic Resonance in Medicine*, 66(6), 1689–1696.

Atuegwu, N.C. et al., 2012. Incorporation of diffusion-weighted magnetic resonance imaging data into a simple mathematical model of tumor growth. *Physics in Medicine and Biology*, 57(1), 225–240.

Baldock, A. et al., 2013. From patient-specific mathematical neuro-oncology to precision medicine. *Frontiers in Oncology*, 3, 62.

Bandula, S. et al., 2013. Measurement of myocardial extracellular volume fraction by using equilibrium contrast-enhanced CT: Validation against histologic findings. *Radiology*, 269(2), 396–403.

Barbolosi, D. et al., 2015. Computational oncology—Mathematical modelling of drug regimens for precision medicine. *Nature Reviews Clinical Oncology*, 13(4), 242–254.

Barnes, S.L. et al., 2015. Correlation of tumor characteristics derived from DCE-MRI and DW-MRI with histology in murine models of breast cancer. *NMR in Biomedicine*, 28(10), 1345–1356.

Barone, D.G., Lawrie, T.A. & Hart, M.G., 2014. Image guided surgery for the resection of brain tumours. *The Cochrane Database of Systematic Reviews*, 1, CD009685.

Berg, W.A., 2004. Image-guided breast biopsy and management of high-risk lesions. *Radiologic Clinics of North America*, 42(5), 935–946.

Bertout, J.A., Patel, S.A. & Simon, M.C., 2008. Hypoxia and metabolism series—Timeline the impact of O(2) availability on human cancer. *Nature Reviews Cancer*, 8(12), 967–975.

Bondiau, P.Y. et al., 2008. Biocomputing: Numerical simulation of glioblastoma growth using diffusion tensor imaging. *Physics in Medicine and Biology*, 53(4), 879–893.

Castell, F. & Cook, G.J.R., 2008. Quantitative techniques in 18FDG PET scanning in oncology. *British Journal of Cancer*, 98(10), 597–601.

Chapman, J.D. & Nahum, A.E., 2015. *Radiotherapy Treatment Planning: Linear-Quadratic Radiobiology*, New York: CRC Press.

Chen, X., Summers, R.M. & Yao, J., 2013. Kidney tumor growth prediction by coupling reaction-diffusion and biomechanical model. *IEEE Transactions on Biomedical Engineering*, 60(1), 169–173.

Chinot, O. et al., 2013. Response assessment criteria for glioblastoma: Practical adaptation and implementation in clinical trials of antiangiogenic therapy. *Current Neurology and Neuroscience Reports*, 13(5), 1–11.

Clatz, O. et al., 2005. Realistic simulation of the 3-D growth of brain tumors in MR images coupling diffusion with biomechanical deformation. *IEEE Trans Med Imaging*, 24(10), 1334–1346.

Corwin, D. et al., 2013. Toward patient-specific, biologically optimized radiation therapy plans for the treatment of glioblastoma. *PLoS ONE*, 8(11), 79115.

Douglas, B.G. & Fowler, J.F., 1976. The effect of multiple small doses of x rays on skin reactions in the mouse and a basic interpretation. *Radiation Research*, 66(2), 401–426.

Einstein, A., 1905. Über die von der molekularkinetischen theorie der wärme geforderte bewegung von in ruhenden Flüssigkeiten suspendierten teilchen. *Annalen der Physik*, 322(8), 549–560.

Eisenhauer, E.A. et al., 2009. New response evaluation criteria in solid tumours: Revised RECIST guideline (version 1.1). *European Journal of Cancer (Oxford, England: 1990)*, 45(2), 228–247.

Eriksson, D. & Stigbrand, T., 2010. Radiation-induced cell death mechanisms. *Tumor Biology*, 31(4), 363–372.

Eschmann, S.M. et al., 2007. Hypoxia-imaging with 18F-Misonidazole and PET: Changes of kinetics during radiotherapy of head-and-neck cancer. *Radiotherapy and Oncology*, 83(3), 406–410.

Folkman, J. et al., 1971. Tumor angiogenesis—Therapeutic implications. *New England Journal of Medicine*, 285(21), 182.

Fujibayashi, Y. et al., 1997. Copper-62-ATSM: A new hypoxia imaging agent with high membrane permeability and low redox potential. *Journal of Nuclear Medicine: Official Publication, Society of Nuclear Medicine*, 38(7), 1155–1160.

Garg, I. & Miga, M.I., 2008. Preliminary investigation of the inhibitory effects of mechanical stress in tumor growth. In *Proc. SPIE*. 69182L–69182L–11.

Gatenby, R.A. et al., 2009. Adaptive therapy. *Cancer Research*, 69(11), 4894–4903.

Goldberg, S.N., 2002. Comparison of techniques for image-guided ablation of focal liver tumors. *Radiology*, 223(2), 304–307.

Gooya, A., Biros, G. & Davatzikos, C., 2011. Deformable registration of glioma images using EM algorithm and diffusion reaction modeling. *IEEE Transactions on Medical Imaging*, 30(2), 375–390.

Grassberger, C. & Paganetti, H., 2016. Methodologies in the modeling of combined chemo-radiation treatments. *Physics in Medicine and Biology*, 61(21), R344–R367.

Guo, Y. et al., 2002. Differentiation of clinically benign and malignant breast lesions using diffusion-weighted imaging. *Journal of Magnetic Resonance Imaging*, 16(2), 172–178.

Harpold, H.L.P., Alvord, E.C.J. & Swanson, K.R., 2007. The evolution of mathematical modeling of glioma proliferation and invasion. *Journal of Neuropathology & Experimental Neurology*, 66(1), 1–9.

Helmlinger, G. et al., 1997. Solid stress inhibits the growth of multicellular tumor spheroids. *Nature Biotechnology*, 15(8), 778–783.

Hogea, C., Davatzikos, C. & Biros, G., 2008. An image-driven parameter estimation problem for a reaction-diffusion glioma growth model with mass effects. *Journal of Mathematical Biology*, 56(6), 793–825.

Hormuth II, D.A. et al., 2015. Predicting in vivo glioma growth with the reaction diffusion equation constrained by quantitative magnetic resonance imaging data. *Physical Biology*, 12(4), 46006.

Hormuth II, D.A. et al., 2017. A mechanically-coupled reaction-diffusion model that incorporates intratumoral heterogeneity to predict in vivo glioma growth. *Journal of the Royal Society Interface*, 14(128), 20161010.

Humphries, P.D. et al., 2007. Tumors in pediatric patients at diffusion-weighted MR imaging: Apparent diffusion coefficient and tumor cellularity. *Radiology*, 245(3), 848–854.

Jaffray, D. et al., 2007. Review of image-guided radiation therapy. *Expert Review of Anticancer Therapy*, 7(1), 89–103.

Jaffray, D.A., 2012. Image-guided radiotherapy: From current concept to future perspectives. *Nature Reviews Clinical Oncology*, 9(12), 688–699.

Jbabdi, S. et al., 2005. Simulation of anisotropic growth of low-grade gliomas using diffusion tensor imaging. *Magnetic Resonance in Medicine*, 54(3), 616–624.

Jiang, X. et al., 2016. In vivo imaging of cancer cell size and cellularity using temporal diffusion spectroscopy. *Magnetic Resonance in Medicine*, 78(1), 156–164.

Johnson, N.F. et al., 1997. DNA damage-inducible genes as biomarkers for exposures to environmental agents. *Environmental Health Perspectives*, 105 Suppl, 913–918.

Joiner, M.C., 2009. Quantifying cell kill and cell survival. In M. C. Joiner & A. J. van der Kogel, *Reds Basic Clinical Radiobiology*. Boca Raton,FL: CRC Press, 41–55.

Koh, D.M. & Collins, D.J., 2007. Diffusion-weighted MRI in the body: Applications and challenges in oncology. *American Journal of Roentgenology*, 188(6), 1622–1635.

Köhler, M.O. et al., 2009. Volumetric HIFU ablation under 3D guidance of rapid MRI thermometry. *Medical Physics*, 36(8), 3521–3535.

Konukoglu, E. et al., 2010. Image guided personalization of reaction-diffusion type tumor growth models using modified anisotropic eikonal equations. *IEEE Transactions on Medical Imaging*, 29(1), 77–95.

Krohn, K.A., Link, J.M. & Mason, R.P., 2008. Molecular imaging of hypoxia. *Journal of Nuclear Medicine*, 49(Suppl 2), 129S–148S.

Kubota, K., 2001. From tumor biology to clinical PET: A review of positron emission tomography (PET) in oncology. *Annals of Nuclear Medicine*, 15(6), 471–486.

Lapi, S.E., Lewis, J.S. & Dehdashti, F., 2015. Evaluation of hypoxia with Cu-ATSM. *Seminars in Nuclear Medicine*, 45(2), 177–185.

Law, M. et al., 2003. Glioma grading: Sensitivity, specificity, and predictive values of perfusion MR imaging and proton MR spectroscopic imaging compared with conventional MR imaging. *American Journal of Neuroradiology*, 24(10), 1989–1998.

Leder, K. et al., 2014. Mathematical modeling of PDGF-driven glioblastoma reveals optimized radiation dosing schedules. *Cell*, 156(3), 603–616. Available at: http://www.ncbi.nlm.nih.gov/pmc/articles/PMC3923371/.

Lencioni, R. et al., 2010. Hepatocellular carcinoma: New options for image-guided ablation. *Journal of Hepato-Biliary-Pancreatic Sciences*, 17(4), 399–403.

Levenberg, K., 1944. A method for the solution of certain non-linear problems in least squares. *Quarterly Journal of Applied Mathematics*, II(2), 164–168.

Lewin, J.S., Nour, S.G. & Duerk, J.L., 2000. Magnetic resonance image-guided biopsy and aspiration. *Topics in Magnetic Resonance Imaging*, 11(3), 173–183.

Li, S.P. & Padhani, A.R., 2012. Tumor response assessments with diffusion and perfusion MRI. *Journal of Magnetic Resonance Imaging*, 35(4), 745–763.

Lima, E.A.B.F. et al., 2016. Selection, calibration, and validation of models of tumor growth. *Mathematical Models and Methods in Applied Sciences*, 26(12), 2341–2368.

Lima, E.A.B.F. et al., 2017. Selection and validation of predictive models of radiation effects on tumor growth based on noninvasive imaging data. *Computer Methods in Applied Mechanics and Engineering*, 327, 277–305.

Liu, Y. et al., 2014. Patient specific tumor growth prediction using multimodal images. *Medical Image Analysis*, 18(3), 555–566.

Marquardt, D.W., 1963. An algorithm for least-squares estimation of nonlinear parameters. *Journal of the Society for Industrial and Applied Mathematics*, 11(2), 431–441 CR–Copyright © 1963 Society for Ind.

Menze, B.H. et al., 2015. The multimodal brain tumor image segmentation benchmark (BRATS). *IEEE Transactions on Medical Imaging*, 34(10), 1993–2024.

Moffat, B.A. et al., 2005. Functional diffusion map: A noninvasive MRI biomarker for early stratification of clinical brain tumor response. *Proceedings of the National Academy of Sciences of the United States of America*, 102(15), 5524–5529.

Morrow, M., Waters, J. & Morris, E., 2011. MRI for breast cancer screening, diagnosis, and treatment. *The Lancet*, 378(9805), 1804–1811.

Mosayebi, P. et al., 2012. Tumor invasion margin on the riemannian space of brain fibers. *Medical Image Analysis*, 16(2), 361–373.

Muzi, M. et al., 2015. [F-18]-Fluoromisonidazole quantification of hypoxia in human cancer patients using Image-derived blood surrogate tissue reference regions. *Journal of Nuclear Medicine: Official Publication, Society of Nuclear Medicine*, 56(8), 1223–1228.

Neal, M.L. et al., 2013. Discriminating survival outcomes in patients with glioblastoma using a simulation-based, patient-specific response metric. *PLoS ONE*, 8(1), 51951.

O'Connor, J.P.B. et al., 2011. Dynamic contrast-enhanced imaging techniques: CT and MRI. *The British Journal of Radiology*, 84(Spec Iss 2), S112–S120.

Omuro, A. & DeAngelis, L.M., 2013. Glioblastoma and other malignant gliomas: A clinical review. *JAMA*, 310(17), 1842–1850.

Padhani, A.R. et al., 2007. Imaging oxygenation of human tumours. *European Radiology*, 17(4), 861–872.

Padhani, A.R. et al., 2009. Diffusion-weighted magnetic resonance imaging as a cancer biomarker: Consensus and recommendations. Neoplasia (New York, N.Y.), 11(2), 102–125.

Painter, K.J. & Hillen, T., 2013. Mathematical modelling of glioma growth: The use of diffusion tensor imaging (DTI) data to predict the anisotropic pathways of cancer invasion. *Journal of Theoretical Biology*, 323, 25–39. Available at: http://www.sciencedirect.com/science/article/pii/S0022519313000398.

Panagiotaki, E. et al., 2014. Noninvasive quantification of solid tumor microstructure using VERDICT MRI. *Cancer Research*. 74(7), 1902–1912.

Pappa, V.I. et al., 1996. Role of image-guided core-needle biopsy in the management of patients with lymphoma. *Journal of Clinical Oncology*, 14(9), 2427–2430.

Rajendran, J.G. & Krohn, K.A., 2015. F-18 fluoromisonidazole for imaging tumor hypoxia: Imaging the microenvironment for personalized cancer therapy. *Seminars in Nuclear Medicine*, 45(2), 151–162. Available at: http://www.sciencedirect.com/science/article/pii/S000129981400124X.

Rockne, R. et al., 2009. A mathematical model for brain tumor response to radiation therapy. *Journal of Mathematical Biology*, 58(4–5), 561–578.

Rockne, R. et al., 2010. Predicting the efficacy of radiotherapy in individual glioblastoma patients in vivo: A mathematical modeling approach. *Physics in Medicine and Biology*, 55(12), 3271–3285.

Rockne, R.C. et al., 2015. A patient-specific computational model of hypoxia-modulated radiation resistance in glioblastoma using (18)F-FMISO-PET. *Journal of the Royal Society Interface*, 12(103), 20141174.

Scherer, H.J., 1938. Structural development in gliomas. *The American Journal of Cancer*, 34(3), 333–351.

Scott, J.G. et al., 2017. A genome-based model for adjusting radiotherapy dose (GARD): A retrospective, cohort-based study. *The Lancet Oncology*. 18(2), 202–211.

Soloviev, D. et al., 2012. [18F]FLT: An imaging biomarker of tumour proliferation for assessment of tumour response to treatment. *European Journal of Cancer*, 48(4), 416–424.

Sugahara, T. et al., 1999. Usefulness of diffusion-weighted MRI with echo-planar technique in the evaluation of cellularity in gliomas. *Journal of Magnetic Resonance Imaging*, 9(1), 53–60.

Sundgren, P.C. et al., 2004. Diffusion tensor imaging of the brain: Review of clinical applications. *Neuroradiology*, 46(5), 339–350.

Swan, A. et al., 2018. A patient-specific anisotropic diffusion model for brain tumour spread. *Bulletin of Mathematical Biology*, 80(5), 1259–1291. Available at: doi:10.1007/s11538-017-0271-8.

Swanson, K.R., Alvord, E.C. & Murray, J.D., 2000. A quantitative model for differential motility of gliomas in grey and white matter. *Cell Proliferation*, 33(5), 317–329.

Swanson, K.R., Harpold, H.L.P. et al., 2008. Velocity of radial expansion of contrast-enhancing gliomas and the effectiveness of radiotherapy in individual patients: A proof of principle. *Clinical Oncology*, 20(4), 301–308.

Swanson, K.R., Rostomily, R.C. & Alvord, E.C., 2008. A mathematical modelling tool for predicting survival of individual patients following resection of glioblastoma: A proof of principle. *British Journal of Cancer*, 98(1), 3–9.

Takayasu, K. et al., 1995. CT diagnosis of early hepatocellular carcinoma: Sensitivity, findings, and CT-pathologic correlation. *American Journal of Roentgenology*, 164(4), 885–890.

Therasse, P. et al., 2000. New guidelines to evaluate the response to treatment in solid tumors. *Journal of the National Cancer Institute*, 92(3), 205–216.

Titz, B., Jeraj, R. & Jeraj, B.T. and R., 2008. An imaging-based tumour growth and treatment response model: Investigating the effect of tumour oxygenation on radiation therapy response. *Physics in Medicine and Biology*, 53(17), 4471.

Trédan, O. et al., 2007. Drug resistance and the solid tumor microenvironment. *Journal of the National Cancer Institute*, 99(19), 441–454.

Vaupel, P. & Mayer, A., 2007. Hypoxia in cancer: Significance and impact on clinical outcome. *Cancer Metastasis Reviews*, 26(2), 225–239.

Villers, A. et al., 2009. Current status of MRI for the diagnosis, staging and prognosis of prostate cancer: Implications for focal therapy and active surveillance. *Current Opinion in Urology*, 19(3), 274–282.

Wang, C.H. et al., 2009. Prognostic significance of growth kinetics in newly diagnosed glioblastomas revealed by combining serial imaging with a novel biomathematical model. *Cancer Research*, 69(23), 9133–9140.

Warner, E., 2011. Breast-cancer screening. *New England Journal of Medicine*, 365(11), 1025–1032.

Weis, J.A. et al., 2013. A mechanically coupled reaction-diffusion model for predicting the response of breast tumors to neoadjuvant chemotherapy. *Physics in Medicine and Biology*, 58(17), 5851–5866.

Weis, J.A. et al., 2015. Predicting the response of breast cancer to neoadjuvant therapy using a mechanically coupled reaction-diffusion model. *Cancer Research*. 75(22), 4697–4707.

Weis, J.A., Miga, M.I. & Yankeelov, T.E., 2017. Three-dimensional image-based mechanical modeling for predicting the response of breast cancer to neoadjuvant therapy. *Computer Methods in Applied Mechanics and Engineering*, 314, 494–512.

Wen, P.Y. et al., 2010. Updated response assessment criteria for high-grade gliomas: Response assessment in neuro-oncology working group. *Journal of Clinical Oncology*, 28(11), 1963–1972.

West, G.B., Brown, J.H. & Enquist, B.J., 2001. A general model for ontogenetic growth. *Nature*, 413(6856), 628–631.

Whisenant, J.G. et al., 2014. Assessing reproducibility of diffusion-weighted magnetic resonance imaging studies in a murine model of HER2+ breast cancer. *Magnetic Resonance Imaging*, 32(3), 245–249.

Wind, J.J. et al., 2012. The role of adjuvant radiation therapy in the management of high-grade gliomas. *Neurosurgery Clinics of North America*, 23(2), 247–258.

Wong, K.C.L. et al., 2015. Tumor growth prediction with reaction-diffusion and hyperelastic biomechanical model by physiological data fusion. *Medical Image Analysis*. 25(1), 72–85.

Wong, K.C.L. et al., 2017. Pancreatic tumor growth prediction with elastic-growth decomposition, Image-Derived Motion, and FDM-FEM Coupling. *IEEE Transactions on Medical Imaging*, 36(1), 111–123.

Woolf, D.K. et al., 2014. Evaluation of FLT-PET-CT as an imaging biomarker of proliferation in primary breast cancer. *British Journal of Cancer*, 110(12), 2847–2854.

Yankeelov, T. et al., 2010. Modeling tumor growth and treatment response based on quantitative imaging data. *Integrative Biology*, 2(7–8), 338–345.

Yankeelov, T.E. & Gore, J.C., 2009. Dynamic contrast enhanced magnetic resonance imaging in oncology: Theory, data acquisition, analysis, and examples. *Current Medical Imaging Reviews*, 3(2), 91–107.

Yankeelov, T.E. et al., 2015. Toward a science of tumor forecasting for clinical oncology. *Cancer Research*, 75(6), 918–923.

Yap, J.T. et al., 2004. Image-guided cancer therapy using PET/CT. *The Cancer Journal*, 10(4), 221–223.

Zaidi, H., Vees, H. & Wissmeyer, M., 2009. Molecular PET/CT imaging-guided radiation therapy treatment planning. *Academic Radiology*, 16(9), 1108–1133.

Zisman, A. et al., 2002. Mathematical model to predict individual survival for patients with renal cell carcinoma. *Journal of Clinical Oncology: Official Journal of the American Society of Clinical Oncology*, 20(5), 368–374.

Looking ahead
Opportunities and challenges in radiomics and radiogenomics

RUIJIANG LI, YAN WU, MICHAEL GENSHEIMER, MASOUD BADIEI KHUZANI, AND LEI XING

22.1 OVERVIEW OF RADIOMICS AND RADIOGENOMICS

The fields of radiomics and radiogenomics have experienced significant growth in recent years. Many studies have identified novel putative imaging signatures that demonstrated promising results in terms of diagnostic, prognostic, or predictive performance, and appeared to improve upon currently used imaging metrics in various oncologic and other clinical applications.[1–3] Radiomics involves the high-throughput extraction of quantitative image features such as shape, histogram, and texture that capture tumor heterogeneity, which are applicable to any type of standard-of-care clinical images such as CT, MRI, or positron emission tomography (PET).[4] Radiomics may be used in a variety of clinical settings including cancer screening/early detection, diagnosis, prediction of prognosis, and evaluation of treatment response. When combined with appropriate statistical or bioinformatics tools, predictive models can be developed that will potentially improve accuracy of clinical outcomes.[5–8] Radiogenomics concerns the study of relations between radiomic features at the tissue scale and underlying molecular features at the genomic, transcriptomic, or proteomic level.[9–16] This may allow identification of the underlying biological basis and molecular underpinnings of clinically relevant imaging phenotypes. Given their potentially enormous clinical impact, there has been significant development in radiomics and radiogenomics, and new paradigms and directions continuously emerge. In particular, deep learning-based radiomics and radiogenomics [17–20] are on the horizon and promise to mitigate many of the bottleneck problems in traditional approaches based on supervised learning.

22.2 EMERGING DIRECTIONS

22.2.1 INTRA-TUMORAL PARTITIONING TO CHARACTERIZE SPATIAL HETEROGENEITY

A major benefit of imaging is that it provides noninvasive depiction of the tumor in its entirety. Currently, the vast majority of radiomic studies follow a standard pipeline where the gross tumor is first segmented and then features are extracted based on the aggregate analysis of the tumor. While texture features of the tumor provide a measure of intra-tumor heterogeneity to a certain extent, this characterization is not complete. Because their calculation is applied to the entire tumor as a whole, this approach implicitly assumes that the tumor is well mixed, and neglects the regional variations within a tumor.

It has been widely recognized that tumors often do not consist of spatially well mixed cell populations, but demonstrate regional variations in genotypes and phenotypes due to clonal evolution.[21,22] Some parts of the tumor may be more biologically aggressive and treatment-resistant than others. Image-based tumor partitioning could reveal aggressive subregions that are more important for determining prognosis and treatment response.[23] To address this issue, Gatenby and colleagues proposed to cascade T1 post-gadolinium MRI with T2-weighted fluid attenuated inversion recovery sequences to divide the whole tumor into multiple regional habitats with distinct contrast enhancement and edema/cellularity.[24] A preliminary study of 32 the cancer genome atlas (TCGA) glioblastoma multiforme patients showed that the distribution of MRI-based habitats was significantly correlated with survival. Cao and colleagues proposed a clustering-based algorithm to identify the significant subvolumes for primary tumors from dynamic contrast-enhanced (DCE) MRI in head and neck cancer.[25] They showed that large poorly perfused subvolumes of primary tumor at baseline and persisting during the early course of chemoradiotherapy might be used to predict local or regional failure, which could potentially stratify patients for local dose intensification.

Earlier studies have used somewhat crude methods to segment intra-tumor subregions, e.g., by thresholding. Wu et al. developed a robust tumor partitioning method by a two-stage clustering procedure and identified three spatially distinct and phenotypically consistent subregions in lung tumors.[26] One subregion associated with the most metabolically active, metabolically heterogeneous, and solid component of the tumor was defined as the "high-risk" subregion. The volume of high-risk intra-tumoral subregion predicted distant metastasis and overall survival in patients with non-small cell lung cancer (NSCLC) treated with radiation therapy.

Tumor partitioning can be combined with radiomic or texture analysis to allow more detailed and refined image phenotyping. Wu et al.[27] showed that the early change of texture features for the intra-tumoral subregion associated with fast contrast-agent washout at DCE MRI predicted pathological complete response to neoadjuvant chemotherapy in breast cancer. Chaudhury et al. showed that texture features of intra-tumoral regions with rapid gadolinium washout were correlated with estrogen receptor status and nodal metastasis in breast cancer.[28] Cui et al.[29] performed radiomic analysis on tumor subregions and defined 120 multiregional image features on MRI in glioblastoma. A 5-feature radiomic signature was identified and independently validated in an external cohort to predict overall survival, which outperformed whole-tumor measurements. Stoyanova and colleagues investigated the association of MRI radiomic features with prostate cancer gene expression profiles from MRI-guided biopsy tissues.[30] They extracted radiomic features for the identified habitats on MRI/3D-ultrasound fusion and found strong associations between radiomic features and gene expression profiles.

Beyond separately analyzing tumor subregions themselves, a recent study further explored the spatial interaction among subregions to better characterize intra-tumoral heterogeneity and investigate their clinical relevance in breast cancer.[31] Wu et al. analyzed a discovery cohort and an external multicenter validation cohort from the Investigation of Serial Studies to Predict Your Therapeutic Response With Imaging and Molecular Analysis (I SPY-1) clinical trial. Each tumor was divided into multiple spatially segregated, phenotypically consistent subregions based on perfusion imaging parameters. A multiregional spatial interaction matrix was defined, based on which 22 image features were calculated. Three intra-tumoral subregions with high, intermediate, and low perfusion were identified and showed high consistency between two cohorts. Network analysis of multiregional image features stratified patients regarding recurrence-free survival (RFS) in both cohorts. Aggressive tumors were associated with larger volume of the poorly perfused subregion as well as interaction between poorly and moderately perfused subregions and surrounding parenchyma. On multivariate analysis, the proposed imaging marker was independently associated with RFS adjusting for age, estrogen receptor (ER), progesterone receptor (PR), human epidermal growth factor receptor 2 (HER2) status, tumor volume, and pathological complete response (pCR). Further, imaging stratified patients for RFS within the ER+ and HER2+ subgroups, and among patients without a pCR after neoadjuvant chemotherapy. These results confirm that breast cancer consists of multiple spatially distinct subregions and showed novel markers of imaging spatial heterogeneity are an independent prognostic factor beyond traditional risk predictors.

Taken together, these studies highlight the benefit of tumor partitioning to identify biologically relevant, aggressive intra-tumoral subregions.[32] This may have significant implications for clinical oncology by identifying important tumor regions for biopsy. In addition, this is particularly relevant for radiotherapy treatment planning and adaptation, because high-risk tumor subregions associated with the aggressive disease can then be targeted with a radiation boost to potentially improve local control and patient survival. This approach is generally applicable to many types of solid tumors that demonstrate intra-tumor heterogeneity at imaging and may lead to identification of additional clinically useful imaging biomarkers.

22.2.2 LEVERAGING COMPLEMENTARY VALUES OF DIFFERENT TYPES OF DATA IN RADIOGENOMICS

Radiogenomics aims to integrate imaging and genomic data with the goal of gaining biological interpretation or improving patient stratification for precision medicine.[9-14,33-38] Historically, there have been two major types of radiogenomic association studies. One approach that most radiogenomic studies so far have adopted is to find imaging correlates or surrogate of a specific genotype or molecular phenotype of the tumor. For instance, CT image features (either semantic or radiomic) were found to be associated with epidermal growth factor receptor (EGFR) or *KRAS* genetic mutations in lung cancer.[20,39-43] MRI radiomic features were correlated with intrinsic molecular subtypes or existing genomic assays in breast cancer.[44-48]

Radiogenomics can also be used to create association maps between molecular features and a specific imaging phenotype to reveal its biological underpinnings. For example, tumors with higher maximum standardized uptake value from fluorodeoxyglucose Positron-emission tomography (FDG-PET) were demonstrated to be associated with the epithelial-mesenchymal transition in non-small cell lung cancer.[49] In another recent radiogenomic study, lung tumors with a higher CT image-based pleural contact index are associated with upregulated pathways such as extracellular matrix remodeling, which is implicated in cancer invasion.[50] In breast cancer, it has been shown that heterogeneous enhancing patterns of tumor-adjacent parenchyma from perfusion MRI are associated with the tumor necrosis signaling pathway and poor survival.[14]

Beyond radiogenomics association studies, one important emerging direction is to leverage the complementary power of imaging and molecular data and integrate them into a unifying model to further improve prediction accuracy of clinical outcomes. This is particularly relevant for improving the value of imaging for precision medicine. Cottereau et al.[51] showed that combination of molecular profile and metabolic tumor volume at FDG-PET imaging improved patient stratification for progression-free and overall survival in diffuse large B-cell lymphoma. Grossmann et al.[38] combined gene expression and CT radiomic signatures to enhance the accuracy of survival prediction in lung cancer. Cui et al.[52] showed that integrating O-6-Methylguanine-DNA Methyltransferase (MGMT) methylation status and volume of the high-risk subregion at multiparametric MRI improved survival stratification in glioblastoma. Recently, Lee et al. developed a CT image-based prognostic signature and validated it in an external cohort of patients with stage I NSCLC.[53] Further, it was shown that a composite imaging and genomic signature improved prognostic accuracy upon either one used alone. While still preliminary, these studies provide the initial evidence that image-based biomarkers can provide additional information beyond molecular analysis alone and integrating both will provide more accurate assessment of prognosis and outcomes for individual patients.

Finally, one interesting area of investigation is to classify tumors into subtypes based on imaging phenotypes rather than molecular features. Itakura et al. identified novel glioblastoma subtypes based on MRI phenotypes that are associated with distinct molecular pathway activities.[12] In a more recent study, Wu et al.[53] discovered three breast cancer subtypes, which were characterized by distinct imaging phenotypes of the tumor: homogeneous intra-tumoral enhancement, minimal parenchymal enhancement, and prominent parenchymal enhancement. They further independently validated the imaging subtypes in multiple cohorts over 1,000 patients, and showed that each of the imaging subtypes is associated with distinct prognoses and dysregulated molecular pathways, and is complementary to known intrinsic molecular subtypes.

22.2.3 New approaches and applications of radiomics and radiogenomics

Radiomics and radiogenomics is a technology intense and clinically oriented discipline. With the rapid technical and clinical advancements, the field is also progressing in an unprecedented speed. Applications specifically designed to meet various demands of new imaging information and/or new therapeutics are being proposed frequently. The radiomics today, for instance, has gone far beyond the tradition CT and MRI data and extended to other existing or emerging imaging modalities, such as PET and other functional imaging techniques. Practically, development of multimodality and multiclassifier radiomics predictive models should help to best utilize the available imaging data to generate analytical features to characterize tumor spatial complexity, elucidate the tumor genomic heterogeneity and composition, and identify subregions in terms of tumor viability or aggressiveness, and response to therapeutics. Ultimately, these radiomics approaches can help to reveal unique information about tumor behavior and lead to better staging, diagnosis, and prediction of outcomes of patient treatments.

Numerous investigations in using the omics approaches to assess the response of traditional chemotherapy and radiation therapy have been carried out in the past few years.[12] It is important to point out that the omics techniques have been applied to some emerging therapeutics. Notably, Sun et al. have reported a radiomics approach to assess tumor-infiltrating cluster of differentiation 8 (CD8) cells and response to anti-programmed cell death protein 1 (PD-1) or anti-PD-L1 immunotherapy.[54] Clinically, immunotherapies with and without radiation therapy have emerged as one of the promising approaches for cancer treatment by exploiting patients' own immune systems to specifically target tumor cells. However, it has been recognized that responses often occur in only a subset of patients in any given immunotherapy. This treatment is also associated with drug toxicity (e.g., cytokine storm) and high cost. As this treatment modality continues to evolve, a significant clinical question that needs to be addressed is to determine which patients would benefit from immunotherapies. In addition, there is increasing need for newer methods to evaluate the efficacy and potential toxicities of the treatment and monitor cancer patients' prognosis. The imaging predictors resulted from radiomics approach may prove to be effective in predicting the immune phenotype of tumors and to infer clinical outcomes for immunotherapy patients.[54]

22.3 CHALLENGES IN OMICS DESIGN AND DATA COLLECTION

Given the growing interest in the field, it is important to highlight some technical and practical challenges associated with radiomics and radiogenomics to maximize the translational potential. The major challenges lie in insufficient data and limited prediction model.

22.3.1 RATIONAL RADIOMIC DESIGN

The lack of reproducibility is probably the biggest challenge toward the clinical translation of radiomics. Several pitfalls can be implicated, including poor experimental design, multiple testing leading to false discovery and model overfitting, and unadjusted biases or confounding factors among others.[55] A rational radiomic design should include robustness evaluation of radiomic features regarding segmentation variability, as well as rigorous model training and testing. Each radiomic analysis step should be documented in complete detail and original codes and data should ideally be accessible, allowing other investigators to replicate the results. Recently, Lambin and colleagues have proposed the radiomics quality score (RQS) as evaluation criteria for radiomic studies.[2] The RQS contains 16 key components that aim to minimize bias and enhance the reproducibility of radiomics models. These recommendations cover the image acquisition protocol, image preprocessing, image feature extraction, and statistical modeling, which establish the reporting guidelines for future radiomic studies.

22.3.2 ISSUES IN RADIOMICS DATA COLLECTION

Big data size is crucial to achieve high accuracy, gain statistical confidence, as well as avoid overfitting.[56] A reasonable rule of thumb is to prepare 10 samples for each feature in a model-based classifier.[4] Furthermore, the incorporation of clinical factors or genomic information into the predictive model puts a higher demand for the size of datasets.

However, it is not easy to collect a large dataset with disease diagnosis or progression information available. For tumor characterization, pathology reports that confirm the malignancy or grade of tumor need to be provided, which are available via surgery or biopsy. For treatment response prediction, the progression data to be collected could be time of progression, survival, or longitudinal change in tumor size. In particular, treatment response prediction faces a bigger challenge on data collection than tumor characterization. Treatment response prediction is typically performed on a cohort of patients that receive the same therapy to treat the same type of tumor so as to provide technical supports for decision making. Due to the variety of tumor type and cancer therapy, the patient number in a cohort is prone to be small. The problem becomes more severe when patients receive novel treatment for rare tumors. In these situations, the prediction of treatment response would be more interesting and provide more insights for patient care. For this reason, some radiomics studies were performed even though the sample size is small (with 20 patients).

In addition to the limited progression data, the availability of data is further reduced by manual annotation. In radiomics, region segmentation is typically performed before feature extraction to localize the region and boundary of tumors for further analysis on texture, shape, or boundary smoothness. Currently, manual contour of structures is commonly adopted by physicians, which is time consuming, therefore reducing the quantity of annotated images. Software tools are being developed for automatic segmentation,[57–59] which, however, need further validation for accuracy or reliability.

The quality of images is also very important to radiomics, where people consider the selection of appropriate imaging modality and the optimization of image quality.

22.3.3 STRATEGY OF DATA SHARING AND QUANTITATIVE IMAGING

Data sharing is an effective way to expand datasets, which, however, is hindered by the inconsistency in image quality caused by diverse imaging protocols. Currently, variation in imaging protocols is allowed in different medical centers with various prescriptions on image resolution, radiation dose, acquisition timing after contrast agent injection, as well as other parameters (such as relaxation time [TR], echo time [TE], flip angle in MRI). These specifications will have impacts on signal to noise ratio (SNR), tissue contrast, and contrast enhancement patterns, which are indicative of characteristics and progression of disease.

It is important to standardize imaging protocol for data sharing. Consensus should be reached on the prescription of imaging protocols. Moreover, quantitative imaging should also be advanced, which alleviates the dependence on imaging parameters. In recent years, there have been much effort in this direction. The National Cancer Institute (NCI) initiated the Quantitative Imaging Network (QIN) to promote the development and validation of imaging methods for the prediction of tumor response to therapies in clinical trial settings. In QIN, the Bioinformatics Working Group (BIWG) defines the common informatics requirements and establishes consensus standards that are generalizable and useful to the imaging community. The Cancer Imaging Archive (TCIA) under the support of NCI provides an open archive of cancer-specific medical images and metadata, where a huge amount of clinical and research images are uploaded. The American Association of Physicists in Medicine (AAPM) provides guidelines for quantitative imaging in terms of modality-dependent reports on imager operation and testing. The Radiological Society of North America (RSNA) sponsors the Quantitative Imaging Biomarkers Alliance (QIBA) to promote quantitative imaging and the use of imaging biomarkers in clinical trials, which has 14 biomarker committees for CT, MR, nuclear medicine, and ultrasound. By successful data sharing, the challenge on data quantity should be alleviated.

As far as the quality of image is considered, it is important to select the appropriate imaging modality in the very beginning, whose advantages will be highly appreciated (high spatial resolution of CT, superior soft tissue contrast of MRI, or physiological information from PET). For a given imaging modality, imaging protocol should be optimized for an optimal tradeoff between SNR, spatial resolution, and scan time. In many cases, spatial resolution, particularly in through-plane direction, is compromised, posing challenge on texture analysis. In this situation, super-resolution as post-processing technique can help.[60,61]

Meanwhile, novel image acquisition techniques should be adopted as much as possible. For example, MR fingerprinting (MRF) technique was recently proposed as a fast imaging technique to provide quantitative multiparametric MRI map. By randomly changing imaging parameters (TR, TE, flip angles), the signal evolution pattern of different tissues is identified in clinically acceptable time. The resultant quantitative T1 and T2 maps are purely determined by tissue properties and independent of imaging parameters, facilitating data sharing. In recent years, compressed sensing has also been widely adopted for the reconstruction of random sampled data,[62–64] which is an optimization algorithm that enforces data consistency while promoting the sparsity of images. In this way, compressed sensing successfully recovers signal from incoherent artifacts and noise, achieving high SNR in MRI, CT, and PET.

In brief, to enlarge the quantity of data, it would be beneficial to establish data sharing mechanism, standardize imaging protocols, promote quantitative imaging, and automate image annotation. By selecting appropriate imaging modality, optimizing imaging protocol, and adopting advanced imaging techniques, image quality can be improved, making contributions for radiomics framework.

22.4 CHALLENGES ON FEATURE-BASED PREDICTION MODELS AND THE NEED FOR DEEP LEARNING-BASED RADIOMICS

22.4.1 ISSUES WITH FEATURE-BASED APPROACH

Existing radiomic features can be divided into four major classes: (a) morphological, (b) statistical, (c) regional, and (d) model-based. Each category quantitatively characterizes specific aspects of a disease. In reality, seamlessly integration of radiomic data with demographic, clinical, imaging, pathological, and genomic information to decode/decipher different types of disease or tissue biology is most challenging. Algorithmically, feature-based models are widely used, such as logistic regression (LR), decision tree, support vector machine (SVM), and k-mean clustering. In these models, clinical outcomes are predicted from image features that are explicitly extracted from radiological images using specified descriptors of tumor texture, tumor shape, heterogeneity, smoothness of tumor boundary, etc.

There are two major issues in feature-based models. The first is how to extract features in an optimal sense. Various feature descriptors have been proposed to extract information on texture, shape, and boundary, but none of them work perfectly. The accuracy of extracted features could be further decreased by the inaccuracy caused by annotation (segmentation). Other features have never been quantified, such as the opacity and location of lung nodules, but known to be indicative of tumor characterization. The lack of effective quantification method hinders the translation of human knowledge to machine interpretable information for more accurate prediction. Moreover, there could be features beyond current human perception, but contributive to outcome prediction.

Another challenge is how to integrate complementary information for comprehensive prediction. Given effective feature descriptors, it is challenging to select most relevant features, assign appropriate weights, explore inter-dependence between features, reduce the dimension of features, integrate image features extracted from multimodal images, and fuse image features and clinical factors into one unified model. With these major issues remaining in conventional feature-based prediction models, a new generation of techniques, deep learning, is appealing.

22.4.2 DEEP NEURAL NETWORKS FOR RADIOMICS AND RADIOGENOMICS

Deep neural networks have been leading to a series of breakthroughs in many applications.[65] The tremendous possibilities they brought to medical physics have triggered a flood of activities in the clinical applications of the technology.[17] Dramatically different from traditional radiomics models, deep learning represents a specific type of machine learning and does not require hand-crafted features to build prediction models. The approach naturally optimizes feature extraction and classification in an end-to-end fashion. In addition, the huge number of parameters and hyper-parameters provide the potential to represent sophisticated nonlinear mappings with high accuracy. It is applicable to a number of tasks in image analysis and quantitative imaging.

22.4.3 CONVOLUTIONAL NEURAL NETWORK AS CLASSIFIER FOR IMAGE PROCESSING APPLICATIONS

The neural network technique is a machine learning method that makes prediction. A neural network can be thought of as a collection of nodes that are positioned in multiple layers and connected with their peers at different layers. At each node, the input passes through a linear and a nonlinear operation, and the generated feature map is forwarded to the subsequent layer. Figure 22.1 illustrates a very simple neural network with three layers. There are different categories of neural network, such as standard fully connected neural network, convolutional neural networks, recurrent neural networks, etc.

A convolutional neural network is widely used for image related applications.[66,67] The use of convolution has two benefits. First, spatial structures are well preserved by 2D or 3D convolution operations. Moreover, the number of model parameters is significantly reduced due to the parameter sharing provided by convolution. An input image would have required a huge number of parameters in a standard fully connected neural network, which is not the case in a convolutional neural network. Convolution kernels that extract features are shared across the whole image, since feature detectors that are useful in one part of the image could be useful in other parts of the image (an edge detector is a good example). No matter how large the input image is, the number of parameters defined in convolutional kernels remains small, relaxing the demand on quantity of training data, and facilitating the training procedure. Therefore, convolutional neural networks are effective and efficient for quantitative image analysis and radiomics applications.

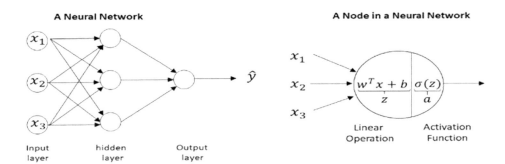

Figure 22.1 Illustration of a neural network.

22.4.4 OPTIMIZED FEATURE DESCRIPTOR

In radiomics, clinical outcomes are predicted based on image features. In conventional classifiers, features are extracted using manually specified feature descriptors, whereas in deep convolutional neural networks, optimal feature descriptors are learned in an end-to-end mapping that integrates feature extraction and classification into a streamline. Therefore, feature extractions in convolutional neural network are adaptive to the final prediction and expected to outperform conventional feature-based models, given a large enough dataset size.

For example, conventionally, texture features are extracted using specified filter banks that correspond to multiple orientations and scales, such as the Leung-Malik (LM) filter bank, Schmid filter bank, and Maximum Response filter bank, as shown in Figure 22.2. Particularly, LM is a set of first and second derivatives of 2D Gaussian functions at six orientations and three scales, coupled with eight Laplacian of Gaussian and four Gaussian functions. If a convolutional neural network is employed for the same purpose, the parameters of the filter bank will be adaptively learned with loss minimized. Therefore, the convolutional neural network model has the potential to outperform hard coded filter banks in texture feature extraction.

22.4.5 UNIFIED CLASSIFIER WITH COMPREHENSIVE FEATURES INCORPORATED

In most cases, features are not individually extracted for texture, shape, or smoothness of region boundary in a deep convolutional neural network. Instead, discriminative features are extracted, representing comprehensive information. Even features that have not been manually quantified or discovered to be relevant are implicated extracted and incorporated.

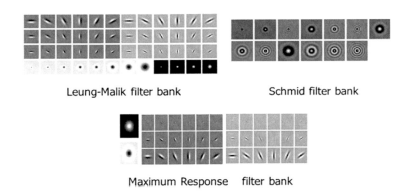

Figure 22.2 Examples of manually specified filter bank, which are candidates for texture descriptor. Each filter corresponds to a specific orientation and scale.

Moreover, a deep convolutional neural network is a unified prediction model. Appropriate weights are naturally assigned to various features. The inter-dependency between different features is implicated, exploited. Information obtained from multimodal radiological images, genomics, and clinical factors are easily integrated into the unified classifier.

22.4.6 DEEP LEARNING AND THE SIZE OF TRAINING DATASET

Transfer learning is a popular approach in deep learning, where a model is pre-trained for one task from one dataset and reused as the starting point for another task given another dataset.[65] The fine-tuning of a pre-trained convolutional neural network requires relatively small data to achieve high performance with little concern of overfitting.

Another approach to increase the size of dataset is to augment the data via applying various elastic deformations to training images.[68,69] Using the deformed images, the network can learn representations that are invariant to transformations, without using the annotated image corpus. Data augmentation in conjunction with a U-Net type network has shown remarkable success for biomedical image segmentation, mainly due to the fact that deformation is the most common variation in tissue, and realistic deformations can be captured extremely well using data augmentation techniques.[66] Data augmentation has also shown successful results for unsupervised feature learning and yields improvement in classification accuracy compared to other competing unsupervised methods.[70]

22.4.7 INTERPRETABILITY OF DEEP LEARNING MODELS

With the growing success of deep neural networks in classification and generative tasks, there is a corresponding need to be able to explain their outputs. A key property is *interpretability*—the ability to *comprehend* and *explain* the model output and to build confidence about detecting model bias. Interpretable machine learning techniques are particularly useful and relevant to address issues that arise when deep networks generate unexpected responses, or give false positives for unrecognizable images. To elucidate the inner structure of neurons in hidden layers, three approaches have been proposed in the literature, namely: (1) feature visualization, (2) saliency maps, and (3) embedding methods. In the sequel, we review these methods.

1. *Feature visualization*: This can be classified into input modification methods and deconvolutional methods. A comprehensive taxonomy of feature visualization methods can be found in Grün et al. (2016).[71] Input modification methods are visualization techniques in which the input is modified, and the resulting changes are measured in the output or intermediate layers of the network. These methods see the network (or all the layers before the layer of interest) as a black-box, and they are designed to visualize properties of the function this black-box represents. The resulting visualizations can be used to analyze the importance and locality of features in the input space.

   Deconvolutional methods based on deconvolutional networks (DeConvNets) reverse the path of excitatory stimuli through the network to reveal which pixels of the input image are responsible for the observed activations. A DeConvNet can be thought of as a convolutional network model that uses the same components such as filtering and pooling, but in reverse. As a result, it is a map from feature space to pixels or voxels. Deconvolutional networks (DeConvNets) were originally proposed as a method for unsupervised feature learning[72] and later applied to visualization.[73]

2. *Saliency maps*: While feature visualization is a useful technique to perceive *what* representations the network learns, it does not explain *how* the network makes inference based on such learned representations. Saliency maps provide a tool to address such questions by establishing the relationships between neurons.[74] This method is derived from the concept of saliency in images. For a given image, the saliency represents how distinctive the image regions are and in what order the eye and the nervous system process them.

3. *Embedding methods*: The embedding methods (a.k.a. dimensionality reduction methods) attempt to visualize the weight parameters of trained neural networks by embedding high-dimensional

parameters into a two or three-dimensional manifold while preserving the pairwise distances of the points. There are many embedding methods that have been developed with the intuition of embedding high-dimensional vectors in a low-dimensional space. Among such embedding methods, t-distributed stochastic embedding (t-SNE) technique[75] proposed by Van der Maaten and Hinton is widely recognized to consistently produce appealing results. t-SNE itself is a variation of stochastic neighbor embedding[76] that is much easier to optimize, and produces significantly better visualizations by reducing the tendency to crowd points together in the center of the map. For some applications of embedding methods to visualize parameters of deep networks see.[65,77]

22.4.8 MATHEMATICAL FOUNDATIONS OF DEEP LEARNING METHODS

Although research into rigorous mathematical foundations of deep convolutional neural networks is ongoing, recent progresses provide a fair understanding of the intrinsic properties of deep convolutional neural networks and the function of its building blocks.

The mathematical analysis of deep convolutional neural networks for feature extraction was initiated by Mallat.[78] Specifically, Mallat studied scattering networks—a cascade of wavelet transforms convolutions with nonlinear modulus and averaging operators. The scattering networks are distinguished from the standard convolutional neural networks in that they lack the max pooling operation, and the convolution in each network layer is followed by the modulus nonlinearity. Mallat has shown that such scattering wavelet networks compute a translation invariant image representation which is stable to deformations and preserves high-frequency information for classification. Wiatowski and Bölcskei have generalized Mallat's results by developing a theory that includes general convolutional transforms, or in more technical parlance, general semi-discrete frames (including Weyl-Heisenberg filters, curvelets, shearlets, ridgelets, wavelets, and learned filters), general Lipschitz-continuous nonlinearities (e.g., rectified linear units, shifted logistic sigmoids, hyperbolic tangents, and modulus functions), and general Lipschitz-continuous pooling operators emulating, e.g., subsampling and averaging (but not max-pooling).[79]

22.5 DEEP NEURAL NETWORKS AS APPLIED TO RADIOMICS

22.5.1 OPTIMIZATION WITH FORWARD AND BACKWARD PROPAGATION

Convolutional neural networks are a category of learning algorithms that form the basis of most deep learning methods. A convolutional neural network is composed of multiple nodes, each having a convolution function and a nonlinear activation. The convolution function is parameterized by weights and biases and may be followed by zero-padding to keep image size constant. Subsequently, a nonlinear activation function is applied, which could be sigmoid, hyperbolic tangent, rectified linear unit (ReLU), or parametric ReLU (PReLU). The last two functions are more frequently utilized and defined as $ReLu = \max(x, 0)$ and $PReLu = \max(x, 0) + \alpha \min(0, x)$. By introducing a new parameter, PReLU avoids zero gradients at the negative part (so that both positive and negative responses of the filters are respected), leading to an improved performance over ReLU.

As a supervised learning method, convolutional neural networks are optimization algorithms in essence. The supervised training aims to find model parameters that best predict the label from the input image, where a label can be a category in a classification problem, or a continuous value in a regression problem. The discrepancy between the predictions and the ground truth, named "loss function," can be defined to be a variety of functions, such as L1 norm, L2 norm, cross-entropy, dice score coefficient, etc. In addition, a regularization term can be incorporated into the loss function, which introduces a tradeoff between the bias and variance to avoid overfitting.

In a convolutional neural network, errors are back-propagated so that the gradient of every parameter can be calculated according to the chain rule.[65] Given the gradients, parameters are updated using stochastic gradient descent algorithms. To speed up convergence and reduce the risk of falling into local minimum, algorithms that support adaptive learning rates are proposed, such as Momentum, RMSProp, and Adam.

In Momentum, the current update is a linear combination of the gradient and the previous update.[80] In RMSProp (root mean square propagation), the learning rate of every parameter is calculated from a running average of recent gradients for that weight.[81] In Adam (adaptive moment estimation), adaptive learning rates are computed from estimates of first and second moments of the gradients.[82] These optimization methods offer faster convergence rate than conventional stochastic gradient descent methods.

22.5.2 ARCHITECTURE OF CONVOLUTIONAL NEURAL NETWORKS

In the past decade, there have been quite a few milestones in the development of convolutional neural network architecture. LeNet[83] and AlexNet[84] were first proposed with few convolutional layers (two to five layers) and relatively large-sized convolutional kernels. Starting from the VGG model,[85] convolutional neural networks became much deeper with smaller size kernels. Deep network architectures help to reduce number of parameters and approach better performance. They generally have a lower memory footprint during inference. Alternatively, a set of convolutions of different sizes composed a network-in-network[86] to replace the conventional mapping, which was used as the building block of GoogLeNet[87] (and known as the inception block).

As neural networks grew deeper, training became more challenging. To facilitate network training, ResNet[88] was proposed, introducing the idea of residual learning. Instead of the mapping function itself, the residual (difference between the prediction and the ground truth) is learned by establishing a shortcut connection between the input and output of a residual block (which is composed of several consecutive network layers). In addition, when identity mappings[89] are used as the shortcut connections and after-addition activation, the forward and backward signals can be directly propagated from one block to any other block. In this way, improved performance and nice convergence behavior were achieved. More architectures with sophisticated shortcut patterns were proposed, such as DenseNet.[90]

As the scale of deep neural network became larger, the prevention of overfitting became more important with limited training data. The dropout technique[91] was developed, which means random dropout of network nodes (and related connections) during the training of neural networks. This prevents network nodes from co-adapting too much. Consequently, the magnitude of weightings in convolutional kernels is limited, which is similar to incorporating a regularization term into a loss function. The dropout technique significantly reduces overfitting and gives major improvements over other regularization methods.

Similarly, batch normalization[92] was proposed to prevent network layers from co-adapting. The inputs of every layer were normalized for each mini-batch, which facilitates higher learning rates in deep neural networks and reduces the attention on initialization. Batch normalization sometimes acts as a regularizer and eliminates the need for dropout.

Multi-stream architectures[93] are natural for convolutional neural networks, since multiple sources of information are easily accommodated in the form of channels presented to the input layer, and channels can be merged at any point in the network. Multistream architectures could be useful for the integration of radiology image, genomic information, and other clinical factors. Additionally, multiresolution architectures[94] were investigated to provide context information for a variety of applications, such as abnormality detection, segmentation, and image reconstruction. Down-scaled representations that support larger context are provided as an additional input channel, offering complementary information to original high resolution patches.

Extended from the multiresolution design, the encoder-decoder architecture was invented for applications whose output is an image (instead of a label), such as segmentation, super-resolution, image reconstruction, and image synthesis. U-net[66] is one of the most popular architectures in this category, where the whole image is processed rather than dealing with numerous image patches using the sliding window or shift-and-stitch patterns.[95] Here, feature maps are first downsampled in an encoder, and then upsampled in a decoder. Additionally, shortcut connections are established between the same levels of both paths to compensate detailed information gradually lost in downsampling. There are several variants of U-net, both in 2D and 3D.[96,97]

22.5.3 APPLICATIONS OF DEEP NEURAL NETWORKS IN RADIOMICS CLASSIFICATION

Given the salient features of deep learning, it is foreseeable that the approach will play an important role in radiomics for tissue characterization and treatment prediction. In addition, deep learning is able to support other related tasks, such as segmentation, super-resolution, and image reconstruction.

Classification: There are different strategies to apply deep learning for tissue characterization or treatment response prediction. Some earlier studies used convolutional neural networks for feature extraction or classification, whereas more recent investigations used the networks to accomplish both feature extraction and classification in an end-to-end mapping.

In a study that classified the knee osteoarthritis severity,[98] discriminative image features were extracted using convolutional neural networks. Subsequently, a linear SVM model was trained on these features and achieved higher classification accuracy in comparison to previous methods. Similarly, in another study that predicted overall survival time of patients with high-grade glioma,[99] image features were extracted from multimodal preoperative brain images (T1 MRI, functional magnetic resonance imaging [fMRI], and diffusion tensor imaging [DTI]) using multiple 3D convolutional neural networks, which were fed into a SVM model to make the final prediction.

Convolutional neural networks were also adopted for pure classification, where image features were extracted using other methods. In a study that predicted Alzheimer's disease from brain MRI,[100] a sparse auto-encoder was used to learn filters for convolution operations, and a 3D convolutional neural network was built for Alzheimer's disease prediction, which took the filters learned with the auto-encoder as the input.

Other investigations employed convolutional neural networks for both feature extraction and classification. In a study that predicted lung nodule malignancy from CT images,[101] separate convolutional neural networks were trained for feature extraction and classification. First, image feature descriptors were learned from a very large discovery set with malignancy likelihood labeled given based on multiple radiologists' assessments. Subsequently, on a smaller diagnosis-definite set with true pathologically proven lung cancer labels, features were obtained from multiple nodules using the feature extractor described above and fed into a multiple instance learning (MIL) enabled convolutional network model for patient level malignancy prediction.

More generally, a single convolutional neural network was established as the unified framework that incorporated feature extraction and classification into an end-to-end mapping. In a study that aimed for pulmonary nodule detection in CT,[102] a multiview convolutional network was developed, as illustrated in Figure 22.3. Initially, a set of 2-D patches were formed for each nodule in different orientations and discriminative features were extracted accordingly. Subsequently, each image patch passed through a corresponding sub convolutional neural network, and the outputs of multistream subnetworks were combined using a dedicated fusion method, leading to reduced false positive rate.

In another study that predicted neuro-developmental outcomes from a brain connectivity graph derived from diffusion tensor MRI,[103] a convolutional neural network was proposed whose layers were composed of different types of convolutional filters (edge-to-edge, edge-to-node, and node-to-graph) so that the topological locality of structural brain networks was leveraged. The network outperformed existing methods in predicting cognitive and motor scores.

In addition to image features, other clinical factors can be seamlessly incorporated into the framework of convolutional neural network. In a study that predicted the damage of radiation to organs at risks (OAR),[8] both 3D dose maps and additional clinical factors were exploited for their association with toxicities. A convolutional neural network was first pre-trained using 3D images of various human organs and then transferred on liver data (125 liver stereotactic body radiation therapy (RT) cases with follow-up toxicity data). Non-dosimetric features, such as patients' demographics, underlying liver diseases, and liver-directed therapies, were inputted to the fully connected neural network layer for more comprehensive prediction. The saliency map of the convolutional neural network was used to estimate the toxicity risks associated with irradiation of anatomical regions of specific organs at risks

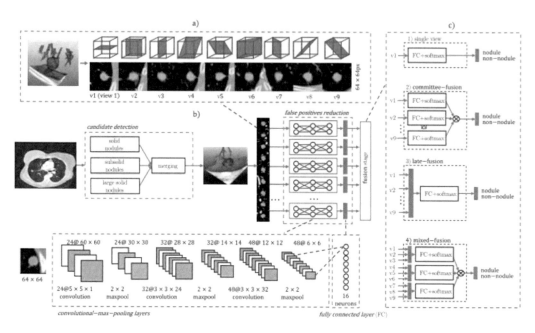

Figure 22.3 (See color insert.) An overview of the pulmonary nodule detection study. (a) An example of extracted 2-D patches from nine symmetrical planes of a cube. The candidate is located at the center of the patch with a bounding box of 50 × 50 mm and 64 × 64 px. (b) Candidates are detected by merging the outputs of detectors specifically designed for solid, subsolid, and large nodules. The false positive reduction stage is implemented as a combination of multiple ConvNets. Each of the ConvNets stream processes 2-D patches extracted from a specific view. (c) Different methods for fusing the output of each ConvNet stream. Gray and orange boxes represent concatenated neurons from the first fully connected layers and the nodule classification output. Neurons are combined using fully connected layers with softmax or a fixed combiner (product-rule). (a) Extracted 2-D patches using nine views of a volumetric object. (b) Schematic of the proposed system. (c) Fusion methods.

and achieved improved results than conventional model-based predictions, as illustrated in Figure 22.4. This type of application (i.e., prediction of therapeutic damage to normal tissue) has attracted more and more attention in radiation therapy.

Segmentation: Typically employed as preprocessing in radiomics, segmentation of structures in medical images provides support to quantitative analysis of parameters related to volume, shape, and texture. Automatic segmentation is valued due to the mitigated inter-observer variation and the possibility to enlarge datasets for radiomics analysis. In the past few years, substantial progress has been made on automatic segmentation using convolutional neural networks.

The most well-known architecture for image segmentation is U-Net, which had a hierarchical architecture with shortcut connections established between the same levels of encoder and decoder. In U-Net, processing was applied to the whole image rather than to individual patches, which not only improved the computation efficiency by avoiding redundant computation between adjacent patches, but also provided valuable context information. Meanwhile, shortcut connections effectively compensate the detailed information lost in down-sampling. For radiological images that are typically acquired volumetrically, 3D U-Net had more attractive performance than 2D U-Net[103] due to the exploitation of continuity in 3D space. Figure 22.5 demonstrates the segmentation result obtained using a volumetric hierarchical deep residual convolutional neural network.

While U-Net offers compelling performance, some patch-based convolutional neural networks also achieved excellent results. Fully convolutional networks were to reduce redundant computation in sliding window-based methods. Additionally, convolutional neural networks were sometimes combined with graphical models such as Markov random fields and conditional random fields. In a study on automatic segmentation of OAR for head and neck radiotherapy,[59] a representative number of

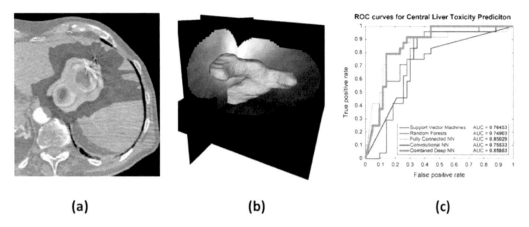

Figure 22.4 (See color insert.) Prediction of radiation toxicity to liver tissue (portal vein). (a) The portal vein (green) on the CT image with superimposed radiation dose map. (b) A 3D dose map delivered to the central hepatobiliary tract (15 mm expansion of the portal vein), which was analyzed using deep learning for hepatobiliary toxicity prediction. (c) Receiver operating characteristic (ROC) curves for the proposed deep dose analysis (combined deep neural network) and alternative toxicity predictors (support vector machine, random forests, fully connected neural network, and convolutional neural network).

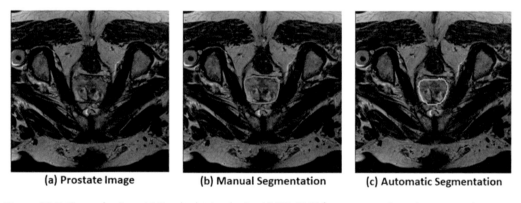

Figure 22.5 (See color insert.) Result obtained using VHDR-CNN for segmentation, where manual segmentation and automatic segmentation are compared for a single case study. (a) prostate image (b) manual segmentation and (c) automatic segmentation. The dice similarity coefficient for prostate delineation was 87% for the entire test set.

positive intensity patches around OAR and negative intensity patches that belonged to the surrounding structures were extracted, both of which were used to train the convolutional neural network. For test images, voxels in the region of interest were classified using the trained network, and the results were smoothed by using Markov random fields.

Moreover, there are smarter ways to define image patch. A super-pixel-based and boundary sensitive convolutional neural network (SBBS-CNN) framework was developed for liver segmentation,[60] as illustrated in Figure 22.6. The CT images were first partitioned into super-pixel regions, where nearby pixels with similar CT number were aggregated. All the super-pixel regions fell into three classes (interior liver, liver boundary, and nonliver background) and sampled with the guidance of an entropy-based saliency map. Hence, more patches were extracted from liver boundary than interior liver or background. Finally, deep convolutional neural network was trained to predict the probability map of the liver boundary, which was integrated into the final segmentation.

Super-resolution and image reconstruction: Improving the quality of radiological images will have significant impacts on radiomics. While there is a tradeoff between spatial resolution, SNR, and data acquisition time, deep learning offers the promise to improve spatial resolution or reduce scan time without

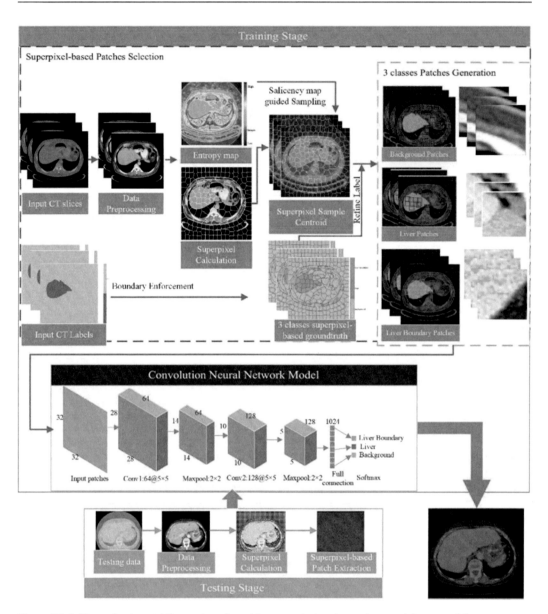

Figure 22.6 (See color insert.) Illustration of sensitive convolutional neural network framework for segmentation. Superpixel patches were extracted (under the guidance of saliency maps) and assigned class labels (interior liver, liver boundary, and nonliver background). A CNN network was trained with given image patches and their labels. The class labels of testing data were predicted, where the image patches that cover all superpixels from individual patients were analyzed and assigned the probability for three classes, and the probability maps were integrated into the final segmentation result.

sacrificing other factors. In reality, image resolution can be improved using super-resolution techniques, which aim to establish an end-to-end mapping between low resolution and high resolution images. In a brain MRI study, for example, a 3D densely connected super-resolution networks (DCSRN) [104] was adopted, whereas in another study investigating super-resolution of CT at multiple organs, a convolutional neural network employing deconvolution operation (SR-DCNN) [68] was developed.

Higher degree of performance improvements can be achieved at the time of data acquisition via sparse sampling and advanced image reconstruction, which provides an end-to-end mapping from undersampled data

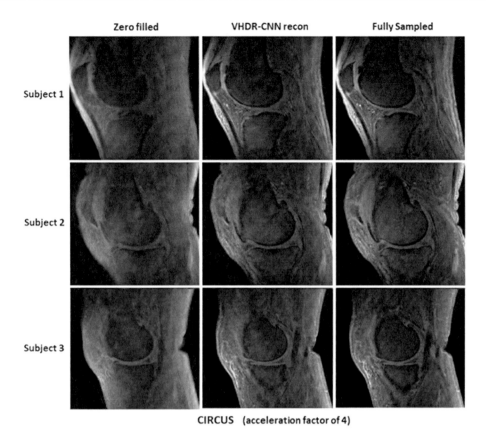

Figure 22.7 Results obtained using VHDR-CNN for image reconstruction. With an acceleration factor of 4 achieved, the zero filled, VHDR-CNN reconstructed, and fully sampled images of three subjects were compared. The micro-structures lost in the zero-filled images were significantly recovered in the VHDR-CNN reconstructed images, which had high fidelity with the ground truth.

to high quality images. Most studies have been conducted to accelerate data acquisition while keeping image quality unaffected. In a cartilage MRI study, k-space was sparsely sampled in a pseudo-random Cartesian trajectory, where a volumetric hierarchical deep residual convolutional neural network (VHDR-CNN) was developed for image reconstruction, obtaining a substantially high acceleration factor without apparent degradation in image quality, as shown in Figure 22.7. Using the same reconstruction algorithm, the reduction in scan time can be traded off to improved SNR and spatial resolution, which would be valuable for the exploitation of heterogeneity.

22.6 THE PATH FORWARD

Radiomics and radiogenomics have been widely investigated for noninvasive acquisition of quantitative textural information from medical imaging data and correlate them to various clinical findings or genomics data. The field has shown a great promise for the discovery of clinically useful imaging markers having diagnostic and prognostic values for variety of clinical applications. It is important to emphasize that, despite the enthusiasm and excitement around this, many radiomics and radiogenomics studies so far are limited by their hypothesis-generating nature, as rigorous validation in independent cohorts has been lacking. Another caveat is that existing biologic knowledge about a certain disease is often not taken into account in many studies. To be of practical value, any new imaging biomarkers should be complementary and add predictive value to known clinical and pathologic factors. It is also useful to point out that an under-explored area of investigation is how radiomics/radiogenomics can be applied to longitudinal imaging scans to better evaluate therapeutic response and monitor disease given the increasing availability of treatment regimens.[105] Although initial studies on simple delta-radiomics are encouraging,[106] alternative approaches to characterizing longitudinal change should also be explored to maximize the information extracted from serial imaging scans.

Advanced machine learning techniques, notably deep convolutional neural networks,[107] are expected to be increasingly used to identify useful image features automatically rather than defined manually.[108] Several successful applications of deep learning in medical imaging have been demonstrated.[109-111] Further research is needed to firmly establish the role of deep learning in radiomics and radiogenomics.[112] While it is known that deep learning approach typically requires extremely large, well annotated datasets in order to train a reliable model, which further highlights the need for curation of high-quality datasets and data sharing, transferred learning techniques,[74,113] and learning with small sample sizes are active areas of research and new progresses are being made routinely. To overcome the same problem of limited size of training dataset, we and other groups sought after distributed learning solutions for both traditional and deep learning-based radiomics. Ultimately, prospective validation in multicenter clinical trials will be required to demonstrate the clinical validity and utility of newly identified imaging markers and truly establish the value of radiomics and radiogenomics in precision medicine.[114-116]

REFERENCES

1. Sala E, Mema E, Himoto Y, Veeraraghavan H, Brenton J, Snyder A, Weigelt B, Vargas H: Unravelling tumour heterogeneity using next-generation imaging: Radiomics, radiogenomics, and habitat imaging. *Clin Radiol* 2017, 72:3–10.

2. Lambin P, Leijenaar RT, Deist TM, Peerlings J, de Jong EE, van Timmeren J, Sanduleanu S, Larue RT, Even AJ, Jochems A: Radiomics: The bridge between medical imaging and personalized medicine. *Nat Rev Clin Oncol* 2017, 14:749.

3. Wu J, Tha KK, Xing L, Li R: Radiomics and radiogenomics for precision radiotherapy. *J Radiat Res* 2018, 59:i25–i31.

4. Gillies RJ, Kinahan PE, Hricak H: Radiomics: Images are more than pictures, they are data. *Radiology* 2016, 278:563–577.

5. Wu J, Aguilera T, Shultz D, Gudur M, Rubin DL, Billy W. Loo J, Diehn M, Li R: Early-stage non–small cell lung cancer: Quantitative imaging characteristics of 18F fluorodeoxyglucose PET/CT allow prediction of distant metastasis. *Radiology* 2016, 281:270–278.

6. Cui Y, Song J, Pollom E, Alagappan M, Shirato H, Chang DT, Koong AC, Li R: Quantitative analysis of FDG-PET identifies novel prognostic imaging biomarkers in locally advanced pancreatic cancer patients treated with SBRT. *Int J Radiat Oncol Biol Phys* 2016, 96:102–109.

7. Ohri N, Duan F, Snyder BS, Wei B, Machtay M, Alavi A, Siegel BA, Johnson DW, Bradley JD, DeNittis A, Werner-Wasik M, El Naqa I: Pretreatment 18F-FDG PET textural features in locally advanced non–small cell lung cancer: Secondary analysis of ACRIN 6668/RTOG 0235. *J Nucl Med* 2016, 57:842–848.

8. Ibragimov B, Toesca D, Chang D, Yuan Y, Koong A, Xing L: Development of deep neural network for individualized hepatobiliary toxicity prediction after liver SBRT. *Med Phys* 2018, 45:4763–4774.

9. Diehn M, Nardini C, Wang DS, McGovern S, Jayaraman M, Liang Y, Alclape K, Cha S, Kuo MD: Identification of noninvasive imaging surrogates for brain tumor gene-expression modules. *Proc Natl Acad Sci USA* 2008, 105:5213–5218.

10. Gevaert O, Xu JJ, Hoang CD, Leung AN, Xu Y, Quon A, Rubin DL, Napel S, Plevritis SK: Non-small cell lung cancer: Identifying prognostic imaging biomarkers by leveraging public gene expression microarray data-methods and preliminary results. *Radiology* 2012, 264:387–396.

11. Segal E, Sirlin CB, Ooi C, Adler AS, Gollub J, Chen X, Chan BK, Matcuk GR, Barry CT, Chang HY: Decoding global gene expression programs in liver cancer by noninvasive imaging. *Nat Biotechnol* 2007, 25:675.

12. Itakura H, Achrol AS, Mitchell LA, et al.: Magnetic resonance image features identify glioblastoma phenotypic subtypes with distinct molecular pathway activities. *Sci Transl Med* 2015, 7:303ra138.

13. Wu J, Cui Y, Sun X, Cao G, Li B, Ikeda DM, Kurian AW, Li R: Unsupervised clustering of quantitative image phenotypes reveals breast cancer subtypes with distinct prognoses and molecular pathways. *Clin Cancer Res* 2017, 23:3334–3342.

14. Wu J, Li B, Sun X, Cao G, Rubin DL, Napel S, Ikeda DM, Kurian AW, Li R: Heterogeneous enhancement patterns of tumor-adjacent parenchyma at MR imaging are associated with dysregulated signaling pathways and poor survival in breast cancer. *Radiology* 2017:162823.

15. Colen R, Foster I, Gatenby R, Giger ME, Gillies R, Gutman D, Heller M, Jain R, Madabhushi A, Madhavan S: NCI workshop report: Clinical and computational requirements for correlating imaging phenotypes with genomics signatures. *Transl Oncol* 2014, 7:556–569.

16. Fehr D, Veeraraghavan H, Wibmer A, Gondo T, Matsumoto K, Vargas HA, Sala E, Hricak H, Deasy JO: Automatic classification of prostate cancer gleason scores from multiparametric magnetic resonance images. *Proc Natl Acad Sci USA* 2015, 112:E6265–E6273.

17. Xing L, Krupinski EA, Cai J: Artificial intelligence will soon change the landscape of medical physics research and practice. *Med Phys* 2018, 45:1791–1793.

18. Giger ML: Machine learning in medical imaging. *J Am Coll Radiol* 2018, 15:512–520.

19. van Griethuysen JJM, Fedorov A, Parmar C, Hosny A, Aucoin N, Narayan V, Beets-Tan RGH, Fillion-Robin JC, Pieper S, Aerts H: Computational radiomics system to decode the radiographic phenotype. *Cancer Res* 2017, 77:e104–e107.

20. Rios Velazquez E, Parmar C, Liu Y, et al.: Somatic mutations drive distinct imaging phenotypes in lung cancer. *Cancer Res* 2017, 77:3922–3930.

21. Yates LR, Gerstung M, Knappskog S, Desmedt C, Gundem G, Van Loo P, Aas T, Alexandrov LB, Larsimont D, Davies H: Subclonal diversification of primary breast cancer revealed by multiregion sequencing. *Nat Med* 2015, 21:751–759.

22. Gerlinger M, Rowan AJ, Horswell S, Larkin J, Endesfelder D, Gronroos E, Martinez P, Matthews N, Stewart A, Tarpey P: Intratumor heterogeneity and branched evolution revealed by multiregion sequencing. *N Engl J Med* 2012, 366:883–892.

23. Gatenby RA, Grove O, Gillies RJ: Quantitative imaging in cancer evolution and ecology. *Radiology* 2013, 269:8–15.

24. Zhou M, Hall L, Goldgof D, Russo R, Balagurunathan Y, Gillies R, Gatenby R: Radiologically defined ecological dynamics and clinical outcomes in glioblastoma multiforme: Preliminary results. *Transl Oncol* 2014, 7:5–13.

25. Wang P, Popovtzer A, Eisbruch A, Cao Y: An approach to identify, from DCE MRI, significant subvolumes of tumors related to outcomes in advanced head-and-neck cancer. *Med Phys* 2012, 39:5277–5285.

26. Wu J, Gensheimer MF, Dong X, Rubin DL, Napel S, Diehn M, Loo BW, Jr., Li R: Robust intratumor partitioning to identify high-risk subregions in lung cancer: A pilot study. *Int J Radiat Oncol Biol Phys* 2016, 95:1504–1512.

27. Wu J, Gong G, Cui Y, Li R: Intratumor partitioning and texture analysis of dynamic contrast—enhanced (DCE)—MRI identifies relevant tumor subregions to predict pathological response of breast cancer to neoadjuvant chemotherapy. *J Magn Reson Imaging* 2016, 44:1107–1115.

28. Chaudhury B, Zhou M, Goldgof DB, Hall LO, Gatenby RA, Gillies RJ, Patel BK, Weinfurtner RJ, Drukteinis JS: Heterogeneity in intratumoral regions with rapid gadolinium washout correlates with estrogen receptor status and nodal metastasis. *J Magn Reson Imaging* 2015, 42:1421–1430.

29. Cui Y, Tha KK, Terasaka S, Yamaguchi S, Wang J, Kudo K, Xing L, Shirato H, Li R: Prognostic imaging biomarkers in glioblastoma: Development and independent validation on the basis of multiregion and quantitative analysis of MR images. *Radiology* 2015, 278:546–553.

30. Stoyanova R, Pollack A, Takhar M, Lynne C, Parra N, Lam LL, Alshalalfa M, Buerki C, Castillo R, Jorda M: Association of multiparametric MRI quantitative imaging features with prostate cancer gene expression in MRI-targeted prostate biopsies. *Oncotarget* 2016, 7:53362.

31. Wu J, Cao G, Sun X, Lee J, Rubin DL, Napel S, Kurian AW, Daniel B, Li R: Intratumoral spatial heterogeneity by perfusion MR imaging predicts recurrence-free survival in locally advanced breast cancer treated with neoadjuvant chemotherapy. *Radiology* 2018, 288: 26–35.

32. O'Connor JP, Rose CJ, Waterton JC, Carano RA, Parker GJ, Jackson A: Imaging intratumor heterogeneity: Role in therapy response, resistance, and clinical outcome. *Clin Cancer Res* 2015, 21:249–257.

33. Bakas S, Akbari H, Pisapia J, Martinez-Lage M, Rozycki M, Rathore S, Dahmane N, O'Rourke DM, Davatzikos C: In vivo detection of EGFRvIII in glioblastoma via perfusion magnetic resonance imaging signature consistent with deep peritumoral infiltration: The φ index. *Clin Cancer Res* 2017, 23: 4724–4734.

34. Smits M, van den Bent MJ: Imaging correlates of adult glioma genotypes. *Radiology* 2017, 284:316–331.

35. Vargas HA, Huang EP, Lakhman Y, Ippolito JE, Bhosale P, Mellnick V, Shinagare AB, Anello M, Kirby J, Fevrier-Sullivan B: Radiogenomics of high-grade serous ovarian cancer: Multireader multi-institutional study from the cancer genome atlas ovarian cancer imaging research group. *Radiology* 2017:161870.

36. Lee J, Cui Y, Sun X, Li B, Wu J, Li D, Gensheimer MF, Loo BW, Diehn M, Li R: Prognostic value and molecular correlates of a CT image-based quantitative pleural contact index in early stage NSCLC. *Eur Radiol* 2017:1–11.

37. Zhu Y, Li H, Guo W, Drukker K, Lan L, Giger ML, Ji Y: Deciphering genomic underpinnings of quantitative MRI-based radiomic phenotypes of invasive breast carcinoma. Sci. Rep. 2015, 5: 17787.

38. Grossmann P, Stringfield O, El-Hachem N, Bui MM, Velazquez ER, Parmar C, Leijenaar RT, Haibe-Kains B, Lambin P, Gillies RJ: Defining the biological basis of radiomic phenotypes in lung cancer. *Elife* 2017, 6: e23421.

39. Liu Y, Kim J, Balagurunathan Y, Li Q, Garcia AL, Stringfield O, Ye Z, Gillies RJ: Radiomic features are associated with EGFR mutation status in lung adenocarcinomas. *Clin Lung Cancer* 2016, 17:441–448. e6.

40. Zhou M, Leung A, Echegaray S, Gentles A, Shrager JB, Jensen KC, Berry GJ, Plevritis SK, Rubin DL, Napel S: Non–small cell lung cancer radiogenomics map identifies relationships between molecular and imaging phenotypes with prognostic implications. *Radiology* 2017, 286:307–315.

41. Yamamoto S, Korn RL, Oklu R, Migdal C, Gotway MB, Weiss GJ, Iafrate AJ, Kim DW, Kuo MD: ALK molecular phenotype in non-small cell lung cancer: CT radiogenomic characterization. *Radiology* 2014, 272:568–576.

42. Rizzo S, Petrella F, Buscarino V, De Maria F, Raimondi S, Barberis M, Fumagalli C, Spitaleri G, Rampinelli C, De Marinis F: CT radiogenomic characterization of EGFR, K-RAS, and ALK mutations in non-small cell lung cancer. Eur. Radiol. 2016, 26:32–42.

43. Hasegawa M, Sakai F, Ishikawa R, Kimura F, Ishida H, Kobayashi K: CT features of epidermal growth factor receptor–mutated adenocarcinoma of the lung: Comparison with nonmutated adenocarcinoma. *J Thorac Oncol* 2016, 11:819–826.

44. Wu J, Sun X, Wang J, Cui Y, Kato F, Shirato H, Ikeda DM, Li R: Identifying relations between imaging phenotypes and molecular subtypes of breast cancer: Model discovery and external validation. *JMRI* 2017, 46:1017–1027.

45. Ashraf AB, Daye D, Gavenonis S, Mies C, Feldman M, Rosen M, Kontos D: Identification of intrinsic imaging phenotypes for breast cancer tumors: Preliminary associations with gene expression profiles. *Radiology* 2014, 272:374–384.

46. Li H, Zhu Y, Burnside ES, Drukker K, Hoadley KA, Fan C, Conzen SD, Whitman GJ, Sutton EJ, Net JM: MR Imaging radiomics signatures for predicting the risk of breast cancer recurrence as given by research versions of mamma print, oncotype DX, and PAM50 gene assays. *Radiology* 2016:152110.

47. Wang J, Kato F, Oyama-Manabe N, Li R, Cui Y, Tha KK, Yamashita H, Kudo K, Shirato H: Identifying triple-negative breast cancer using background parenchymal enhancement heterogeneity on dynamic contrast-enhanced MRI: A pilot radiomics study. *PLoS One* 2015, 10: e0143308.

48. Sutton EJ, Dashevsky BZ, Oh JH, Veeraraghavan H, Apte AP, Thakur SB, Morris EA, Deasy JO: Breast cancer molecular subtype classifier that incorporates MRI features. *J Magn Reson Imaging* 2016, 44:122–129.

49. Yamamoto S, Huang D, Du L, Korn RL, Jamshidi N, Burnette BL, Kuo MD: Radiogenomic analysis demonstrates associations between 18F-Fluoro-2-Deoxyglucose PET, prognosis, and epithelial-mesenchymal transition in non–small cell lung cancer. *Radiology* 2016:160259.

50. Lee J, Cui Y, Wu J, Sun X, Li B, Wu J, Li D, Gensheimer M, Billy W. Loo J, Diehn M, Li R: Prognostic value and molecular correlates of a CT image-based quantitative pleural contact index in early stage NSCLC. *Eur Radiol* 2018, 28:736–746.

51. Cottereau AS, Lanic H, Mareschal S, Meignan M, Vera P, Tilly H, Jardin F, Becker S: Molecular profile and FDG-PET/CT total metabolic tumor volume improve risk classification at diagnosis for patients with diffuse large B-Cell lymphoma. *Clin Cancer Res* 2016, 22:3801–3809.

52. Cui Y, Ren S, Tha KK, Wu J, Shirato H, Li R: Volume of high-risk intratumoral subregions at multiparametric MR imaging predicts overall survival and complements molecular analysis of glioblastoma. *Eur Radiol* 2017, 27:3583–3592.

53. Lee J, Li B, Sun X, Cui Y, Wu J, Zhu H, Yu J, Gensheimer M, Billy W. Loo J, Diehn M, Li R: A quantitative CT imaging signature predicts survival and complements established prognosticators in stage I non-small cell lung cancer. *Int J Radiat Oncol Biol Phys* 2018, 102: 1098–1106.

54. Sun R, Limkin EJ, Vakalopoulou M et al.: A radiomics approach to assess tumour-infiltrating CD8 cells and response to anti-PD-1 or anti-PD-L1 immunotherapy: An imaging biomarker, retrospective multicohort study. *Lancet Oncol* 2018, 19: 1180–1191.

55. Chalkidou A, O'Doherty MJ, Marsden PK: False discovery rates in PET and CT studies with texture features: A systematic review. *PLoS One* 2015, 10:e0124165.

56. Deng J, Xing L: *Big Data in Radiation Oncology*. Abingdon, UK: Taylor & Francis Group, 2018.

57. Ibragimov B, Korez R, Likar B, Pernus F, Xing L, Vrtovec T: Segmentation of pathological structures by landmark-assisted deformable models. *IEEE Trans Med Imaging* 2017, 36:1457–1469.

58. Ibragimov B, Xing L: Segmentation of organs-at-risks in head and neck CT images using convolutional neural networks. *Med Phys* 2017, 44:547–557.

59. Qin W, Wu J, Han F, Yuan Y, Zhao W, Ibragimov B, Gu J, Xing L: Superpixel-based and boundary-sensitive convolutional neural network for automated liver segmentation. *Phys Med Biol* 2018, 63:095017.

60. Liu H, Xu J, Guo Q, Ibragimov B, Wu Y, Xing L: Learning deconvolutional deep neural network for high resolution (HR) medical image reconstruction. *Infor Sci* 2018, 468:142–154.

61. Wu Y, Zhao W, Mistretta C, Du J, Xing L: Incorporating prior knowledge via volumetric deep residual network to optimize the reconstruction of sparsely sampled MRI. *Med Phys* 2019:in press.

62. Choi K, Xing L, Koong A, Li R: First study of on-treatment volumetric imaging during respiratory gated VMAT. *Med Phys* 2013, 40:040701.

63. Choi K, Wang J, Zhu L, Suh TS, Boyd S, Xing L: Compressed sensing based cone-beam computed tomography reconstruction with a first-order method. *Med Phys* 2010, 37:5113–5125.

64. Mardani M, Gong E, Cheng J, Vasanawala S, Zaharchuk G, Alley M, Thakur N, Han S, Dally W, Xing L, Pauly J: Deep generative adversarial networks for compressed sensing Automates MRI. *IEEE Trans Med Imaging* 2019, 38:167–179.

65. LeCun Y, Bengio Y, Hinton G: Deep learning. *Nature* 2015, 521:436.

66. Ronneberger O, Fischer P, Brox T: U-net: Convolutional networks for biomedical image segmentation. *International Conference on Medical Image Computing and Computer-Assisted Intervention*, Springer, 2015. pp. 234–241.

67. Dong C, Loy CC, He K, Tang X: Image super-resolution using deep convolutional networks. *IEEE Trans Pattern Anal Mach Intell* 2016, 38:295–307.

68. Zhao W, Han B, Yang Y, Hancock S, Bagshaw H, Buyyounouski M, Xing L: Visualizing the invisible in prostate radiation therapy: Markerless prostate target localization via a deep learning model and monoscopic kV projection X-ray image 2018. *Annual Meeting of ASTRO*. San Antonio, TX, 2018.

69. Zhao W, Shen L, Wu Y, Han B, Yang Y, Xing L: Automatic marker-free target positioning and tracking for image-guided radiotherapy and interventions. *Proceedings Volume 10951, Medical Imaging 2019: Image-Guided Procedures, Robotic Interventions, and Modeling*, 2019: 109510B, SPIE Medical Imaging, San Diego, CA. doi:10.1117/12.2512166.

70. Dosovitskiy A, Springenberg JT, Riedmiller M, Brox T: Discriminative unsupervised feature learning with convolutional neural networks. Adv Neural Inf Process Syst 2014. pp. 766–774.

71. Grün F, Rupprecht C, Navab N, Tombari F: A taxonomy and library for visualizing learned features in convolutional neural networks. arXiv preprint arXiv:160607757 2016.

72. Zeiler MD, Krishnan D, Taylor GW, Fergus R: *Deconvolutional Networks*. IEEE Computer Society Conference on Computer Vision and Pattern Recognition, CVPR, 2010.

73. Zeiler MD, Fergus R: Visualizing and understanding convolutional networks. *European Conference on Computer Vision*. Springer, 2014. pp. 818–833.

74. Yuan Y, Qin W, Ibragimov B, Han B, Xing L: RIIS-DenseNet: Rotation-invariant and image similarity constrained densely connected convolutional network for polyp detection. *MICCAI 2018: Medical Image Computing and Computer Assisted Intervention - MICCAI*, 2018. pp. 620–628.

75. Maaten Lvd, Hinton G: Visualizing data using t-SNE. *J Mach Learn Res* 2008, 9:2579–2605.

76. Hinton GE, Roweis ST: Stochastic neighbor embedding. *Adv Neural Inf Process Syst* 2003: 857–864.

77. Zhu B, Liu JZ, Cauley SF, Rosen BR, Rosen MS: Image reconstruction by domain-transform manifold learning. *Nature* 2018, 555:487.

78. Bruna J, Mallat S: Invariant scattering convolution networks. *IEEE IEEE Trans Pattern Anal Mach Intell* 2013, 35:1872–1886.

79. Wiatowski T, Bölcskei H: A mathematical theory of deep convolutional neural networks for feature extraction. IEEE IEEE Trans Inf Theory 2018, 64:1845–1866.

80. Qian N: On the momentum term in gradient descent learning algorithms. *Neural Netw* 1999, 12:145–151.

81. Hinton G, Srivastava N, Swersky K: Neural networks for machine learning lecture 6a overview of mini-batch gradient descent. *Cited on* 2012:14.

82. Kingma DP, Ba J: Adam: A method for stochastic optimization. arXiv preprint arXiv:14126980 2014.

83. LeCun Y, Bottou L, Bengio Y, Haffner P: Gradient-based learning applied to document recognition. *Proceedings of the IEEE* 1998, 86:2278–2324.

84. Krizhevsky A, Sutskever I, Hinton GE: Imagenet classification with deep convolutional neural networks. *Adv Neural Inf Process Syst* 2012. pp. 1097–1105.

85. Simonyan K, Zisserman A: Very deep convolutional networks for large-scale image recognition. arXiv preprint arXiv:14091556 2014.

86. Lin M, Chen Q, Yan S: Network in network. arXiv preprint arXiv:13124400 2013.

87. Szegedy C, Liu W, Jia Y, Sermanet P, Reed S, Anguelov D, Erhan D, Vanhoucke V, Rabinovich A: Going deeper with convolutions. *Proceedings of the IEEE Conference on Computer Vision and Pattern Recognition*, 2015. pp. 1–9.

88. He K, Zhang X, Ren S, Sun J: Deep residual learning for image recognition. *Proceedings of the IEEE Conference on Computer Vision and Pattern Recognition*, 2016. pp. 770–778.

89. He KM, Zhang XY, Ren SQ, Sun J: Identity mappings in deep residual networks. *Lect Notes Comput Sci* 2016, 9908:630–645.

90. Huang G, Liu Z, Van Der Maaten L, Weinberger KQ: Densely connected convolutional networks. *CVPR*, 2017, 3:4700–4708.

91. Srivastava N, Hinton G, Krizhevsky A, Sutskever I, Salakhutdinov R: Dropout: A simple way to prevent neural networks from overfitting. *J Mach Learn Res* 2014, 15:1929–1958.

92. Ioffe S, Szegedy C: Batch normalization: Accelerating deep network training by reducing internal covariate shift. arXiv preprint arXiv:150203167 2015.

93. Kamnitsas K, Ledig C, Newcombe VF, Simpson JP, Kane AD, Menon DK, Rueckert D, Glocker B: Efficient multi-scale 3D CNN with fully connected CRF for accurate brain lesion segmentation. *Med Image Anal* 2017, 36:61–78.

94. Farabet C, Couprie C, Najman L, LeCun Y: Learning hierarchical features for scene labeling. *IEEE Trans Pattern Anal Mach Intell* 2013, 35:1915–1929.

95. Long J, Shelhamer E, Darrell T: Fully convolutional networks for semantic segmentation. *Proceedings of the IEEE Conference on Computer Vision and Pattern Recognition*, 2015. pp. 3431–3440.

96. Milletari F, Navab N, Ahmadi S-A: V-net: Fully convolutional neural networks for volumetric medical image segmentation. 3D Vision (3DV), *2016 Fourth International Conference on: IEEE*, 2016. pp. 565–571.

97. Çiçek Ö, Abdulkadir A, Lienkamp SS, Brox T, Ronneberger O: 3D U-Net: Learning dense volumetric segmentation from sparse annotation. *International Conference on Medical Image Computing and Computer-Assisted Intervention.* Springer, 2016. pp. 424–432.

98. Antony J, McGuinness K, O'Connor NE, Moran K: Quantifying radiographic knee osteoarthritis severity using deep convolutional neural networks. *Pattern Recognition (ICPR), 2016 23rd International Conference on: IEEE*, 2016. pp. 1195–2000.

99. Nie D, Zhang H, Adeli E, Liu L, Shen D: 3D deep learning for multi-modal imaging-guided survival time prediction of brain tumor patients. *International Conference on Medical Image Computing and Computer-Assisted Intervention.* Springer, 2016. pp. 212–220.

100. Payan A, Montana G: Predicting Alzheimer's disease: A neuroimaging study with 3D convolutional neural networks. arXiv preprint arXiv:150202506 2015.

101. Shen W, Zhou M, Yang F, Dong D, Yang C, Zang Y, Tian J: Learning from experts: Developing transferable deep features for patient-level lung cancer prediction. *International Conference on Medical Image Computing and Computer-Assisted Intervention.* Springer, 2016. pp. 124–131.

102. Setio AAA, Ciompi F, Litjens G, Gerke P, Jacobs C, Van Riel SJ, Wille MMW, Naqibullah M, Sánchez CI, van Ginneken B: Pulmonary nodule detection in CT images: False positive reduction using multiview convolutional networks. *IEEE Trans Med Imaging* 2016, 35:1160–1169.

103. Kawahara J, Brown CJ, Miller SP, Booth BG, Chau V, Grunau RE, Zwicker JG, Hamarneh G: BrainNetCNN: Convolutional neural networks for brain networks; towards predicting neurodevelopment. *NeuroImage* 2017, 146:1038–1049.

104. Chen Y, Xie Y, Zhou Z, Shi F, Christodoulou AG, Li D: Brain MRI super resolution using 3D deep densely connected neural networks. Biomedical Imaging (ISBI 2018), 2018 *IEEE 15th International Symposium on: IEEE*, 2018. pp. 739–742.

105. Pollom EL, Song J, Durkee BY, Aggarwal S, Bui T, von Eyben R, Li R, Brizel DM, Loo BW, Le QT, Hara WY: Prognostic value of midtreatment FDG-PET in oropharyngeal cancer. *Head Neck* 2016, 38:1472–1478.

106. Fave X, Zhang LF, Yang JZ, Mackin D, Balter P, Gomez D, Followill D, Jones AK, Stingo F, Liao ZX, Mohan R, Court L: Delta-radiomics features for the prediction of patient outcomes in non-small cell lung cancer. *Sci Rep* 2017, 7.

107. Tajbakhsh N, Shin JY, Gurudu SR, Hurst RT, Kendall CB, Gotway MB, Liang J: Convolutional neural networks for medical image analysis: Full training or fine tuning? *IEEE Trans Med Imaging* 2016, 35:1299–1312.

108. Litjens G, Kooi T, Bejnordi BE, Setio AAA, Ciompi F, Ghafoorian M, van der Laak J, van Ginneken B, Sanchez CI: A survey on deep learning in medical image analysis. *Med Image Anal* 2017, 42:60–88.

109. Esteva A, Kuprel B, Novoa RA, Ko J, Swetter SM, Blau HM, Thrun S: Dermatologist-level classification of skin cancer with deep neural networks. *Nature* 2017, 542:115–118.

110. Kermany DS, Goldbaum M, Cai W, et al.: Identifying medical diagnoses and treatable diseases by image-based deep learning. *Cell* 2018, 172:1122–1131 e9.

111. Ehteshami Bejnordi B, Veta M, Johannes van Diest P, et al.: Diagnostic assessment of deep learning algorithms for detection of lymph node metastases in women with breast cancer. *JAMA* 2017, 318:2199–2210.

112. Greenspan H, van Ginneken B, Summers RM: Guest editorial deep learning in medical imaging: Overview and future promise of an exciting new technique. *IEEE Trans Med Imaging* 2016, 35:1153–1159.

113. Yuan Y, Qin W, Hancock S, Buyyounouski M, Ibragimov B, Han B, Xing L: Prostate cancer classification with multiparametric MRI transfer learning model. *Med Phys.* 2019, 46:756–765.

114. Yankeelov TE, Mankoff DA, Schwartz LH et al.: Quantitative imaging in cancer clinical trials. *Clin Cancer Res* 2016, 22:284–290.

115. Yankeelov TE, Abramson RG, Quarles CC: Quantitative multimodality imaging in cancer research and therapy. *Nat Rev Clin Oncol* 2014, 11:670–680.

116. O'Connor JP, Aboagye EO, Adams JE et al.: Imaging biomarker roadmap for cancer studies. *Nat Rev Clin Oncol* 2017, 14: 169.

Index

Note: Page numbers in italic and bold refer to figures and tables, respectively.

Printed and bound by CPI Group (UK) Ltd, Croydon, CR0 4YY

24/10/2024

01778309-0008